The

Chemical Engineering

Guide to Corrosion

CHEMICAL ENGINEERING BOOKS

Sources and Production Economics of Chemical Products
Calculator Programs for Chemical Engineers—Volumes I and II
The *Chemical Engineering* Guide to Compressors
The *Chemical Engineering* Guide to Corrosion
The *Chemical Engineering* Guide to Pumps
The *Chemical Engineering* Guide to Heat Transfer—Principles
The *Chemical Engineering* Guide to Heat Transfer—Equipment
The *Chemical Engineering* Guide to Valves
Controlling Corrosion in Process Equipment
Effective Communication for Engineers
Fluid Movers: Pumps, Compressors, Fans and Blowers
Industrial Air Pollution Engineering
Industrial Wastewater and Solid Waste Engineering
Microcomputer Programs for Chemical Engineers
Modern Cost Engineering: Methods & Data—Volumes I and II
Practical Process Instrumentation & Control—Volumes I and II
Process Energy Conservation
Process Heat Exchange
Process Piping Systems
Process Technology and Flowsheets—Volumes I and II
Safe and Efficient Plant Operation and Maintenance
Selecting Materials for Process Equipment
Separation Techniques 1: Liquid-Liquid Systems
Separation Techniques 2: Gas / Liquid / Solid Systems
Skills Vital to Successful Managers
Solids Handling
Supplementary Readings in Engineering Design
You and Your Job

BOOKS PUBLISHED BY CHEMICAL ENGINEERING

Fluid Mixing Technology: James Y. Oldshue
Physical Properties: Carl L. Yaws
Pneumatic Conveying of Bulk Materials: Milton N. Kraus

The Chemical Engineering Guide to Corrosion

Edited by
Richard W. Greene
and
the Staff of Chemical Engineering

McGraw-Hill Chemical Engineering

McGraw-Hill Publications Co., New York, N.Y.

Library of Congress Cataloging in Publication Data
Main entry under title:

The Chemical engineering guide to corrosion.

Includes index.
1. Chemical plants—Equipment and supplies—Corrosion.
I. Greene, Richard, 1944– . II. Chemical engineering.
TP155.5.C47 1985 660.2'83 85-19772
ISBN 0-07-606927-3
ISBN 0-07-024309-3 (Book Company)

1221 Avenue of the Americas, New York, New York 10020

1234567890 EDW/EDW 8932109876

ISBN 0-07-024309-3

Printed in the United States of America.

CONTENTS

Section III PAINTS AND COATINGS

Section IV MATERIALS TO RESIST CHLORIDES AND CHLORINE

Section V METALS AND ALLOYS

INTRODUCTION

Yearly costs of corrosion to industry are in the billions of dollars. What should the chemical engineer do in such situations? Pipes rust, tanks develop pinholes, vessel walls stress crack. Corrosion problems range from discoloration of structures to catastrophic failure of machinery and equipment. Also, plants must be built to handle new processes and existing facilities must be changed to accommodate process modifications.

Here are guidelines to help cope with a variety of corrosion and materials-selection problems. THE CHEMICAL ENGINEERING GUIDE TO CORROSION contains articles that appeared in *Chemical Engineering* during the past five years. It covers theory and applications in a variety of areas that will help you to make the proper materials selection and to solve many problems you may face. The book is divided into seven sections:

Section I—Basics on corrosion prevention, monitoring and water-treating. Includes topics on stopping stress corrosion-cracking, preventing hydrogen embrittlement, performing acoustic-emission testing, and controlling biofouling in cooling water.

Section II—Materials selection and identification. Tells how to identify the various metals and alloys found in the typical chemical-process-industries (CPI) plant, and offers tips on picking the best ones for a specific use.

Section III—Paints and coatings. Details the various formulations available for the CPI, when to use them, how to inspect them, and more.

Section IV—Materials to resist chlorides and chlorine. Includes high-performance stainless steels, alloys for desalting, and alloys to resist chlorine.

Section V—Metals and alloys. Covers wear and corrosion, high-temperature effects, steels, and refractory metals and alloys.

Section VI—Fiberglass-reinforced plastics and other polymers. Includes FRP, plastic pumps and piping, and, in general, how to select plastics.

Section VII—Refractories, insulation and ceramics. Details selection, installation and failure analysis.

These seven sections contain a wealth of practical information that design, project, process and research engineers can use to eliminate materials failures. Numerous charts, graphs and guidelines are presented to simplify the often-complex task of materials selection. This book offers expert advice and practical solutions—information that will remain current for years to come.

The
Chemical Engineering
Guide to Corrosion

Section I
BASICS ON CORROSION PREVENTION, MONITORING AND WATER-TREATING

Stop stress-corrosion cracking

A network of branched cracks in equipment or piping indicates the presence of this type of corrosion. Preventing it requires selection of the proper metal or alloy.

James L. Gossett, Fisher Controls International, Inc.

☐ The wrong combination of environment, material, and minimum tensile strength will produce stress-corrosion cracking (SCC) in any metal or alloy: steel, stainless steel, nickel-based alloys, aluminum, titanium—even gold.

Many of these failures occur only under the most severe laboratory conditions and in the past were usually not encountered in the field. However, such conditions now are found commercially with greater frequency as chemical-processing, oil-refining and power plants are using higher pressures and temperatures to increase efficiencies. The table lists common environments known to cause SCC.

Knowing how stress-corrosion cracking occurs and how to avoid it will help you to select equipment and piping that will not crack in process and utility applications.

Identifying SCC

SCC is a progressive type of failure that produces cracking at stress levels that are well below those of a material's tensile strength. The break or fracture appears brittle, with no localized yielding, plastic deformation or elongation. SCC has a characteristic appearance, which in many cases makes it easy to identify (see Fig. 1). Rather than a single crack, a whole network of fine, feathery, branched cracks will form (see Fig. 2).

Severe general corrosion will not be found where SCC develops. The rate of general attack will always be quite low. If the rate were high, metal would be removed faster than a crack could penetrate the material. Thus, a crack would have no chance to begin and grow.

Pitting is frequently seen, and will often serve as a stress concentrator to initiate cracking. One or more cracks will grow from the pit, eventually leading to failure.

The oldest known form of SCC is "season cracking" of brass sleigh bells and cartridge cases. Shortly after the use of brass cartridges became common in the mid-1800s, cracking was often noted. The shells always cracked on the crimped end, where residual stresses were highest, and usually during the rainy season when shells were stored in barns. Thus, the name season cracking.

The three required elements of SCC are all found here. The brass used for the cartridge is a susceptible material. Crimping deforms the brass, and creates a stress above the minimum level for SCC. Finally, the barnyard provides the environment. Decaying manure and other organic material produce ammonia, which is known to stress-crack brass. Moisture is required to absorb the ammonia.

Environment and materials

A typical stress-cracking environment is an aqueous solution containing certain ions. A minimum ion concen-

A typical network of branched cracks on a U-bend sample of Type 316 stainless steel **Fig. 1**

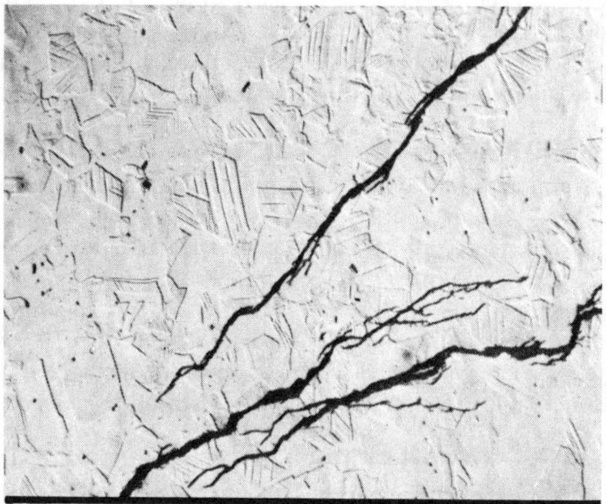

Photomicrograph of Type 316 shows multiple lacy, branched cracks (magnified 440 ×) **Fig. 2**

tration is required to produce SCC. As the concentration increases, the environment becomes more severe, reducing the time to failure. Although the minimum ionic concentration may not be present throughout the solution, SCC may still develop, however. The reason for this is that ions can collect in crevices and pits, concentrating to very high levels. Concentrations of 10, 100, or even 1,000 times the overall average can easily develop.

Ion concentration is a particular problem with Type 316 stainless steel in dilute chloride solutions. Solutions with just 10-100-ppm chlorides can pit Type 316. The concentration in the pits will rise to the 1,000-10,000-ppm range required for chloride SCC. Thus, one must be conservative in selecting materials for such environments.

Temperature also is a factor. In general, the likelihood of SCC increases with increasing temperature. A minimum threshold temperature exists for most systems, below which SCC is rare. This holds for most types of SCC, except for hydrogen embrittlement, which is most severe at ambient temperatures.

Almost all common metals and alloys are susceptible to SCC. A material's susceptibility is dependent upon various metallurgical conditions, including cold-work, residual stresses, machining stresses, heat-treating, hardness and strength level. At higher hardness levels, failure occurs extremely fast. As the level drops, the time to failure becomes longer and longer, until a level of immunity is reached. This immune level is about Rockwell 22 for carbon steel, alloy steel, and the 400 series stainless steels, in several environments, including sour gas or H_2S.

Stress level

Surface tensile stresses are the final requirement for SCC. These stresses can be caused by cyclic loads, applied tensile loads, or residual stresses from the original forming operations (i.e., heat treatment, casting or welding). A threshold stress level exists below which SCC will not occur. For different materials, this threshold may vary anywhere from 10 to 100% of a material's yield strength. SCC of Type 316 has occurred where the stress was from bending a 1/4-in.-thick plate to a 4-ft radius.

Although it is seldom practical, design stresses can be made to follow these threshold levels in special cases. This is risky, however, as some stresses, such as residual ones, may exceed expected levels. Undetected internal defects—inclusions, porous areas or shrinkage—act as stress risers and produces unexpectedly high stress levels. Let us look at the basic forms of SCC.

Season cracking

Ammonia and amine compounds will produce season cracking of all copper, brass and bronze alloys. The required stress and ion-concentration levels are low.

Season cracking can occur at temperatures from ambient on up. The use of copper in systems containing any concentration of ammonia is not recommended. Alloying copper with nickel, such as cupro-nickel alloys or Monel* alloys produces immunity to season cracking.

Caustic embrittlement

Caustic embrittlement is generally encountered in boilers or steam systems where sodium or calcium scales form on heated surfaces. The susceptibility of a welded steel structure to SCC is a well-defined function of sodium hydroxide concentration and temperature. As-welded steel is acceptable at all NaOH concentrations below 120°F, and at low concentrations to 180°F. Stress-relieving carbon steel will extend the service range another 30-70°F. The 300 series stainless steels resist caustic embrittlement to high temperatures at low NaOH concentrations. Nickel 200 and Inconel 600 alloys are the best for caustics.

*Monel (and Inconel) are registered trademarks of Huntington Alloys, Inc.

These environments are known to cause stress-corrosion cracking in certain alloys and metals

Alloy/metal	Environment
Aluminum-base	Chlorides Humid industrial atmospheres
Copper-base	Ammonium ions Amines
Gold	Potassium permanganate
High-strength stainless and alloy steels	Hydrogen sulfide Chlorides Humid industrial atmospheres
Low-alloy steels	Boiling, concentrated hydroxides Boiling, concentrated nitrates
Nickel-base	Hot, concentrated hydroxides Hydrofluoric acid vapors
Stainless—300 series	Hot chlorides Boiling, concentrated hydroxides
Stainless—400 series	Hydrogen sulfide Chlorides Reactor cooling-water
Titanium-base	Chlorides Methyl alcohol Solid chlorides above 550°F

Chloride SCC

Chloride stress-corrosion cracking is the most widely encountered and most extensively studied form of SCC. Much is still unknown, however, about its mechanism. One theory says that hydrogen, generated by the corrosion process, diffuses into the base metal in the atomic form and embrittles the lattice structure.

A second, more widely accepted theory proposes an electrochemical mechanism. Stainless steels are covered with a protective oxide film. The chloride ions rupture the film at weak spots, resulting in anodic (bare) and cathodic (film-covered) sites. The galvanic cell produces accelerated attack at the anodic sites, which when combined with tensile stresses produces cracking.

The minimum stress level required for chloride SCC for 300 series stainless is extremely low. Most of the commonly used fabrication techniques (welding, cold-drawing, rolling, bending or crimping) will produce residual stress exceeding the minimum value. Annealing to relieve the residual stresses is seldom effective. Heating to 1,000-1,200°F is required to stress-relieve, within a reasonable time period. Heating in this temperature range, however, severely reduces corrosion resistance.

This is because chromium carbides precipitate in the grain boundaries, producing a "sensitized" structure, which is very undesirable in most corrosive environments. Heating to 1,700-1,900°F prevents sensitization, but may cause other problems. The likelihood of warpage is greatly increased, and new residual stresses may result if the part is cooled too fast.

Alloys to use

A minimum solution temperature of 140°F is usually required for chloride SCC of the 300 series. As the temperature increases above 140°F, the time to failure will decrease. An extreme example of this is a boiling solution on a heated surface. Very severe extensive cracking can develop in hours. A surface concentration of only 1,000 ppm of chloride ions is all that is required for SCC of the 300 series. Most natural waters contain chlorides of at least such levels.

For all practical purposes, all the 300 series stainless steels have the same susceptibility to SCC. The corrosion resistance does improve, however, starting from 302 and 304 and going to 316 to 317. However, the improvement is so small that a stress-cracked 304 part should never be replaced with a 316 one.

Carpenter 20 CB-3* and Incoloy† 825 specialty stainless steels have a higher alloy content than does 316, and both resist chloride SCC. These two are highly economical SCC-resistant alloys. Although both can stress-crack in extremely severe laboratory tests, experience has found them to be immune in field applications.

The Inconel and Monel alloys and titanium are immune to chloride SCC. They are typically used where additional general corrosion resistance is required.

E-Brite‡ 26-1 and 29-4 alloys are relatively new, improved versions of the 400 series stainless steels, such as 410. All hardened E-Brite alloys readily resist cracking. The E-Brites are only slightly more expensive than Type 316.

Sulfide stress-cracking

Hydrocarbon streams are termed "sour" when they contain hydrogen sulfide. Careful material selection is required to prevent sulfide stress-cracking. NACE Standard MR-01-73, Sulfide Stress Cracking Resistance Material for Oil Field Equipment, is a widely accepted guide for sour service, and has even been adopted as law in some states. The characteristics of sulfide SCC and the methods used to prevent it are similar to those for hydrogen embrittlement. The ideas discussed here can be applied to both situations, and hydrogen embrittlement will not be discussed separately.

As little as 750 ppm H_2S in 65-psia methane can cause sulfide stress-cracking. It is most severe in the 20-120°F range, and is almost never seen below 20°F. Above 120°F, stress cracking may still occur, although it will probably be chloride SCC, not sulfide stress-cracking. Most deep sour wells contain chlorides in addition to H_2S. Therefore, materials must be carefully selected.

Resistant alloys

Carbon and alloy steels will resist sulfide SCC provided their hardness is less than Rockwell 22. Post-weld heat treatment (PWHT) is generally recommended. These same requirements apply to cast steels such as types WCB and LCB. Although the base metal may be softer than Rockwell 22, "hard spots" are nearly always present in the weld heat-affected zone. PWHT will eliminate such spots.

The restrictions for Type 410 are similar to restrictions on the wrought and cast carbon steels, except that the maximum hardness allowed is Rockwell 25. The composition of Type 416 is the same as that of 410, except for the addition of sulfur to produce free-machining characteristics. This alloy is not, however, acceptable for sour service. As a matter of fact, no free-machining alloys are acceptable.

Annealed 304, 316 and 317 are three acceptable austenitic stainless steels. Components made from these materials must be free of cold-work and be softer than Rockwell 22. Standard, cold-worked 316 parts should not be used unless they have been annealed.

Some parts, particularly noise- and cavitation-reducing trim in valves, are made of 17-4PH (precipitation-hardening) stainless steel. When heat-treated in H900 condition, 17-4PH is quite brittle and extremely susceptible to stress-corrosion cracking. For H_2S service, the NACE specification requires that one use a double H1150 heat treatment to achieve maximum hardness of Rockwell 32. The double H1150 should also be considered for other corrosive services to prevent some other form of SCC where H_2S is not present. If the double treatment cannot be used, H1075, H1100, or H1150 (single) should be considered. These treatments reduce hardness, strength, wear, galling, cavitation and erosion resistance. If a reduction in these properties is unacceptable, use other materials.

Two other high-strength stainless steels acceptable for sour service are A-286 and S20190 (this is the standard designation for Nitronic§ 50 and Carpenter 22Cr-13Ni-5Mn alloys). A-286 is a precipitation-hardening alloy with excellent resistance to SCC and general corrosion.

S20190 in the high-strength condition has recently been approved by NACE for sour service. The maximum allowable hardness is Rockwell 35. Where the choice exists, S20190 should be selected over 17-4 PH for even mildly corrosive or sour applications.

The various nickel-based alloys are highly resistant to most forms of stress-corrosion cracking. This makes them very attractive engineering alloys, but they are more expensive than the stainless steels. Monel, Monel K-500, Inconel 600, 625, 718, and X-750 and Hastelloy‖ C and B alloys are used most commonly. Standard MR-01-75 limits all these alloys to Rockwell 35 maximum.

* Carpenter 20 CB-3 is a registered trademark of Carpenter Technology Corp.
† Incoloy is a registered trademark of Huntington Alloys, Inc.
‡ E-Brite is a registered tradename of Allegheny Ludlum Steel Corp.
§ Nitronic is a registered trademark of Armco Steel Corp.
‖ Hastelloy is a registered trademark of Stellite Div., Cabot Corp.

The author

James L. Gossett is engineering specialist, Fisher Controls International, Inc., Technical Center, P.O. Box 11, Marshalltown, IA 50158. Tel: (515) 754-3011 x2275. His areas of expertise include corrosion, stress-corrosion cracking, fractography, scanning electron microscopy, failure analysis, coatings, welding, and wear-resistant materials. He holds two degrees in metallurgical engineering—a B.S. from Purdue University and an M.S. from Stanford University. He is active in the Amer. Soc. for Metals and the Natl. Assn. of Corrosion Engineers.

How to prevent stress-corrosion cracking in stainless steels—I

Stress-corrosion cracking, or fear of it, is a major factor limiting the use of stainless steels in refineries and chemical plants today. However, it is not inevitable; it can be delayed, even eliminated, in many cases.

Dale R. McIntyre, *Battelle Memorial Institute*

☐ Stainless steels as a class offer the chemical engineer many attractive features: good corrosion resistance, weldability, fabricability and reasonable cost. However, any engineer hoping to take advantage of these benefits must consider the possibility of stress-corrosion cracking.

Stress-corrosion cracking (SCC) is an interaction between tensile stress and corrosion, which results in localized cracking. The cracking can take place at very low stresses and in environments where general corrosion, as measured by reduction in wall thickness, is negligible.

In austenitic-stainless-steel piping and vessels, SCC usually results in leakage, not catastrophic failure. However, severe strength loss and catastrophic failure are possible in extreme cases. Fig. 1 presents a photograph of a 6-in., Schedule 5, Type 304 stainless-steel transfer line that was inadvertently left half-full of brackish hydrotest water with the steam-tracing on. After one year, a 150-psi hydrotest caused a jagged 2-ft rupture due to the extensive stress-corrosion cracking in this line. Normal burst pressure of 6-in., Schedule 5, Type 304 stainless-steel pipe is 2,130 psi, so its strength was reduced 93% by SCC.

Stress-cracking agents in process streams are not the only cause of SCC; many failures take place due to traces of SCC agents in the air. Fig. 2 shows a Type 304 stainless-steel heat-exchanger flange so badly weakened by *external* SCC, from the breakdown of polyvinyl chloride (PVC) dust in the air, that pieces could be broken off by hand.

Conditions that cause SCC

SCC of stainless steels results from a combination of three conditions:

- A susceptible alloy-environment combination.
- Tensile stress.
- Elevated temperatures.

"Stainless steel" is a rather loose term denoting all iron-base, low-carbon alloys having a chromium content in excess of 11%. Since stainless steels vary widely in strength, structure and composition, it is not surprising that they also vary widely in resistance to SCC. Table I lists a number of environments in which the stainless steels are known to crack. However, Table 1 is meant as a guide only; stress-corrosion cracking is not inevitable, even in solutions in which a given stainless is listed as being nonresistant.

Environments that contain stress-cracking agents may be regarded as of three types:

- Process streams having high bulk concentrations of a

The strength of this Type 304 stainless-steel pipe was reduced 93% by SCC — Fig. 1

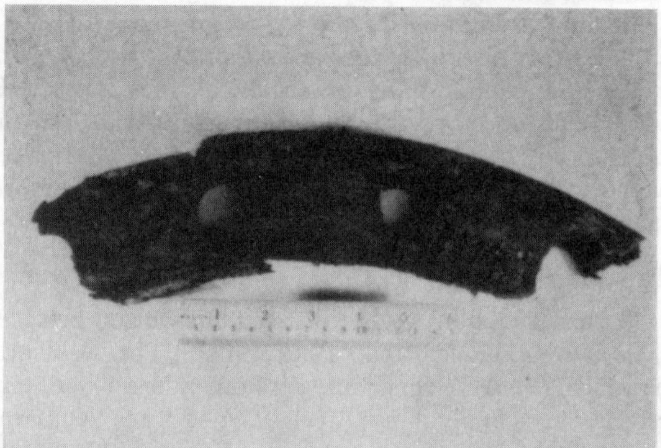

Type 304 stainless-steel heat-exchanger flange weakened by SCC due to PVC dust — Fig. 2

Originally published April 7, 1980

Aqueous environments that can cause SCC in stainless steels **Table I**

Material	Class	Cl^-, acid	Cl^-, neutral	Cl^-, oxidizing	Br^-	I^-	OH^-	F^-	$S^=$	$S^=/Cl^-$	$S_4O_6^=$	$SO_3^=$	$SO_4^=$	$CrO_4^=$	NO_3^-	NH_3	Ultra-pure $H_2O + O_2$	Seawater
		Environment																
405	Ferritic		1				1		4					1	5	1	1	X
18-2	Ferritic	4	1	4			3		1	1						1		X
26-1	Ferritic	4	1	4			4		2				5	1	1	1		X
26-1S	Ferritic	4	1	4			4		2				5	1	1	1		X
29-4	Ferritic	1	1	1			4		2					1	1	1		4
430	Ferritic		1						1	1				1	1	1		X
434	Ferritic		1							3								
431	Martensitic, quenched and tempered or precipitation-hardened	5	5	5			4	2	5				2	1	1	1		X
410	Martensitic, quenched and tempered or precipitation-hardened	5	2		5	5	2	5	5	5				1	2	1	2	X
CA-6NM	Martensitic, quenched and tempered or precipitation-hardened		5						5	2								X
440 A, B, C	Martensitic, quenched and tempered or precipitation-hardened	5	5	5					5									X
PH15-7Mo	Martensitic, quenched and tempered or precipitation-hardened	5	4					1	5				5	1	1	1		X
17-4PH	Martensitic, quenched and tempered or precipitation-hardened	5	4					1	5	5			5	1	1	1		X
PH13-8Mo	Martensitic, quenched and tempered or precipitation-hardened	5	4						5									X
17-7PH	Martensitic, quenched and tempered or precipitation-hardened	5	5					1	5									X
Custom 450	Martensitic, quenched and tempered or precipitation-hardened	4	1	4			4	2	2	2			2	1	1	1		X
Custom 455	Martensitic, quenched and tempered or precipitation-hardened	4	1	4			4	2	2				2	1	1	1		X
202	Austenitic	5	5	5		1	4	1	2		3	3	1	1	1	1		X
216	Austenitic	5	5	5		1	4	1	2		3	3	1	1	1	1		X
216L	Austenitic	5	5	5		1	4	1	2		3,4	3,4	1	1	1	1		X
Nitronic 50	Austenitic	5	4	5		1	4	1	2		3	3	4	1	1	1	1	4
Nitronic 60	Austenitic	5	5	5		1	4	1	2		3	3	1	1	1	1		X
304	Austenitic	5	5	5		1	4		2	5	3	3	4	1	1	1	3	X
304L	Austenitic	5	5	5		1	4	1	2	5	3,4	3,4	4	1	1	1	3,4	X
309S	Austenitic	5	5	5		1	4	1	2		3	3	1	1	1	1	3	X
310S	Austenitic	5	5	5		1	4	1	2		3	3	1	1	1	1	3	X
316	Austenitic	5	5	5	4	1	4	1	2	5	3	3	1	1	1	1	3	X
316L	Austenitic	5	5	5			4		2		3,4	3,4	1	1	1	1	3,4	X
317	Austenitic	5	5	5		1	4	1	2		3	3	1	1	1	1	3	X
317L	Austenitic	5	5	5		1	4	1	2		3,4	3,4	1	1	1	1	3	X
321	Austenitic	5	5	5		1	4	1	2	5	1	1	1	1	1	1	1	X
347	Austenitic	5	5	5		1	4	1	2	5	1	1	4	1	1	1	1	X
329	Duplex	4	5	5		2	4	2					1	1	1	1		X
3RE60	Duplex	4	1	5			3		4	4			1	1	1	1	3	4
18-18-2	Duplex	4	4	5		1	4	1					1	1	1	1		X

Code:
1 Resistant
2 Resistant unless cold-worked or hardened
3 Resistant unless sensitized
4 Resistant except at high temperatures and concentrations
5 Nonresistant
X Not recommended for this environment

Portions of this table have been taken from Ref. *1*.

stress-cracking agent (such as a chloride or sulfide).

- Process streams having low levels of SCC agents that concentrate at leaks, under deposits, and so on.
- External atmospheric conditions that contain low levels of an SCC agent.

Without doubt, the most common stress-cracking agent is the aqueous chloride ion. (SCC is an electrochemical process, and water is necessary to allow electron flow; completely dry chloride compounds are not normally cracking agents.) Common in brackish river water, seawater and coastal atmospheres, the chloride ion can cause SCC of austenitic stainless steels even at extremely low concentrations. Failures have been reported in steam condensate having as little as 0.5 ppm chlorides. Such low levels are not normally dangerous to stainless. However, in spots where evaporation and concentration raise the local level of chloride—such as crevices, deposits and liquid-vapor interfaces—cracking can still take place. Under these conditions, the only safe level of chlorides is zero.

Caustic environments may also crack stainless steels, and are perhaps the second most common cause of unexpected SCC failures. The austenitic stainless steels find many uses in such environments at temperatures below 150°C. Above that temperature, however, cracking can take place. Many high-pressure steam systems have low levels of caustic present that may concentrate at flange leaks, stainless valve stems, etc. and cause cracking.

The hardenable stainless steels, whether quenched and tempered or precipitation-hardened, encounter their worst problems in environments that involve exposure to ionic hydrogen. Stress-corrosion cracking takes place in these alloys via a hydrogen-embrittlement mechanism. As a result, their serviceability in corrosive environments depends on the generation and absorption of ionic hydrogen. Hardenable stainless steels crack most readily in environments containing ions, such as sulfides and arsenic, that "poison" the hydrogen recombination reaction. Coupling a hardenable stainless steel to a less-noble material, such that the stainless becomes the cathode in an electrolytic cell, frequently results in rapid stress-corrosion cracking of the stainless.

Tensile stresses are necessary for the propagation of stress-corrosion cracks. However, these stresses need not be applied ones; residual tensile stresses from forming, welding and heat-treating have the same effect. This is important to remember in pressure vessels and piping. Applied stresses are usually quite low, but welding and fabrication stresses are often at or beyond the yield point.

Stress-corrosion cracks usually require a certain amount of initiation time before they appear, even if all conditions required for cracking are present. In austenitic stainless steels, initiation time is often controlled by the pitting tendency, since cracks usually propagate from pits. Consequently, molybdenum-bearing alloys such as 316 and 317 will show longer initiation times than alloys such as 304.

Stress level plays an important role in the initiation time for SCC; the higher the stress, the shorter the initiation time. Hardenable stainless steels often exhibit a minimum critical-stress intensity, K_{ISCC}, below which cracks will not propagate (see Table II). Parts designed with stresses and geometries such that K_{ISCC} is not exceeded should not fail

Critical stress intensities for some hardenable stainless steels **Table II**

Material	Heat-treat condition	Critical stress intensity, ksi $\sqrt{in.}$ Air	Seacoast	3.5 NaCl %
15-5PH	H900	71.8	35.9	32.3
	H1150	75.7	>72.0	>72.0
AM355	SCT850	36.6	>18.3	5.5
	SCT1000	70.0	>35.0	28.0
431	TS = 125 ksi	75.7	>37.8	43.0
	TS = 200 ksi	79.2	<39.6	11.9
PH13-8Mo	H950	62.6	>31.3	>45.9
	H1050	87.8	>43.9	>65.8
PH15-7Mo	RH950	30.6	15.3	10.1
	RH1050	40.7	20.1	12.1
17-7PH	RH1050	47.0	11.7	9.4

H900 - Solution annealed and precipitation-hardened at 900 °F; RH - Solution annealed, refrigerated and precipitation-hardened; SCT - Sub-zero cooling transformation; TS - Hardened to this tensile strength. Data from Ref [10]. ksi = 1,000 psi.

Effectiveness of SCC prevention methods **Table III**

Prevention method	Ferritic	Austenitic	Martensitic	Duplex
Materials selection	1,2,3	1,2,3	1,2,3	1,2,3
Barrier coatings	3	3	4	3
Eliminating the stress-cracking agent	1,2	1,2	1,2	1,2
Adding inhibitors	NA	2	NA	NA
Thermal stress relief	4	4	NE	4
Heat treatment	*	*	1,3	NE
Shot peening	1,2	1,2	4	1,2
Cathodic protection	X	1, not 2	X	1, not 2
Lowering the temperature	2,3	2,3	NE	2,3
Design techniques	2	2	2	2

Code:

1 Eliminates SCC in process streams having high SCC-agent concentration.

2 Eliminates SCC in process streams having concentration effects (wetting and drying).

3 Eliminates external cracking from atmospheric SCC agents.

4 Will delay the onset of SCC.

NE Not effective for these alloys.

X Not effective, may accelerate cracking.

NA No industrial application reported.

* Effective only for intergranular SCC of sensitized material.

in service. Austenitic stainless steels also exhibit the K_{ISCC} behavior, but K_{ISCC} is so low in many chloride solutions (~6 ksi $\sqrt{in.}$) that it is impractical to design equipment made out of these steels for such conditions [2].

Temperature is the third element normally required to produce stress-corrosion cracking. Each alloy/environment system usually has a temperature range in which cracking occurs. Austenitic stainless steels, for example, seldom crack in chlorides below about 50°C, and cracking in caustic solutions is rare below 150°C.

Prevention of SCC

Knowledge of the necessary conditions for SCC suggests techniques for prevention. These are:

- Changing the alloy/environment combination.
- Eliminating tensile stresses.
- Cathodic protection.
- Lowering the temperature.
- Design techniques.

These options will be outlined more fully in the sections below. Table III presents a summary of the different methods for SCC prevention, and their observed effectiveness on different environments and alloy classes.

Changing the alloy/environment

Changing the alloy/environment combination can take the following forms: materials selection; barrier coatings; eliminating the stress-cracking agent; adding inhibitors.

Materials selection, which simply means the use of materials resistant to the environment in question, is without doubt the most effective method for controlling SCC. However, it is not always the most economical. Materials-selection options include: resistant alloys or nonmetals; and composite construction with bimetallic tubing or plate.

As Table I shows, several of the "lean alloy" ferritics such as 405, 409 and 430 are not susceptible to chloride SCC. However, these alloys should be used with caution since they pit readily in chloride-bearing streams. A leak from a pit is just as troublesome as a leak from a crack.

Several of the new low-interstitial ferritic and austenitic-ferritic alloys are serviceable in areas where the regular 300 and 400 stainlesses are not. In heat-exchanger tubing especially, 18-2, 26-1 and 3RE60 stainless steels have proven their worth in many hot chloride-bearing waters and process streams that quickly crack 304 and 316. For severe applications, titanium or the high-nickel alloys such as Hastelloy* C or G often are used.

As with any corrosion-control technique, the use of specialty stainless steels or more-exotic alloys must be justified economically, since these materials impose penalties of higher cost, more-difficult fabrication and longer delivery times. Ref. 3 outlines methods for calculating annual costs based on anticipated life.

Bimetallic tubes can prevent water-side SCC in heat exchangers [4]. Such tubes require ferrules at the tubesheet to effect a roll joint (Fig. 3). Bimetallic tubes with deoxidized high-residual-phosphorus (DHP) copper lightly drawn over 304 stainless steel currently have lower prices and better delivery times than many of the SCC-resistant specialty stainless steels [5].

For vessels, bimetallic plate, with a thin layer of

*Hastelloy is a trademark of Cabot Corp., Stellite Div.

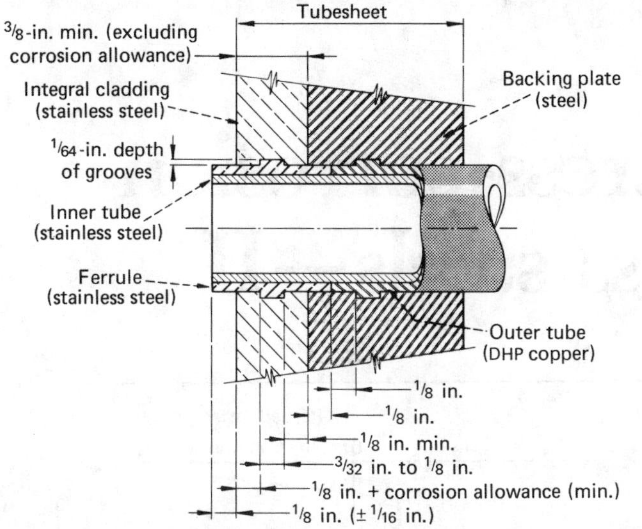

Typical joint for duplex tube and tubesheet in heat exchanger **Fig. 3**

stainless steel roll-bonded or explosion-clad to a carbon-steel backing plate, is sometimes used in environments that would stress-crack solid stainless. Even if the stainless cladding should crack, SCC will stop at the steel, and SCC cracks are usually too tight to allow significant amounts of corrodent to attack the steel. In thick sections, clad plate has the additional advantage of being about 25% cheaper than solid stainless. This approach should not be used in environments that stress-crack or severely corrode steel.

Materials selection for SCC resistance does not always mean more-expensive alloys. For instance austenitic-stainless-steel valve-bonnet bolts are particularly prone to SCC because bonnet leaks or atmospheric chlorides can concentrate under the bolt heads. Using B7 steel bolts coated with Xylan* resin over a zinc primer will eliminate this problem. Compared with stainless steel, the coated bolts are more resistant to SCC, as well as cheaper, stronger and less prone to galling.

Barrier coatings will prevent SCC in mild environments. Such coatings are often used to preclude external SCC under insulation, a widespread problem in coastal areas. Paint simply prevents chlorides from coming in contact with the stainless surface. For most coastal environments, a modified silicone paint is adequate [*6*]. In more-rigorous industrial environments, epoxy-phenolic coatings have been used. Insulated lines or vessels should be completely painted; uninsulated equipment need be painted only under slip-on flanges.

No paint system ever goes on without defects, so there is a possibility of SCC at pinholes and coating flaws. However, the incidence of external SCC can be greatly reduced with this method.

Eliminating the stress-cracking agent (which is often an unwanted impurity in a process stream) may not affect the process or the product, but will eliminate SCC.

Example—A 304L stainless-steel column stripped organic chlorides from an 80°C solvent containing 0.2% water. After 18 mo in service, wet chloride-bearing process residues had accumulated in the column and the trays, causing extensive SCC.

A new 304L column was installed, and the process was revised to remove the water upstream of the column. Chloride-bearing process residues still accumulate, but examination of stressed samples exposed in the column indicates that no further SCC has taken place.

All too often, however, stress-cracking agents cannot be eliminated economically. Chlorides are so widespread that it is virtually impossible to ensure their absence.

In some cases, inhibitors are effective in counteracting stress-cracking agents. In particular, alkaline washes are used to prevent polythionic acid SCC during shutdowns in oil refineries [*7*]. Laboratory studies have shown that 3% $NaNO_3$ will prevent stress-corrosion cracking of 304 stainless steel in boiling 42% $MgCl_2$, and that 0.005M Na_2CrO_4 will prevent SCC of 304 stainless steel in 20% NaOH [*8*]. No large-scale industrial application of these inhibitors has been reported.

Oxygen is apparently necessary for the SCC of austenitic stainless steels, and laboratory researchers have observed that mechanical or chemical deaeration will prevent SCC. In the field, however, failures have been reported in chloride-bearing water fully deaerated with an excess of sodium sulfite [*9*].

*Xylan is a trademark of Whitford Corp.

References

1. Dillon, C. P., Guidelines for Control of Stress-Corrosion Cracking in Nickel-Bearing Stainless Steels and Nickel-Base Alloys, MTI Manual No. 1, June 1979, Materials Technology Institute of the Chemical Process Industries.
2. Speidel, M. O., Stress-Corrosion Crack Growth in Austenitic Stainless Steel, *Corrosion* (Houston), Vol. 33, No. 6, June 1977.
3. Direct Calculation of Economic Appraisals of Corrosion Control Measures, Standard RP-02-72, Natl. Assn. of Corrosion Engineers, Katy, Texas.
4. Ashbaugh, W. G., and Doughty, S. E., Bimetallic Heat Exchanger Tubes Solve Dual Corrosion Problem, *Met. Prog.*, Dec. 1960, pp. 115-132.
5. Davisson, R. L., Elder, G. B., and Montero, A. J., internal memo, Union Carbide Corp., Nov. 8, 1979.
6. Ashbaugh, W. G., External Stress-Corrosion Cracking of Stainless Steel Under Thermal Insulation", *Mater. Prot.*, May 1965, pp. 18-23.
7. Protection of Austenitic Stainless Steels in Refineries Against Stress-Corrosion Cracking by Use of Neutralizing Solutions During Shutdown, Standard RP-01-70, Natl. Assn. of Corrosion Engineers, Katy, Texas, 1970.
8. Park, Y. S., Agrawal, A. K., and Staehle, R. W., Inhibitive Effect of Oxyanions on the Stress-Corrosion Cracking of Type 304 Stainless Steel in Boiling 20N NaOH solution, *Corrosion* (Houston), Aug. 1979, pp 333-339.
9. Dillon, C. P., Stress-Corrosion Cracking of Austenitic Stainless Steels, internal memo, Union Carbide Corp., Apr. 17, 1961, File No. MF: 760:17.
10. Sprowls, D. O., Shumaker, M. B., Walsh, J. D., and Coursen, J. W., Evaluation of Stress-Corrosion Cracking Susceptibility Using Fracture Mechanics Techniques, Alcoa Laboratories, Report on NASA Contract NAS8-21487.

The author

Dale R. McIntyre, formerly of Union Carbide Corp., is currently employed by Battelle Memorial Institute, 2223 West Loop South, Houston, TX, telephone (713) 877-8034. Mr. McIntyre received a B.S. in Metallurgical Engineering from Oklahoma U. in 1972 and an M.S. in Metallurgical Engineering from the U. of Missouri-Rolla (St. Louis Extension Center). He is a registered Professional Engineer in the state of Texas, and a specialist in corrosion and materials engineering. His published works include papers on metallurgical-failure analysis, electron fractography, hydrogen embrittlement, stress-corrosion cracking and corrosion control.

How to prevent stress-corrosion cracking in stainless steels—II

Stress-corrosion cracking in stainless steels can be eliminated by removing tensile stresses, providing cathodic protection, lowering the temperature, and using intelligent design techniques.

Dale R. McIntyre, ** Battelle Memorial Institute*

☐ In Part I of this two-part series [*1*], we looked at conditions that cause stress-corrosion cracking (SCC) and at various ways of changing the alloy-environment combination to prevent it. Here are some other ways of stopping this phenomenon, which is a major factor in limiting the use of stainless steel in the chemical process industries.

Eliminating tensile stresses

This is a very effective method for preventing stress-corrosion cracking. In process vessels and piping, stresses due to internal pressure are usually quite low, typically only 25% of the ultimate tensile strength. Most SCC failures are due to residual stresses from welding and fabrication [*2*].

The three most common methods for SCC prevention by eliminating tensile stresses are thermal stress relief, shot peening, and heat treatment.

Thermal stress relief for the 300 series stainless steels should be done above 1,600°F; lower temperatures will not completely eliminate fabrication stresses [*3*]. Thermal stress relief eliminates only fabrication stresses; obviously, service stresses from fit-up and internal pressure are still present. Still, the lowered stress levels may make the austenitic stainless steels economical by extending the initiation time.

Example—A flash tank handled a 120°C stream of water and organic chlorides. Rolled-in 316L stainless-steel tubes in the reboiler were failing by stress-corrosion cracking every six months. A seal-welded 316L stainless-steel tube bundle was stress-relieved at 1,650°F and installed in the reboiler. This bundle lasted four years before replacement became necessary.

Thermal stress relief will sensitize regular carbon grades of stainless steel, regardless of structure. In some environments, sensitization leads to accelerated intergranular attack (IGA), so this must be considered whenever stress-relieved equipment is used. However, a moderately high, predictable rate of IGA often is preferable to rapid, unpredictable SCC.

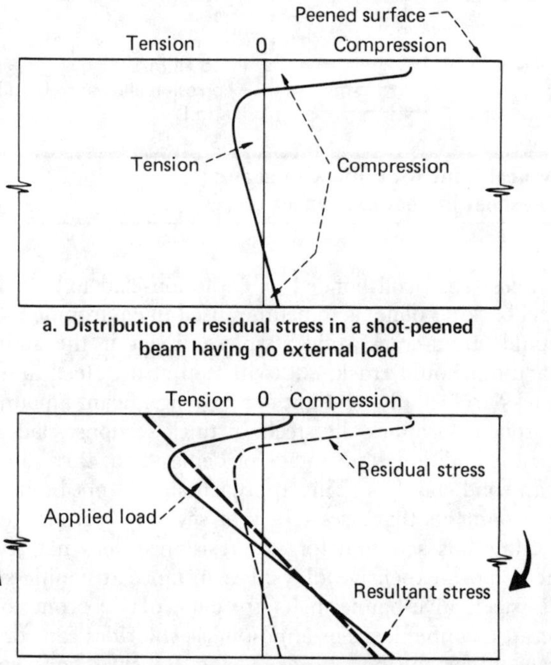

a. Distribution of residual stress in a shot-peened beam having no external load

b. Resultant distribution of stress in the same beam with external load applied

Drawing courtesy of Metal Improvement Co., Teaneck, N.J.

Stress distribution in a shot-peened item **Fig. 1**

Shot peening is one of the most promising methods for prevention of SCC. During shot peening, the wetted surfaces of the item are cold-worked with steel shot under carefully controlled conditions to produce a thin layer of metal with a net residual compressive stress [*4*]. Stress-corrosion cracks cannot propagate through compressive stresses; therefore, the SCC problem is eliminated.

Ordinary shot blasting, as used to clean steel surfaces before painting, is *not* equivalent to controlled shot peening and is *not* a reliable SCC prevention method. In controlled shot peening, shot size, shape and velocity and the intensity of cold-work are carefully and continuously monitored. No such controls are used in shot blasting.

*To meet the author, see p. 7.

Originally published May 5, 1980

Extensive SCC in the head of a 316 stainless-steel centrifuge handling a hot salt slurry Fig. 2

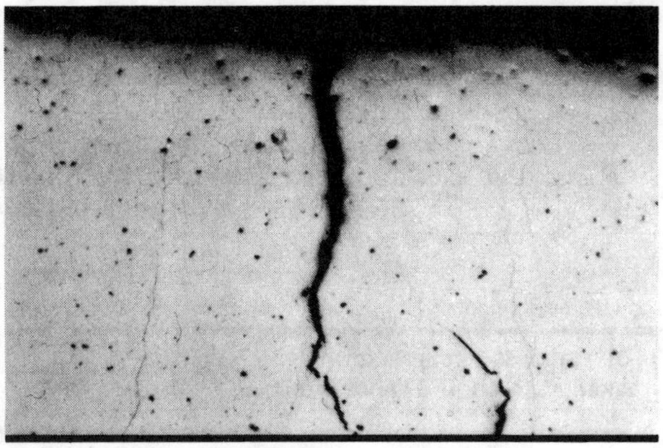

Cross-section of 304 stainless-steel check valve cracked by SCC after 3 wk in a 7% NaOH line Fig. 3

Done properly, shot peening produces a layer of material about 0.02 in. thick that has residual compressive stresses of yield-point magnitude. Not only are welding stresses cancelled out, but subsequent application of service stresses should still leave the surface with net compressive stresses. This is an important advantage over stress relieving, which may delay but not necessarily eliminate SCC. Theoretically, shot peening should eliminate SCC entirely.

For maximum effectiveness, shot peening should be the last manufacturing operation before the item is placed in service. No welding or torch heating should be performed on an item that has been shot-peened unless the heat-affected area is re-peened after heating. On process vessels, shot peening should be performed after the final hydrostatic test. The deliberately severe hydrotest stresses would reduce the magnitude of compressive stresses on a previously shot-peened surface.

Shot peening an item that is already stress-cracked will *not* retard further cracking and may, in fact, accelerate it. The residual compressive stresses induced on the surface by peening are balanced by subsurface residual tensile stresses of equal magnitude (Fig. 1). Peening an item already cracked may accelerate the crack growth, due to the presence of these added tensile stresses.

To be effective, peening must be done to at least 100% coverage (100% coverage is the time taken to completely cover the entire surface with shot; any additional time is referred to as being in excess of 100% coverage). Some applicators spread fluorescent dye over the surface before peening. After peening, black-light inspection quickly highlights any areas missed.

It must be remembered that the protective layer of compressive stresses is relatively thin (0.02 in. maximum). If pitting or general corrosion proceeds to the point where this layer is penetrated, SCC can then take place.

If the above-mentioned cautions are observed, dramatic savings can be obtained by the judicious use of shot-peened stainless instead of high-alloy equipment.

Example—A Type 316 stainless-steel centrifuge separated solids from a 60°C organic chloride stream. After one year in operation, extensive stress-corrosion cracking had taken place on all the wetted parts (Fig. 2). A replacement centrifuge made out of Hastelloy C, which would have been fully resistant to SCC, was estimated at $450,000. Instead, a new 316 stainless-steel centrifuge was purchased for $150,000. Shot peening added only $2,000 to the cost. The shot-peened 316 stainless-steel centrifuge was installed and, when last inspected after 18 mo in service, showed no evidence of SCC.

Shot peening has been used successfully on pump shafts, piping and process vessels. Shot-peening service centers have recently been able to shot-peen small-diameter heat-exchanger tubing; the cost makes 304 and 316 competitive with more-expensive specialty stainless steels for brackish-cooling-water service.

Heat treatment: Hardenable stainless steels are usually quenched from high temperatures. Quenching induces internal stresses from thermal contraction and structural changes that are great enough to propagate SCC. These stresses are reduced by tempering. Normally, the higher the temperature, the lower the internal stresses. Therefore, martensitic stainless steels often show marked variations in SCC susceptibility, due to differences in heat treatment. In general, the lower-strength heat treatments show better SCC resistance. For instance, 410 stainless steel shows its best cracking resistance when quenched and tempered to less than Rockwell "C" hardness 22 [*5*] and the martensitic precipitation-hardened stainlesses show markedly improved SCC resistance in the over-aged conditions [*6*]. This is unfortunate since a principal reason for using hardenable stainless is the need for high strength.

A solution-annealing heat treatment will prevent intergranular SCC of regular-carbon austenitic stainless steels that have been sensitized by welding. The high solution-annealing temperature (1,850-2,050°F) and the need for a water quench make this method impractical for many field applications. Solution-annealing is not effective in preventing the transgranular SCC normally observed in high-chloride environments.

Cathodic protection

Stress-corrosion cracking normally takes place in a fairly narrow range of potential. As a result, adjusting the potential with cathodic protection (CP) can prevent SCC. The following incident gives an example of cathodic protection on stainless steel that was totally inadvertent but, oddly enough, effective.

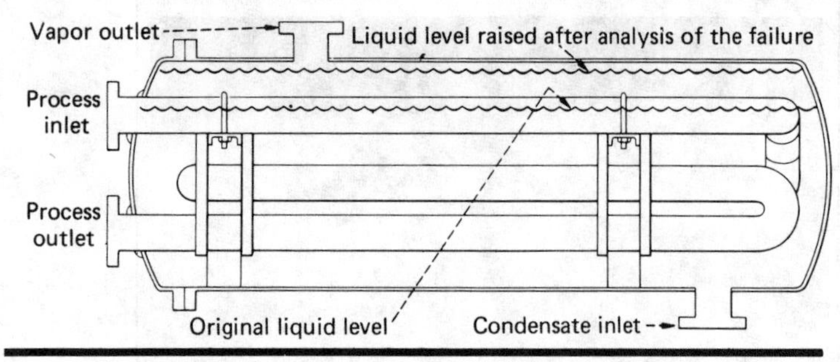

316L stainless-steel reactor in which the tubes cracked at the liquid-vapor interface **Fig. 4**

Example—A feed/tails intercooler cooled a 120°C tubeside organic stream with 50°C scrubber water on the shell. The scrubber water had about 700 ppm chlorides and 40-50 ppm organic acid, and was oxygen-saturated. The intercooler was originally installed with 316 stainless tubes, a 316 tubesheet and a Monel* shell. After four yr in service, an inspection revealed serious corrosion of the shell but, surprisingly, no damage to the 316 stainless tubes. A new shell of 316 stainless was ordered. Within a year, one-third of the tubes were leaking in the U-bends. Examination revealed classical transgranular stress-corrosion. Apparently the Monel shell and the copper ions in solution provided a form of cathodic protection to the tubes. When the Monel was removed, rapid SCC of the stainless resulted.

In general, galvanic coupling of heat-exchanger tubes with a less-noble shell material is not a reliable method for preventing SCC. The complex geometry of a heat-exchanger bundle normally results in some areas being "shadowed," or cut off from current distribution. However, the above example confirms laboratory studies indicating that cathodic protection is a valid option for SCC prevention.

A potential shift of at least 100 mV cathodic is required [7]. This can be accomplished either by impressed currents from a rectifier, or by galvanic coupling to sacrificial anodes of less-noble metals. Zinc, aluminum, lead and magnesium have all been used successfully for cathodic protection of Series 300 stainlesses. Iron is less successful, since the potential shift is not as large.

Cathodic protection should not be used on martensitic or precipitation-hardened stainless steels. Cathodic polarization or coupling to less-noble metals produces hydrogen at the surface of the stainless, which will greatly *accelerate* cracking in stainless having martensitic structure.

Cathodic protection works best for austenitic stainless immersed in neutral or near-neutral chloride solutions. As the pH of the solution becomes more acidic, larger and larger currents are required to produce the needed potential shift. The current demand for strongly acid solutions may make this method impractical [8].

Many SCC failures are due to concentration of low levels of chlorides under deposits, behind flanges or in vapor spaces above a liquid level. In such cases, impressed-current CP or sacrificial anodes would not be effective, since they require electrical continuity to function. In some cases, sacrificial coatings such as zinc or solder have been applied. However, the life of many such coatings in chloride-bearing waters is short, and once the coating is consumed, the stainless is left unprotected.

*Monel is a trademark of International Nickel Co.

Another concern with sacrificial coatings using lead, aluminum, zinc or cadmium is the relatively low melting points of these active metals. During welding or fire exposure, contact with such liquid metals may cause catastrophic fluxing attack on the stainless (in the case of lead and aluminum) or intergranular cracking (with zinc and cadmium) [9].

Temperature reduction

As stated earlier, stress-corrosion cracking usually takes place above some threshold temperature. Exposure to the same environment below that threshold temperature will not cause cracking.

For austenitic stainless steels in chloride solutions, the threshold temperature is approximately 50°C. In caustic solutions, the threshold temperature is about 150°C. Where process constraints permit, lowering the temperature below threshold can render a stress-cracking environment innocuous.

Lines carrying products with high viscosity or high freezing points must be heat-traced to keep the products fluid. All too often, such lines are steam-traced at temperatures far in excess of those required to keep the products moving. Rapid SCC can result.

Example—A run of 304 stainless steel tubing was installed to carry a 7% NaOH solution to a cooling-water treatment pump. This line was heat-traced to keep the NaOH liquid during the winter. The handiest source of heat was a nearby superheated 200-psi steam header that normally operated at 290°C (550°F). After three weeks in service, the line cracked at several valves, fittings and bends in the tubing (Fig. 3). All the cracks were found to be due to stress-corrosion. 70-psi steam (132°C, or 270°F, which is below the threshold temperatures for cracking) was substituted for the high-temperature steam; after a year, no further failures have been reported.

Whenever possible, electric tracing is preferred to steam tracing, mainly because of the lower temperatures normally employed.

Design practices

A high percentage of SCC failures take place in aqueous streams that have harmlessly low chloride levels in the bulk composition. Stress-corrosion cracking results from local concentration to high levels due to crevices, evaporation, intermittent immersion, and so forth.

Thus, the most fruitful area for improved design is in the avoidance of places where chlorides can concentrate.

Example—A 316-L stainless-steel serpentine reactor used condensate on the shell side to control the reaction temperature at 190°C. By chance, the liquid level was normally maintained in the middle of the top bank of tubes (Fig. 4). Condenser leaks occasionally resulted in organic chlorides being introduced into the unit condensate header in small amounts (the average chloride-ion concentration in the condensate was 13 ppm).

After nine months in service, the reactor began leaking; examination indicated extensive stress-corrosion in the top

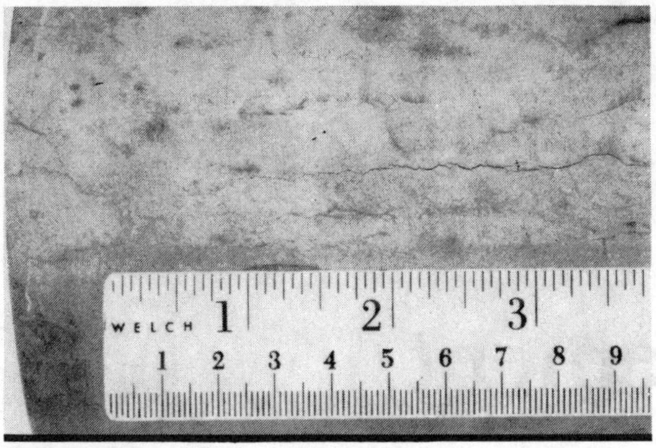

Outer surface of 316-L stainless-steel reactor tube **Fig. 5**

row of tubes (Fig. 5 and 6). No tubes below the liquid level were cracked.

Alternate wetting and drying on the partly exposed surface of the top row of tubes resulted in concentration of the low level of chlorides in the condensate to levels high enough to cause SCC. The number of condenser leaks was reduced, thus reducing chloride levels in the condensate. However, it was considered uneconomical to render the condensate completely free of chlorides.

In this case, the problem was solved by simply raising the liquid level in the shell about a foot. This completely covered the top row of tubes and eliminated the vapor space where chlorides could concentrate. To date, two and a half years later, no further SCC failures have occurred.

As stated earlier, the initiation time for SCC is often controlled by the pitting tendency, since cracks usually start at pits. Pitting is aggravated by stagnant conditions and low flow rates. Consequently, designing for high flowrates on stainless equipment (>3 ft/s in brackish river water) minimizes pitting under deposits and hence minimizes the possibility of SCC.

When designing heat exchangers, putting the cooling water on the tube side will often give velocities high enough to prevent pitting and thus minimize stress-corrosion cracking. For instance, 316 stainless steel is not recommended for seawater, yet some successes have been reported using 316-tubed condensers with seawater on the tube side [*10*]. However, the tubes in these cases were carefully cleaned of all deposits on a weekly or monthly schedule.

Repair-welding stress-cracked vessels

Attempts to weld up the cracks in a stress-corrosion-cracked stainless steel vessel often meet with failure, especially if welding heat is applied directly to the cracks. Thermal stresses and corrodents trapped in the cracks combine to cause the cracks to propagate ahead of the welding arc. Arc-gouging out the cracks, or grinding them out, will also cause the cracks to run ahead of the heated zone, and should therefore be avoided.

A procedure that has been used with some success:

- Clean the vessel wall down to bare metal.
- Define the crack as precisely as possible, using dye penetrant testing.

Photomicrograph of cracked reactor tube, showing branching transgranular cracking **Fig. 6**

- Mark out a rectangle enclosing the crack with at least 6 in. of apparently sound metal on all sides.
- Apply torch heat around the boundary of the rectangle. Any tight, previously unnoticed stress-corrosion cracks will open up under the heat.
- Dye-check the heated area.
- If no cracks are observed, proceed by torch-cutting out the marked area and welding in a patch.
- If new cracks show up in the heated area, however, move back another 6 in. all around and try again.

If still more cracks appear on the second attempt, SCC is widespread, and weld repair is impractical.

References

1. McIntyre, D. R., How to Prevent Stress-Corrosion Cracking in Stainless Steels—I, pp. 4–7 in this book.
2. Izumiyama, M., Problems Concerning Stress-Corrosion Cracking of Austenitic Stainless Steel Used in Chemical Plants, *Corrosion* (Houston), Jan. 1979, pp. i, ii.
3. Sangdahl, G. S., "Stress-Relieving of Austenitic Stainless Steel," Metals Handbook, 8th ed., Vol. 2, pp. 253-254.
4. Friske, W. H., Shot-Peening to Prevent the Corrosion of Austenitic Stainless Steels, Report No. AI-75-52, Rockwell Intl. Corp., Atomics Intl. Div., Sept. 15, 1975.
5. Burns, D. S., Laboratory Test for Evaluating Alloys for H_2S Service, *Mater. Perform.*, Jan. 1976, pp. 21-28.
6. Armco 17-4PH Precipitation-Hardening Stainless Steel, Armco Steel Corp. Product Bull. S-6C.
7. Mears, R. P., Brown, R. H., and Dix, E. H., A Generalized Theory of Stress-Corrosion of Alloys, Proc., Symp. on Stress-Corrosion Cracking of Metals, ASTM/AIME, 1944, p. 323.
8. Laliberté, L. H., Mueller, W. A., and Sharp, W. B. A., Cathodic Protection of Stainless Steels in Bleach Plant Filtrates: A Laboratory Investigation, presented at Corrosion/78, sponsored by the Natl. Assn. of Corrosion Engineers, Mar. 6-10, 1978, Houston, Tex.
9. Sadigh, S., The Effect of Molten Zinc on Some Commercial Alloys, presented at Corrosion/79, sponsored by the Natl. Assn. of Corrosion Engineers, Mar. 12-16, 1979, Atlanta, Ga.
10. Deverell, H. E., and Maurer, J. R., Stainless Steels in Sea Water, *Mater. Perform.*, Mar. 1978, pp. 15-19.

Guarding against hydrogen embrittlement

Four common forms of this potentially catastrophic phenomenon are discussed.

R. S. Treseder, Consultant

☐ Hydrogen embrittlement is defined as the loss of ductility in metals resulting from absorption of hydrogen. The small size of the hydrogen atom permits it to permeate metals in the atomic (not molecular) form. Absorbed hydrogen can react irreversibly with some constituents of the metal and reduce ductility, or it can have a reversible effect that influences ductility only as long as the hydrogen is present.

Further complexities arise from differences in metals and in sources of the absorbed hydrogen. Thus the many phenomena that come under the general heading of hydrogen embrittlement are complex in both the mechanistic and the engineering sense. This leads to semantic problems, since specialists differ in the terms they use to describe the various phenomena. We shall try to avoid such problems by defining each hydrogen phenomenon prior to discussing it.

Though this article will discuss four forms of hydrogen embrittlement of ferrous alloys that are of practical interest to the chemical process industries, there are problems with other alloy systems as well. For example, hydrogen embrittlement is a potential problem with both titanium and tantalum.

High-temperature attack

Above about 430°F (220°C), hydrogen atoms permeating carbon steel can gradually react with iron carbide in the steel to form methane. The effect is two-fold: (1) decarburization of the steel with resultant loss of strength; and (2) fissuring of the steel by the pressure of methane formed at the grain boundaries.

In mild cases, it may take years for the mechanical properties of the steel to be seriously affected. However, in an extreme case of hydrogen damage, a steel actually lost 60% of its tensile strength; its ductility, as measured by elongation in 2 in., was reduced from 30% to nil.

The source of hydrogen atoms can be the dissociation of molecular hydrogen under conditions of high pressure and high temperature, as in hydrogenation processes, or the generation of hydrogen atoms on the surface resulting from a high-temperature corrosion reaction involving water, as in high-temperature steam generation. Severity of the hydrogen attack increases with increasing temperature and with increasing hydrogen pressure.

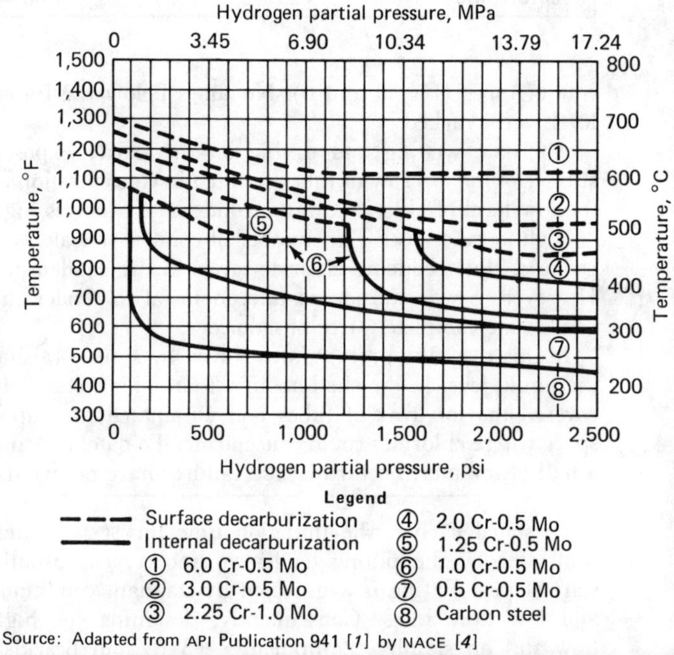

Source: Adapted from API Publication 941 [1] by NACE [4]

Fig. 1 Operating limits for steels in hydrogen service

Fortunately, a practical engineering solution to the above problem has been available for many years. It involves use of low-alloy steels containing carbide stabilizers such as chromium, molybdenum, tungsten, vanadium, titanium and niobium. These reduce the reactivity of carbon with the absorbed hydrogen. Experience has resulted in a definition of the acceptable limits of temperature and hydrogen partial pressure for each of the common low-alloy steels generally considered for construction of hydrogenation equipment. These limits are expressed graphically in the well-known Nelson curves [1], a simplified version of which is shown in Fig. 1. (These curves are revised periodically by a committee of the American Petroleum Institute.)

Although materials selection problems for conventional hydrogenation processes can be handled with the Nelson curves, there is a need for improved mechanistic understanding of the phenomena so that predictions

Originally published June 29, 1981

can be made for new steel/environment combinations through extrapolation of short-term tests.

Hydrogen stress cracking

This form of hydrogen embrittlement is defined as a failure process that results from the initial presence or absorption of hydrogen in metals in combination with residual or applied stress. It occurs most frequently with high-strength alloys. Primarily a problem at temperatures below about 250°F (160°C), the effects are often most severe around room temperature.

For those cases where the hydrogen is initially present in the steel, the effect has been called *delayed fracture* or *static fatigue,* since failure usually occurs with some time delay after the steel is put in service. The hydrogen initially present can be from several sources:

1. A manufacturing or welding operation in which reaction of moisture with molten metal results in hydrogen entry into the steel.
2. A prior process condition, such as in a hydrogenation, where hydrogen enters the steel as a result of high-pressure, high-temperature conditions, and is retained in the steel during subsequent cooling.
3. An electroplating or other electrolytic process, where the steel was cathodic in an operation where hydrogen was being generated on the steel surface.

If the hydrogen is absorbed after the steel is put in service, other terms are used to describe the effect. If the absorption results from contact with high-pressure hydrogen gas, the term hydrogen stress cracking or hydrogen embrittlement is used. If the hydrogen is absorbed in connection with a corrosion reaction, the definition problem becomes controversial. Some prefer to consider it as a form of stress-corrosion cracking, while others prefer to call it hydrogen stress cracking or hydrogen embrittlement. If the corrosion reaction involves hydrogen sulfide, the commonly accepted term is sulfide stress cracking; because of its engineering significance, this phenomenon is discussed in a separate section.

Contributing to the above semantic confusion is the possibility that in some steel/environment combinations cracking failures may result from either of two mechanisms: absorbed hydrogen as the dominant factor in growth of the crack (hydrogen stress cracking); or an anodic process being the dominant factor in crack growth (anodic stress-corrosion cracking).

One characteristic of the various forms of hydrogen stress cracking is the effect of temperature. The tendency for cracking decreases with increasing temperature, with significant changes occurring above about 150°F (70°C). Another characteristic is the tendency for cracking to be associated with strength of the steel, stronger steels being much more susceptible to cracking. Except where the hydrogen source is a corrosion reaction involving either hydrogen sulfide or hydrofluoric acid, hydrogen stress cracking is not considered to be an engineering problem with most steels having yield strengths below about 150 psi (1,000 MPa). With the two acids named, this limit drops to about 80 psi (550 MPa).

Control of hydrogen stress cracking is generally achieved by use of low-strength steels in preference to high-strength steels. If the source of hydrogen is a corrosion reaction, then it is obvious that stopping the corrosion reaction will prevent hydrogen stress cracking. However, the rate of corrosion required for failure may be well below the values normally considered acceptable from a weight-loss criterion. The amount of absorbed hydrogen needed to cause cracking is low, a few ppm or less.

For those cases where hydrogen is initially present, cracking can be prevented by heating the steel to about 200°C prior to exposure to stress. The higher diffusion rate of hydrogen at elevated temperature will permit the hydrogen to escape in a relatively short time. This practice is used to avoid hydrogen stress cracking of electroplated items. The same idea has been incorporated in shutdown procedures for hydrogenation equipment where there is concern about possible hydrogen stress cracking of high-strength steel components.

Sulfide stress cracking

This effect is defined as brittle failure by cracking under the combined action of tensile stress and corrosion in the presence of water and hydrogen sulfide. It is a problem of great economic importance in the petroleum industry, being encountered in the production and transportation of sour natural gas and sour crude oil, and in oil refining. When steel is corroded by aqueous solutions of hydrogen sulfide, a much higher percentage of the hydrogen generated at the cathodic areas enters the steel as atomic hydrogen than is the case with other weak acid systems such as carbon dioxide and water. This is attributed to a catalytic effect of the iron sulfide film formed on the metal surface. A typical failure is shown in Fig. 2.

As noted above, susceptibility to sulfide stress cracking increases with increased strength of the steel, with the critical strength level being about 80 psi (550 MPa) yield strength. A number of metallurgical factors influence this critical strength level, including steel composition, heat treatment and degree of cold working. Quenched and tempered steels are preferred to normalized steels. Cold working has a pronounced adverse effect. Free-machining steels are more susceptible to sulfide stress cracking than other steels at the same strength level.

Materials selection is the major method of controlling sulfide stress cracking. Because of the influence of many metallurgical factors, acceptable steels must be defined carefully. The oil industry and its suppliers have done this in a standard prepared by a National Assn. of Corrosion Engineers committee, based on the accumulated experience of the industry. This standard is known as NACE Standard MR-01-75; the latest revision is dated 1980 [*2*]. For most carbon and low-alloy steels this standard calls for a maximum hardness of HRC 22. Hardness is used as the criterion for acceptability because it is a nondestructive measure of strength that can be used on finished equipment. For those applications where higher-strength materials are needed, the standard lists a number of nonferrous alloys as acceptable up to a hardness of HRC 35.

The above standard indicates sulfide stress cracking to be a potential problem in natural-gas environments with a hydrogen sulfide partial pressure greater than

Sulfide stress cracking of a centrifugal compressor impeller **Fig. 2**

Hydrogen blistering of steel equipment **Fig. 3**

0.05 psia. However, lower values are often selected as control limits when this factor is used as a means of preventing sulfide stress cracking. Temperature control has also been used as a means of preventing sulfide stress cracking, taking advantage of the observation that susceptibility to cracking of many of the steels used in oil production decreases dramatically above about 150°F (65°C).

Hydrogen blistering

This effect is defined as subsurface voids produced by hydrogen absorption in (usually) low-strength alloys with resultant surface bulges. It results from the recombination of absorbed hydrogen atoms to form molecular hydrogen at very high pressure at internal discontinuities such as inclusions. A typical example is shown in Fig. 3. This effect was experienced first in storage tanks and vessels handling sour gas or sour oil, and resulted from corrosion by the combination of moisture and hydrogen sulfide. Later, hydrogen blistering became a serious problem in process vessels in petroleum refining processes where the process stream contained moisture plus hydrogen sulfide, ammonia and hydrogen cyanide. The general solution to the problem has been by control of the environment, often by addition of a corrosion inhibitor. The inhibitor commonly used for the refinery case is polysulfide ion, either added as such or formed *in situ* by addition of oxygen (air) to the process stream.

Recently a related problem has been encountered with some pipelines made of welded pipe that has been exposed to moisture and hydrogen sulfide. This effect has been characterized by an internal crack pattern in which cracks along parallel planes in the rolling direction of the plate used to make the pipe are joined by short transverse cracks in such a way as to affect the integrity of the pipe. This variation of hydrogen blistering is called stepwise cracking by some; others use the term hydrogen-induced cracking.

Study of the problem has concentrated on the effects of steelmaking practices, particularly those practices that control the nature and distribution of inclusions [*3*]. A committee of the NACE is examining laboratory test methods that could be used to evaluate the resistance of pipeline steels to stepwise cracking. The laboratory test currently favored consists of exposing a test specimen to an aqueous solution of hydrogen sulfide. The specimen is then given a visual and metallographic examination to determine the number, size and distribution of cracks resulting from the blistering reaction. These results are then correlated with actual service experience.

Monitoring

In-plant monitoring of process conditions that might lead to hydrogen embrittlement effects is being done by use of hydrogen probes. These are of two types: one consists of a thin-walled, closed-end steel tube inserted into the process stream, the other is in the form of a patch on the outside of the pipe or vessel. In each case the permeation rate of hydrogen through the steel is measured. These data can be correlated with plant conditions, and used to monitor mitigation measures such as inhibition.

References

1. API Publication 941, Steels for hydrogen service at elevated temperatures and pressures in petroleum refineries and petrochemical plants, second ed. 1971.
2. NACE Standard MR-01-75 (1980 Revision), Sulfide stress cracking resistant material for oil field equipment.
3. Bruno, T. V., and Hill, R. T., Stepwise cracking of pipeline steels—A review of the work of Task Group T-1F-20, presented at NACE Annual Conference, Chicago, March 1980.
4. Treseder, R. S., Ed., "Corrosion Engineer's Reference Book," NACE, 1980.

The author

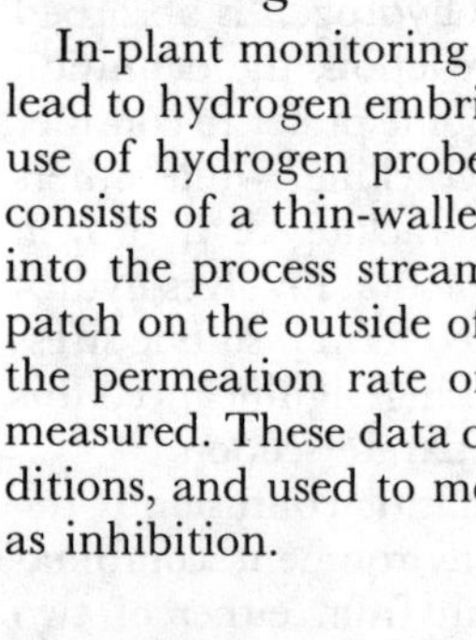

Richard S. Treseder is a consultant in corrosion engineering, specializing in process industries and oil production corrosion. He is located at 6272 Girvin Drive, Oakland, CA 94611. Tel.: 415/531-9560. He spent 34 years with Shell Development Company at its Emeryville Research Center. Mr. Treseder holds a B.S. in chemical engineering from the University of Utah. He has been active in the National Association of Corrosion Engineers and the American Society for Metals. He is a Fellow of ASM and was given the NACE Speller Award in 1966.

Preventing unscheduled shutdowns

Through the use of online monitoring devices, dangerous corrosive conditions can be detected and corrected without interrupting production.

C. F. Britton, Corrosion Monitoring Consultancy

☐ In selecting materials for process plants, the main priority is strength, coupled with the confidence that this will not be affected for the lifetime of the plant. Corrosion can cause a loss of strength.

Several methods for obtaining online corrosion information from process plants are available. These need to be supported by other corrosion-monitoring methods such as coupons or ultrasonics. Eventually, plants will be equipped with automated corrosion control (such as inhibitor addition) triggered by means of online monitoring devices.

Materials selection is generally based on information supplied by manufacturers, which is based on corrosion tests carried out in the laboratory. Corrosion tests are remarkably prone to error, and the prediction of long-term behavior in laboratory accelerated tests is difficult. Of course, if the plant to be built is identical to an existing plant and is using the same technology, this background of experience will provide considerable corrosion information, especially if inspection experience is considered.

Variance in process conditions, such as temperature, pressure and flowrate, can lead to corrosion problems, as can increased product output.

Also important is the presence of particular corrodents in the plant. The tables of corrosion data supplied by metal manufacturers give corrosion rates in chemicals at specified concentrations, purities and temperatures. In practice, however, alloy compositions can cover a wide range of tolerance, and foreign substances can often be present. The unintentional presence of other metals in the plant (such as carbon steel in a stainless-steel reaction vessel) can have an adverse effect on the corrosion behavior of stainless steel.

The presence of unaccounted-for chemical ions, arising as impurities or from upsets in operation, can also seriously increase corrosion, as can the production of intermediate chemical compounds in many processes.

The above are some examples whereby corrosion is caused by factors not taken into account at the design stage. The prevention of unscheduled plant shutdowns and consequent product loss, and avoidance of danger to personnel, are continuous and onerous responsibilities of the inspection department.

The aim of the inspection engineer is to obtain a measurement of wall thickness and to compare this with previous measurements. Unfortunately, since corrosion rates are not measured with the preciseness of other plant parameters, several methods of corrosion monitoring must be used together and the results compared. Corrosion is a complex phenomenon that can occur in many forms. It is unrealistic to expect one method to detect all of these; in fact, there are specific forms of corrosive attack that cannot yet be monitored.

The inspection engineer relies on several monitoring methods: coupons (for baseline data); appropriate online electronic sensors; and ultrasonics, a form of nondestructive testing (NDT). Data from these methods will give information on general corrosion and to a lesser extent on localized or pitting corrosion. Other techniques are based on radiography, eddy currents and acoustic emissions.

Of course, measurement of the corrosivity of the process stream provides a criterion to assess the efficiency and optimum dosage of chemical corrosion inhibitors. This type of measurement is extremely useful and is widely used in recirculating cooling-water systems and oil-production waterfloods (used for secondary recovery of oil). Some sensing systems produce real-time electronic signals; thus, process conditions can be correlated with "corrosivity" signals. If process conditions are upset, corrective action can be taken without delay.

It is important to correlate all the factors that relate to corrosion and keep the necessary records on hand. The microprocessor will obviously make its contribution here, as well as the computer used for plant control. By these means, a corrosion signal can be used as a process-control parameter and recorded in the control-room for eventual correlation with records obtained by other methods. Let us examine the available methods in more detail.

Coupons

This method is the baseline for any corrosion inspection program. Coupons are specimens of the metal of

Originally published June 1, 1981

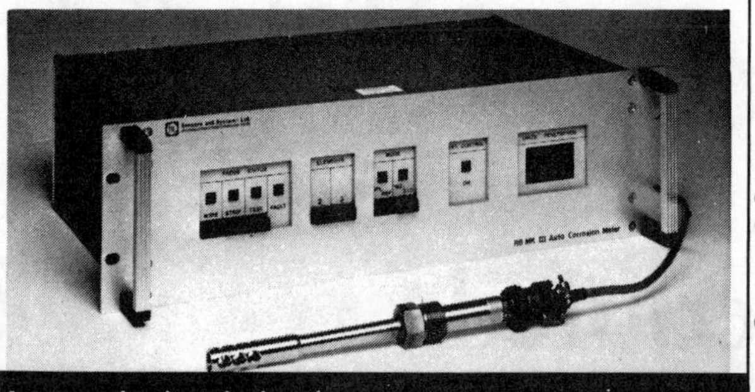

Automatic electrical-resistance corrosion monitor and sensor Fig. 1

Courtesy: Sensors & Systems, Ltd.

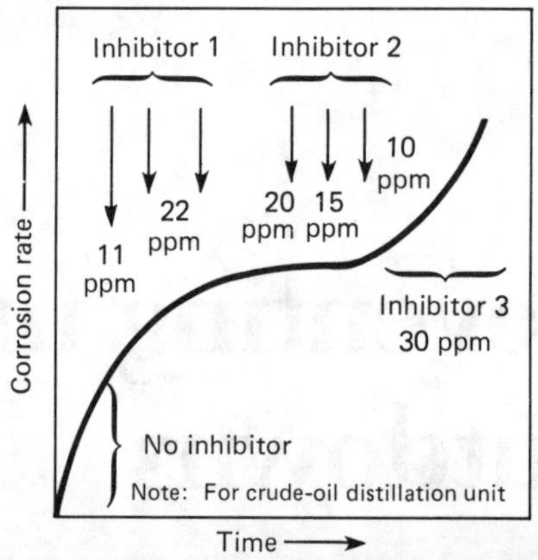

Online electrical-resistance corrosion data correlated with operating conditions Fig. 2

interest (generally, 76.2 x 25.4 x 3.2mm) drilled with a supporting hole, surface treated and mounted on a suitable holder for exposure to the process stream. Before exposure, the coupons are weighed and the surface area recorded. After exposure, the coupons are chemically cleaned (in an acid or electrolytic-bath treatment) and reweighed. The weight loss is converted into a penetration rate expressed as either mils/yr (mil = 0.001 in.) or mm/yr. Procedures for coupon preparation, cleaning and reporting of results are published by the National Assn. of Corrosion Engineers and the American Soc. for Testing and Materials.

Accuracy of results depends on good management regarding the laboratory treatment of the coupons and eventual data presentation. The results obtained are an average assessment and do not inform the process engineer that an upset to the system occurred resulting in serious corrosion at a specific hour on a certain shift. Coupons can be inserted and retracted from plant processes by use of special fittings and retrievers.

Electrical resistance (ER)

A metal loop (similar to a hairpin), made of the metal of interest, is mounted in a holder called a probe. The probe is fitted so that the loop is exposed to the process stream. As the loop loses cross-sectional area due to corrosion, its electric resistance increases. This change is monitored by a portable instrument or one permanently connected to control-room or data-handling equipment (Fig. 1). Equipment is available that can handle 30–40 probe channels. Other elements are available, such as strips, flush-strips and tubes. The elements chosen depend on the particular application, sensitivity required and duration of test.

The system can be used in any liquid or gaseous environment and has the advantage that online data are obtained that can be correlated with process conditions, upsets, or any methods used to mitigate the corrosion (Fig. 2). The method, though, is still averaging the corrosion rate, similar to coupon exposure. The increase in electrical resistance can be related to diameter reduction of the element due to corrosion, which can be expressed as a penetration rate in the same units as given for coupons.

Polarization resistance (LPR)

This method is very interesting because it can respond instantaneously to a change in corrosivity. A sensor or probe represents, on a macro scale, the actual corrosion cell on the metal surface of interest. By making potential/current measurements, an estimate of the actual corrosion current can be made. This method is a simplification of laboratory electrochemical techniques used for studying corrosion reactions. There are several versions available, such as D.C. and A.C., for liquids only, and A.C. impedance. Using A.C. impedance, measurements can be made in more resistive corrodents such as crude oil or even concrete. This method is still under development.

Successful application of LPR techniques has been reported in multiphase systems such as oil/brine; however, relevance to actual corrosion rate is sometimes doubtful.

The system approach is similar to the electrical-resistance method, in that a probe is exposed to the process stream and read either by a portable instrument or one permanently connected to the control room. Multichannel instrumentation is available. Portable instruments are particularly useful for surveys of corrosion problems in many industrial situations—a good example being measurement in a cooling-water circuit.

Potential measurement

Measurement of potential is a very useful technique for assessing the conditions of a number of metals used for plant construction. The information can indicate whether the metal is passive (noncorroding) or active (corroding), and if combined with LPR techniques can provide more-detailed information.

Equipment is simple, only needing an appropriate reference electrode (exposed to the corroding medium) and a voltage-measuring instrument, such as a pH meter or electronic voltmeter. The readings can be continuously recorded in the control room with the other process parameters and can be used as an indicator as to the passivity or activity of the metal of interest. Penetration rates are not obtained.

Laboratory investigation into the actual plant situation can often provide the criteria basis for the plant operators. A drawback to potential monitoring (except when used with cathodic protection systems) is that spe-

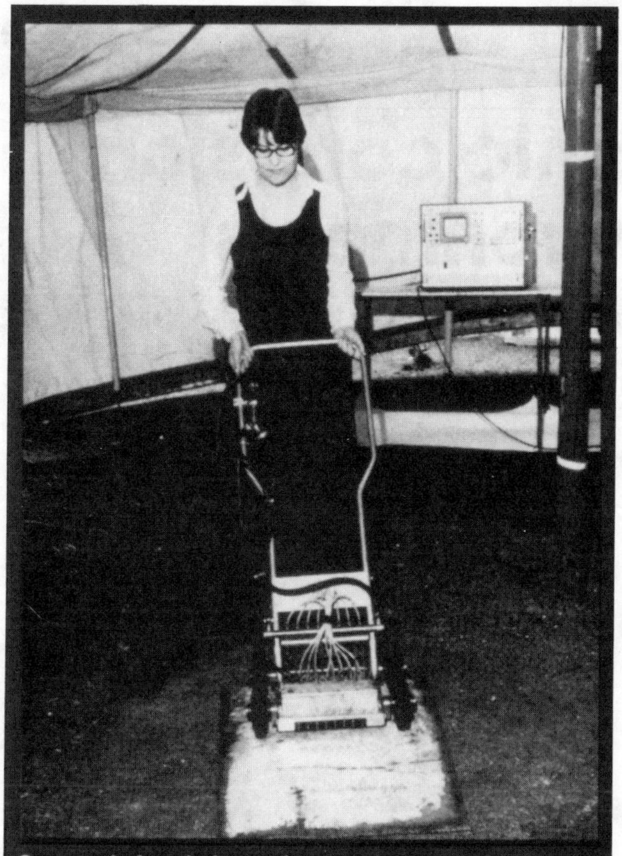

Custom-designed ultrasonic corrosion-monitoring system for storage-tank floors Fig. 3

cialized corrosion knowledge is required to set up the installation and interpret the information obtained.

Galvanic corrosion

This type of measurement assesses the possibility of corrosion between two different metals. A probe containing two electrodes, one of each of the metals in question, is exposed to the corroding liquid. A zero-resistance ammeter is used to measure the current produced by the galvanic cell. The magnitude of the current indicates the tendency for galvanic corrosion to occur if the two metals are coupled together and exposed to a corroding liquid. Penetration rates are not obtained.

Hydrogen monitoring

Hydrogen is a product of a corrosion reaction and can be measured by a hydrogen probe mounted in the plant. Atomic hydrogen diffuses through a thin wall of the probe, is combined into molecular hydrogen, and is registered on a Bourdon gage mounted on top of the probe.

A recent development has been the use of electrochemical cells for hydrogen measurement. The advantage is that the hydrogen present can be related to an electric current for recording/alarm purposes. There is no need to penetrate the plant wall, as the cells can be strapped to the outside wall and are easily moved from location to location. Information is transmitted to the control room. These methods for hydrogen measurement are extremely useful, as the hydrogen produced can be dangerous due to hydrogen embrittlement.

Nondestructive testing (NDT)

NDT consists of a number of methods, of which ultrasonics, eddy currents, radiography, acoustic emissions and thermography are the most important. The most common method is ultrasonics, which is frequently used to detect metal loss in industrial processes. A beam of ultrasound is transmitted through the pipe wall. The beam is reflected at the inner surface and is picked up by the receiver part of the probe. The "time of flight" can be related to metal thickness, using the appropriate factor for the metal in question. The thickness measurement can be displayed in digital form.

Problems associated with this technique are: (a) there is difficulty in coupling the probe to the metal surface (because acoustic waves are reflected by a metal/air or liquid interface) if oxide layers (rust) are present, and (b) high temperatures can destroy the piezoelectric properties of the crystal used in the probe. Results can also be displayed on a CRT, which is generally used for more-specialized requirements, such as the detection of pits or other forms of localized corrosion. A number of special-application ultrasonic systems are now available (see Fig. 3).

Other NDT methods are used for special situations as they occur. Eddy currents can detect corrosion inside condenser tubes (offstream) in nonferrous materials (ultrasonics can also be used) and radiography is used for corrosion detection in pipes and process vessels. Special instrumentated devices called "pigs" are passed through pipelines for corrosion inspection. These use both eddy currents and ultrasonics, as well as caliper measurements.

References

1. Britton, C. F., Corrosion Monitoring and Chemical Plant, "Corrosion," 2nd Ed., Chapter 20.3, Newnes-Butterworth, London, 1976.
2. Industrial Corrosion Monitoring, Department of Industry, HMSO, London, 1978.
3. Corrosion Testing and Monitoring, *Proc.,* Institution of Corrosion Science and Technology, London, 1977.
4. On-Line Surveillance and Monitoring of Process Plant, *Proc.,* Society of Chemical Industry, London, 1977.

Acknowledgement

This paper was first presented at the Eurochem 80 Conference. Proceedings of the conference are available from the Conference Secretariat, Runfold (025 18) 2066, England.

The author

Colin F. Britton is the President of Corrosion Monitoring Consultancy (CMC), St. Pirans, Denchworth Road, Wantage, Oxon, OX12 9AU, England. Tel: 02357 3867. CMC provides both consultancy and training courses in all aspects of corrosion inspection and measurement. Mr. Britton began his career in corrosion at the Atomic Energy Research Establishment at Harwell, England. In 1970, he joined an American manufacturer of inhibitors and monitoring instrumentation. Mr. Britton is a member of the National Assn. of Corrosion Engineers.

Preventing galvanic corrosion in marine environments

When dissimilar metals are placed together in seawater, the least-noble can suffer accelerated corrosion. Proper selection of alloys or use of cathodic protection can prevent this from becoming a problem.

T. S. Lee, LaQue Center for Corrosion Technology, Inc.

Structures and components for seawater use are typically made up of several materials. Because requirements for mechanical properties and corrosion tolerances can vary for different parts of a structure, use of more than one material proves economical.

However, when several metals are in physical or electrical contact with each other in seawater, galvanic corrosion can take place. Ways to prevent such corrosion will be presented here. But first, to understand how prevention works, the mechanism of galvanic corrosion will be summarized.

Galvanic corrosion

This has been defined as the "accelerated corrosion of a metal because of an electrical contact with a more-noble metal or nonmetallic conductor in a corrosive electrolyte"[*1*]. While the more-active metal (i.e., the anode) suffers accelerated corrosion, this process is usually slowed down for the more-noble metal (i.e., the cathode). Such changes in corrosion rates are related to how each metal would corrode in the electrolyte were it alone.

When the individual metals are immersed in a corrosive electrolyte, a corrosion potential is established, due to differences in local anodic (oxidation) and cathodic (reduction) reactions on the metal surfaces. If reactions take place simultaneously and at equal rates, the specific sites where these local reactions occur may change during exposure, and uniform corrosion may result. On the other hand, localized corrosion occurs at a given site if the site remains anodic.

The corrosion potential is the potential that a metal shows in an open circuit. However, upon coupling with another metal, a circuit is created, and electrons flow through the external circuit from anode to cathode. The result is that the anodic and cathodic reaction rates increase at the respective surfaces.

Thus, the metals will no longer be at their equilibrium corrosion potentials. Such a deviation from the corrosion potential due to the net current is called polarization. The relationship between the current to polarize a given metal and the resultant potential is the polarization curve. The shape of the curve and the magnitude of the current required for polarization will determine the rate of corrosion.

Assessing galvanic corrosion

Whether galvanic corrosion is likely to occur when two dissimilar metals are coupled is indicated by comparing their individual corrosion potentials in the solution they are in.

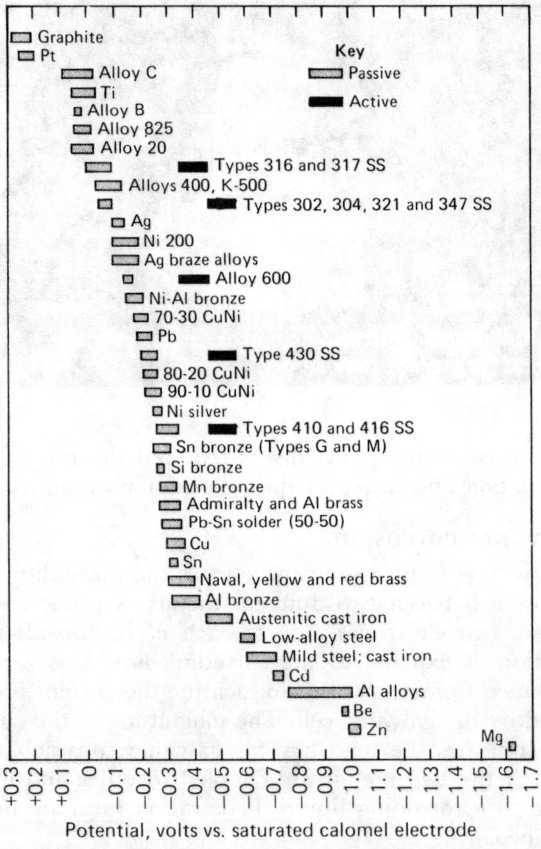

Figure 1 — Galvanic series of metals/alloys in flowing seawater is useful in picking materials

Table I — Changing the relative areas of cathode and anode will affect corrosion rate

	Effect of increasing anode area* on:		Effect of increasing cathode area† on:	
Type of control	Total anode mass-loss	Intensity of anodic corrosion	Total anode mass-loss	Intensity of anodic corrosion
Anodic	Increased	None	None	None
Mixed	Increased	Decreased	Increased	Increased
Cathodic	None	Decreased	Increased	Increased

*With constant cathode area.
†With constant anode area.

From Ref. [*2*].

Originally published April 1, 1985

Table II — Galvanic compatibility of metals in seawater helps to select materials

S—Small Metal A to Metal B area ratio.
E—Equal area ratio.
L—Large Metal A to Metal B area ratio.
● Unfavorable—Galvanic acceleration expected.
x Uncertain—Variable direction and/or magnitude of galvanic effect.
o Compatible—No galvanic acceleration expected.

Metal A ↓ / Metal B →		Magnesium alloys	Zinc	Aluminum alloys	Cadmium	Mild steel, wrought iron	Cast iron	Low-alloy high-strength steel	Brasses, Mn bronze	Copper, Si bronze	Lead-tin solder	Tin bronze (Types G and M)	90-10 copper-nickel	70-30 copper-nickel	Nickel-aluminum bronze	Silver braze alloys	Types 302, 304, 321 and 347 stainless steel	Alloys 400, K-500	Types 316 and 317 stainless steel	Alloy 20, alloy 825	Titanium, alloys C, C-276 and 625	Graphite, graphitized cast iron
Magnesium alloys	S		●	●	●	●	●	●	●	●	●	●	●	●	●	●	●	●	●	●	●	●
	E		●	●	●	●	●	●	●	●	●	●	●	●	●	●	●	●	●	●	●	●
	L		●	●	x	●	●	●	●	●	●	●	●	●	●	●	●	●	●	●	●	●
Zinc	S	●		●	●	●	●	●	●	●	●	●	●	●	●	●	●	●	●	●	●	●
	E	●		x	x	●	●	●	●	●	●	●	●	●	●	●	●	●	●	●	●	●
	L	●		o	o	o	o	o	●	●	●	●	●	●	●	●	●	●	●	●	●	●
Aluminum alloys	S	●	o		●	●	●	●	●	●	●	●	●	●	●	●	●	●	●	●	●	●
	E	●	x		●	●	●	●	●	●	●	●	●	●	●	●	●	●	●	●	●	●
	L	●	●		x	x	x	x	●	●	●	●	●	●	●	●	x	●	x	x	x	●
Cadmium	S	x	o	x		●	●	●	●	●	●	●	●	●	●	●	●	●	●	●	●	●
	E	●	x	●		●	●	●	●	●	●	●	●	●	●	●	●	●	●	●	●	●
	L	●	●	●		o	o	o	●	●	●	●	●	●	●	●	●	●	●	●	●	●
Mild steel, wrought iron	S	●	o	x	o		●	●	●	●	●	●	●	●	●	●	●	●	●	●	●	●
	E	●	●	●	●		o	x	●	●	●	●	●	●	●	●	●	●	●	●	●	●
	L	●	●	●	●		o	o	o	o	o	o	o	o	o	o	o	o	o	o	o	●
Cast iron	S	●	o	x	o	o		x	●	●	●	●	●	●	●	●	●	●	●	●	●	●
	E	●	●	●	●	o		x	●	●	●	●	●	●	●	●	●	●	●	●	●	●
	L	●	●	●	●	●		●	x	x	x	x	x	x	x	x	x	o	o	o	o	x
Low-alloy high-strength steel	S	●	o	x	o	o	●		●	●	●	●	●	●	●	●	●	●	●	●	●	●
	E	●	●	●	●	x	x		●	●	●	●	●	●	●	●	●	●	●	●	●	●
	L	●	●	●	●	●	x		o	o	o	o	o	o	o	o	o	o	o	o	o	●
Brasses, Mn bronze	S	●	●	●	●	o	x	o		●	●	●	●	●	●	●	●	●	●	●	●	●
	E	●	●	●	●	●	●	●		x	x	x	x	x	x	x	●	●	x	x	x	●
	L	●	●	●	●	●	●	●		x	x	x	x	x	x	x	●	o	x	o	o	●
Copper, Si bronze	S	●	●	●	●	o	x	o	x		●	●	●	●	●	●	●	●	●	●	●	●
	E	●	●	●	●	●	●	●	x		x	x	x	x	x	●	●	x	x	x	x	●
	L	●	●	●	●	●	●	●	●		x	x	x	x	x	x	●	o	●	●	o	●
Lead-tin solder	S	●	●	●	●	o	x	o	x	x		●	x	x	x	●	●	●	●	●	●	●
	E	●	●	●	●	●	●	●	x	x		x	o	o	o	x	●	●	●	x	x	●
	L	●	●	●	●	●	●	●	●	●		x	x	x	x	x	●	o	o	o	o	●
Tin bronze (Types G and M)	S	●	●	●	●	o	x	o	x	x	x		x	x	x	●	●	●	x	x	x	●
	E	●	●	●	●	●	●	●	x	x	x		o	x	x	●	●	o	x	o	o	●
	L	●	●	●	●	●	●	●	●	●	●		x	o	o	x	●	o	x	o	o	●
90-10 copper-nickel	S	●	●	●	●	o	x	o	x	x	x	x		●	x	x	●	●	●	●	●	●
	E	●	●	●	●	●	●	●	x	x	o	o		o	o	x	●	x	x	x	x	●
	L	●	●	●	●	●	●	●	●	●	x	x		o	o	x	●	o	o	o	o	●
70-30 copper-nickel	S	●	●	●	●	o	x	o	x	x	x	o	o		o	x	●	●	●	●	●	●
	E	●	●	●	●	●	●	●	x	x	o	x	o		o	x	●	x	x	o	o	●
	L	●	●	●	●	●	●	●	●	●	x	x	●		o	x	●	o	o	o	o	●
Nickel-aluminum bronze	S	●	●	●	●	o	x	o	x	x	x	o	o	o		x	●	●	●	●	●	●
	E	●	●	●	●	●	●	●	x	x	o	x	o	o		x	●	x	x	x	x	●
	L	●	●	●	●	●	●	●	●	●	x	x	x	o		x	●	o	o	o	o	●
Silver braze alloys	S	●	●	●	●	o	x	o	x	x	x	x	x	x	x		●	x	x	x	x	●
	E	●	●	●	●	●	●	●	x	●	x	●	x	x	x		●	x	x	x	x	●
	L	●	●	●	●	●	●	●	●	●	●	●	x	x	x		●	x	x	x	x	●
Types 302, 304, 321 and 347 stainless steel	S	●	●	x	●	o	x	o	●	●	●	●	●	●	●	●		●	●	●	●	●
	E	●	●	●	●	●	●	●	●	●	●	●	●	●	●	●		●	●	●	●	●
	L	●	●	●	●	●	●	●	●	●	●	●	●	●	●	●		x	●	●	●	●
Alloys 400, K-500	S	●	●	●	●	o	o	o	o	o	o	o	o	o	o	x	x		x	x	x	●
	E	●	●	●	●	●	●	●	●	x	●	o	x	x	x	x	●		x	x	x	●
	L	●	●	●	●	●	●	●	●	●	●	●	●	●	●	x	●		x	x	x	●
Types 316 and 317 stainless steel	S	●	●	x	●	o	o	o	x	●	o	x	o	o	o	x	●	x		x	x	●
	E	●	●	●	●	●	●	●	x	x	●	x	x	x	x	x	●	x		x	x	●
	L	●	●	●	●	●	●	●	●	●	●	x	●	●	●	x	●	x		x	x	●
Alloy 20, alloy 825	S	●	●	x	●	o	o	o	●	●	o	o	o	o	o	x	●	x	x		x	●
	E	●	●	●	●	●	●	●	x	x	x	o	x	o	x	x	●	x	x		x	●
	L	●	●	●	●	●	●	●	●	●	●	x	●	●	●	x	●	x	x		x	●
Titanium, alloys C, C-276 and 625	S	●	●	x	●	o	o	o	o	o	o	o	o	o	o	x	●	x	x	x		x
	E	●	●	●	●	x	x	x	x	x	x	o	x	o	x	x	●	x	x	x		x
	L	●	●	●	●	●	●	●	●	●	●	x	●	●	●	x	●	x	x	x		x
Graphite, graphitized cast iron	S	●	●	●	●	●	x	●	●	●	●	●	●	●	●	●	●	●	●	●	x	
	E	●	●	●	●	●	●	●	●	●	●	●	●	●	●	●	●	●	●	●	x	
	L	●	●	●	●	●	●	●	●	●	●	●	●	●	●	●	●	●	●	●	x	

Generally, the greater the difference in corrosion potentials, the greater the probability of corrosion. However, this difference neither predicts the kinetics and corrosion rates nor accounts for the degree of polarization.

Fig. 1 shows corrosion-potential ranges in seawater for various metals and alloys. The series was developed from tests that were run for 30–600 d in flowing seawater. The temperature varied from 10° to 26°C, with flowrates ranging from 2.4 to 3.9 m/s. Since potentials change as films of corrosion products build up, each metal or alloy shows a range, rather than a single potential. For some alloys, active and passive states are shown.

This galvanic series helps to tell which metal in a galvanic couple may corrode. Fig. 2 shows how the corrosion current, which indicates the rate, can be estimated. E_a and E_c are the anodic and cathodic corrosion potentials, respectively.

In Fig. 2 (left), both metals polarize equally. The potential of the galvanic couple, E_{galv}, is midway between the anodic

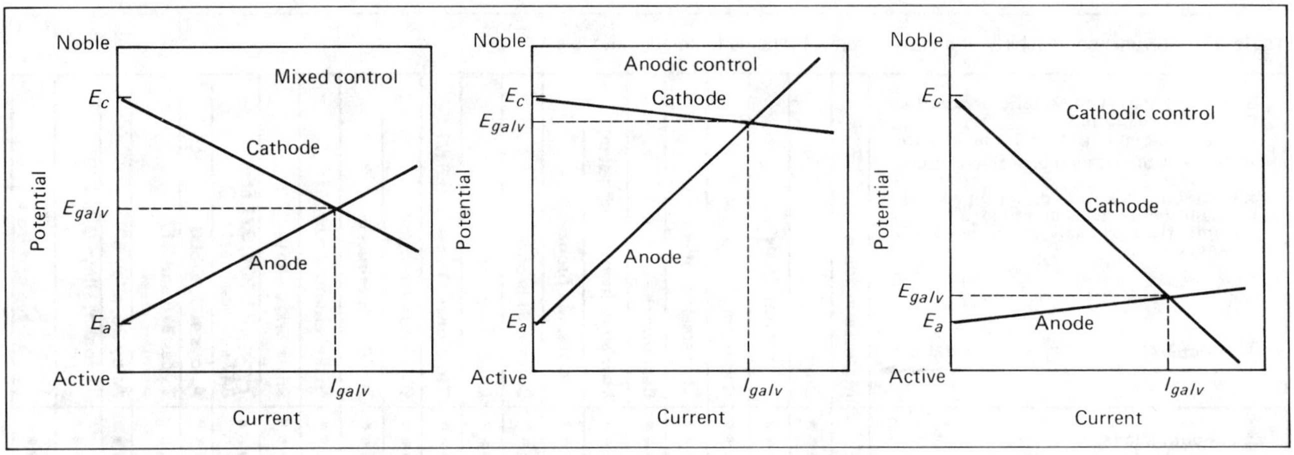

Figure 2 — Galvanic systems can be under these different types of control

and cathodic potentials. The corrosion rate of the anode is determined by the current at this point, I_{galv}. This system is under mixed control, since both members of the couple show equivalent polarization.

In Fig. 2 (center), polarizations are unequal; this system is under anodic control. The anode readily polarizes; the cathode does not. The reverse of this is cathodic control; see Fig. 2 (right). In both cases, the potential of the couple is closer to that of the less-polarized metal than it is to the midpoint.

To estimate galvanic corrosion rates for a given service, experiments should simulate actual service conditions. For some materials, polarization behavior may be significantly affected by variables such as temperature and flow characteristics (including degree of aeration). Formation or removal of corrosion products over time also affects polarization. So do the resistance of the metallic/electrical circuit and that of the electrolyte. The electrolytic resistance reflects that of corrosion films as well as that through the electrolyte.

Where it is impractical to set up a test that accurately simulates the real situation, the component that might corrode can be identified from a galvanic series. The magnitude of any adverse effects can be assessed by looking at basic corrosion mechanisms, plus relevant field experience.

Factors influencing galvanic corrosion

Effect of anode and cathode area—The examples in Fig. 2 assumed equal anode and cathode areas. In practice, this is unlikely. Table I summarizes the effects on the degree of corrosion of changing the anode or cathode area for the three types of control [*2*]. Except for anodic control, increasing the cathode area also increases the corrosion rate.

Effect of environment—As noted, the conductivity of the environment affects galvanic corrosion. A low-conductivity electrolyte results in the intensity of corrosion being greatest near the anode/cathode junction, and decreasing with distance from the junction. For a high-conductivity electrolyte, such as seawater, the distance effect is less pronounced and more-uniform corrosion results [*2*].

Such corrosion can also take place in the atmosphere if films of condensed moisture are present. The films usually are low in conductivity. This is so even in marine environments, where the film contains chlorides. Galvanic corrosion produced by atmospheric exposure usually occurs near the anode/cathode junction. For example, a steel core was surrounded by magnesium and exposed to sea air for 9 years. Corrosion of the Mg was within about 5 mm of the steel's.

Guidelines to minimize corrosion

Since it is often impractical to build a system of one material, galvanic corrosion can be minimized in several ways:

- Select materials that, through experience, have been found compatible. Table II offers guidance on this for a variety of metals and alloys in flowing seawater at ambient temperatures.
- Avoid unfavorable area ratios. Thus, key components such as fasteners, valve seats, welds, etc., should be made from the more-noble material.
- Where possible, insulate dissimilar metals by *complete* electrical and physical separation.
- If coatings or paint are specified, coat or paint both the anodic and cathodic members. Defects in coatings on the anode will provide sites for intensified corrosion if large, uncoated cathodic areas exist. Alternatively, coat or paint only the cathode, thus reducing the effective cathodic area.
- If, in a couple, increased corrosion of an anodic material is expected, design the structure for easy replacement of it, or use thicker sections of it.
- Use cathodic protection by adding a third material that is anodic and will polarize the components, making them cathodic. Zinc or aluminum anodes are often used here.

Galvanic corrosion need not limit the service life of structures or components. Recognizing this and adopting preventive measures can overcome service problems.

References

1. ASTM, Standard Definitions of Terms Relating to Corrosion and Corrosion Testing, ASTM Designation G15, in "1983 Annual Book of ASTM Standards," Vol. 03.02, ASTM, Philadelphia.
2. Wesley, W. A., and Brown, R. H., Fundamental Behavior of Galvanic Couples, in "Corrosion Handbook," H. H. Uhlig, ed., John Wiley & Sons, New York, p. 481, 1948.

The author

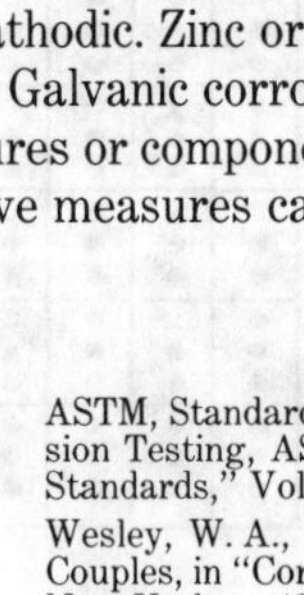

T. S. Lee is vice-president of research and engineering at the LaQue Center for Corrosion Technology, Inc., P.O. Box 656, Wrightsville Beach, NC 28480. Tel: (919) 256-2271. He is responsible for LCCT's corrosion-research program. Prior to joining LCCT, he was a senior engineer in corrosion testing for Cabot Corp. He holds B.S. and M.S. degrees in materials science and engineering from the University of Florida, and is a member of Alpha Sigma Mu, Sigma Tau, the Electrochemical Soc. and NACE.

Measuring corrosion as it happens

A number of corrosion-monitoring techniques now available provide valuable information and help to solve corrosion problems.

G.P. Rothwell, National Corrosion Service (U.K.)

☐ Corrosion often causes catastrophic failure. But one need not sit back and wait for disaster. By logging the rate at which materials are corroding, it is possible to develop a maintenance program free of surprises.

Techniques available

Many techniques are available for corrosion monitoring, and those more commonly used are summarized in Table I, classified in terms of their origins. Some measure the rate of attack, or allow it to be inferred; others only measure total loss of thickness, or thickness of metal remaining. Some techniques indicate the distribution of attack; others monitor the corrosion regime in which the plant or probe is operating, indicating the possible occurrence of uniform corrosion, passivity, pitting, etc.

The techniques also differ in other respects. Some make measurements directly on the plant, while others require special probes and the information they provide must be applied to the plant with caution. Many of the techniques may be used while the plant is online, but a few are only applicable in offline situations. The environment itself may limit the range of techniques available; e.g., a nonconducting environment rules out the use of electrochemical techniques.

The expertise required to interpret information varies from method to method, as does the technological background required. The nature of the problem will generally dictate the most suitable techniques. Many of them are complementary, and it is often advisable to employ more than one method to provide confirmation, calibration or a greater breadth of data.

A check-list of the characteristics of corrosion-monitoring techniques is given in Table II, and a more-detailed analysis in Table III. Fig. 1 and 2 show photographs of corrosion-monitoring equipment.

Description of techniques

The simplest method of monitoring is systematic *visual inspection*, though this is only qualitative, or at best semiquantitative when the eye is aided by a variety of gages, closed-circuit TV or other instruments. The method is generally limited to accessible areas of plant that may be examined on shutdown. A more flexible approach is to install sample *coupons* in the plant that can be removed

Photo: Rohrback Instruments

```
*******************************
PROBE NUMBER:               4
AVERAGE FOR LAST:    16 HOURS
PRESENT RATE IS:     13.7 MPY
ALARM RATE SET AT:        OFF
CONTROL RATE SET AT: 12.0 MPY
------CONTROL  CONDITION------
LIFETIME RATE IS:    11.7 MPY
PROBE STYLE:        ALL METAL
INITIAL THICKNESS: 10.0 MILS
CORROSION LIFE:     4.66 MILS
TOTAL CORROSION:    4.87 MILS
PERCENT USED IS:      104.2 %
-----USEFUL LIFE EXCEEDED-----
INITIAL PERCENT USED:   6.8 %
CHECK READING IS:
```

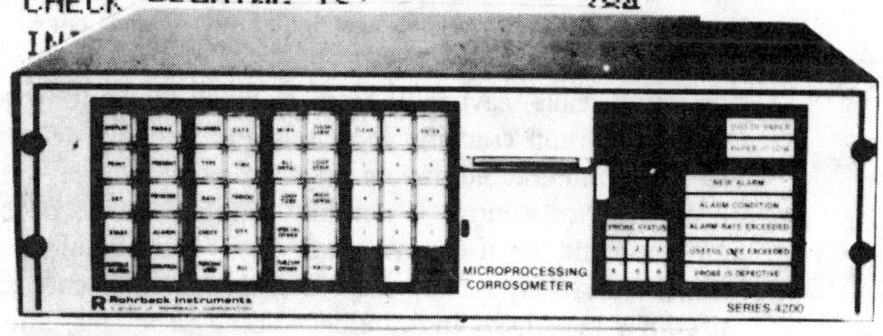


```
PROBE NUMBER:               5
AVERAGE FOR LAST:     2 HOURS
PRESENT RATE IS:        0 MPY
ALARM RATE SET AT:   10.0 MPY
CONTROL RATE SET AT:      OFF
LIFETIME RATE IS:     .09 MPY
PROBE STYLE:       HIGH SENSE
INITIAL THICKNESS:  1.00 MILS
CORROSION LIFE:     .451 MILS
TOTAL CORROSION:    .092 MILS
PERCENT USED IS:       28.1 %
INITIAL PERCENT USED:   9.8 %
CHECK READING IS:         293
INITIAL CHECK WAS:        291
TIME PROBE STARTED:     08:43
DATE PROBE STARTED:   08MAR78
PROJ. END OF LIFE:    17MAR83
*******************************
PROBE NUMBER:               6
PROBE IS NOT IN SERVICE.
*******************************
```

A microprocessing corrosometer based on the electrical resistance method **Fig. 1**

Originally published May 7, 1979

and examined at intervals. Unless bypass lines or online access fittings are used, the inspection interval may be restricted to the period between shutdowns. Either method will give information on the type of corrosion as well as allow corrosion rates to be estimated. If prior warning of reduction to the minimum allowable thickness is the main requirement, the use of *sentinel holes*, drilled through the design thickness at selected points, will indicate by leaking when the corrosion allowance has been lost. The leak can be plugged until an appropriate time to repair or replace. The method is convenient in some cases, although it does not meet with universal approval.

A number of methods based on nondestructive-testing techniques may be used for monitoring. *Radiography* indicates the distribution or depth of attack, and *ultrasonic* examination shows metal thickness, cracking or pitting, and may often be used online, provided surface temperatures are not excessive. *Eddy-current* measurements may be used to reveal surface defects, but are not normally possible except during shutdown. None of these techniques shows particularly rapid response to changes in operating conditions, and the techniques are of most value in predictive-maintenance applications.

Two rather specialized techniques have useful applications in some particular corrosion problems. *Thermography* (infrared imaging) is used on refractories, insulation and furnace tube inspections to check surface temperature patterns as an indicator of the physical state of the material. This can aid in the detection of local corrosion problems. *Acoustic emission* is a technique capable of detecting leaks, cavitation, corrosion fatigue, pitting and stress corrosion cracking in vessels and lines by detection of the sound emitted during their propagation.

If *analytical* control is normally applied to the process environment, or if analyses can easily be obtained, the information may be of value. The presence of corrosion products may itself indicate developing problems, and in some cases the ratio of metallic ions in a process liquor may even give a suggestion of the components that are corroding. If some information is available on the conditions likely to cause corrosion problems, analytical data will warn when the environment moves into a dangerous condition.

If prior information is available on the system, simple electrochemical techniques, such as *potential* or *galvanic-current* measurement, may indicate the corrosion regime in which the plant, or probe, is operating and allow estimates to be made of the probable rate of attack and whether this is likely to be uniform or localized.

Onstream monitoring

The final grouping in Table I includes two techniques that have been developed especially for onstream monitoring, and it is the growth of applications in this area that has drawn increased attention to the whole field of corrosion monitoring.

The first of these is the *electrical resistance method*, simply based on the fact that as a sample corrodes its cross-section decreases and its electrical resistance is therefore increased. By a suitable design of probe and a bridge-type detector, a device can be made that will instantaneously indicate the thickness of metal remaining, which can be sufficiently sensitive to change to enable

Some common techniques of corrosion monitoring, classified in terms of their origins — Table I

Direct observation	Visual/optical techniques Coupon testing Sentinel holes
Nondestructive testing methods	Radiography Ultrasonics Eddy-current measurements Thermography Acoustic emission
Conventional chemical and electrochemical methods	Analytical measurements Potential measurements Galvanic measurements
Onstream monitoring techniques	Electrical-resistance measurements Polarization-resistance measurements (complex impedance methods)

Checklist for characteristics of corrosion-monitoring techniques — Table II

- ☐ What does the technique measure?
- ☐ In what range of environments can it operate?
- ☐ Does it give information about plant or probe?
- ☐ How rapidly can a reading be obtained?
- ☐ How rapidly does it respond to change?
- ☐ Are specialist skills required for interpretation?
- ☐ Is sophisticated technological backup required?

corrosion rates to be determined in relatively short test periods. See Fig. 1.

Probes can be prepared of a variety of materials to suit most practical corrosion rates; and with varied geometries to suit pipeline and tank applications, atmospheric and buried corrosion investigations, packaging, and so on. The electrical-resistance technique has a major advantage in that it does not require a conducting medium for its application; so it can be used in the air, buried, or in organic fluids. However, it is sensitive to temperature fluctuations, and careful compensation is required. And the technique only indicates how aggressive a specific environment is to a probe that is broadly similar to the plant in question. Successful correlation of probe behavior with plant performance is important in establishing the value of the electrical resistance technique in specific applications.

The second technique in this group is the linear polarization or *polarization resistance technique*. Under certain conditions, the slope of the corrosion potential/current curve for a corroding specimen is virtually linear for small perturbations from the free corrosion potential, and the gradient is proportional to the corrosion rate. It is relatively simple to devise a two-electrode probe and appropriate instrumentation to determine the gradient; and knowledge or estimation of the proportionality constant gives the corrosion rate. The reading can be

Characteristics of corrosion-monitoring techniques — **Table III**

Technique	Type of information	Type of corrosion	Environment	Relation to plant	Time for measurement	Response to change	Ease of interpretation
Visual/ optical	distribution (rate of attack)	localized	any	accessible areas on plant	slow	slow	easy
Coupons	average rate; type of attack	general or localized	any	probe	long exposure	slow	easy
Sentinel holes	"go/no go" on thickness	general	any, gas or vapor preferred	localized on plant	slow	slow	easy
Radiography	distribution, thickness	pitting; possibly cracking	any	localized on plant	fairly slow	slow	generally easy
Ultrasonics	remaining thickness; presence of cracks, pits	general or localized	any	localized on plant	fairly rapid	fairly slow	needs experience, otherwise easy
Eddy current	cracks, pits	localized	any	localized on plant	fairly rapid	fairly slow	needs experience
Thermography (infrared)	distribution of attack	localized	any; temperatures warm or subambient	localized on plant	rapid	slow	easy
Acoustic emission	cracking, cavitation, leaks	–	any	generally on plant	instantaneous	rapid	generally easy
Analytical and related	state; total corrosion; item corroding	general	any	generally on plant	normally fairly rapid	normally fairly rapid	generally easy, needs plant knowledge
Potential measurement	state; indirect indication of rate of attack	general or localized	electrolyte	probe or plant in general	instantaneous	rapid	generally easy, needs experience
Galvanic (zero-resistance ammeter)	state; galvanic effects	general or unfavorable conditions localized	electrolyte	probe or, occasionally, plant in general	instantaneous	rapid	generally easy, needs experience
Electrical resistance	integrated corrosion	general	any	probe	instantaneous	moderate	generally easy
Polarization resistance	rate	general	electrolyte	probe	instantaneous	rapid	generally easy

virtually instantaneous, and commercial instruments are available, frequently calibrated directly in corrosion rates (see Fig. 2), although such a direct approach is clearly hazardous. It shares with the electrical resistance technique the drawback that the information is necessarily obtained on probes rather than on the plant itself, and it can only be used in conducting environments. However, with calibration and understanding it may be employed over a wide range of conditions with sufficient accuracy for most industrial applications, and it is an extremely useful tool.

A development based on the polarization resistance technique is the measurement of complex polarization impedance, which removes many of the limitations of the simpler approach. Although interpretation is more complex, this promises to be a much more powerful method when fully developed.

Applications

There are five main ways in which corrosion monitoring can be applied:

■ *Diagnosing a problem.* The use of online monitoring methods to correlate plant conditions with corrosion behavior is one of several techniques available to the materials engineer to diagnose corrosion problems. It is a great help to know precisely the conditions that exist within an operating plant when a problem arises.

■ *Monitoring a solution.* If a particular technique has led to a problem being solved, it is a logical extension to continue to use the same technique, or a modification of it, to ensure that the solution is successful and remains so.

■ *Monitoring plant operation.* If there is a clear correlation between process variables and corrosion behavior, it is often possible to monitor the process conditions and to use the information to control the plant in such a way as to maintain acceptable corrosion behavior. A similar approach can be used during optimization trials, where rapid measurement of corrosion rates onstream enables the effects of process changes to be assessed before damage to the plant can occur. Simple examples are the use of potential measurements to monitor the operation of cathodic and anodic protection systems, and the instrumental methods that are available to monitor water-treatment procedures.

■ *Controlling plant operation.* Since a monitoring system can provide information to guide the operation of a plant, the monitoring signals themselves can often be used to control the plant, through suitable feedback systems. This is already being done with cathodic and anodic protection installations and with some water-treatment plants. The increasing use of computers for process-plant control may be expected to significantly extend the range of possible applications.

■ *Providing management information.* Corrosion moni-

Photos: Rohrback Instruments

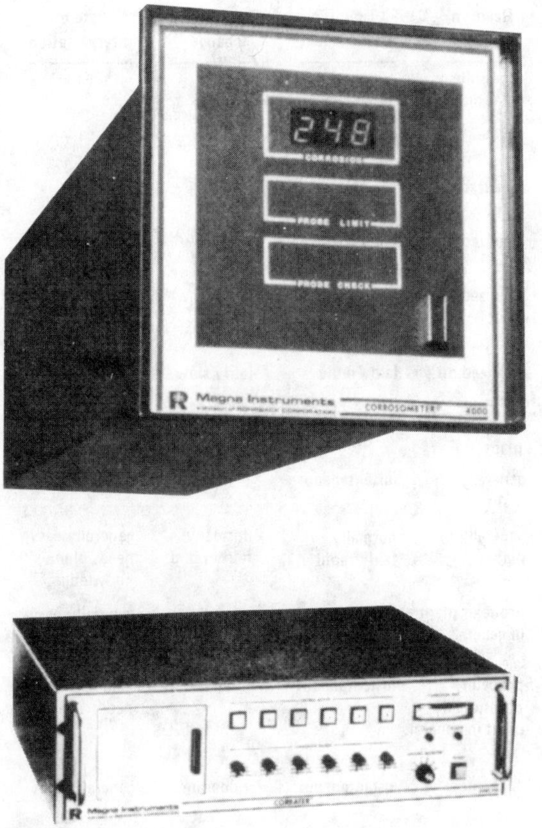

Corraters, portable and bench-or-rack design, based on the polarization resistance method **Fig. 2**

toring is one of a number of techniques—including vibrational analysis, lubricant analysis and ferrography—that can be used to provide onstream inspection and continuous analysis of the condition of the plant. As an aid to predictive maintenance, these techniques permit peak-shaving of inspection activity during shutdown, and better scheduling of maintenance work. Advantages of this approach are maximum plant availability and better use of manpower, even though it involves higher inspection costs and frequently the need for more highly trained personnel. This application in particular is one that requires baseline data for maximum benefit. While it may seem to be counterproductive to spend hours making measurements on new and uncorroded plant before it goes onstream, the absence of this information seriously reduces the value of any subsequent monitoring scheme.

Examples

There are many impressive case histories to be found, particularly in the petroleum, chemical and power-generation industries. But too great an emphasis on the glamorous is often counterproductive since it implies that this approach to corrosion control is only applicable in industries that are technologically very sophisticated, which is far from the case. It will be more useful in the present context to summarize some of the areas in which corrosion monitoring techniques find regular use in a number of industries.

Potential measurement finds wide use in monitoring and controlling cathodic and anodic protection systems. It is valuable for confirming that materials that may show active/passive behavior do, in fact, remain in the passive state. Also, it can be used to ensure that the correct dosage of certain types of corrosion inhibitor is maintained. Galvanic probes may be incorporated into chimney cladding and critical insulation to indicate the ingress of moisture. They are used to warn of the onset of impingement attack in flowing systems and of unacceptably high oxygen levels in oilfield water flood systems.

Analytical measurements are widely used to maintain acceptable water quality both in boiler feed water and in plant cooling-water systems, as indeed in any water-handling plant. A variety of parameters may be measured. Avoidance of scaling is just as important as control of corrosion in boiler and cooling-water heat-transfer systems. On the other hand, measurement of dissolved iron content in the entrained water from gas wells is used to monitor corrosion of downhole equipment, and acceptable inhibitor dosage.

Electrical-resistance probes are frequently used to indicate corrosion rates as well as total metal loss, particularly in systems where the rate is not expected to vary rapidly with time. These probes have also been used in packaging applications, in atmospheric and buried corrosion tests, and even to simulate the corrosion of reinforcing bars in concrete structures.

If the corrosion rate is expected to fluctuate fairly rapidly, the direct measurement of rate by the *polarization resistance technique* may be preferable, and this technique is used in many plant monitoring and control systems. It is useful in monitoring inhibitor additions in acid pickling baths and in cooling-water systems. Commercial packages are available that not only monitor corrosion rate, pH and total dissolved solids in cooling-water circuits but also control water-treatment dosing and blowdown to optimum levels without operator intervention.

References

1. "Controlling Corrosion—Monitoring," Dept. of Industry, Committee on Corrosion (1977).
2. "Handbook of Industrial Corrosion Monitoring," Her Majesty's Stationery Office (1978).
3. Corrosion Testing and Monitoring, Institution of Corrosion Science and Technology, Conference, Westfield College, London, March/April 1977, Chameleon Press Ltd. (1977).
4. On Line Surveillance and Monitoring of Process Plant, Soc. of Chemical Industry Symposium, City University, London, September 1977.

The author

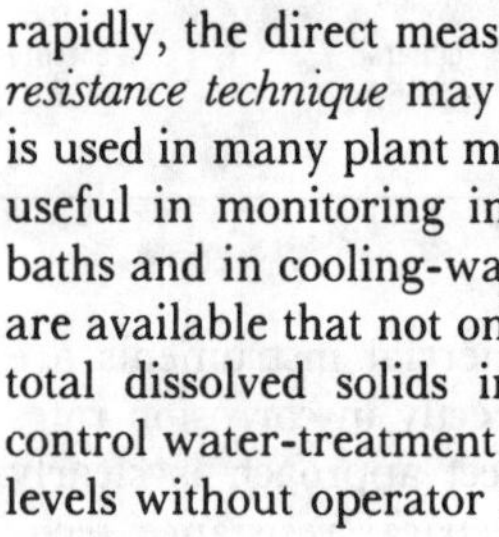

Peter Rothwell is Head of the National Corrosion Service of the U.K. Dept. of Industry, a post that he has held since the Service was formed in 1975. Prior to this, he was Assistant Director of Research in the Dept. of Metallurgy and Materials Science, University of Cambridge, and a Fellow of Downing College. He graduated from Jesus College, Cambridge, in 1959, and took his doctorate in 1962. He is a Chartered Engineer and Chartered Chemist, a member of the Institution of Metallurgists, a fellow of the Royal Institute of Chemistry and of the Institution of Corrosion Science and Technology, and is a member of the Natl. Assn. of Corrosion Engineers.

Welding practices that minimize corrosion—I

The author tells how to lessen the vulnerability of welds in process equipment slated for corrosive service.*

Frank C. Brautigam, Sharples-Stokes Div., Pennwalt Corp.

☐ Inadequate welding techniques used during fabrication often degrade the inherent corrosion resistance of the austenitic stainless steels (300 series) and the chromium-bearing, high-nickel alloys. In extreme environments, even if the proper parent metal and filler metal have been specified, one or more of the following factors can cause weldments to corrode:

- Weldment design
- Fabrication technique
- Welding practice
- Welding sequence
- Sensitization of heat-affected zone
- Moisture contamination
- Organic and metallic contamination
- Sulfur and chloride contamination
- Oxide films and scale
- Weld slag and spatter
- Iron contamination after welding
- Final surface finish

Weld design

The ideal design allows the welder to first fit up and tack a fabrication, and then finish with the weld in a flat position. Vertical-position welding and overhead welding require greater skill and boost the chances of poor bead-formation or slag entrapment.

Proper weld-joint geometry (given in standards published by alloy makers) permits the welder to manipulate his electrode or filler wire in such a way as to deposit relatively flat beads of low profile without slag entrapment or weld dropthrough. The good designer will anticipate the field conditions under which the weld is to be made.

A poor design can leave crevices, stagnant areas, and areas of impingement or turbulent flow where corrosion is most likely to take place. A welder may be faced with the problems of unaligned butt joints, overlapping corner joints, too wide or narrow a root-gap, or a space limitation that prevents him from manipulating his electrode properly.

Joints that are too wide require excessive filler metal; in nickel alloys this fosters overheating that results in hot cracking, severe undercutting, or sensitization (usually chromium depletion) of the parent alloy. Sensitized, heat-affected zones may be subject to selective attack when exposed to a corrosive medium. Too narrow a gap makes welding difficult and increases the possibility of slag inclusions, lack of penetration and lack of fusion.

Surface preparation

Other than grossly poor welding, there is probably no factor that can produce immediately more critical defects on stainless steel and high-nickel alloy welds than contamination. Such defects are unacceptable mechanically and offer sites for corrosive attack in aggressive environments. The presence of moisture on the parent metal, or in electrode coatings, can lead to porosity and extremely poor welds. The high-nickel alloys are particularly prone to cracking or embrittlement due to contamination from sulfur, lead and other elements in grease-pencil markings, paint, machine oil, drawing compounds, or shop dirt. Contamination from organic materials causes carbon pickup, or carburization, and a heavy oxide-film left by thermal cutting may introduce excessive chrome-oxide into the weld.

A contamination source often overlooked is dirt, oil, grease or moisture on bare wire-filler metal or covered electrodes. Filler metals are produced and packaged under clean conditions, but are sometimes handled with oily or wet gloves, or stored in areas exposed to shop dirt or chemicals.

Although proper cleaning is quite simple, it should be noted that the cleaning method itself can inadvertently be a trouble source. For example, sandblasting can do an excellent job of removing scale or oxide film, but may pound a grease mark into the surface. Cleaning with a sandblaster, grinding wheel or file that is contaminated with iron, copper or other metals may leave the surface in poorer condition than before. Solvent- or chemical-cleaning, grinding, filing, and scrubbing with clean stainless-steel scouring pads or brushes are all excellent surface preparation methods, used individually or in combination. Contaminants should be stripped from both sides of the weld area and from adjacent surfaces. Information on the cleaning and descaling of stainless steel parts, equipment, and systems is included in ASTM Standard A-380.

Influence of the welding process

Shielded metal arc-welding (SMAW) with covered electrodes and gas tungsten arc-welding (GTAW) are

* This article and an upcoming one are part of a series of reports initiated by the Welding Research Council's Subcommittee on Corrosion of Weldments.

Originally published January 17, 1977

the most common techniques used for corrosive service. In recent years, gas metal arc-welding (GMAW) has gained wider acceptance, whereas submerged arc welding (SAW) has been used more selectively. Little information is available on the corrosion resistance of welds made by newer processes such as electron-beam, laser-beam, plasma, and friction welding.

Shielded metal arc-welding—This process has a major advantage in that experienced welders are generally available for manual welding with covered electrodes. The techniques are not difficult for a qualified welder of steel, although added training and practice may be needed for welding the high-nickel alloys. If properly done, manual arc-welding of high-nickel alloys introduces less heat into the weldment than do the GTAW or GMAW processes, especially in multiple-pass welds. Improper technique and poor electrode manipulation, however, can result in undercutting, entrapped slag, lack of fusion, cratering and porosity.

Gas tungsten arc-welding—Also known as TIG welding, this process demands a higher degree of welder skill than does SMAW welding with covered electrodes. Although it produces the cleanest and soundest welds if done correctly, it can also cause overheating and undercutting in multiple-pass welds. A high heat-input reduces the post-weld cooling rate and fosters sensitization in the heat-affected zone of susceptible stainless steels or high-nickel alloys. Gas shielding with argon or helium prevents contact with air that can result in oxidation, nitridation and porosity.

Gas metal arc-welding—In this process (sometimes called MIG welding), the electrode is a bare metal wire that is fed automatically through the welding gun. Transfer of metal from the electrode to the work is governed by the wire feed-speed, which in turn regulates the current. With a nonpulsing current, the mode of transfer will be globular at low current settings, and spray at higher currents. When a pulsing current is used, spray transfer can be obtained at low average currents; this technique—pulsed-spray welding—works in all positions.

Spray transfer offers excellent arc stability, good penetration depth, and high deposition rates, but is accompanied by high heat input. It is also limited to flat welding.

Transfer can also be achieved by a short-circuiting mode in which the molten tip of the electrode contacts the molten pool. This mode has the advantage of low heat input and may be used for thin material. It is also useful for out-of-position welding. Problems associated with GMAW include porosity, lack of fusion, overlapping, poor bead formation, crater-cracking, and oxide or nitrogen contamination. These defects are most often caused by improper current settings, improper gas-shielding and poor welder technique.

Welder technique (SMAW)

Although both a.c. and d.c. power sources are available, most stainless-steel SMAW is done with reverse-polarity d.c. because it produces a steady arc, low spatter, good penetration and consistent fusion.

Current settings vary with material type and thickness, electrode size and welding position. Many welding machines have no ammeter or voltmeter but have variable settings that control arc current. The welder must therefore judge the current setting by visual observation, and "feel" for the proper arc length. For stainless steels and high-nickel alloys, the shortest arc length (lowest voltage) consistent with complete fusion, good arc stability and proper weld-puddle formation should be selected. Short arc length minimizes alloy loss through the arc and reduces porosity and spatter. Low current reduces heat input and prevents excessive weld dilution when dissimilar metals are welded. Conversely, an excessively high voltage and a long arc length promote uneven deposits, weld spattering, high heat input, loss of alloying elements and a detrimental influx of atmospheric elements.

Electrodes

The AWS has set up a numerical code system to identify stainless-steel and high-nickel-alloy electrodes. (Some of the latter, however, may be identified by the trade name of their manufacturer.) In the AWS stainless-steel code, the first three digits indicate the electrode composition, and the last two digits, separated by a dash, indicate the current to be used.

For example, "316-15" indicates a Type 316 stainless electrode material to be used with reverse-polarity d.c. (Use with alternating current is sometimes attempted, but "15" electrodes are not intended to qualify for use with a.c.) A "16" suffix indicates that the electrode may be used with either a.c. or reverse-polarity d.c. Both 15 and 16 electrodes are usable in all orientations, in sizes up to and including 5/32 in. Sizes of 3/16 in. and larger are intended to be used in the flat and horizontal-fillet positions (ASW Specification A5.4).

The coverings of 15 electrodes are referred to as "lime" type, because they contain a high proportion of alkaline earth materials, especially calcium. The 16 electrodes use either lime or "titania" coverings, the latter predominating.

When used with reverse-polarity d.c., both electrodes yield satisfactory welds on virtually all stainless steels. Each type has certain innate characteristics that are introduced by the proprietary electrode-coating the manufacturer applies to meet AWS requirements. These characteristics may favor one brand over another for specific welding conditions.

The 15 electrode is characterized by uniform fusion and high depth of penetration, and is often preferred for tack-welding and out-of-position welding, especially on pipe. It produces a convex, rough-appearing nugget, which when deposited, is accompanied by more spatter than the 16 electrode. The latter produces a smooth, uniform, concave deposit with less spattering and easier slag-removal. In fillet welds, the concave surfaces produced by the 16 electrode are more readily polished for sanitary applications. However, this type electrode achieves the shallower weld penetration, and is more suited for single-pass welds on thin materials. The differences between these electrodes are minor, though, and satisfactory welds may be made with either type.

High-nickel alloy electrodes are also used with reverse-polarity d.c. The electrode maker should be contacted for specific information on weld characteristics.

Moisture pickup

The function of the electrode covering is to form a gas (primarily CO_2) to shield the molten metal as it is transferred across the arc. It also protects and insulates the molten pool from the surrounding atmosphere and prevents excessive contamination by oxygen, hydrogen and nitrogen. Stainless-steel and high-nickel-alloy coatings may contain limestone, fluorides, titania, potassium compounds and other minerals in various proportions. They act as scavenging agents in the weld puddle and, upon cooling, form a slag around the solidifying weld nugget to protect it from oxidation. These coatings also have a major effect on arc stability, on fluidity of the slag and molten metal, and thereby on the final contour of the weld bead.

Alloying elements are sometimes added to the coating as a means of modifying or controlling the chemistry of the deposited weld metal, if required.

A major problem of covered electrodes is excessive moisture pickup by the electrode coating after it is removed from its sealed storage container. The 16 electrodes, particularly the titania type, will exhibit increased surface porosity that can be readily seen, especially at weld-bead starts. With 15 electrodes, the porosity problem may be more subtle, since excessive moisture encourages subsurface porosity.

Compact baking- or dry-rod-ovens are available for storing loose electrodes before use; oven temperatures of 100°F or more above ambient will deter moisture absorption. Moisture that has already been absorbed may be driven off by baking the rods at temperatures from 350°F (1 h) for stainless steel, to 600°F (1 h) for high-nickel alloys. Specific supplier recommendations as to the temperature and duration of heating should be consulted.

Slag removal during welding

Difficulties with entrapped slag can crop up with both electrode types, particularly when multiple-pass welds are performed. Slag from a.c./d.c. titania-coated electrodes (type 16) usually breaks off in large pieces and is easily removed. These electrodes, however, are also more likely to trap slag in the molten weld pool. Slags from lime-coated electrodes (type 15) rise quickly to the molten surface and crumble into particles when solidified. Small particles are difficult to remove from the rough surface produced by lime-coated electrodes.

With all multiple-pass welds intended for corrosive or high-temperature service, it is essential that the welder remove slag after each pass with a power grinder or power chipping tool.

Electrode manipulation

For the best corrosion resistance in welded stainless steels, a stringer-bead technique using the smallest possible electrode size compatible with the job is recommended. A straight weld-pass made without electrode oscillation or weaving keeps heat input at a minimum and reduces cracking and undercutting.

There are instances, however, when some weaving may be essential to ensure proper fusion in the welding of high-nickel alloys, or in the vertical welding of austenitic stainless steels. For stainless steels, the oscillation should not be more than 2.5 times the electrode diameter. For nickel alloys, any weaving should be kept at a minimum consistent with a sound deposit.

In all cases, the welder should inspect for possible crater cracks at the end of each weave and adjust his technique so as to eliminate cracks and still produce a fully fused weld. The crater that is formed upon breaking the arc at the end of a stringer bead can be a potential trouble source for cracking, porosity, or slag entrapment. To avert these difficulties, an experienced welder will either pause at the end of a pass to fill in the crater, or back-weld into the crater when initiating the arc for a continuing pass.

In joints for services involving severe corrosive attack, weld starts and stops should be ground to remove slag, porosity, or fissures before a new weld pass is begun. Care should also be taken to avoid contact between electrode and puddle.

The joining procedures used for the stainless steels are generally applicable to the various grades of high-nickel alloys. Rigorous surface cleaning, low-amperage reverse-polarity d.c., covered electrodes, short arc-length, stringer-bead technique, and thorough slag removal are all essential. The high-nickel alloys, however, demand closer control of welding parameters to overcome a lack of fluidity, a tendency to undercut, and a sensitivity to cracking, especially as the weld puddle begins to solidify.

Welders experienced in welding steel or stainless steel find the lack of fluidity of the high-nickel alloys a major problem. They often attempt to compensate for the apparent sluggishness by increasing amperage to obtain better metal flow. This increase may result in excessive spatter and transverse cracking with covered electrodes, or excessive undercutting, fissures and poor-quality welds with inert-gas welding. Undercut areas along the edge of the weld are attacked by aggressive media. Attempts to cover the undercut by reinforcing with another weld often produce a second and more severe undercut.

The welding sequence used for multiple-pass welds is also important in reducing heat sensitization and welding stresses. Staggered- or skip-welding, and cooling between passes are procedures often used when such conditions are anticipated. In some cases, it is also advisable to deposit the final weld-pass on the surface of the weldment that is exposed to the corrodent, so that this pass may be relatively free from sensitization.

The author

Frank C. Brautigam is manager of metallurgy and materials engineering for the Sharples-Stokes Div. of Pennwalt Corp., 955 Mearns Rd., Warminster, PA 18974, where he directs the metallurgical activities of the company's centrifuge manufacturing plants. A graduate of the University of Rochester, he holds a B.S. degree, and has done advanced work in welding and corrosion engineering at various institutions. Mr. Brautigam is accredited as a corrosion specialist by the Natl. Assn. of Corrosion Engineers and belongs to the Welding Research Council, American Soc. for Metals, American Welding Soc., and American Soc. for Testing and Materials.

Welding practices that minimize corrosion—II

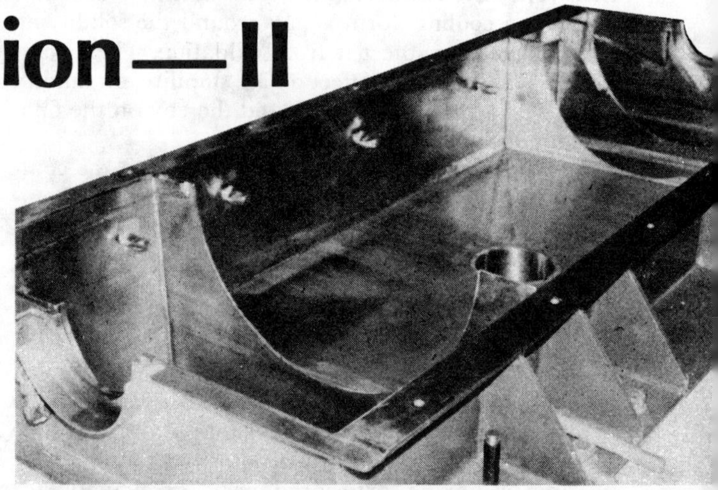

Interior of centrifuge casing, showing all welds and surfaces ground, polished and passivated

This second of two articles covers some common modes of weld corrosion, as well as preferred methods of cleaning and finishing weld surfaces.

Frank C. Brautigam, Sharples-Stokes Div., Pennwalt Corp.

☐ The first article described techniques used for welding austenitic stainless steels (300 series) and chromium-bearing, high-nickel alloys. Recommendations pertained to weld design, surface preparation, weld process, electrode type and welder technique — all of which must be chosen carefully if the inherent corrosion resistance of these metals is to be preserved.

Improper practices can lead to weld cracking, and may foster such forms of attack as intergranular corrosion, stress-corrosion cracking and high-temperature corrosion. Equally important is weld surface-finishing. For maximum corrosion resistance, the final surface should be smooth and free from slag, scale, metallic contaminants and other foreign matter.

Intergranular corrosion

During the welding process, portions of the joint close to the weld deposit (heat-affected zone) are heated to 900–1,500 F. At these temperatures, available carbides are precipitated at the grain boundaries (Fig. 1). A marked depletion of chromium occurs at the edge of each grain, causing stainless and high-nickel alloys to become sensitized. Sensitization makes the part vulnerable to severe attack by chemical environments in which these alloys would normally be inert, or only mildly active. Fig. 2 shows a sensitized, stainless-steel casting in which whole grains have dropped out, causing it to be weakened.

Fig. 3 shows another extreme case of heat sensitization. Severe attack has occurred at the weld interface and heat-affected zone of a weld in which a hard-surfacing alloy was deposited by gas tungsten arc-welding (GTAW). Heat generated by multiple-passes, coupled with the carbon content of the parent metal, induced the formation of a readily attacked carbide network at the grain boundaries.

In the 300-series stainless steels, and in some of the high-nickel alloys, the destructive carbide network (as in Fig. 1) can be completely eliminated by a solution annealing treatment; typically this involves heating to 2,000°F, then quenching in water.

When solution annealing is impractical, intergranular corrosion may be averted or reduced by the use of austenitic alloys (Types 304 and 316) having a low carbon analysis (0.03% C max.), or alloys stabilized with titanium or columbium. Avoiding overheating during welding will also lessen intergranular attack.

Stress-corrosion cracking

Almost all alloy systems are subject to this form of attack. Although stress-corrosion cracking is associated with specific corrodents for each system, tensile stresses, either residual or applied, must be present. Residual stresses most often stem from the welding process, or from extreme cold working. Applied or operating stresses can also be influential if they are cyclic in nature.

Fig. 4 shows stress-corrosion cracks along the exterior surface of a vertical weldseam in a hot reactor column

* Jan. 17, 1976 issue, p. 145. To meet the author, see p. 147.

Originally published February 14, 1977

Carbide precipitation at grain boundaries (120×) Fig. 1

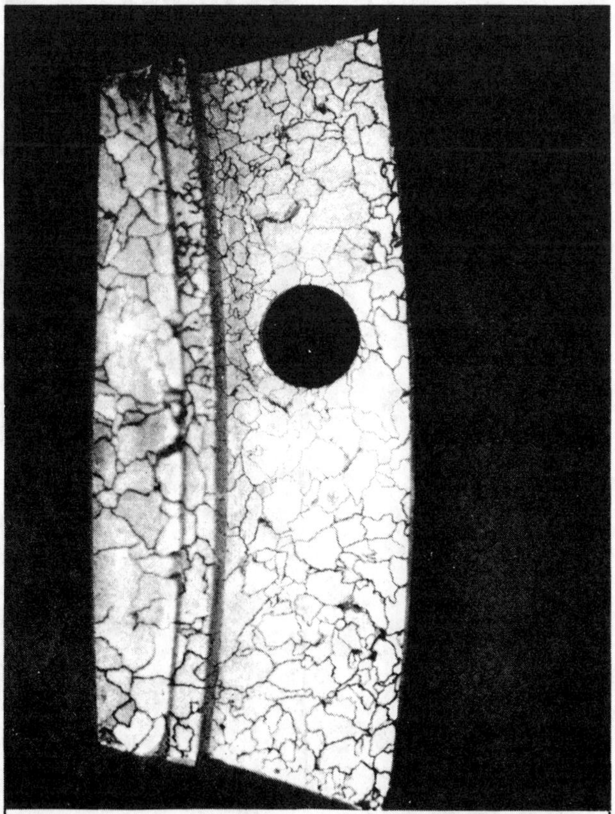

Intergranular corrosion in a heat-sensitized, stainless-steel casting (0.6×) Fig. 2

constructed of Type 316 stainless. The corroding medium contained chlorides which concentrated at the column surface. Cracking was initiated by residual welding stresses, but was propagated by expansion and contraction stresses developed during startup and shutdown of the equipment over a 2.5-yr period.

High-temperature corrosion

Service at high temperatures entails a different set of corrosive mechanisms that can be extenuated by the presence of oxidizing, carburizing or sulfur-bearing gases; ammonia; fluorides; and fuel-ash mixtures. Thermal cycling between wide temperature ranges, and alternating chemical environments (e.g., oxidizing and carburizing), are difficulties typical of extreme corrosive service.

Under these conditions, slag, inclusions, porosity, undercutting and surface irregularities encourage premature failure; the removal of these flaws is essential.

Fig. 5 shows sections taken from opposite ends of a high-nickel-alloy tube subjected to thermal cycling in an alternately carburizing and oxidizing environment. All areas exposed to the cycling were attacked but rupture initiated at the welds. Note the excessive graingrowth, carburization and intergranular corrosion at the heated end of the tube. Black areas are carbonaceous deposits forced into spaces where grains have dropped out.

Weld cracking

Cracks or fissures in weld deposits are of major concern in all corrosive environments. In high-temperature service accompanied by thermal cycling, and in stress-corrosion-cracking environments, these defects can cause premature failure. Cracking frequently occurs at the start of a weld deposit, where the arc is struck, or at the end of a weld-pass, where the arc is broken.

Striking the arc on a piece of metal similar and adjacent to the parent metal, or on top of a previous weld-pass, is recommended for establishing the proper weld heat and arc length. The area where the strike has been made can be ground later on to remove defective or porous metal. Crater cracking at the end of a weld-pass may be prevented by reducing the size of the weld pool before breaking the arc. In manual weldments, this is sometimes accomplished by boosting travel speed. Backwelding in the opposite direction will also forestall crater formation.

Similar procedures can be employed with GTAW. Improved accuracy is possible with machines equipped with a pedal-controlled current regulator. In multiple-pass welds, cracks that develop in one pass must be removed before another pass is deposited.

Weld surface-finishing

Cleaning and finishing specifications should be a part of every fabricating procedure. Where possible, the weld deposit should be inspected visually immediately after welding. This will often disclose surface defects or other substandard conditions that can be repaired immediately by the welder. Maximum corrosion resistance demands a smooth non-oxidized surface, free from irregularities and foreign particles.

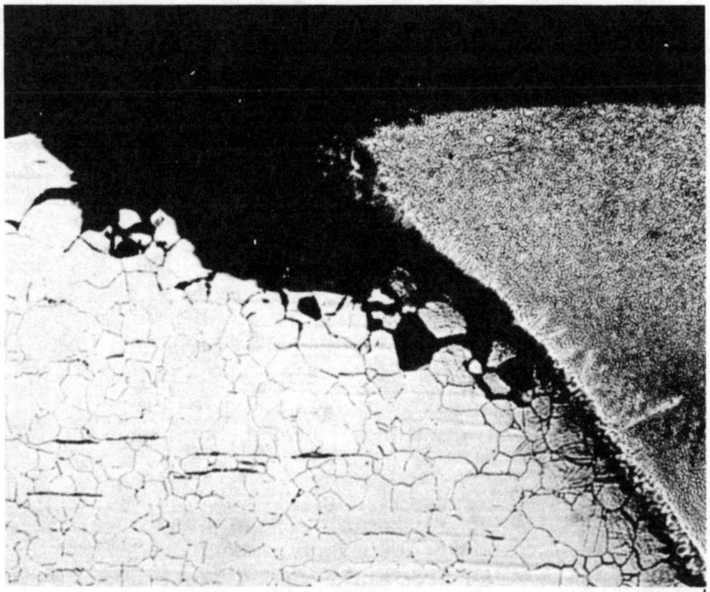

Intergranular corrosion and grain dropout in a heat-affected zone (40×) **Fig. 3**

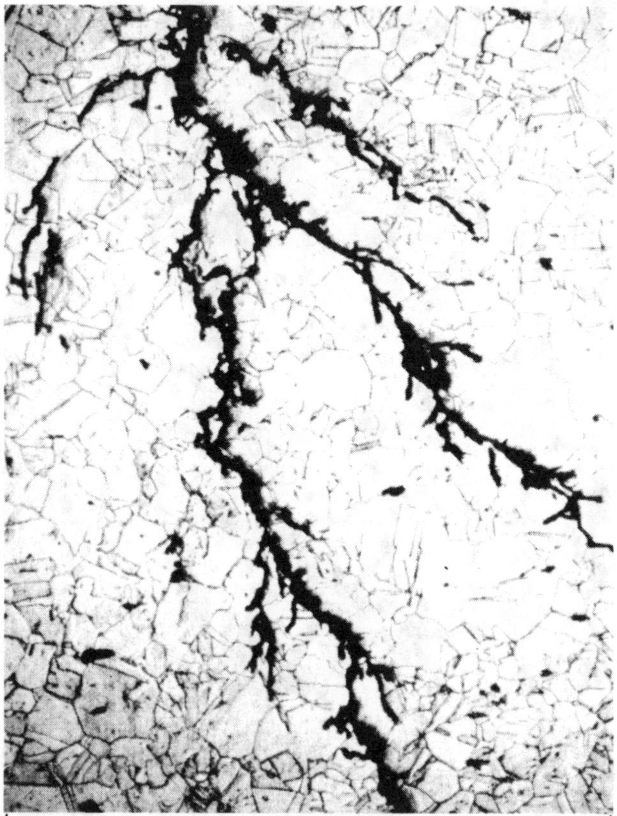

Stress-corrosion cracking initiating at a welded area (120×) **Fig. 4**

Various weld procedures produce characteristic conditions on the weld surface. Slag will always be present on shielded metal arc-welds and, in multiple-pass welds, may be entrapped to hinder fusion at the weld/parent-metal interface.

In even mildly corrosive media, slag, scale, and iron particles can set up active corrosion cells. Slag is particularly deleterious in high-temperature service (1,000°F). It reacts with oxidizing environments to foster fluoride attack on stainless-steel and nickel alloys. In reducing environments, slag absorbs sulfur (even if it is present at low concentrations), resulting in sulfidation of the underlying surface, which leads to failure.

Deposits applied with covered electrodes vary greatly in roughness and in the degree of weld spatter present on adjacent areas. For such deposits, grinding is the preferred finishing method, and chipping or wire-brushing should be reserved for very smooth deposits. All finishing equipment must be free from iron contamination.

GTAW does not present slag problems, and deposits are generally very smooth. Moreover, the good penetration characteristics of this weld process favor GTAW for high-temperature piping work, especially for the root pass. Adequate cleaning can usually be performed by wire brushing and a minimum of grinding. Care must be taken, however, to remove oxidized metal from the back of weld-joints or corner-joints that may not have been properly shielded.

Iron contamination

Particles of free iron or steel on surfaces of stainless-steel and high-nickel-alloy weldments can become the focus of pitting attack. Although lightweight iron dusts may be scoured away by the process fluid, heavier embedded particles become sites for concentration or galvanic cells where the normally passive surface of the alloy is broken down.

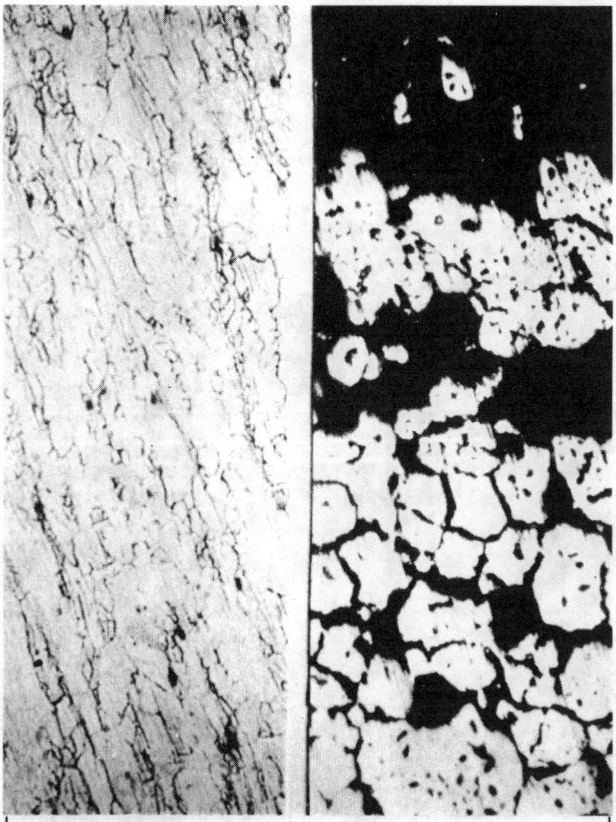

Unaffected and corroded areas in a high-temperature reactor tube (120×) **Fig. 5**

Sources for iron pickup include contaminated pickling solutions, grinding wheels, steel tools, welding fixtures, and sandblasting operations. Dust from manufacturing operations, airborne sand and iron-oxide particles, and chips or particles from chipping and filing operations can also contribute.

The presence of iron can be quickly detected by swabbing or spraying a ferroxyl test solution onto the weld. The appearance of a blue stain within 15 seconds reveals the presence of iron.

After the test has been run, the solution should be removed by rinsing the surface thoroughly with water. (This test is not recommended for food-processing equipment unless extreme care is exercised to remove all traces of the test solution.)

Ferroxyl test solutions are prepared by adding nitric acid to distilled water and dissolving potassium ferricyanide salts in the following proportions:

Distilled water	94 wt%	1,000 cm^3
Nitric acid (70%)	3 wt%	20 cm^3
Potassium ferricyanide	3 wt%	30 g

As with other acid solutions, care should be exercised during handling, and the use of rubber gloves and plastic face-shields is recommended, particularly when the spray technique is used. (ASTM A-380).

Cleaning weld surfaces

Mechanical methods—No cleaning method can match the effectiveness of grinding the weld deposit and adjacent areas. Both grinding and polishing are often stipulated for welds in vessels and piping slated for high-temperature, nuclear, or sanitary food service. Final polishing to a 120-grit finish provides a surface that is easily cleaned to prevent a buildup of solid particles from the process stream.

Abrasive cleaning with glass beads, alumina, sand or silica will also effectively remove surface slag, oxides and scale. Most blasting machines, however, eventually become contaminated with iron or steel fines, unless used exclusively for the high-nickel alloys. Even then, the surface should be checked for iron contamination. And care must be taken to make sure that the surfaces are free from shop dirt, oil, and grease marks prior to blasting.

Acid pickling—Treatment with a nitric-hydrofluoric acid solution will remove scale, metallic contaminants, and welding oxides from stainless steel and some of the high-nickel alloys. Since this solution will attack both the scale and the clean parent metal, it should be used as a last resort, and overexposure should be avoided. The solution may also attack the heat-affected zone of weldments that have been sensitized by welding or other thermal treatments. Short exposures to the pickling solution remove surface oxides and loosen scale so that it can be removed by hand with stainless-steel brushes or scouring pads.

The acid solution is prepared by adding 15–25% (volume) nitric acid and 1–4% (volume) hydrofluoric acid to distilled water, at temperatures from ambient up to 140°F. The acid solutions can cause severe burns, and strict safety precautions during preparation and handling must be followed. Workers should wear protective gloves, clothing and goggles. The work area should also be adequately ventilated.

Passivation and iron decontamination

When stainless-steel and high-nickel alloys are exposed to an oxidizing medium, such as the atmosphere, they very quickly develop a protective, passivating film. However, slag, scale or iron particles adhering to the surface of these alloys can prevent the formation of the oxide film, and create a small galvanic cell or crevice where corrosive attack is initiated.

Embedded particles can be removed by acid cleaning or grinding. Nitric-acid cleaners and passivating solutions are then applied annually by swabbing or immersion. Since these solutions do not effectively remove grease, oil, and shop-dirt, the weldment must first be cleaned with a solvent, hot alkaline cleanser, or mild phosphoric acid. During field construction—or when parts are too large for tank immersion—it is often convenient to identify and treat only the contaminated areas. This can be done by administering the ferroxyl test, and then swabbing the part with a 20–30 wt% aqueous solution of nitric acid. At 120–140°F, immersion times run 15 to 30 min. Cold immersion and hand application require longer periods. Care must be taken to prevent acid depletion and buildup of metallic-salt concentrations, which accompany prolonged treatments.

A more comprehensive discussion covering the cleaning and descaling of stainless steel is given in ASTM A-380, "Cleaning and Descaling Stainless Steel Parts, Equipment and Systems." The user should contact the makers of high-nickel alloys and stainless steels for data on specific welding and cleaning problems.

Acknowledgements

The author wishes to thank the Welding Research Council's Sub-Committee on Corrosion Resistance of High Alloy Weldments for their many helpful comments and recommendations during the preparation of this work.

References

1. Wilcox, W. L., How to Use Covered Electrodes, *Metals Progress*, August 1967.
2. Norcross, J. E., "Seven Common Troubles Found in Welding Stainless and Other Alloy Steels," Arcos Corp. Bulletin, Philadelphia, Pa.
3. Lackey, J. Q., and Streicher, M. A., Detection and Removal of Iron Contamination from Stainless Steel Surfaces, ASTM Reprint (1968) from *Journal of Materials*, Vol. 3, No. 4.
4. Committee of Stainless Steel Producers, "Maintenance Welding of Stainless Steel," American Iron and Steel Institute, New York, 1971.
5. Technical Bulletin T-2, "Joining," Huntington Alloys, Inc., div. of International Nickel Co., Huntington, W. Va., 1972.
6. Technical Bulletin T-20, "Heating and Pickling," Huntington Alloys, Inc., div. of International Nickel Co., Huntington, W. Va., 1968.
7. "Carpenter Stainless No. 20cb-3," Carpenter Technology Corp., Reading, Pa., 1970.
8. "ASTM A-380 Recommended Practice for Cleaning and Descaling Stainless Steel Parts, Equipment and Systems," ASTM Vol. 3, Philadelphia, Pa., 1974.
9. "Filler Metal Specification AWS A5.4-69, Corrosion Resisting Chromium and Chromium-Nickel Steel Covered Welding Electrodes," American Welding Soc., Miami, Fla.
10. "Welding Handbook," 6th ed., American Welding Soc., Miami, Fla. 1972.
11. Henthorne, M., Corrosion Testing of Weldments, *Corrosion*, Vol. 30, No. 2, Natl. Assn. of Corrosion Engineers, Houston, Tex., March 1975.
12. Brautigam, F. C., Selective Corrosion of Weld Metal in High-Nickel Alloys and Stainless Steels, *Corrosion*, Vol. 31, No. 3, Natl. Assn. of Corrosion Engineers, Houston, Tex., March 1975.

Controlling external underground corrosion

The proper use of cathodic-protection systems and coatings will guard underground pipes, tanks and other such structures against attack.

Joseph S. Dorsey, Consultant

□ In designing a chemical-process-industries plant, one of the first considerations should be corrosion prevention for underground metal structures. Such plants typically have extensive underground pipelines, tanks, electrical grounds and other setups, which may corrode or result in corrosion of other constructions.

Corrosion problems must be assessed and preventive measures taken while the plant is in the design stage. To help the engineer do this, we shall detail the principal causes of underground corrosion and the methods to eliminate it.

Types of underground corrosion

Soil corrosion

In general, corrosion is caused by a direct current that leaves a metal and enters an electrolyte. For underground pipelines, soil is the electrolyte. At the place where the current leaves the pipeline or other structure to enter the soil, corrosion occurs. If the soil is the sole cause of corrosion, then the corrosion rate depends upon the electrical resistance and driving potential of the electrical circuit.

The resistance varies with the moisture content in the soil (see Fig. 1, from Ref. [*1*]). No corrosion will occur in completely dry soil; however, nearly all soils have enough moisture to sustain corrosion. Some are so dry, however, that the rate is very slow.

The main factor that causes corrosion on pipelines is differential aeration of the soil [*2*]. If a pipeline is located in soil that is high in oxygen at one location and low at another, an oxygen-concentration cell will be established. A direct current results, with corrosion occurring at the low-oxygen area where the current enters the soil. The current flows back onto the pipe at a high-oxygen area, providing cathodic protection there. The anode and cathode may be inches or hundreds of feet apart, but in either case, serious corrosion may result.

The corrosion mechanism is different in oxygen-rich soils than it is in oxygen-poor ones. When oxygen is plentiful, the initial rate is high, but it is slowed by the corrosion products that adhere tightly to the pipe. These products have a higher resistivity than does the surrounding soil, thus reducing the corrosion rate [*3*]. Fig. 2 shows examples of an oxygen-concentration cell.

It should not be assumed that the corrosion products form a coating that is even remotely comparable to an insulating pipeline-coating. A hard scale is formed, but it does not adhere tightly to the pipeline in all places. In other places, it is thin, and in still other places there is no scale at all. Oxygen-concentration cells will exist in this well-aerated area, and corrosion will continue. Even so, the scale formed by the corrosion products increases the electrical resistance and often reduces the overall corrosion rate significantly. Sometimes, it quickly drops to a low value.

While the corrosion rate in a well-aerated area decreases in the first few months, this is not so in a poorly aerated location. The corrosion products produced in heavy moist soil differ from those formed in well-aerated earth. The corrosion products produced in oxygen-deficient soil do not adhere tightly to the pipe, and do not form an insulating barrier that reduces current flow.

Thus, corrosion continues unabated. The effect of this

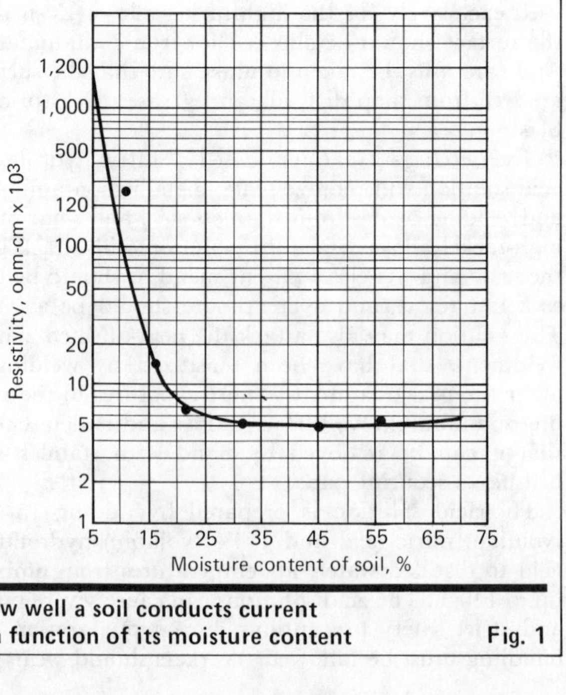

How well a soil conducts current is a function of its moisture content **Fig. 1**

Originally published March 5, 1984

continuing corrosion is cumulative, and the total amount over a few years is much greater than in well-aerated soil.

In these heavy, moist soils, pitting corrosion can worsen the problem. Pits grow rapidly, and a pipe can be penetrated quickly. Often, in such cases, even though more than 95% of the pipe surface shows little corrosion, a section will have to be replaced because of a few deep pits.

Bacterial corrosion

Most of this type of corrosion is attributed to *Desulfovibrio desulfuricans*, an anaerobic sulfate-reducing bacterium. This organism is often found in heavy, moist soils, such as the clays, and thrives where there is a lot of organic matter [*4*].

D. desulfuricans may cause severe corrosion on underground pipelines. Where there are high numbers of sulfate-reducing bacteria in the soil around the pipeline over extended periods, a typical pattern of corrosion is often seen: Deep, craterlike pits develop and the pipe wall may be penetrated in a short time. However, other factors also create this pattern, so it is not always an indication of sulfate-reducing bacteria at work.

Galvanic corrosion

This type of corrosion occurs when two dissimilar metals are connected together in an electrolyte, such as soil. For instance, copper water pipes are often run to steel pipes or tanks—a powerful corrosion cell is set up and serious corrosion can result in a short time.

Any two metals in the galvanic series will cause a current to flow when connected together in an electrolyte. The less-noble metal will corrode.

The use of insulators eliminates galvanic couples. Also, a sacrificial anode (e.g., one made out of magnesium or zinc) can be connected to a steel pipe to cathodically protect the steel.

Chemical corrosion

When a plant is expanded or a new one is built and land is scarce, marginal land may be used. This includes dumps, landfills and abandoned sewage-treatment plants. Chemical dumps are often adjacent to some of the older chemical-processing plants—just the locations where new plants are likely to be built.

Van Eck [*5*] warns that all of these marginal areas may contain corrosives, and their presence must be determined by chemical tests. Soil samples should be collected at enough locations and depths to provide a good assessment of conditions. Often, however, only a small part of an area will contain corrosives. Most plants will not be built where the soil is contaminated, but some soil tests and chemical analyses should be made anyway.

In most cases, if there is enough foreign chemical in the soil to damage pipelines, there will be some indication of it from a cursory examination. Appearance, odor and pH will usually flag a warning. A conventional chemical analysis may also indicate the presence of corrosives. If a plant that is manufacturing or using chemicals is nearby, then soil tests should be made for the chemicals found at that plant.

Chemicals that leak into the soil or are applied to it (e.g., fertilizers) can be corrosive. If the concentration is high enough, fertilizers can corrode steel. In most cases where chemicals have damaged substructures, care has simply been inadequate.

Most chemical damage is in the form of pitting corrosion—a costly problem. Stress-corrosion cracking (SCC) may also occur, and it could result in catastrophic failure of a pipeline. Even a small amount of material can cause SCC.

Sulfides, carbonates, nitrates, chlorides and caustics have all been reported to cause SCC in steel [*6*]. Although these groups are normally found in soils, they are typically a problem only when they occur in large concentrations, i.e., when they come from manmade chemicals.

The use of coatings, SCC-resistant steels, and cathodic protection have reduced the number of accidents and problems caused by this form of corrosion. Nonetheless, contaminated soil is an undesirable environment for steel structures. If such soil cannot be removed, the pipeline should have one or two layers of pipeline coating applied over the standard coating already on it. Cathodic protection should be used, too.

Tests to determine soil corrosivity

The most common method of doing this is by measuring electrical resistivity. Many factors affect the resistivity of a soil (e.g., moisture, temperature), so this measurement only provides an approximation. Further, the resistivity of a soil may change from time to time.

For example, rainfall may differ from one season to the next. In Southern California, tests made in desert

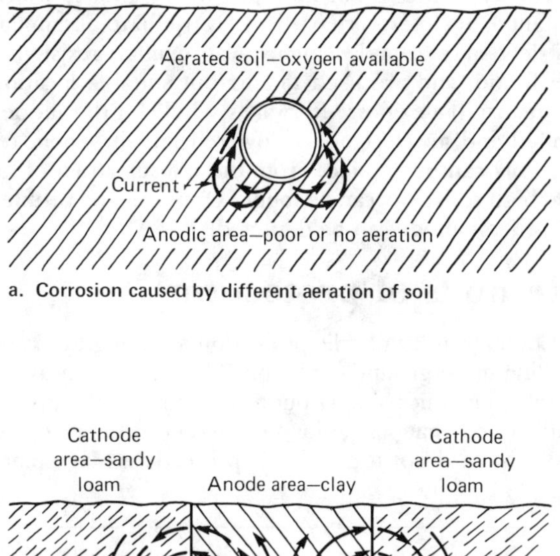

a. Corrosion caused by different aeration of soil

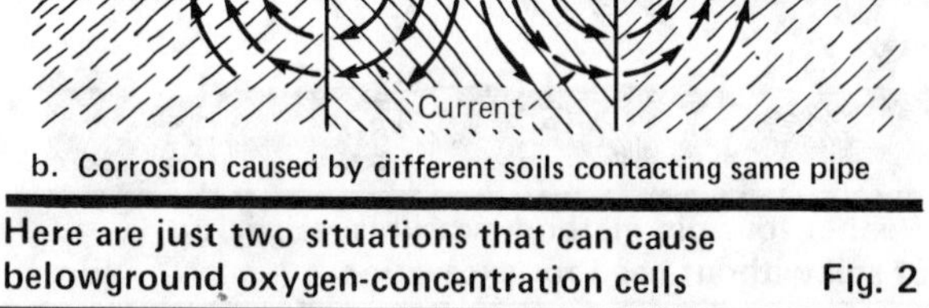

b. Corrosion caused by different soils contacting same pipe

Here are just two situations that can cause belowground oxygen-concentration cells Fig. 2

areas during summer often yield values of 50,000-100,000 ohm-cm. During the rainy winter, readings may drop to less than 5,000 ohm-cm.

An excellent booklet on earth-resistance testing is published by the James G. Biddle Co. [7]. As described in the booklet, most resistivity tests use the Wenner four-pin method. It has the advantage that only surface measurements need be made; no excavation or hole boring is necessary.

Fig. 3 shows a typical setup for making this test. Current is put into the soil through pins C_1 and C_2; potential is measured at pins P_1 and P_2. By spacing the pins at different distances, average resistivity can be found at various depths. For instance, with a 10-ft spacing, resistivity at a 10-ft depth will be found. If the soil is uniform from the surface down to 10 ft, then the measurement will be correct for all depths down to 10 ft.

In most cases, readings are made at several depths, and it is possible to see from these the changes in resistivity with depth. The resistivity at the bottom of the hole where the pipe will be placed can then be estimated. Usually, such a method suffices to determine resistivity.

Resistivity is not the only important measurement; the soil pH has a significant effect on the corrosion rate. Recognizing this, Stratfull [8] developed a test that incorporated pH and soil resistivity.

The reasons for making soil tests have changed. Federal law now requires that interstate pipelines carrying flammables be coated and cathodically protected. Many state and local governments have statutes for pipelines and other structures not covered by federal laws.

So today, tests are done primarily to determine the resistivity of the soil at locations where anodes will be installed. Such tests are particularly useful to obtain data for deep-anode emplacements. Approximate resistivities are found, and are used to sort out suitable locations from unsuitable ones. Afterwards, a hole for the anode is drilled and an electrode is lowered into it. Then, test currents can be applied. From these tests, strata of low and high resistivity can be found. Once this soil profile is known, an anode can be designed.

Methods of protection

Coatings and cathodic protection are integral to controlling underground corrosion. This was not always so. Some companies simply put a bare pipe in the ground and hoped that it would not corrode. Others coated pipes but did not use cathodic protection. Many plants located near the sea coast installed bare pipes and flooded them with cathodic protection. This was because power was cheap and the coastal soils have low resistivity, thus not wasting much power.

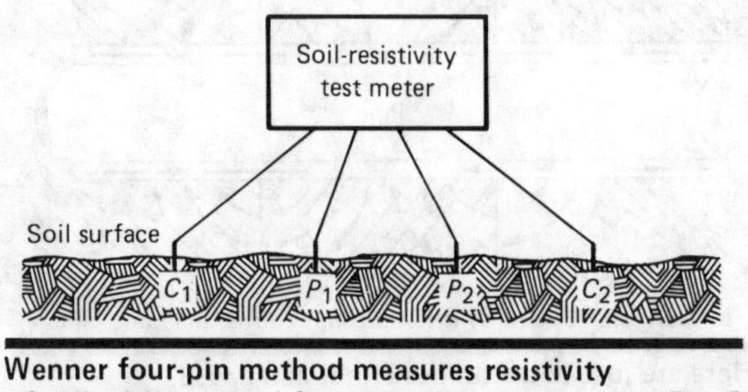

Wenner four-pin method measures resistivity of soil without need for excavation **Fig. 3**

However, none of these methods is practical for new construction. Coatings and cathodic protection are needed. Sometimes, law requires this. Also, if there is concern about leakage of noxious liquids, a firm not using both systems might be found negligent if leakage were to occur. If cathodic protection is properly designed and installed, it will be nearly 100% effective.

Coatings

Selecting coatings for plant piping differs from doing it for a cross-country pipeline. Plant structures and belowground plant piping are usually ordered factory-coated, while pipelines are coated "over the ditch."

Such differences preclude some types of coatings from being used for plant facilities. Also, some coatings are difficult to use, and some are difficult to obtain in certain locations.

For example, the bitumens are good coatings, but they are messy to apply, and small-diameter pipe coated with them is often unavailable. Typical factory-applied coatings:

1. *Heat-cured plastics.* Liquids that are applied to the pipe and heat-cured. Mainly, these are epoxies and phenolics.

2. *Extruded plastics.* Polyethylene is commonly used for extruded coatings. An adhesive, usually butyl rubber, is applied to the pipe and then the polyethylene is extruded over it. The pressure due to extrusion forces the adhesive into good contact with the plastic and the pipe. This process is limited to such shapes as pipes and bars.

3. *Heat-cured powder resins.* Included are epoxies, which are applied to a hot pipe (or other metal object), which quickly cures the resin. Such curing yields a hard, dense plastic coating that has good metal-adherence. The coating is relatively thin, about 12-25 mils. The manufacturer supplies compatible materials for repairs; its directions for field joint-covering *must* be followed.

In selecting coatings, talk to local coatings applicators. They know the types most in demand, and can point out their features. They can also provide information on cost and availability [9].

Various organizations have prepared standards and specifications for coatings used in underground service. Such properties should be reviewed and applicable ones included in a purchase order.

Most coatings withstand temperatures to about 160°F; above that, special coatings must be used. These are often more expensive and require more care during application.

Some structures may be coated at the plantsite. These include tanks that will be installed underground and tank bottoms that will rest on the ground or on concrete pads. Such coatings must be self-curing, to provide ease of application. Coal-tar epoxies are a favorite here. Many other materials also perform well.

Cathodic protection

Cathodic protection is not simply a matter of pouring current onto a pipeline. To make it effective, coatings, test leads, and so on are needed. Insulators are used, when needed, to electrically isolate substructures that are to be cathodically protected. A cathodic-protection system shoud be included at the earliest stage of plant design.

At that stage, many decisions have to be made. For example, it must be decided whether to protect all underground metal structures as a unit, or to insulate those that do not need cathodic protection. For an existing facility that has not had cathodic protection before, insulating may not be feasible, and it is better to protect the whole plant as one unit. New plants usually employ insulators, and protect only those structures that need it. When an area is isolated and protected, it is referred to as the basic system. Such isolation eliminates most cathodic interference problems and saves power costs. But more care is needed and technicians must be better skilled.

The basic cathodic system

The basic system comprises all underground, coated, metal structures, including pipelines and tanks. It does not include reinforcing steel in slabs of foundations, copper ground-rods or mats, or copper water-pipes.

When structures are to be isolated, insulators will have to be installed, and continual diligence will be required to make sure that they are not breached. Tests should be made routinely to ensure that insulators are not shorted and that no piping or wiring has been connected to them. Also, the plant staff involved in maintenance, new construction or plant modifications should preserve this electrical isolation during such operations.

Care must be taken that electrical controls not be so installed that they provide a path around an insulator.

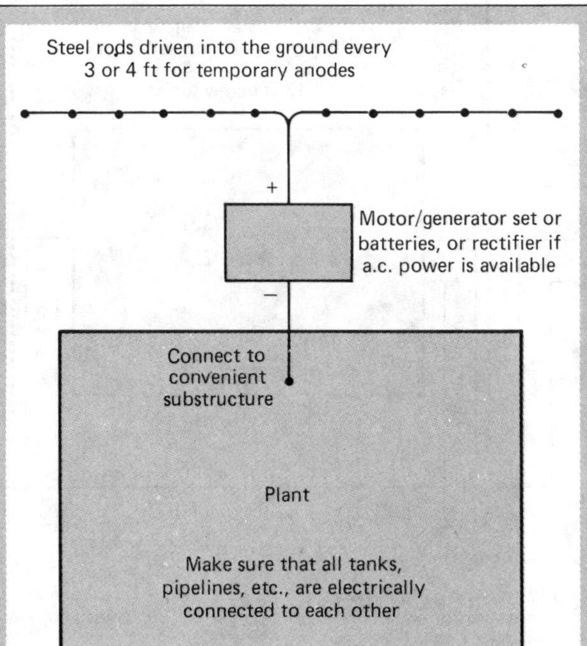

A temporary cathodic-protection station is used to apply a test current **Fig. 4**

Similarly, be careful of small-diameter pipes. They are plentiful in many plants, and any of them may accidentally be connected around an insulator, shorting it. When this happens, it sometimes takes considerable testing to determine the cause of the short.

Whole-plant protection

In some plants, maintaining electrical isolation will be too difficult, and all substructures will therefore be cathodically protected, whether they need it or not. While this system is simpler since no insulators are needed, the current required is greater, and this is expensive. This is somewhat compensated for by eliminating the costly process of searching for shorts.

When the plant is protected as one unit, the copper electrical grounds are tied to the underground steel structures. A galvanic couple is set up between copper and steel. This couple could cause corrosion if the cathodic-protection current that reached such areas were not strong enough to overcome it. To guard against this, a magnesium anode should be connected to the steel near each ground rod. The anode will protect the steel and improve the plant's electrical grounding system.

The other drawback to whole-plant protection is cathodic interference, which will be covered later on. Large protection currents may be needed in some plants. Some of this current may cause corrosion on pipelines and cables nearby.

A procedure will be given for finding the proper amount of current needed. It applies whether the whole plant or just the basic system is to be protected. The values shown are for a typical basic system of a very large petrochemical plant, assuming that there would be some damage to the coatings during construction.

Amount of current required

After plant construction is completed, tests are needed to design the cathodic-protection system. First, the amount of current needed must be determined. To do this, test current is applied to the substructures to be protected. Pipe-to-soil potentials are measured at strategic plant locations with the current on and off.

Fig. 4 shows how a test current is supplied by a temporary cathodic-protection station. The ideal amount of current would create a minimum pipe-to-soil potential—anywhere in the plant—of 1.0 V, and a change in the off-and-on potential of about 0.5 V.

If there is insufficient current to do this, then the resistance of the temporary anode should be lowered by watering it or by adding more rods to it. If it is impossible to bring the pipe-to-soil potential to 1.0 V, then it can be extrapolated. If the off-potential is 0.6 V, and when 10 A is applied the potential is 0.8 V, then 20 A should produce about 1.0 V.

For this plant, 20 A is assumed to offer excellent protection. However, most new plants would not initially require as much current unless the coatings were substantially damaged or improperly applied. But in this hypothetical plant, there is a fence around the property line. If nothing were done, the bottom of the steel fence posts would corrode due to cathodic interference. To prevent this, bond wires are connected from the fence to the rectifiers.

Impressed current vs. anodes

Now, the current source must be selected. Should galvanic anodes or impressed current be used? Galvanic anodes are usually made of zinc and, more often, magnesium, due to the latter's higher driving potential. In most cases when the impressed current is above 3-4 A, an impressed-current system is usually the less expensive of the two.

Further, such a system provides greater flexibility: The current can be adjusted easily, while with galvanic anodes more of them must be added to increase the current. Also, the impressed-current system makes testing easier. The current can be turned on and off with a switch, whereas anodes must be disconnected and reconnected.

Impressed current can be supplied in several ways. A rectifier can change a.c. into d.c.; solar systems or fuel cells can produce d.c. directly; and gas engines can drive d.c. generators. However, when commercial electric power is available, it is almost always chosen. It is reasonably efficient, troublefree and cheap.

More than one rectifier can be used. With large plants, particularly old ones, a large amount of current might be required, possibly 100 A or more. Better current distribution would be obtained if four or five cathodic-protection stations were installed, rather than having the current come from a single source.

In a plant where only 20 A is needed, the decision is not so clear. With the substructures coated, the current would probably be evenly distributed around the plant. Another rectifier costs money to install and maintain. It does, however, provide a means of balancing potentials should the need arise. Also, with two rectifiers, partial protection will be afforded if one of them is out of service. Weighing all these aspects, two cathodic-protection stations are chosen to yield the best protection.

Anode design

For simplicity and to show both types of installation, with two cathodic-protection stations, assume that one anode will be a shallow horizontal type and the other, a deep vertical design. Each will have an initial output of 10 A.

Fig. 5 shows the location of the plant and some features that affect anode design. Adjoining plants are too close to put anodes on the sides of the plant. Pipelines, cables, etc., in the street rule out the use of shallow anodes in the front of the plant. At the rear, borings taken prior to construction showed a layer of rock about 17 ft below the surface. This dictated that any anode installed here would have to be shallow.

Soil tests will be made in the front and rear of the plant, using the Wenner four-pin method. At the rear, soil samples will be taken from the surface down to a depth of 17 ft. Resistivities of these samples will be measured in the laboratory.

The type of anode must be selected. Graphite or high-silicon cast-iron rods are the most common. Both are usually surrounded with coke breeze (undersized coke screenings about ⅝ in. and smaller) for 4-5 in. If installed in a round hole, the diameter would be 11-12 in. Steel pipes were once a favorite for anodes, but are now rarely used. Today's anodes are more expensive and are specifically suited for use in certain soils [*10*].

For the two cathodic-protection stations, graphite rods with coke breeze are chosen. For the station at the rear of the plant, holes, 10 ft apart, will be drilled in a straight line parallel to the rear lot line. The holes will be 14 ft deep, with the anode occupying 12 ft, allowing 2 ft of soil cover. Each hole will be of 12-in. dia. and will contain a single graphite rod, 3 × 60 in., surrounded by coke breeze.

If the soil resistivity is assumed to be equal to 4,800 ohm-cm, then the anode's resistivity is 2.2 ohms, as calculated by a formula given by Dwight [*11*]. With a 10-A output, the mimimum required voltage is found to be 22 V.

The deep anode, at the front of the plant, will be a single unit, 70 to 150 ft deep. The soil resistivity from 70 to 150 ft deep averages 5,600 ohm-cm. Graphite rods are used; this anode will also have an output of 10 A.

Seven graphite rods will be installed in a 12-in.-dia. hole and be surrounded by coke breeze. The anode will be 80 ft long, with its top 70 ft below the surface. A vent pipe will be installed from the anode to the surface to carry away any gases that might form in the anode and raise its resistance. This pipe also serves as a means of watering the anode if the soil around it becomes too dry. The resistance is calculated to be 2.0 ohms; the required voltage is 20 V.

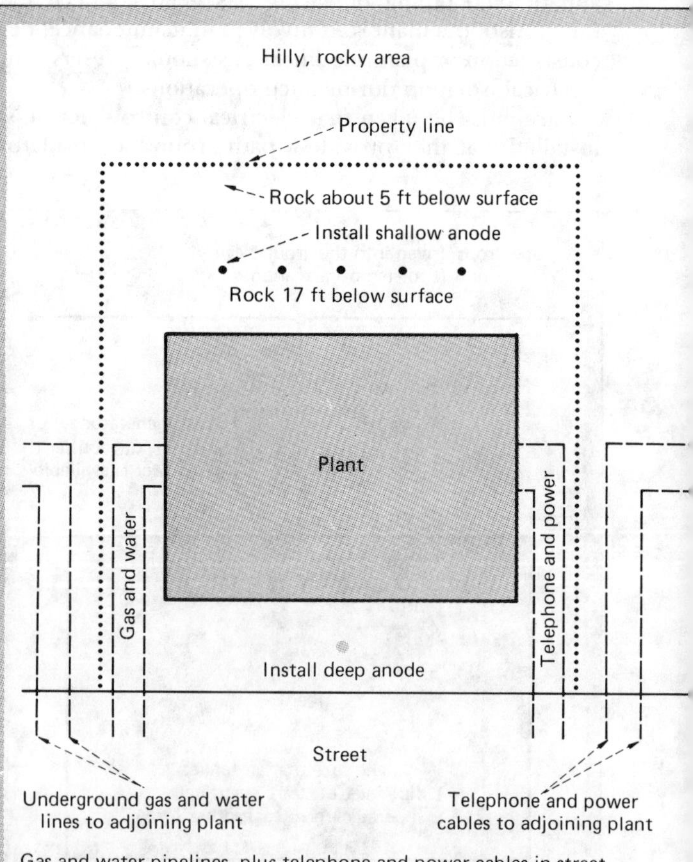

Gas and water pipelines, plus telephone and power cables in street, make it impractical to install shallow anode in front of plant

A rough, typical plant layout shows factors affecting cathodic-protection system design **Fig. 5**

Installation and startup

Some engineers delay ordering rectifiers until the anodes have been installed. Then, additional tests are made to see exactly how much voltage is needed to supply the required current. It it not always possible, particularly with deep anodes, to accurately calculate anode resistances. Measuring the resistances after anode installation is more accurate. But even using this method, rectifiers should not be sized precisely to match these resistances.

Although the shallow anode has a calculated resistance of 2.2 ohms, measurement shows it to be 2.0 ohms. A driving voltage of at least 20 V is needed. But rectifiers are always ordered with extra capacity for current and voltage; how much extra capacity is a matter of judgment. Here, the rectifier will be ordered with an output of 40 V and 15 A. This will provide extra voltage in case the anode dries out, and extra current if the coating is damaged or additions are made to the plant.

For the deep anode, the measured resistance is 2.3 ohms. The rectifier's output will also be 40 V and 15 A.

Once installed, the rectifiers are turned on and set at 10 A. Pipe-to-soil potentials are measured throughout the plant. Although the ideal protection potential is 1.0-1.2 V, some potentials may fall out of this range. Outputs of the rectifiers will be adjusted to bring all of the low spots to at least 1.0 V, and the high spots down to as close to the 1.0-1.2 V range as possible. This may result in an imbalance of currents from the two rectifiers, but this is not a matter of concern since the pipe-to-soil potential is the governing factor.

Even after the currents have been adjusted, some of the potentials may fall outside of the desired range. If they are low, and if it is a matter of one or two spots with potentials of no lower than 0.9 V, a few magnesium anodes can be installed at these spots.

Potentials that are slightly higher than the desired ranges at a few spots are tolerated by most corrosion engineers. However, it is advisable to get the coating manufacturer's recommendations in regard to this. Some coatings have withstood high potentials for long periods, but a few corrosion engineers think that some disbonding may occur if potentials are as high as 1.0 V. If necessary, zinc anodes can be installed at the "high" spots to lower the potential.

Corrosion can result if the potential is too high. The coating will disbond, and then moisture will penetrate under it. The cathodic-protection current will not travel far under a disbonded coating and will not reach all spots. Corrosion occurs and the cathodic-protection system does nothing to stop it.

When oxygen is absent, the corrosion process under disbonded coatings may be complex. Anaerobic bacteria may play a role in this [*12*]. Disbonding of a coating is just about the worst thing that can happen to it. Most new pipeline coatings can withstand potentials to about 1.6 V without much disbonding. But to be safe, potentials less than this should be the norm.

Notices and permits

The local cathodic-protection committee should be notified that a system will be installed. So should managers of nearby plants. In some cases, they may suggest changes. Notification may avoid trouble later if a cathodic interference problem develops at an adjacent plant. The same people should be sent a notice once the system is in operation. This final notice should state the operating current and voltage of each rectifier.

There are nearly fifty cathodic-protection committees in the U.S. and Canada. These groups were devised so that companies owning pipelines or other underground facilities would be informed when cathodic-protection stations were being installed.

Such groups do not maintain local offices, and their officers change every year or two. NACE—the Natl. Assn. of Corrosion Engineers (Houston)— has tried to maintain a mailing list of these committees, but unsuccessfully since they have no fixed addresses. Nonmember companies may send representatives to their meetings. For instance, a plant engineer with a specific interference problem can attend a session and get expert advice.

These groups often use formal notification forms when a cathodic-protection station is being installed. Also, they have developed guidelines for handling cases of cathodic interference. NACE has published a booklet that contains procedures developed by various cathodic-protection committees [*13*].

Many states and local governments are concerned about pollution of underground aquifers. Some require permits for any hole drilled deeper than 50 ft. California is one, and has written specific guidelines for drilling and installing deep anodes. When adoped by cities and counties in that state, these guidelines become law [*14*].

For our example, the deep anode will require a permit. In addition, it is usually required that the anode hole be sealed for the top 50 ft or more with concrete or similar material. This keeps out surface water. Sometimes, much deeper sealing is called for to prevent interchange of water between aquifers.

Some local governments require building permits for cathodic-protection stations using commercial power.

Routine tests and recordkeeping

After the cathodic-protection system is in operation, pipe-to-soil measurements should be made throughout the entire plant. Minor rectifier adjustments might be needed to keep the potential within the desired range.

After about a month of operation, the pipe-to-soil potentials should be measured again. If they are satisfactory, a routine test schedule should be set up. Rectifier outputs should be read at least once a month. For the first year, pipe-to-soil potentials should be measured every three months.

If the potentials are stable during the first year, the time between measurements can be lengthened, but the risk of corrosion must be balanced against the money saved. Many companies adopt a modified schedule whereby tests are made only at key points every three months. A complete set of potentials is measured annually. Routine test records should be kept so that results can be compared.

Cathodic interference

There are two types of cathodic interference, one from current, the other from voltage. Current interference is by far the worse (see Fig. 6). Fig. 6 shows a cathodic-protection station with a rectifier and a ground bed. Normally, such a setup will cause several amperes to flow through the soil to the protected pipeline.

Some current collects on a nearby foreign pipeline and travels on it for some distance. As the current nears the cathodically protected pipeline, it discharges into the soil, causing corrosion. If the foreign pipeline is coated, as most are, the interference current leaves the pipeline at holidays. This can cause rapid, severe corrosion.

Fig. 7 illustrates voltage interference, in which an excessive voltage may be created between the foreign pipeline and the surrounding soil. If the foreign pipeline is coated, disbonding may occur. The excessive voltage and high current density at the holidays make the soil quite alkaline.

This alkalinity destroys the coating's bond. If the voltage is high enough, hydrogen gas will be formed at the pipe wall and will penetrate the coating, lifting it. Considerable damage can be done to a coating.

Equally bad are the effects on uncoated amphoteric-metal structures, e.g., lead. Such metals are subject to corrosion in highly acidic or alkaline environments. Alkaline products may be formed if lead cable-sheaths are made excessively negative, especially in soils containing appreciable amounts of sodium or potassium salts [*15*].

Another amphoteric metal is aluminum. It is not used much underground, although some pipelines are made of it. Amphoteric metals may suffer much more corrosion than other metals when affected by cathodic interference. Extra precautions must be taken.

And remember that interference can be caused on a plant's substructures by the operation of cathodic-protection stations of other companies.

Avoiding interference

Whether or not a cathodic-protection station will cause interference depends primarily on the current output of the anode and the proximity of the anode to other substructures. Galvanic anodes produce small currents, usually less than 200 mA. Rectifiers may have outputs of over 100 A.

However, if enough galvanic anodes are installed and spread out over a long pipeline, they could produce a substantial current. Usually, this is not so, because rectifiers are more economical above a certain threshold total current value.

Locating the anode as far away as possible from any substructure that might pick up current will reduce interference. A cardinal error in corrosion engineering is to place the anode so that a foreign substructure is between it and the protected structure.

The second-worst error is to place the anode close to a foreign substructure that offers a good path for the current to follow in its effort to get to the protected pipeline.

Designing to avoid interference

Substructure maps for the streets around the plants should be examined. These are kept by most large cities and some county offices. At this time, the engineer should contact nearby companies to determine whether their plants might be affected by cathodic interference, and to arrange for cooperative efforts to prevent any damage. Finally, the local cathodic-protection committee should be notified.

In avoiding cathodic interference, prime responsibility rests with the company installing the cathodic-protection system. Once the system is in, the responsibility shifts to any company that installs a new pipeline or other new substructure that might suffer interference from the system.

The owner of the cathodic-protection system should be notified. Joint tests would then be arranged between the two companies, and a decision made as to how to solve the interference problem.

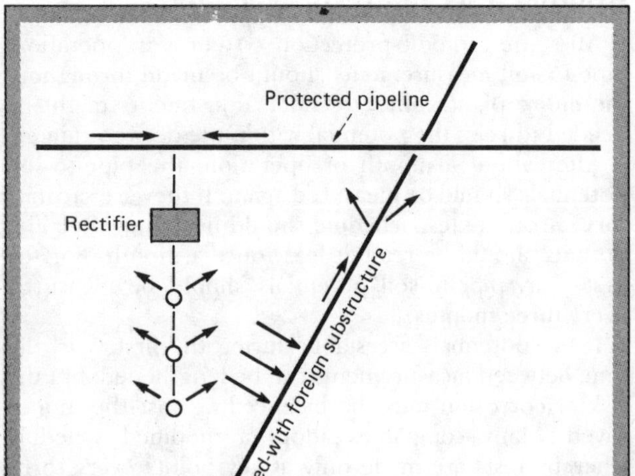

Current interference is by far the worst kind of cathodic interference Fig. 6

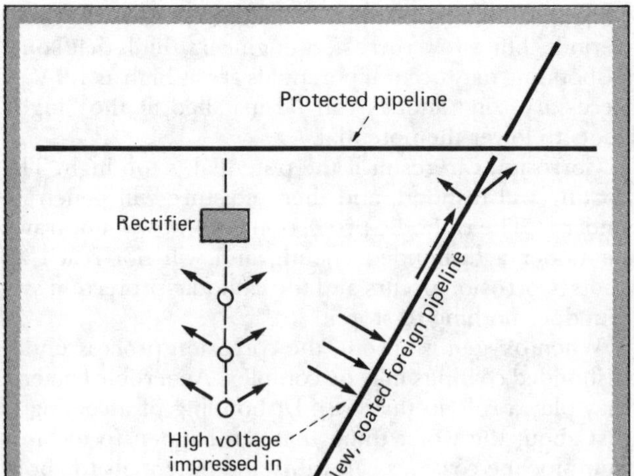

Voltage interference can result in coating disbondment and corrosion Fig. 7

Legal aspects

Some consideration should be given to the legal aspects of cathodic interference [*16*]. The corrosion engineer should keep records of tests and designs of cathodic-protection systems, especially those pertaining to interference.

It is difficult to assign blame in cases of interference, because of the natural tendency of underground pipelines to corrode. It is hard to determine how much corrosion would occur were there no interference. But, in many cases, losses amount to hundreds of thousands of dollars, and lack of documentation may lose a case for a company.

Effects of a.c.

Alternating current may cause corrosion on underground steel structures in large plants such as oil refineries and chemical-processing facilities. Electrical grounds are usually connected through some metallic paths to underground tanks, pipelines, etc.

As a result, these steel structures often carry several amperes of a.c. While a.c. is not normally considered to cause significant corrosion, this may not be true with underground steel. Fuchs et al. [*17*], in laboratory tests, found a relatively high corrosion rate with high current densities. Other investigators have come up with mixed results.

However, even in plants where large amounts of a.c. flow on underground structures, there is no need for concern if these are cathodically protected. Bruckner [*18*] states that cathodic protection will prevent a.c. corrosion, but cautions that magnesium anodes may not be satisfactory. This is because a reversal in potential of the anodes may sometimes occur. He suggests that impressed-current systems be used.

Other corrosion engineers have recommended that the pipe-to-metal potentials of cathodically protected steel structures be maintained from 0.1-0.2 V higher than would be needed if there were no a.c. on the structures. This seems to be a reasonable precaution.

Insulators

In most plants, steel substructures will inadvertently be connected by metallic paths to electrical grounds, reinforced steel in building slabs, and various kinds of pipelines and electrical cables coming into the plant. If cathodic protection is installed without using insulators, then the reinforcing steel, copper ground-rods, etc., will soak up large quantities of protective current.

Shielding and test connections

In large plants, substructures are often crowded together, and shielding may occur so that a sufficient cathodic-protection current does not reach all parts of a protected structure [*19*]. In older plants, holidays may develop in coatings, and shielding may begin to cause corrosion. High-resistivity soil tends to make shielding worse [*20*].

A good way to provide needed current at shielded locations is use magnesium anodes. During plant design, locations where shielding might occur can be surmised by examining the plans. If a large pipe is in front of a small one, the small pipe might not get enough current to prevent corrosion. One or more magnesium anodes connected to the small pipe and placed between the two pipes would offer protection.

A related precaution: One should install a permanent copper sulfate electrode at each shielded location to provide for future measurements of pipe-to-soil potentials. All wires should be brought to a surface test-box. A few other permanent copper sulfate electrodes should be installed at strategic locations throughout the plant. These locations are at remote spots where the level of protection cannot be estimated with reasonable accuracy.

A final note: Cathodic protection is extremely effective in controlling corrosion, but it must be designed and maintained properly.

References

1. U.S. Dept. of Commerce, National Bureau of Standards, Underground Corrosion, Circular 579, 1957, p. 155.
2. Schaschl, E., and Marsh, G., Some New Views on Soil Corrosion, *Materials Protection*, Nov. 1963, pp. 8-17.
3. Schiff, M. J., What Is Corrosive Soil?, paper presented at the NACE [Natl. Assn. of Corrosion Engineers, Houston] Western States Corrosion Seminar, 1969.
4. Uhlig, H., "The Corrosion Handbook," John Wiley and Sons, New York, 1955, pp. 474-476.
5. Van Eck, W. A., What Is Soil?, *Pipeline News*, Mar. 1964, pp. 17-22.
6. LaQue, F. L., and Copson, H. R., "Corrosion Resistance of Metals and Alloys," Reinhold Publishing Corp., New York, 1963, pp. 315-319.
7. Getting Down to Earth, booklet published by James G. Biddle Co., Plymouth Meeting, Pa.
8. Stratfull, R. F., A New Test for Estimating Soil Corrosivity, paper presented at NACE Western Region Conference, 1960.
9. Sloan, R. N., How to Select Mill-Applied Coatings, *Pipeline and Gas J.*, Feb. 1982, pp. 40-48.
10. Stephens, R. W., Deep Ground Beds—Material Selection and Economics, paper presented at NACE's Corrosion/83, Apr. 18-22, Anaheim, Calif.
11. Dwight, H. B., Calculation of Resistances to Ground, *Electrical Engineering*, Dec. 1936.
12. Peabody, A. W., "Control of Pipeline Corrosion," NACE, 1967, p. 175.
13. Bulletin IV of the report of the Correlating Committee on Cathodic Protection, NACE.
14. State of California, Dept. of Water Resources, Cathodic Protection Well Standards, Bulletin No. 74-1, 1973, with revisions.
15. Uhlig, *op. cit.*, p. 603.
16. Hatley, H. M., Pipeline Corrosion: The Legal Aspects, *Anti-Corrosion*, Nov. 1971, pp. 6-7.
17. Fuchs, W., et al., Corrosion of Iron by Alternating Current with Relation of Current Density and Frequency, *Das Gas- und Wasserfach*, Jan. 1952.
18. Bruckner, W. H., The Effects of 60-Cycle Alternating Current on the Corrosion of Steel and Other Metals Buried in Soils, *Bulletin*, Vol. 62, No. 32, 1964, University of Illinois.
19. Orton, M. D., Fundamentals of Cathodic Protection, paper presented at NACE Western States Corrosion Seminar, 1978.
20. Martin, B., Cathodic Protection Shielding of Pipelines, *Materials Performance*, Feb. 1982, pp. 17-21.

The author

Joseph S. Dorsey is a consulting engineer in California, and resides at 9 Osprey, Irvine, CA 92714. Tel: (714) 552-3075. He specializes in corrosion control, electrical effects on pipelines, and electrical grounding problems. He started in private practice as a consulting engineer in 1973, after serving for many years as the corrosion engineer for Southern California Gas Co. The author of several technical papers, he holds a B.S. degree from the University of New Hampshire and is a registered electrical, corrosion and metallurgical engineer in California.

Acoustic emission testing for chemical plants

Used mainly to test welds, and fiberglass-reinforced plastic piping and equipment, this nondestructive test method can help to eliminate costly internal inspections.

W. Donald Treleaven,
Monsanto Fibers and Intermediates Co.

☐ Briefly, acoustic emission (AE) testing is a way to determine the structural integrity of a piece of equipment, a weld or a section of piping by analyzing the elastic waves that are generated by it when it is stressed. Sensors are placed on the outside of the item to be tested, and signals from the sensors are sent to an analyzer, enabling a structural determination to be made. Tests are usually conducted by trained engineers, since the results require careful interpretation.

Here, we will discuss the AE test method, its advantages and disadvantages, and where it can be used. Detailed test procedures will not be given; these have been published and are usually only needed by those engineers who perform the tests.

What is AE?

The Amer. Soc. for Testing and Materials (ASTM) Specification E610-77 [*1*] defines acoustic emission as "the class of phenomena whereby transient elastic waves are generated by the rapid release of energy from a localized source or sources within a material, or the transient elastic wave so generated."

As examples, consider either a tree limb cracking or a glass breaking as it hits the floor. The crack tip running through the tree limb or the glass is the localized source of the rapid energy release. The transient elastic waves are the sound waves that travel from the crack tip to the ear.

For the tree limb, at the crack tip, energy is released by the breaking of molecular bonds and through other mechanisms, including friction. Released energy is in the form of heat and mechanical energy. The mechanical energy is dissipated as stress waves that travel through the limb and into the air.

Fig. 1 shows what the stress wave may look like. At time t_0, enough stress is applied to make the crack run a short distance. At time $t_0 + t$, no further energy is being released. The stress wave travels to the ear, and is converted to an electrical impulse that is sent to the brain for analysis. Then, one can decide on a course of action. An AE system works analogously.

A piezoelectric transducer is placed on the limb close to the crack. As the stress wave travels through the limb and strikes the face of the transducer, an electrical signal similar to the one shown in Fig. 1 is generated. This signal is amplified and analyzed by the AE equipment. The results are transmitted to an appropriate receiver, which may include an alarm system or an automatic shutdown device.

However, this is a rough analogy. The ear hears airborne sounds in a frequency range of 0.1 to 10 kHz, and has a sensitivity as low as 20 μPa. AE sensors can measure surface displacements within the frequency range of 50 to 500 kHz, with pressures as low as 1 μPa.

The term acoustic emission is therefore a misnomer: The stress waves transmitted by the sensors are not in the acoustic range—in fact, they are far below the threshold of hearing. AE testing is much more sensitive than the human ear, and can be used to detect stress waves of a fairly small magnitude.

For the tree, AE methods would likely warn of an impending failure several minutes before the ear could. By the time the stress waves would be large enough to

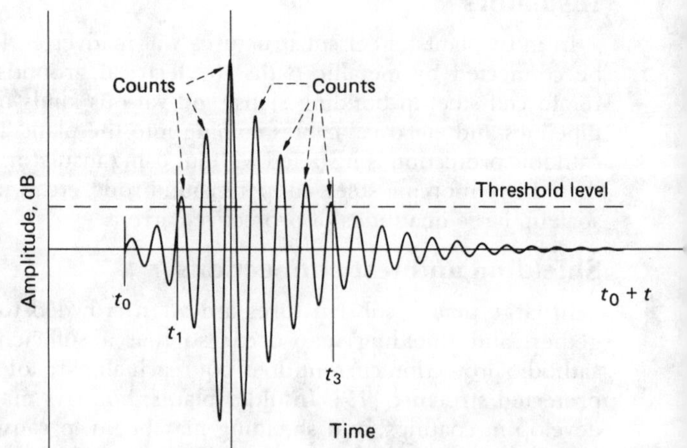

How a typical acoustic-emission signal looks, and the parameters used to characterize it **Fig. 1**

Originally published February 7, 1983

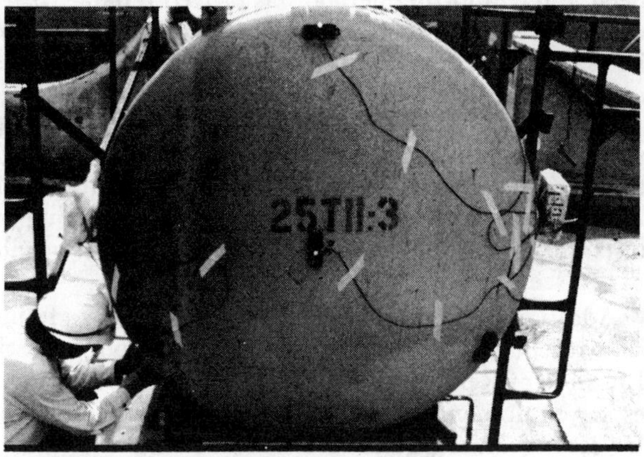

Four sensors on this steel tank are set in a triangle to locate damage Fig. 2

be detected by the ear, a person on the limb would already be falling.

Some basic AE terms are defined in Fig. 1. At time t_1, the signal rises to a decibel level above the threshold value, which is set to eliminate background noise (static). Time t_1 is not the true beginning of the signal. The "end" of the signal at t_3 occurs when the signal falls below the threshold value again. The signal received between t_1 and t_3 is called an event. Totaling the number of events that occur during a period of time will give some idea of how the crack is progressing.

Each event has several measurable parameters. The duration of the event is $t_3 - t_1$. The duration can be used to separate real events from false ones that are caused by electromagnetic interference and have short durations. The number of counts for each event is equal to the number of times the signal rises above the threshold value. In Fig. 1, the event has seven counts.

The total counts obtained during a test are a measure of the amount of emissions. Emissions can originate from various sources, including crack growth, yielding at the crack tip, crack friction, breaking of molecular bonds, intergranular cracking, raindrops falling on the equipment, and pump noise.

The peak amplitude is the highest signal obtained during an event, and is independent of the threshold setting. In Fig. 1, the peak amplitude occurs on the third count. The rise time is the time from the first threshold crossing to the peak amplitude, $t_2 - t_1$.

Equipment needed for AE testing

An acoustic-emission test requires only a few pieces of equipment: several sensor/preamplifiers and an analyzer. The sensor/preamplifier receives the stress waves that travel through the material, converts them into electrical signals, and amplifies the signals so that they can be sent over some distance. This is done with little or no distortion.

The piezoelectric sensors can generate signals as low as about 100 μV. The preamplifier amplifies this signal so that it can be sent to the analyzer. Further amplification may be done by the analyzer.

The size and capabilities of the analyzer depend upon the application and the type of information needed. For example, consider a unit designed to monitor a flange or valve for a high-pressure leak. The unit sounds an alarm if it receives a signal above a specified amplitude. Such a unit could fit into one's hand.

Alternately, a unit used to monitor a large metal pressure-vessel would require data-storage and retrieval systems, detailed signal-analysis functions, source-location capabilities and data-display systems. This unit would be large enough to require a small office or a corner of the laboratory for storage.

Due to increased reliability and advances in miniaturization, AE testing has been moving from the R&D department into the plant. For example, Fig. 2 shows four sensors attached to a steel tank. The preamplifiers are together, to the right. Newer devices have sensor/preamplifiers the size of just the sensors in Fig. 2. The analyzer has been scaled down, too: A unit capable of monitoring 16 sensors is the size of a large briefcase.

Advantages of AE testing

AE testing provides several advantages. The major one is that the test can be performed from the exterior of the piece to be tested. For example, during a hydrostatic test of a vessel, AE sensors are placed on the surface of the vessel, then the pressure is increased.

Since the test is done outside of the vessel, it can be conducted with the process fluid and at operating temperature, provided that appropriate sensors are used. By performing the test at service conditions, a better evaluation can be made. This is especially true for high-pressure, high-temperature service and for fluids denser than water. (For such fluids, the weight in the tank will be greater at the same liquid level, and hence the stress will be greater.)

With AE testing, costly internal inspections that require decontamination and entry may be delayed. Rather than visually inspecting a vessel every three years, this may be necessary only every six years if an AE test is done at two-year intervals. However, AE testing is not a substitute for internal inspection.

In some cases, AE testing can be done online, without any loss of production, and this can be done continuously. If a vessel is jacketed or insulated, only the areas where the sensors have to be placed need be uncovered. For a jacketed vessel, special provisions can be made in the construction of the vessel to accommodate sensors. Sensors can be mounted permanently on vessels, if the cost of scaffolding or insulation removal is high.

By monitoring each sensor, the source of emission can be located within a certain area. If three sensors are affected by an event, the source can be located by triangulation. In a chemical-processing plant, however, much of the accuracy of triangulation is lost because of the curvature and irregularities of process vessels.

Fig. 2 shows four sensors placed on the head of a vessel. These are sufficient to locate the source of a problem; other nondestructive methods can be used for more-detailed analyses. Using AE will, in some cases, reduce or eliminate the need for 100% inspection by X rays or dye penetrants.

For some applications, such as spot welding and leak detection, the AE test is used as a go/no-go procedure that requires no interpretation by the operator. During spot welding, an acoustic emission is generated by expulsion of gases and slag from the weld. Once expulsion takes place, no further current is necessary for welding. Thus, the AE signal is used to interrupt the current flow as soon as expulsion occurs.

In leak detection, a large amount of AE noise is generated at the site of a leak by an escaping fluid. (This assumes that there is a significant pressure drop across the leak.) Once a sensor detects the noise, a signal is generated. This signal can sound an alarm, set off a warning device or shut down a piece of equipment.

Problems with AE testing

Many of the disadvantages of AE testing are because it is a new technique. For testing vessels in a chemical-processing facility, one needs skills and knowledge of both AE and the material being investigated.

Each material has its own properties and reacts differently during testing. Defects in fiber-reinforced plastic (FRP) materials are "noisy" under applied stresses, but attenuation is high and the signal does not travel far. Thus, many sensors must be placed on the vessel.

Most metal defects are much less noisy than are FRP ones but, in metal, the signal is not attenuated as much, and fewer sensors are required. One drawback to AE testing is that not all materials have been studied for their response to the various types of cracking or overload. Interpretation of test data is difficult—or impossible—without knowing a material's response pattern.

Another problem with AE testing is that once an emission has occurred, the defect will not emit again until either a higher stress is applied or the material is weakened in some way. Also, materials do not always generate AE signals under fatigue conditions. Close to failure, an AE signal is usually generated on every cycle, but during earlier parts of the cycle, an AE is not always produced.

FRP and other such materials (graphite-reinforced and other plastic composites) produce emissions during refilling or repressurization. Normally, once a part has been stressed to a certain point and the stress is reduced, there will not be any emissions until that stress level is reached again. However, if there is damage, emissions will be generated at a level somewhere between 90 and 95% of the original value. This is known as the Felicity effect, and can be used as a measure of significant damage to a vessel. Lack of such emissions, however, indicates that the vessel has not been weakened since its last test.

The largest problem in AE testing is signal definition. Real emissions can be masked or eliminated by background noise from agitators, pumps, rain, fill-nozzle bubbling, gas spargers and many other sources. However, certain parameters of the signal can be used to identify and eliminate background noise. For large vessels or structures, interference from local radio stations can pose a problem, and it may be necessary to use special filters to eliminate certain frequencies.

Applications

In the chemical process industries, acoustic-emission testing has been used on FRP structures for the past several years. Vessels, storage tanks and piping have been tested, and the method has proved reliable. If AE tests and visual inspections are done regularly, defects in FRP vessels can be found and repaired before failure results.

A significant improvement in the quality of FRP vessels can be made using the results of AE tests to improve their design, and by performing tests at the equipment fabricator's shop. The theory and practice of using AE tests on FRP vessels is described by Fowler and Scarpellini [*2*], see p. 39.

Additional information on FRP testing can be found in a recent publication of the Committee on Acoustic Emission of Reinforced Plastics [*3*]. It describes the necessary equipment and techniques. Procedures are given for testing atmospheric, pressure and vacuum vessels; criteria are presented to perform an evaluation. With some modification, these tests can be applied to all FRP structures.

The use of AE tests for metal vessels is limited in chemical plants. One such application is the testing of ammonia storage tanks. Under certain conditions, ammonia can cause stress-corrosion cracking of carbon steel. Large ammonia tanks, especially those used for low-temperature service, are difficult and expensive to inspect internally. Bell [*4*] describes a vessel that was AE tested, and gives the results of subsequent visual and magnetic-particle inspections.

In addition to the already mentioned leak-detection and spot-welding tests, AE is also used on bearings. Stress waves emitted during early stages of deterioration of a bearing can be detected. This is because metal-to-metal contact generates acoustic emissions. Continuous monitoring or periodic tests can be used here.

Utilities are using AE testing of FRP booms for bucket trucks, and ASTM is developing a specification to cover this application. The same procedures could be used for manlifts in chemical plants and, with modification, for testing metal crane-booms.

References

1. Amer. Natl. Standards Inst./Amer. Soc. for Testing and Materials (ASTM), Specification E610-77, Standard Definitions of Terms Relating to Acoustic Emission, pub. by ASTM, Philadelphia, 1977.
2. Fowler, T. J., and Scarpellini, R. S., Acoustic Emission Testing of FRP Equipment, *Chem. Eng.*, Oct. 20, 1980, p. 145 (Part I), and Nov. 12, 1980, p. 293 (Part II).
3. Committee on Acoustic Emission of Reinforced Plastics, The Soc. of the Plastics Industry, New York, "Recommended Practice for Acoustic Emission Testing of Fiberglass Tanks/Vessels," Jan. 1982.
4. Bell, H. V., Shutdown, Inspection and Startup of a 15,000-ton Cryogenic NH_3 Storage Tank, paper presented at Safety in Ammonia Plants and Related Facilities Symposium, Montreal, Oct. 1981.

The author

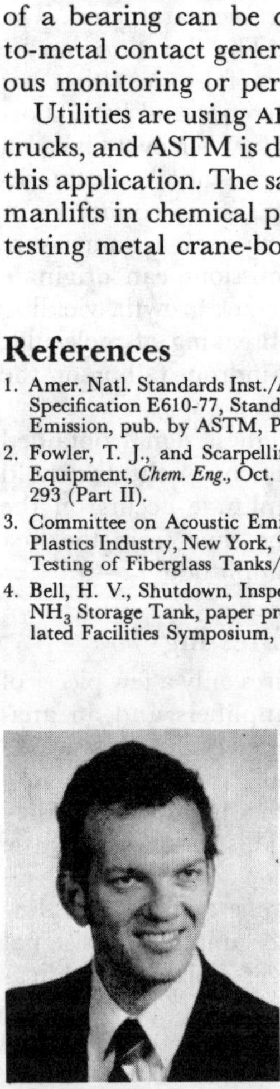

W. Donald Treleaven is senior materials engineer for Monsanto Fibers and Intermediates Co., P.O. Box 1311, Texas City, TX 77590. Tel: (713) 942-3580. He has been with Monsanto since 1978; prior to that, he worked as a process engineer for Diamond Shamrock Corp. Treleaven holds a B.S. degree in chemical engineering from Carnegie-Mellon University. He is a member of the Natl. Assn. of Corrosion Engineers, the Amer. Soc. for Metals, the Amer. Welding Soc., and the Committee on Acoustic Emission of Reinforced Plastics (CARP) of the Soc. of the Plastics Industry.

Corrosion control in steam and condensate lines

Several chemical treatments can be used to control oxygen-based and acidic corrosion. A combination of chemicals affords the best protection.

Paul F. Pelosi and **Carl J. Cappabianca,**
Drew Industrial Div., Ashland Chemical Co.

Corrosion of steam and condensate lines is one of the most costly problems facing plants. A few years ago, the total cost of corrosion in the U.S. was found to be in excess of $70 billion annually [*1*]. Since steam is widely used in process plants, afterboiler corrosion certainly makes up part of this total.

If corrosion of steam and condensate lines amounts to only 1% of the total, this figure still amounts to nearly $1 billion annually.

In the plant, afterboiler corrosion may incur cost penalties by requiring repairs to be made to the steam/condensate system, with resultant lost production. Energy losses through steam leaks and through deposition of insulating corrosion products in the boiler prove costly as well.

Repair costs — These depend on the piping material used and on local labor rates. Such costs have been escalating and will continue to do so. Still, the largest expense is not repairing the pipe; it is the loss of production during repair.

Steam leaks — These are often overlooked in considering the costs of steam-line corrosion. Usually, steam leaks are not serious enough to cause a boiler shutdown, but they do drain a system of valuable latent heat. Also, condensate is lost to the atmosphere.

Table I — Steam leaks are costly; even a small leak in a relatively low-pressure line can waste thousands of dollars annually

Pressure, psig	Dia. of leak, in. 1/16	1/8	1/4	1/2	1
	Steam wasted, lb/h				
100	18.15	72.57	290.33	1,167.27	4,645.08
200	33.93	135.67	542.80	2,171.07	–
400	65.70	261.88	1,047.72	–	–
850	136.53	545.76	–	–	–
	Cost, $/yr at $7/1,000 lb/steam*				
100	1,113	4,450	17,803	71,209	284,836
200	2,081	8,319	33,285	133,130	–
400	4,017	16,059	64,246	–	–
850	8,372	33,466	–	–	–

*Based on 24 h/d, 7d/wk, 52 wk/yr operation.

Table I (from Ref. [*2*]) estimates the cost of steam leaks. This table accounts only for the increased fuel usage needed to compensate for the energy lost through the leak. Costs of replacing the lost steam or condensate should be considered also.

Corrosion products — In afterboiler corrosion, return of corrosion products (i.e., metal oxides) to the boiler water via the condensate is often overlooked. These metal oxides have low solubilities and, therefore, can deposit on boiler heat-transfer surfaces.

Such products have poor thermal conductivity and can bind nonadherent boiler-water sludges to surfaces. The figure shows energy losses due to the low thermal conductivity of iron oxide in boiler deposits.

Causes of corrosion

There are two major causes of corrosion in steam and condensate lines: oxygen, which results in pitting; and a low pH, which gives rise to generalized thinning of piping.

In afterboiler systems, oxygen-caused pitting is rapid. Oxygen is a depolarizer of the corrosion-cell cathode. This electrochemical reaction is quick because of the high temperatures present. Oxygen is found in the afterboiler be-

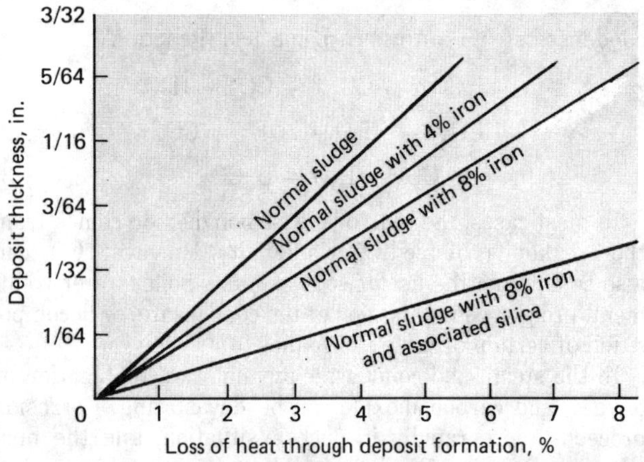

Figure — Iron oxide in boiler-scale cuts down heat transfer

Originally published June 24, 1985

cause (1) it is in the feedwater and subsequently is flashed from the boiler with the steam, and (2) it enters when air enters the afterboiler section.

Usually, formation of a low-pH condensate is due to carbonic acid. When bicarbonate and carbonates break down in the boiler, one of the products is carbon dioxide (see Reactions 1 and 2, which are thermal decompositions). Carbon dioxide flashes with the steam and combines with the condensed steam to form carbonic acid (Reaction 3).

Table II — For neutralizing amines, the distribution ratio indicates the type of system in which the amine will work best

Neutralizing inhibitor	Typical distribution ratio	Most-applicable condensate system
Morpholine	0.4:1	Short
Diethylaminoethanol	1.7:1	Medium
Cyclohexylamine	4:1	Long or branched
Ammonia	10:1	Long or branched

Table III — Combinations of various inhibitors provide optimal corrosion protection for steam and condensate lines

Components	Action
Blended neutralizing amines	Low-pH protection throughout steam and condensate lines
Blended neutralizing amines and catalyzed hydrazine	Low-pH protection throughout steam and condensate lines Oxygen-pitting protection Passivated-metal surfaces
Blended neutralizing amines and ESA	Low-pH protection throughout steam and condensate lines Oxygen-pitting protection Better ESA distribution
Catalyzed hydrazine and ESA	Low-pH protection throughout steam and condensate lines Oxygen-pitting protection Passivated-metal surfaces Surface preparation for ESA
Blended neutralizing amines, catalyzed hydrazine and ESA	All of the above advantages Backup protection

$$2HCO_3^- \rightarrow CO_3^{-2} + CO_2 + H_2O \quad (1)$$

$$CO_3^{-2} + H_2O \rightarrow 2OH^- + CO_2 \quad (2)$$

$$CO_2 + H_2O \rightarrow H_2CO_3 \quad (3)$$

In most cases, the bulk of the carbon dioxide comes from the breakdown of feedwater alkalinity. However, CO_2 can also come from the use of soda ash as a boiler-water treatment, process contamination of the condensate, or decomposition of certain organic compounds [*3*].

In the steam and condensate system, the combination of oxygen and carbon dioxide can be devastating. Corrosion proceeds more rapidly in such a situation, and the final reaction (Reaction 5) releases carbon dioxide, which makes the process self-perpetuating.

$$Fe^0 + 2H^+ + 2HCO_3^- \rightarrow Fe(HCO_3)_2 + H_2 \quad (4)$$

$$4Fe(HCO_3)_2 + O_2 \rightarrow 2Fe_2O_3 + 4H_2O + 8CO_2 \quad (5)$$

Steam/condensate-line treatments

Afterboiler corrosion control through the use of various chemicals is well established. There are four types of treatment: oxygen scavengers, neutralizing inhibitors, filming inhibitors, and combinations of these.

Oxygen scavengers

Hydrazine (N_2H_4) is the most popular and effective oxygen scavenger used in steam and condensate lines. Since it is volatile, it can be fed directly to the afterboiler section or be flashed with the steam and later condensed, to help protect condensate lines from oxygen pitting.

$$N_2H_4 + O_2 \rightarrow N_2 + 2H_2O \quad (6)$$

Neither the reactants nor the products are solids. Treatments such as sulfites cannot be used in steam lines because the sulfites add solids that may lead to afterboiler deposition. Since hydrazine adds no solids, it is quite attractive for use in steam lines.

In addition to being an excellent oxygen scavenger, hydrazine also passivates iron and copper surfaces, rendering them, of course, less susceptible to corrosion. Metal surfaces are chemically reduced from higher to lower oxidation states:

$$N_2H_4 + 6Fe_2O_3 \rightarrow 4Fe_3O_4 + N_2 + 2H_2O \quad (7)$$

$$N_2H_4 + 4CuO \rightarrow 2Cu_2O + N_2 + 2H_2O \quad (8)$$

Addition of a catalyst to the hydrazine mixture ensures completion of the oxygen-scavenging and metal-passivating reactions. Organic catalysts can be used to speed things up, particularly at temperatures found in condensate systems. Depending on temperature, pH and concentration, catalyzed hydrazine can react 10 to 100 times faster than the uncatalyzed compound.

Neutralizing-type inhibitors

These are generally neutralizing-type amines. They react with the carbonic acid in the condensate to form neutral amine salts, thus raising the pH. The most popular such amines are morpholine, cyclohexylamine and diethylaminoethanol. Although not an amine, ammonia can be used, since it elevates the pH through neutralization.

Neutralizing-type amines are volatile compounds that flash with the steam and later condense in the afterboiler. Each amine has its own volatility, which dictates where in the system it will condense the most, thus raising the pH. This phenomenon is determined by the distribution ratio, which is the ratio of the amine in the vapor (steam) phase to that in the liquid (condensate) phase. The higher this ratio, the more volatile the amine. The ratio then indicates in which system the amine is most applicable (see Table II, from Ref. [*4*]).

Neutralizing-type amines can reduce concentrations of iron and copper that return to the boiler in condensate. This lowers the potential for forming insulating iron- and copper-bound sludges. Also, these amines reduce the risk of system shutdown due to corrosion, as well as the risk of energy-robbing steam leaks.

Filming-type inhibitors

These are generally known as filming amines. They form a barrier film on condensate lines that prevents corrosives from reaching the metal surface. This provides protection from corrosion caused by oxygen and by low pH.

Whereas neutralizing amines raise the pH of the condensate, filming amines have little effect on it. Thus, these chemicals cannot be monitored by pH testing. Actually, they are not independent of pH; severe swings into the highly acidic or alkaline ranges can strip the film from condensate lines.

Two popular filming amines are octadecylamine (ODA) and ethoxylated soya amine (ESA). The use of ODA is an older technology and has some drawbacks, which include: inability to effectively form a film over existing pits; relatively poor distribution in long or branched systems; and action as a boiler-sludge binder if overcycled or overfed. ESA minimizes these pitfalls. Also, it is compatible with other boiler-water treatment compounds and is more soluble in water, making it easier to feed to the system.

Table IV — For each treatment option, there are various monitoring methods that can be applied to ensure proper operation

Treatment	Most-appropriate monitoring techniques	Backup monitoring techniques
Catalyzed hydrazine	Hydrazine test	Oxygen concentration Corrosion-product concentration Corrosion coupons
Neutralizing inhibitors	pH	Corrosion-product concentration Corrosion coupons
Filming inhibitors	Corrosion coupons	Corrosion-product concentration
Neutralizing and filming inhibitors	Corrosion coupons	pH Corrosion-product concentration
Neutralizing inhibitor and catalyzed hydrazine	pH Hydrazine test	Corrosion coupons
Catalyzed hydrazine and filming inhibitor	Corrosion coupons Hydrazine test	Corrosion-product concentration
Neutralizing and filming inhibitors plus catalyzed hydrazine	Corrosion coupons	Hydrazine test Corrosion-product concentration pH

Combination treatments

Advanced steam and condensate treatments combine the benefits of the classes of chemicals already mentioned, to give better protection and minimize corrosion (Table III [*5*]).

Neutralizing-amine blends that have multiple volatilities (such as a morpholine-cyclohexylamine combination) have been used for years to provide low-pH protection throughout an entire afterboiler system. Adding catalyzed hydrazine to such a neutralizing combination provides protection against oxygen pitting as well. Additionally, metal surfaces are rendered more resistant to corrosion. This addition reduces the amount of iron or copper that is returned to the boiler, providing a valuable state-of-the-art treatment.

Neutralizing amines and ESA form an excellent combination to protect against low-pH- and oxygen-caused corrosion. The volatility of the neutralizing amines improves the distribution of the filming amine. Catalyzed hydrazine and ESA also provides protection against oxygen pitting and acidic corrosion. The catalyzed hydrazine improves the strength of the ESA film by yielding a smooth, passivated metal surface on which a more durable and complete film can form.

The combination of organically catalyzed hydrazine, neutralizing amines, and ESA can offer the optimum afterboiler treatment, minimizing both forms of corrosion. The catalyzed hydrazine and neutralizing amines provide surface preparation and distribution of the filming amine. Such a multicomponent system affords built-in backup protection, should there be condensate contamination or a system upset.

Monitoring methods

Methods are available to monitor the effectiveness of any afterboiler treatment-program. Table IV summarizes the most applicable techniques for monitoring the various treatment options. In Table IV, the hydrazine test is a colorimetric determination that is very specific to that compound.

References

1. Bennet, L. H., others, Economic Effect of Metallic Corrosion in the United States, paper presented at Corrosion/'79, meeting held March 12–16, 1979, in Atlanta by the Natl. Assn. of Corrosion Engineers (Houston).
2. "Principles of Industrial Water Treatment," pub. by Drew Chemical Corp., Boonton, N.J., 1979 (data were adjusted to reflect current steam costs).
3. Gelosa, L. R., and McCarthy, J. W., Latest Chemical Treatment Controls Corrosion in Condensate Systems, *Power*, January 1979, p. 78.
4. Cappabianca, C. J., Energy Conservation Through Advanced Boiler Water Treatment, *Proc. of the Sixth World Energy Engineering Congress*, held in Atlanta by Assn. of Energy Engineers (Atlanta), Nov. 29 – Dec. 2, 1983.
5. Internal files, Drew Industrial Div., Ashland Chemical Co., div. of Ashland Oil, Inc.

The authors

Paul F. Pelosi is product manager of the boiler-water and fuel-treatment product lines for Drew Industrial Div., Ashland Chemical Co. Address: One Drew Chemical Plaza, Boonton, NJ 07005. Tel: (201) 263-7600. He also aids in technical field support. He holds a B.S. in chemistry from Lafayette College (Pa.) and an M.B.A. from Seton Hall University (N.J.). Pelosi is an author of the "Drew Handbook on Paper Industry Utilities." He belongs to ACS and the Natl. Assn. of Corrosion Engineers (NACE).

Carl J. Cappabianca is senior product specialist of the boiler-water and fuel-treatment product lines for Drew Industrial Div., at the above address. He also provides worldwide technical support for a variety of Drew's products. He has presented numerous technical seminars, and has authored technical papers on topics such as advanced boiler-water treatments. Cappabianca received a B.S. in biology and chemistry from Fairfield University (Conn.) and is currently pursuing an M.B.A. at Fairleigh Dickinson University (N.J.). He is a member of NACE, the Air Pollution Control Assn., and the Oil Mill Superintendents Assn.

Quick way to determine scaling or corrosive tendencies of water

All one need do is to perform a few simple laboratory tests, read a figure, and make a simple calculation.

Krishna V. Mayenkar, *Harza Engineering Co.*

☐ Here is a simple way to see whether a particular water has a tendency to form scale or corrode equipment and piping. The method was developed for use by those unfamiliar with water chemistry, but familiar with such basic concepts as pH. Also, the method allows a water-systems operator to easily understand that scaling or corrosion can be minimized (which one of these would be a problem depends upon the water) by adjusting the water's pH to be close or equal to an "ideal" value. (This ideal value will be explained later on.)

Scaling/corrosiveness indicators

The Langelier index and the Ryznar stability index are two such commonly used indicators. These indexes both assign a number that shows scaling or corrosiveness tendency. They are calculated by determining the pH when the water is saturated with calcium carbonate, for a water of given calcium content and alkalinity. Then, one finds the deviation of the actual pH of the water from the pH of saturation.

The method used here to predict water behavior is based on the Ryzner stability index. This technique calculates an ideal pH, pH_I.

The pH_I is that value that produces a Ryznar stability index equal to 6.0. Using the Ryznar stability index formula:

$$pH_I = 2\ pH_s - 6$$

where pH_s is the pH at saturation for the calcium carbonate.

Comparison of pH_I with the actual pH shows whether the water will cause scaling or corrosion. The pH_I depends upon the calcium content and alkalinity of the water. This method predicts scaling and corrosion due to calcium carbonate and bicarbonates, but does not account for the effects of chlorides or sulfates. Separate chemical and corrosion tests are necessary to determine the activity of these ions.

In this approach, waters are categorized, and dissolved solids are eliminated as a variable.

Most commonly encountered waters will fall into one of these three categories — **Table I**

Category	Alkalinity, mg/L, as $CaCO_3$	Calcium, mg/L, as Ca	Total dissolved solids, mg/L
I	200 – 500	40 – 120	300 – 1,000
II	50 – 200	10 – 40	100 – 300
III	20 – 50	4 – 10	Less than 100

Water categorization

To categorize waters, the quality has been reviewed for several ground and surface (river and lake) waters, using data published by the State of Illinois [*1*]. No attempt has been made to perform any specific analyses.

It has been found that waters commonly have an alkalinity (as $CaCO_3$) of less than 500 mg/L, a calcium concentration less than 120 mg/L, and a total dissolved solids concentration less than 1,000 mg/L. The waters are categorized as shown in Table I. These classifications in most cases cover ground, surface and softened waters.

Important characteristics

Characteristics that affect to scaling or corrosive tendency of water:

- Methyl orange alkalinity (a measure of the total alkalinity of the water, including carbonate, bicarbonate and hydroxyl ions).
- Calcium concentration (as straight calcium).
- pH.
- Temperature.
- Dissolved solids.

(Information on how to perform water-chemistry tests and what they mean is found in Ref. [*2*].)

Among the above indicators, the variation of dissolved solids has a minimal effect on scaling or corrosion tendency. If the other variables are held constant, only a substantial change in dissolved solids concentration will noticeably affect water behavior in regard to scale or corrosion.

Thus, a single value for total dissolved solids concentration has been picked for each category. Based on the commonly encountered ranges found in Table I, the

Originally published May 30, 1983

Here is a numerical indication of the scaling or corrosive tendency of water **Table II**

Actual pH – pH_I	Water quality
0 to 0.5	Little or no scaling
0.5 to 1.0	Little or light scaling
1.0 to 2.0	Little to significant scaling
Greater than 2	Significant to heavy scaling
0 to −0.5	Little or no corrosion
−0.5 to −1.0	Little to light corrosion
−1.0 to −2.0	Light to significant corrosion
Less than −2.0	Significant to heavy corrosion

following values are assumed: 650 mg/L for Category I; 200 mg/L for Category II; and 50 mg/L for Category III.

Ideal pH

For each water category, pH_I values have been calculated for various combinations of calcium and alkalinity. These are based on the average total dissolved solids and 50°F, and are shown in the figure.

Thus, by knowing only the calcium and alkalinity, and by establishing the water category, pH_I can be determined. This value is compared with the actual pH. If the actual pH is greater than pH_I, the water is scale-forming; if the actual pH is less, the water is corrosive. The greater the difference between actual and ideal pHs, the more severe will be the scaling or corrosion. The qualitative analysis of this is shown in Table II.

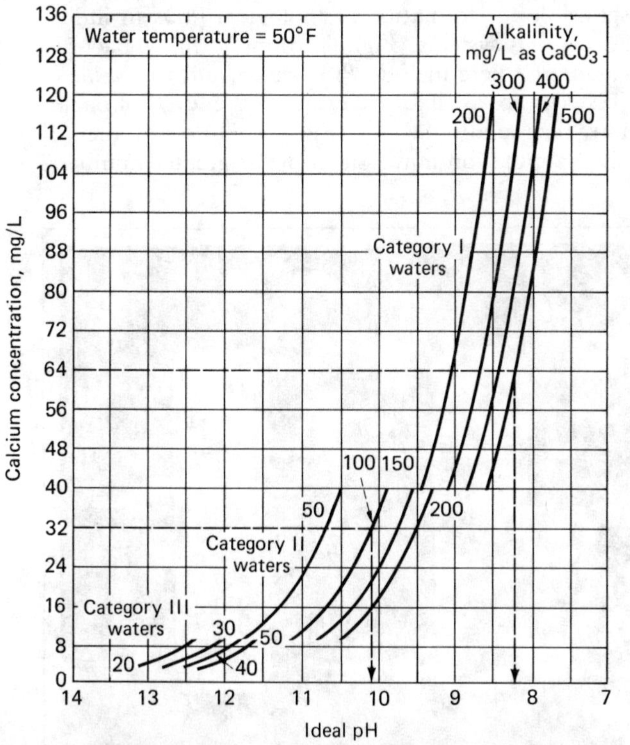

The ideal pH is determined, based on the category, alkalinity and calcium content

Temperature corrections

Corrections can be made if the water temperature is not at 50°F. The pH_I at any temperature can be found by multiplying the temperature differential (t − 50°F) by −0.0167. Add this product to pH_I. The higher the temperature, the lower the value of pH_I for a given water quality.

If one encounters a water with alkalinity or calcium concentrations greater than those shown in the figure, the value of pH_I can be found by extrapolating the curves on the graph.

As the levels of alkalinity and calcium content become higher, the tendency for scale formation extends over a greater pH range. Conversely, as those levels become low, the water becomes corrosive over a wider pH range.

Two examples will illustrate use of the method:

Example I

Determine whether the following waters are scale-forming or corrosive:

	Water A	Water B
Methyl orange alkalinity, mg/L (as $CaCO_3$)	500	100
Calcium, mg/L	64	32
Total dissolved solids, mg/L	950	250
Temperature, °F	50	50
pH	9	9

From the figure, the pH_I value for Water A is 8.25, and the deviation of the actual from the ideal pH is 0.75. This water will be slightly scale-forming. The pH_I for Water B is 10.1, and here the deviation is −1.1. This water will be significantly corrosive.

Example II

Repeat Example I for the same waters, but at 120°F.

$\Delta t = 120 - 50 = 70°F$
$\Delta pH_I = 70 \times -0.0167 = -1.17$
For Water A: 8.25 − 1.17 = 7.08
For Water B: 10.1 − 1.17 = 8.93

At 120°F, Water A will cause heavy scaling, while Water B will cause little or no scaling.

Acknowledgment

The author wishes to express his appreciation to A. L. Lagvankar for assistance in preparing this paper.

References

1. Illinois State Sanitary Water Board in cooperation with the Illinois Dept. tof Public Health, "Summary of Data, 1969 Illinois Water Quality Network," Vol. I and II, Springfield, Ill.
2. Amer. Water Works Assn., Water Pollution Control Federation, and Amer. Public Health Assn., "Standard Methods for the Examination of Water and Wastewater," 14th ed.

The author

Krishna V. Mayenkar is head of the industrial process section in the environmental engineering department of Harza Engineering Co., 150 S. Wacker Dr., Chicago, IL 60606. Tel: (312) 855-7000. He supervises design and development of processes for industrial wastewater and water. He has 12 years' experience as a consultant in this field, plus six years in project management. He received a B.Tech. degree in chemical engineering from the Indian Inst. of Technology (Bombay) and an M.Ch.E. degree from the University of Louisville. He belongs to AIChE and its Environmental Div., the Water Pollution Control Federation, the Natl. Soc. of Professional Engineers, and the Illinois Soc. of Professional Engineers.

Cooling-water system biofouling

Bacteria and algae form films and create slime on heat-transfer piping surfaces, increasing pressure drop and reducing heat flow. Biocides and mechanical cleaning are used to solve these problems.

Denise S. Richardson, Mogul, div. of The Dexter Corp.

☐ Biofouling, the undesirable formation of a biological film, or growth of microscopic fouling organisms, on inanimate surfaces, is a problem receiving increasing attention. Industrial recirculating-cooling-water systems are particularly affected by this phenomenon. Recognition and proper treatment of biofouling can reduce operating costs and improve equipment performance. Here, we will discuss the biofouling process, and methods for treatment.

The fouling process

Biofouling is caused by the attachment of macro- and microorganisms to surfaces with the subsequent production of extracellular products. Fouling proceeds in stages [*1,2*]. Initially, surfaces are conditioned, as nutrients for bacteria are adsorbed from the liquid phase and are concentrated on them. Next, the surfaces attract bacteria themselves. Chemotactic (moving in relation to chemical agents), motile (capable of moving) bacteria migrate toward surface areas high in nutrients. This is then followed by an attachment process, which is at first reversible and then becomes irreversible. The resulting film of organisms is considered the primary bacterial population.

Later, secondary populations attach themselves to the primary film—these are either other bacteria or macroscopic organisms (such as protozoans or shellfish). These secondary organisms grow on top of the primary film, and may graze upon it.

Organisms adhere to surfaces principally because there are advantages in doing so [*3,4*]. Extracellular polymers are produced by the organisms. These polymers encase the organism and provide protection against shear forces and toxic agents in the water. The flowing water brings a supply of nutrients to the organisms and carries away the toxic end-products of their metabolism.

Slime, which is produced by bacteria, consists largely of water and polysaccharides, and can act as an insulator, thereby reducing heat transfer across heat-exchange surfaces. Slime also can attract clay, sand and silt that plug cooling-system tubes and other equipment.

Biofouling organisms in cooling systems

Bacteria and algae are the main types of organisms responsible for formation of biological films in industrial cooling systems (Fig. 1). Slime-forming bacteria encountered here include *Pseudomonas* and *Galionella.*

Test kits are available to estimate the levels of bacteria in recirculating water (Fig. 2). However, the attached population may consist of far greater numbers,

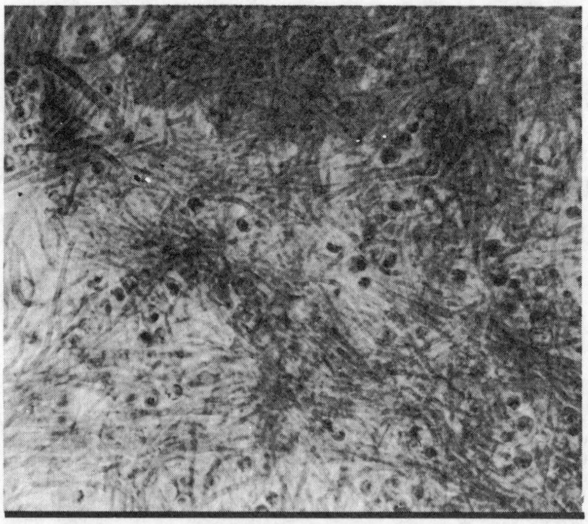

Filamentous and unicellular algae typical of those found on a cooling-tower deck **Fig. 1**

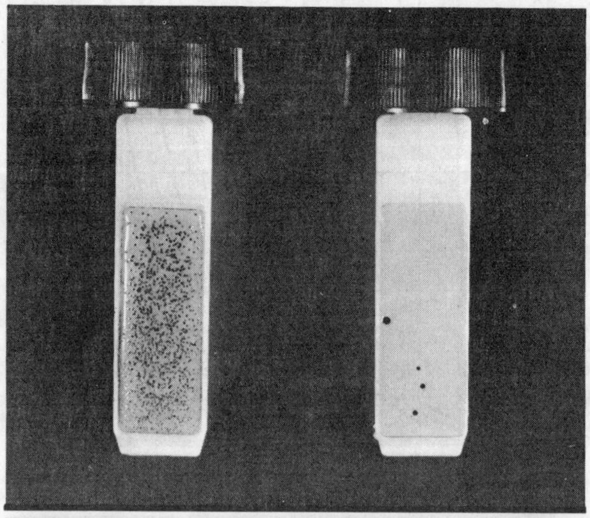

Test kit shows different bacterial densities; left vial is 10^6/mL, right is 10^3/mL **Fig. 2**

Originally published December 13, 1982

and such sessile (permanently attached) organisms are not counted by the kits.

Among the most troublesome algae are the blue-green *Ulothrix* and the green genera *Spirogyra* and *Chlorella.* In general, cellulolytic fungi are responsible for the biodeterioration of wood. Soft rot and pocket rot caused by fungi can severely weaken wooden components of a cooling system.

A macroorganism that frequently causes fouling in cooling systems is the Asiatic clam *Corbicula.* Once established in a system, this shellfish is difficult to remove. It can sense toxicants—including biocides used to remove them—in the water and literally "clam up." When this happens, a costly system shutdown may be required so that shells can be mechanically scraped from surfaces.

Cooling-tower biofouling

In a recirculating cooling system, warm water flows to the top of the cooling tower and through distribution holes in the top of the deck. The holes distribute the flow evenly over the fill material. As the water hits the tower fill, it loses its latent heat of vaporization.

Additional heat (sensible heat) is lost if the temperature of the air flowing through the tower is lower than the water temperature. Air is drawn upward through the tower fill, as the water falls down. The cooled water is pumped to heat exchangers, where it absorbs heat.

Once water evaporates from the tower, it must be replaced. Makeup water is added, and a small amount of the recirculating water is bled off, in order to maintain the proper ion concentration in the water. Thus, as the tower operates, it concentrates nutrients, dissolved gases, dirt and microorganisms in its water, and provides a warm medium for the growth of bacteria, algae and fungi.

Fouling can occur in many parts of the tower. If the tower deck is uncovered and open to sunlight, algae can attach themselves to it and grow. If these organisms are not controlled, they can eventually plug the distribution holes in the deck. Algae growing on the tower fill interfere with droplet formation. This results in less-efficient heat removal. Algae may also provide nutrients for bacteria.

Access to the system's piping and heat exchangers can often be difficult, and biofouling may go unrecognized until an increased pressure drop is observed. The decreased flow and increased pressure drop caused by a thin biofilm are greater than would be expected from just a slight reduction in pipe diameter (as caused by corrosion) [*5*]. The insulating layer of biofilm reduces heat transfer and increases energy costs. Atwood [*6*] projected savings of over $100 million annually for utilities if backpressure could be reduced by 0.1 in. Hg.

Biofouling can also be a problem in pharmaceutical, pulp and paper, and marine facilities. Fouling by aquatic macroorganisms, such as barnacles, has been a continual problem for marine structures. The efficient and economical production of energy by the new technology of ocean thermal energy conversion has been hampered by biofouling on the heat exchangers used in the process.

It is difficult to determine biofouling in industrial settings and any decrease in its effect due to the use of control measures. This is because biological films are relatively thin (tens to hundreds of μm) and generally not visible. Changes in system performance due to biofouling may also be difficult to distinguish from process variations.

Methods to measure biofilms include direct observation, with a light or electron microscope, and indirect observation—noting the pressure in a pipe or rate of heat transfer across a surface. Indirect methods work best in commercial plants.

Methods of control

Control of biofouling organisms has been carried out historically by chemical or mechanical means. Commercially available microbiocides are classified as oxidizing or nonoxidizing. Among the oxidizing types are chlorine gas, chlorine dioxide and chlorine donors, such as calcium hypochlorite. Characklis et al. [*7*] have presented data indicating that chlorine removes the biofilm. Chlorine is inexpensive and has widespread use, but its biocidal properties are pH-dependent. In a pH range of about 4–6, chlorine gas hydrolyzes to form hypochlorous acid and small amounts of hydrochloric acid. Hypochlorous acid has more biocidal activity.

However, many cooling systems use corrosion inhibitors that perform in the alkaline range; chlorine is a relatively poor biocide above a pH of about 7.5. In addition, discharge regulations for chlorine are more stringent than those for other biocides. Chlorine is also under increasing scrutiny, as it is a possible former of trihalomethanes, which cause environmental problems.

Nonoxidizing biocides include the isothiazolins, dibromonitriloproprionamide (DBNPA), methylene bisthiocyanate, quarternary ammonium chlorides, carbamates and organotin compounds. Studies have shown that the isothiazolin-based microbiocides are particularly effective against organisms in biofilms [*8*]. DBNPA is relatively effective at low levels, and decomposes into products that are less toxic than it is.

Mechanical cleaning is an alternative to chemical treatment. Either brushes or abrasive-surfaced balls can be forced through piping, scouring away deposits.

References

1. Corpe, W. A., and Winters, H., "Condenser Biofouling Control," ed. J. F. Garey, et al., Ann Arbor Science Pub., Ann Arbor, Mich., pp. 29–42. 1981.
2. Mitchell, R., paper presented at Amer. Soc. for Microbiology meeting, Atlanta, Mar. 10, 1982.
3. Geesey, G. G., *Amer. Soc. for Microbiology News,* Vol. 48, pp. 9–14, 1982.
4. Costerton, J. W., et al., *Scientific American,* Vol. 238, pp. 86–95, 1978.
5. Norman, G., et al., *Developments in Industrial Microbiology,* Vol. 18, pp. 581–590, 1977.
6. Atwood, K. E., in Garey, *op. cit.,* pp. 43–45.
7. Characklis, W. G., et al., Fundamental Considerations in Biofouling Control, paper presented at Cooling Tower Inst. meeting, Houston, Jan. 1980.
8. Rueska, I., et al., *Technology,* pp. 253–264, Mar. 1982.

The author

Denise S. Richardson is group leader, microbiology, Mogul Div., The Dexter Corp., P.O. Box 200, Chagrin Falls, OH 44022. She has had nine years' experience in microbiology, and has written several scientific papers. She is involved with water-management products for heating and cooling systems. Richardson holds a B.S. in microbiology from Purdue University and an M.A. in biology from Boston University. She is a member of the Amer. Soc. for Microbiology and the Amer. Soc. for Testing and Materials.

Controlling microorganisms in cooling-water systems

Microbes can clog systems and hinder performance. Often, chlorination is the preferred treatment, with continuous application preferred to shock doses.

Robert L. Wetegrove *and* ***Frances C. Pocius,***
Nalco Chemical Co.

☐ Microorganisms find cooling-water systems a natural and convenient place to live. These systems comprise a variety of habitats, so many different types of troublesome microbes can grow in them. Controlling microbe growth is essential to proper operation.

Here, we will discuss various chemical treatments for microbial control—mainly chlorination. First, however, we will briefly cover the basic types of organisms and the problems that they can cause.

Basic types of microbes

Algae are photosynthetic organisms that require only traces of nitrogen and phosphorus to grow. These organisms are most often found as a green mat on open-deck cooling towers or as a green film on tower slats. Algae can plug water-distribution devices, flow-control valves, and heat-exchanger tubes. The waste products of these organisms provide food for bacteria and fungi.

Bacteria are microscopic life forms that can grow on any organic contaminant at any pH, temperature, or oxygen level found in a cooling system. Held in place by a slime layer, bacteria grow on condenser, pipe and cooling-tower surfaces. A layer of bacterial slime less than 100 μm thick dramatically impedes heat transfer in an exchanger, increases fluid friction and pumping costs, fosters corrosion, acts as a nucleation site for scale formation, and impedes performance of water-treatment chemicals (i.e., those for scale and corrosion).

Fungi are filamentous microbes, commonly found on and in the lumber used in cooling towers. Fungal growth can destroy wooden tower fill or weaken structural members, especially in the plenum section.

Chlorination

More cooling-system failures are due to the lack of proper microorganism control than to any other single cause. Proper use of biocides and biodispersants can prevent undesirable growth of microbes in cooling systems, and eliminate such failures.

Chlorine is the most commonly used and economical biocide for typical industrial cooling-water systems. The gaseous form is readily available, inexpensive, effective against all types of microbes, and useful for controlling microorganisms at any pH.

When safe handling is essential, liquid or solid chlorine-releasing chemicals can be used. These include sodium hypochlorite, calcium hypochlorite, chlorinated isocyanurate and chlorinated hydantoin. Whatever the

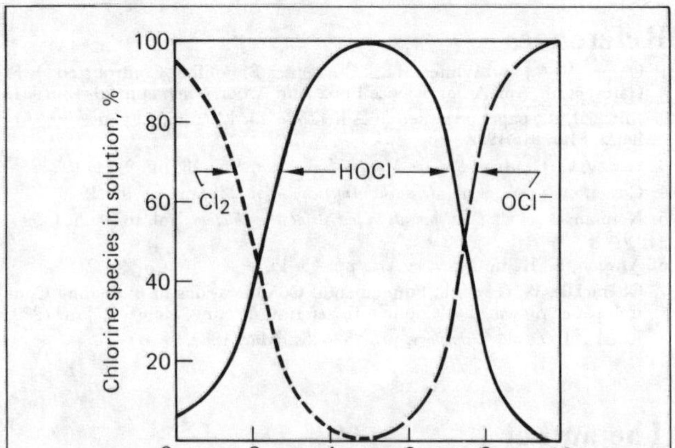

For chlorination, the species present changes as the pH of water varies **Fig. 1**

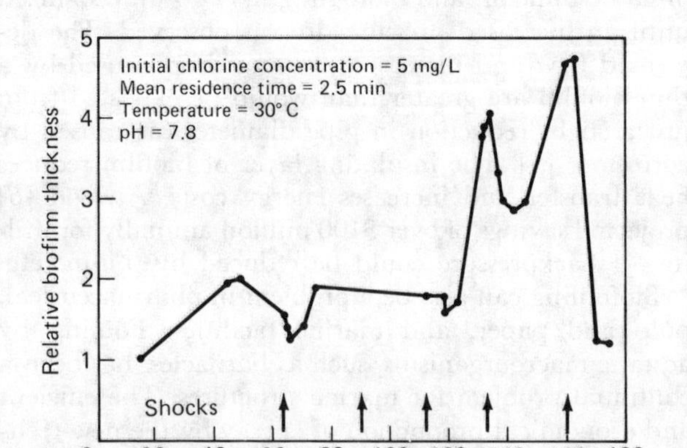

Experiments show that slug treatment is followed by rapid regrowth of microbes **Fig. 2**

From Jolley et al.; used with permission of publisher

Originally published October 31, 1983

form of chlorine used, the active killing forms are hypochlorous acid or hypochlorite ion.

Hypochlorous acid is a weak acid, and undergoes a partial dissociation to H^+ and OCl^- at pH values between 5 and 10. As shown in Fig. 1, the proportion of OCl^- increases as the pH increases.

HOCl has a more rapid biocidal activity than does the hypochlorite ion. The ion, which predominates above a pH of 7.5, penetrates microbial cells less rapidly, thus longer contact times are needed. In continuous low-level chlorination, the contact time is essentially infinity. Here, chlorine is equally effective at all pH values.

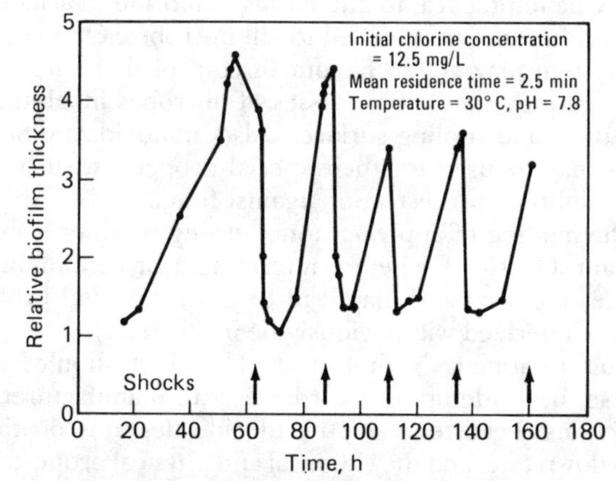

Cycling occurred when a biofouled system was treated with shocks of chlorine **Fig. 3**

From Jolley et al.; used with permission of publisher

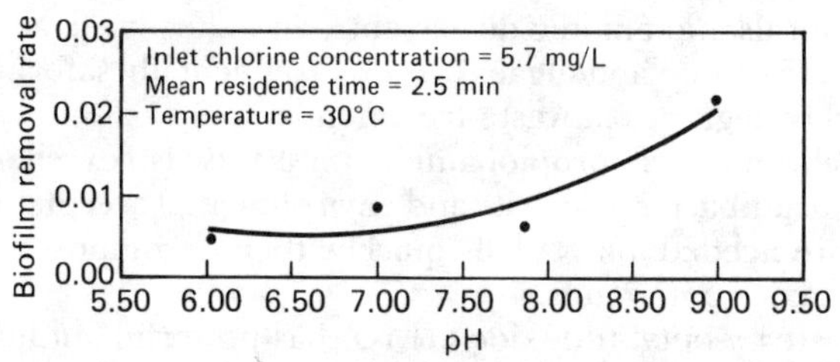

Biofilms are removed more quickly as the water becomes more basic **Fig. 4**

From Jolley et al.; used with permission of publisher

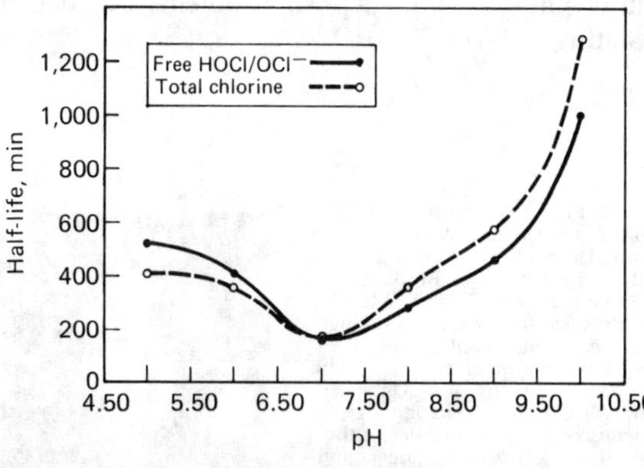

At higher values of pH, chlorine remains in the cooling water for a longer time **Fig. 5**

Reactions with other chemicals

Both the acid and the ion form react with nonbiological cooling-water contaminants in ways that affect chlorine's performance, and the amount of it needed to maintain microbe control. A major such reaction is the one between hypochlorous acid and ammonium ions to form chloramines.

Chloramine formation depends on the pH of the solution and the chlorine-to-nitrogen ratio. The chloramines thus formed are oxidizing biocides, although they are not as effective as hypochlorous acid or hypochlorite. Therefore, the chloramines—which are referred to as combined chlorine—are often considered in determining chlorine residuals. The total disinfecting capacity of a chlorinated system is the sum of combined chlorine (monochloramine and dichloramine) and free chlorine (HOCl and OCl^-).

The amount of chlorine needed will also be increased by the presence of oxidizable inorganics and organics in the cooling water. Chlorine is consumed by reaction with inorganic ferrous, manganous, nitrite and sulfide ions. Nonliving organics are oxidized by chlorine—oil, grease, pollen, leaves and process chemicals can dramatically increase the chlorine demand of a cooling water. A process leak of an organic may require an increase in chlorination rate to compensate for both chlorine consumption and the increase in microorganisms feeding on the leaking process-chemical.

Most carbon compounds react slowly with chlorine, and reaction rates depend on pH and chlorine concentration. Also, overall water quality and other chemical species that may be present strongly affect chloro-organic formation. If hypochlorous acid is the main oxidizing agent, the formation of chlorinated organics will be proportional to the hypochlorous acid concentration. However, if ammonia is present, fewer chlorinated organics will be formed; chloramines form more rapidly.

Ultraviolet rays present in sunlight also increase chlorine demand. The chlorine residual from a shock dosage may persist for hours in the dark, but will disappear in less than 1 hour in an open-deck tower on a bright day.

Although understanding the chemistry of water chlorination is important, control cannot be maintained without the proper method of applying the chlorine to the water. Now, we will discuss continuous and slug chlorination.

Continuous vs. slug treatment

Intermittent application of chlorine typically consists of 1–3 daily treatments of 1–3 h each, with maximum free chlorine residuals to 1.0–2.0 ppm. This type of treatment minimizes the duration of the chlorine discharge. Such a practice may be necessary where system blowdown is discharged directly into a river and a permit given by the U.S.'s National Pollutant Discharge Elimination System limits chlorine discharges. However, intermittent treatment has no performance advantages.

Intermittent chlorination allows for long periods between doses, in which microbial buildup can occur. Characklis et al. investigated the effects of daily shock chlorination (Characklis, W. G., et al., Chapter 32 in "Water Chlorination: Environmental Impact & Health Effects," Vol. 3, ed. by Robert L. Jolley, et al., Ann Arbor Science Publishers, Inc., 1980). A dose of 5 ppm free chlorine resulted in temporary control; however, this was followed by rapid growth to a biofouled state in 5–12 h (see Fig. 2). Doses were 24 h apart.

Starting with a biofouled system, a chlorine dose of 12.5 ppm free chlorine resulted in an even more dramatic cycling of film destruction and rapid regrowth (see Fig. 3). Both studies show that shock chlorination temporarily removes some biological deposits, but regrowth occurs within a few hours.

Intermittent chlorination also has the disadvantage that large amounts of gaseous chlorine must be added to the system in a short time to overcome system demand and achieve a biocide-free residual concentration. High chlorine levels can damage the wooden parts of the cooling tower, and the HCl formed when chlorine is added can result in significant pH excursions.

Reliability of both methods

Regarding operation and maintenance, slug chlorination is less reliable, mainly because an operator must ensure that correct dosage (duration and chlorine level) is administered at a certain time. Sometimes, operators become distracted by other duties or accidentally allow damaging overchlorination to take place.

The continuous method provides a low-level feed to maintain a 0.2–0.5 ppm free residual level. Maximum microbiological control can be ensured by using a biodispersant. Laboratory studies have shown as much as a sevenfold difference in microbe removal using a biodisperant, versus using a biocide only.

Using a continuous chlorine feed eliminates deposit buildups, as well as pH excursions. Also, with the continuous method, process contamination can be detected early on when the chlorine residual drops.

Effect of pH

Characklis et al. investigated how pH influences the rate of biofilm removal (see Fig. 4). Removal rates by chlorine were two to four times greater at a pH of 9 than at one of 6–7.

Also, industrial experience has shown that continuous chlorination is effective in waters having a pH from 5 to 9.5. Further, in the alkaline range, less chlorine may be required to maintain the free residual necessary for the control of microbes, regardless of the proportion of hypochlorous acid to hypochlorite ion. Fig. 5 show the results of a Nalco laboratory study demonstrating that less chlorine is stripped at higher pHs.

Other treatment chemicals

Frequently, a nonoxidizing or organic biocide may be needed, rather than chlorine. For example, if there is a process leak of a chemical with a high chlorine demand, it may be impractical to add enough chlorine to achieve the residual amount needed to kill microbes. Or a cooling system may get so far out of control that chlorine alone cannot handle the masses of microbes inhabiting the tower and cooling surfaces. Also, nonoxidizing biocides may be used for their special properties, such as their ability to protect wood against fungi.

The method of applying nonoxidizing biocides is important: Overfeed, whether intermittent or continuous, is uneconomical and may lead to environmental problems. Underfeed will obviously be ineffective.

Slug treatment is preferred. The feed should, of course, be made up in accordance with manufacturer's directions. Persistence of the biocide depends on the blowdown rate, and the chemical and physical properties of its active ingredients. Reapplication depends on the success of the treatment and the rate of regrowth.

In many cases where cooling systems become overgrown with microbes, the nonoxidizing biocide isothiazolin is the best choice to use to regain control. It is effective at low doses, compatible with chlorination chemicals and anionic dispersants, and effective against bacteria, fungi and algae. It is environmentally safe and readily degrades in waste-treatment systems.

Dibromonitrilopropionamide (DBNPA) is also excellent when antibacterial activity and environmental acceptability are needed. DBPNA kills quickly, then decomposes to nontoxic compounds.

Bis(tri-*n*-butyl-tin)oxide (TBTO) has powerful antifungal and antialgal activity, as well as a strong affinity for wood, making it beneficial to fight fungi that attack wooden cooling-tower components.

Whichever biocontrol agent is used, it is essential to routinely monitor the microorganism count and the biocide feed. Microbe growth is logarithmic, and enormous numbers of them can appear in a few days. The resulting plugging and deposit problems can become devastating.

The authors

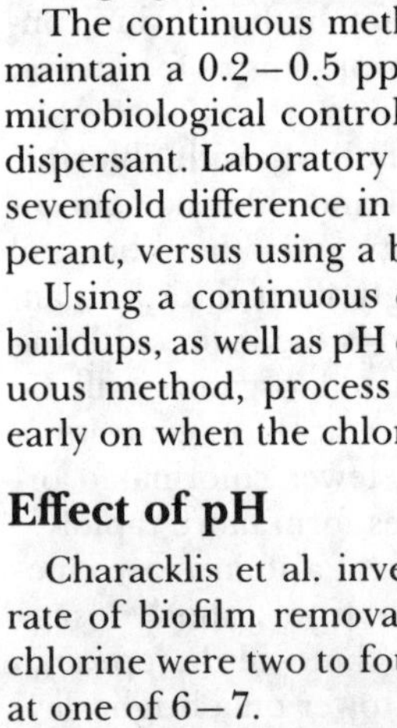

Robert L. Wetegrove is a group leader in the Cooling Water Research Dept. of Nalco Chemical Co., Technical Center, 1801 Diehl Rd., Naperville, IL 60566. Tel: (312) 961-9500. He has been with Nalco since 1978, holding various product-development-research and consulting assignments. He holds B.A., M.E. and Ph.D. degrees in microbiology and chemistry from the University of Texas at Austin. He is a member of the Amer. Soc. for Microbiology, Sigma Xi, the Water Pollution Control Federation and the Microbiologically Influenced CorrosionTask Group of NACE.

Frances C. Pocius is a product specialist in the Cooling Water Marketing Dept. of Nalco Chemical Co., 2901 Butterfield Rd., Oak Brook, IL 60521. Tel: (312) 887-7500. She has been with Nalco for six years, working first in research microbiology, and then moving to cooling-water marketing. She holds a B.A. degree from the University of Illinois at Urbana in microbiology and chemistry. She is a member of the Natl. Assn. of Corrosion Engineers and the Cooling Tower Inst.

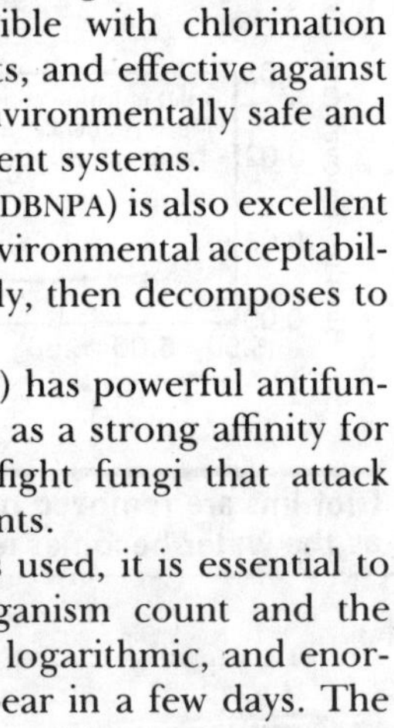

Chelant/phosphate treatment for boiler water

Combining phosphate and chelant treatment for zeolite-softened boiler water will reduce sludge and minimize corrosion.

Leonard Freedman, *Hercules Inc., Water Management Chemicals*

☐ More and more, the combined use of chelants and phosphates for high-purity-feedwater, high-pressure systems is solving critical boiler problems. Classically, phosphates have been added to remove the possibility of caustic corrosion, and chelants have been added to reduce the deposition of various metal ions. More recently, chelants have been proposed as the primary water treatment, with phosphates added to reduce overall corrosion [*1-3*].

Combining chelants and phosphates results in a significant improvement over straight phosphate-based treatment and eliminates the critical control usually necessary when chelants are employed alone. Owing to the latitude built into a combined program, it is applicable in almost any boiler system using zeolite-softened makeup.

Feedwater quality

Most boiler systems will use zeolite-quality makeup, which may or may not be preceded by other pretreatment steps, such as clarification, hot or cold lime softening, etc., to provide a makeup quality low in total hardness. Zeolite softeners do not always accomplish this objective. The reasons may be associated with the resin itself, the equipment, the operating/regeneration procedures, or changes in raw-water quality, and can include:

- Resin fouling.
- Resin loss.
- Leaking valves.
- Poor water/regenerant distribution.
- High or low flows (regenerant, rinse, or service).
- Wrong brine concentrations.
- Intermittent operation.
- Increased hardness in influent (changes in water supply).
- Polyphosphate in influent.

These conditions can lead to some background hardness in the effluent and, therefore, in the boiler feedwater [*4,5*]. Process or condenser leaks could also allow hardness in the feedwater. According to the cause of the problem, hardness leakage may either be at some constant level or fluctuate between two points during the service cycle.

When leakage does occur, the less tenaciously held ions are released first. For instance, cation resins have a greater affinity for calcium than for magnesium [*6*]. As a result, the hardness in a zeolite-softened effluent is primarily magnesium, with far less calcium. When phosphate/caustic or chelant treatments are used alone, this distinction is not important, since "hardness" is generally treated as one problem.

Regardless of the treatment program or the level of hardness in the feedwater, the leakage must be continuously treated. Otherwise, calcium and magnesium form hard, insulating scales that cannot be tolerated in today's boilers. It is the unexpected inleakage of hardness that creates the problem, especially if the program does not cover this possibility.

Traditional approaches

Two approaches have traditionally been used to treat this type of feedwater: precipitating-type treatments (phosphates) and solubilizing treatments (chelants). The phosphate-based programs encourage the precipitation of hardness salts in more easily handled forms:

$$10Ca^{+2} + 6PO_4^{-3} + 2OH^{-1} \rightarrow \underset{\text{hydroxyapatite}}{3Ca_3(PO_4) \cdot Ca(OH)_2}$$

$$3Mg^{+2} + 2SiO_3^{-2} + 2OH^{-1} \rightarrow \underset{\text{serpentine}}{2\,MgSiO_3 \cdot Mg(OH)_2}$$

Enough phosphate is added to precipitate the calcium as hydroxyapatite, and enough alkalinity is maintained to ensure that all reactions go to completion [7]. Almost all water supplies contain more than enough silica, in the presence of reduced levels of magnesium, to guarantee the formation of serpentine. Residual phosphate and alkalinity are maintained to compensate for minor levels of upset from pretreatment or condensate systems.

Even relatively small quantities of hardness in the feedwater can generate large volumes of sludge in a year's time. With a constant 1.0 ppm calcium (as Ca) in the feedwater, each 100,000 lb/h of flow will create more than a ton of hydroxyapatite sludge in a year's time. If that 1 ppm were all magnesium, almost 3,200 lb of sludge would be formed.

Supplemental feed of sludge conditioners remains a standard practice; it is intended solely to remove as much formed and infiltrated suspended solids as possible. The newer, synthetic polymers employ several different mechanisms (i.e., dispersion, sequestration or crystal

Originally published June 11, 1984

modification), and can remove 75% [*8*] or more [*9*] of the solids. Removal of even 90% of the formed solids can still leave substantial quantities of sludge.

When good feedwater quality is maintained, this approach can produce excellent boiler openings. However, where upsets occur, sludge loading can become excessive, and dirty boilers may result.

The other traditional treatment technique involves the use of chelant—EDTA (ethylenediaminetetraacetic acid) or NTA (nitrilotriacetic acid)—as a solubilizing agent for the hardness salts. With this approach, a slight excess of chelant is added beyond that required for the stoichiometric reactions [*10*].

In the boiler, the silica and hydroxide that are present compete with the chelant for the magnesium. As a result, some of the magnesium will precipitate as serpentine, even in the presence of a full chelant treatment. It is reported that as much as 50% of the magnesium will precipitate with an EDTA program, and 90% or more with an NTA program [*11*] Proper control of water chemistry can help to moderate this effect. This ability to solubilize most of the hardness represents a significant improvement over precipitation. Sludge conditioners are still required to prevent any unchelated suspended matter or precipitated hardness from depositing,

Chelants represent virtual state-of-the-art technology for boiler water treatment. It is doubtful whether many improvements in cleanliness will be possible over those achieved with a well-run pretreatment system, an effective return-line treatment program, and a tightly controlled chelant program. However, we will look for improvements in control and ability to handle upsets.

Chelant residuals usually run below 5 ppm, leaving little room for upsets. With composite feedwater hardness as the control, minor swings can be handled. In many cases, though, this method can be too cumbersome, especially in the small-to-medium-sized steam plants. Also, major upsets could cause some deposition that would be difficult to remove with higher chelant feedrates at a later time [*12*].

Each type of program has advantages and disadvantages. For instance, phosphates are less expensive to apply and are easier to feed and control than are chelants. Except for the contribution that it makes to boiler water solids, and its possible effect on cycles, an overfeed of phosphate generally has no negative effect on the system. Finally, with the proper control of external and internal treatment, very good boiler openings can be expected.

Against this is the very significant concern that phosphate programs work by creating a sludge. Upsets can generate conditions that lead to the deposition and baking on of sludges [*13*].

Chelants, on the other hand, are generally more expensive to apply, and usually require a good deal more operational control. Residual testing is often unreliable and subject to many interferences. Feedwater hardness and material balance is a satisfactory way to control most chelant programs, but the method requires commitment and close attention. This can partially compensate for upsets that cause a higher feedwater hardness. Finally, the cleanliness possible with a well-run chelant program is unsurpassed by that of any other approach [*14*].

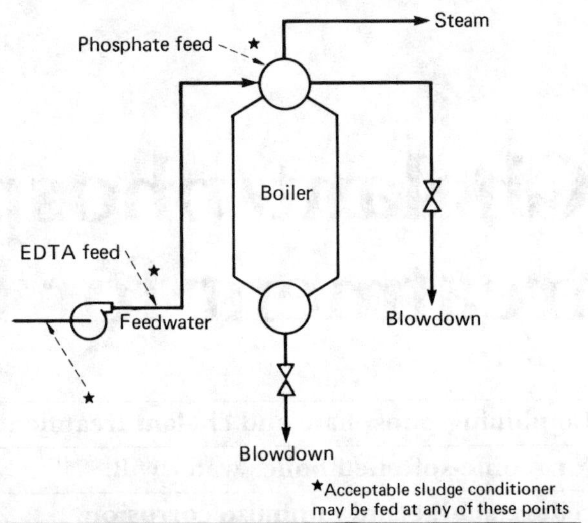

Chemical feedpoints for chelant/phosphate additions

Boilers that are subject to hardness leakage could have reduced the level of potential deposition considerably with the use of chelants. But, costs, control difficulties, and historical fear of chelant corrosion [*15*] restrict their use in these applications. In effect, chelants are rarely considered for systems that they could benefit most.

A new approach

Used together, chelants and phosphates can produce important benefits. Such use thus far has primarily been in high-purity feedwater systems where chelants have been employed to maintain clean boiler surfaces and phosphates have been added to eliminate caustic corrosion potential. In most other reported applications, a similar approach has been described—that is, the program is run as a chelant treatment, and is fed to a slight excess. The phosphate is always employed to provide additional protection, usually from effects of excessive hardness leakage [*16*].

Conceptually, it should be possible to start with a phosphate-based treatment and, by supplementally feeding EDTA, effect a reduction in sludge loading. EDTA is used instead of NTA because of its superior ability to solubilize hardness in the presence of competing anions [*17*]. The mechanics of setting up such a program are relatively simple.

The chelant feedrate would vary according to the degree of sludge reduction desired and the economics, but the chelant would never be fed to a calculated excess. A thorough historical review of the external treatment and condensate systems should provide a fairly accurate estimate of the anticipated levels of hardness inleakage, both high and low. This would serve as the basis for chelant feed. The chelant would be superimposed on a standard phosphate-treatment program. Sludge conditioners would still be required.

Since the chelant is not fed in excess of demand, no further control test is needed for this component of the program. Operator attention is still needed, but it is not as critical as it would be in a straight chelant program.

The phosphate, alkalinity, conductivity and other boiler-water testing would be the same as with a phosphate-based program.

The chelant feed would require all of its normal feed and handling precautions—that is, completely deaerated feedwater, stainless-steel injection nozzle, etc. Recommended feed points for all treatment constituents are shown in the figure.

With such feed of chelant, some hardness would remain in solution, resulting in a reduction of the quantity of sludge that would have to be handled by the system. This would be true even though competitive anions were present for both calcium and magnesium. Even though the sludge reduction would not be directly equivalent to the chelant feed, it would still be significant.

With the chelant being base-fed, the effects of hardness swings can be considered. When excess hardness enters the system, there is always sufficient phosphate present to ensure that it is tied up as a treatable sludge. The chelant still reduces some level of the sludge that would have been created. On the other hand, if the hardness level drops to near zero, there will be an excess of chelant. Fortunately, it has been well established that phosphate acts to inhibit virtually any corrosion potential that might exist with a slight excess of chelant [*18*].

This program is a natural intermediate step between straight phosphate and straight chelant treatments, from the standpoint of results, control, etc. It can, and should, be considered in any system now facing sludge problems. at inspection periods.

An example

A Midwest industrial plant operates three 550-psig boilers for power generation and process use. These field-erected 2-drum units have been in service since the 1940s. External treatment consists of a hot-lime/zeolite softener. Control of these units is erratic, and makeup quality varies from excellent to very poor. Because of a direct product/steam contact, condensate-return-line treatments are not used. Further, the condensate is subject to the infiltration of raw water at irregular intervals. Hence, feedwater quality can fluctuate wildly.

The original treatment was standard phosphate/caustic/sludge-conditioner. Initially, excellent results were achieved and only superficial water washing was required during turnarounds. Owing to deteriorating condensate quality, and to aging pretreatment equipment, boiler openings became progressively worse until, in the mid-1960s, the units had to be turbined annually. Very heavy sludge buildup became typical.

Since the main problem appeared to be contaminated condensate, a sodium zeolite softener was added to the system as a polisher. At the same time, on the assumption that better-quality feedwater would be available, the treatment program was switched from phosphate to EDTA. Unfortunately, this did not eliminate the need for annual turbining of equipment, as had been hoped. Even with the polisher, the return condensate occasionally contributed up to 15 ppm of total hardness, far beyond the ability of the calculated excess of chelant to handle. Further, since the condensate was not treated for corrosion, iron was consistently present. In the absence of a moderating sludge, the deposits that did form in the boiler contained a much higher percentage of iron oxides, which were much more tenacious.

In 1974, on a trial basis, one of the boilers was fed EDTA to an assumed feedwater hardness somewhat below the historical average, supplemented with a phosphate residual in the boiler (10-15 ppm) and an effective sludge conditioner. The treatment was controlled with phosphate, alkalinity, and conductivity readings. No chelant tests were run, though occasional material balances were checked.

After four months of operation, the boiler surfaces were so clean that the other boilers were immediately put on the same treatment regimen. Since that time (1974), the boilers have opened clean, and no further turbining or acid washing has been required.

References

1. Chagnard, H. A., others, An In-Depth Field Evaluation of Iron Transport in 1400 PSI Boilers with Various Water Treatments, 41st International Water Conf., 1980.
2. Venesky D. L., and Dausuel, R. L, EDTA as a Boiler Water Treatment Program at 1200 PSIG, 40th Annual Meeting, International Water Conf., 1979.
3. Freedman, L., Boiler Treatment for High Purity Feedwater, *Hydrocarbon Process.*, Jan. 1983.
4. Ketten, R. E., Proper Maintenance of Ion Exchange Systems, 40th Annual Meeting, International Water Conf., 1979.
5. Kunin, R., Helpful Hints in Ion Exchange Technology, *Rohm and Haas Bull.*, Philadelphia, Pa.
6. Mark, H. F., others, "Encyclopedia of Polymer Science and Technology," Vol. 7, pp. 693-740, John Wiley & Sons, New York, 1967.
7. Herman, K. W., Internal Boiler Water Treatment, 38th Annual Meeting, International Water Conf., 1977.
8. Gaylor, L. A., and Sprague, S. J., Boiler Scale Inhibitors, 42nd Annual Meeting, International Water Conf., 1981.
9. Denman, W. L., and Salutsky, M. L., Boiler Scale Control, *Power*, Sept. 1968.
10. McCoy, J. W., "The Chemical Treatment of Water," Chemical Pub. Co., New York, 1981.
11. Stephens J., and Walker, J., Predicting Chelate Performance in Boilers, *Ind. Water Eng.*, July/Aug. 1973.
12. Seels, F. H., others, Deposit Formation in a 600 psi Boiler Using Chelate Based Water Treatment, 42nd Annual Meeting, International Water Conf., 1981.
13. Bullis, H L., Water Treatment for High Makeup Industrial Steam Generation Systems, 38th Annual Meeting, International Water Conf., 1977.
14. Venesky, D. L., The Ups and Downs of Chelants as Boiler Water Additives, National Assn. of Corrosion Engineers, *Corrosion/80*, 1980.
15. Deal, R. L., others, Role of Chelating Agents in Boiler Corrosion, 17th Annual Meeting, International Water Conf., 1966.
16. Lux, J. A., Chelating Agents for Boiler Treatment—a Manufacturer's Viewpoint, American Power Conf., 1965.
17. Walker J. L., and Stephens, J. R., A Comparative Study of Chelating Agents: Their Ability to Prevent Deposits in Industrial Boilers, 34th Annual Meeting, International Water Conf., Oct. 1973.
18. Tvedt, T. J., and Dawson, J., Fifteen Years of Operating Experience with EDTA in High Pressure Industrial Boilers, American Power Conf. 40th Annual Meeting,. 1978.

The author

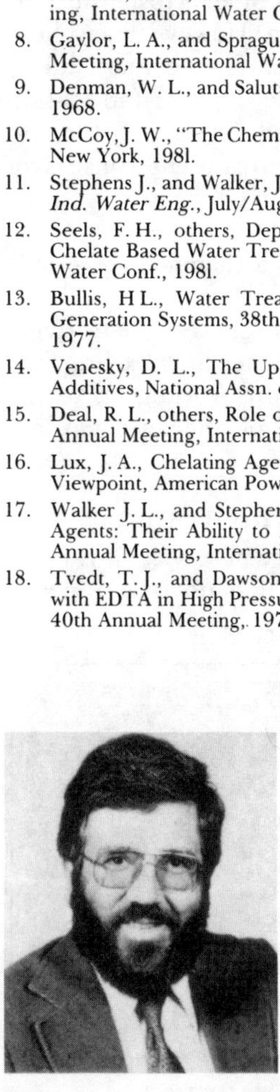

Leonard Freedman is Manager of Technology, Water Management Chemicals, Hercules Inc., Hercules Plaza, Wilmington, DE 19894, Tel.: (302) 992-7522. He is responsible for coordinating technical activities within Water Management Chemicals, including application technology, R&D and customer service. He received his B.S. in chemistry from St. Joseph's College, Philadelphia, Pa.

Pretreating mild-steel water-cooled heat exchangers

Here is a method for cleaning and passivating new mild-steel exchangers. This will help to eliminate waterside corrosion and fouling.

Raymond M. Pasteris, Mobil Oil Corp.

☐ Water-cooled heat exchangers made of mild steel are subject to corrosion and deposits of corrosion products on tube surfaces. However, by using proper pretreatment procedures, one can minimize such problems and guarantee long life and efficient heat transfer.

Mild-steel corrosion can be controlled effectively by the rapid and uniform formation of a corrosion-inhibitor film on a clean metal surface. However, new heat exchangers are not clean. Their surfaces contain lubricants and mill scale, which is the result of fabrication [*2*]. These contaminants must be removed before establishing a protective film. Pretreatment consists of two steps:

- Cleaning the metal surface.
- Passivating the surface with a chemical corrosion-inhibitor [*1*].

We will outline two simple procedures to effectively pretreat new mild-steel cooling-water heat exchangers. Before this, let us review the events leading up to the investigation and development of the two-step exchanger pretreatment program.

A common but ineffective pretreatment

Before placing a new exchanger in service, it is commonly flushed with water. Flushing removes extraneous materials such as dirt and loose rust, but does nothing to get rid of oils and tightly adhering mill scale.

After flushing, a corrosion inhibitor is added to the cooling water at a higher-than-normal concentration to establish corrosion protection. However, our experience has shown that even on clean surfaces such levels may be insufficient to provide good protection, and that on dirty surfaces the film cannot be applied uniformly. A nonuniform film results in localized pitting-attack [*1*].

This conventional method had been employed with success for years at one of our refineries. However, the tubes it was being used on were made of admiralty brass. A problem started to develop when mild steel began to replace the admiralty tubes, owing to process and cost considerations. Previously, there had been no corrosion or fouling, since admiralty brass has excellent resistance to corrosion caused by water. But now, an effective pretreatment program was needed, because the mild-steel units were corroding and fouling—despite good cooling-water treatment.

Fig. 1 shows the effects of corrosion and fouling on a shell-and-tube device, using the old pretreatment process. An analysis of the deposits appears in Table I.

Bench-scale study

A bench-scale study was done to develop an effective procedure. Various pretreatments were tried on mild-steel test-sample tubes, using actual refinery cooling water. Two sample tubes were used in the trials [*2*].

Test exchanger tube A was pretreated with a chromate-zinc corrosion-inhibitor solution at 1,000 ppm (as CrO_4^{-2}) at 130–140°F. The solution was mildly agitated for 1 h. This pretreatment simulated the conventional full-scale one-step procedure used on new mild-steel exchangers.

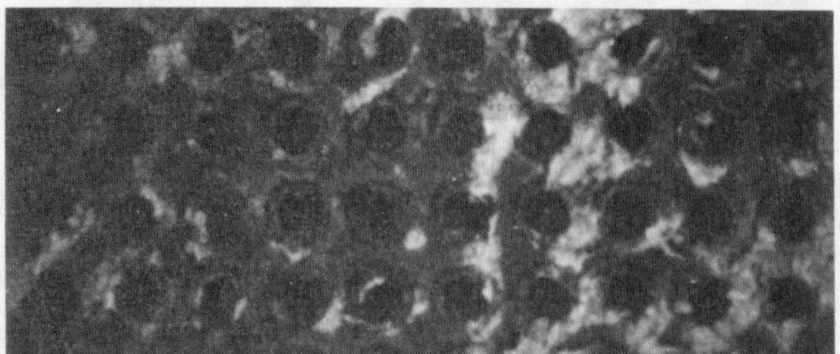

Fouling of mild-steel exchanger after two years' service using old pretreatment **Fig. 1**

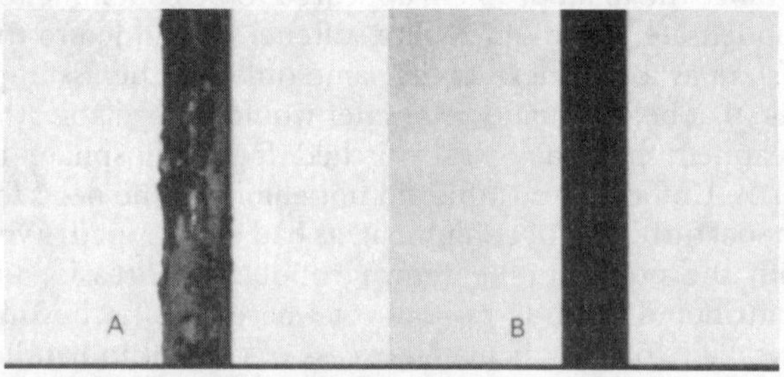

Two-step method reduces corrosion: Tube A had no initial cleaning; B did **Fig. 2**

Originally published November 16, 1981

Deposit analysis for Fig. 1 **Table I**

Metal - mild steel. Deposit appearance - brown

	Percent
Loss at 105°C	4
Loss at 800 °C	20
Inorganic analysis	
	Percent
Iron	71
Chromium	13
Phosphorus	5
Silicon	4
Sulfur	2
Aluminum	1
Zinc	1

Typical analysis for cooling water used in test of two-step treatment **Table II**

	ppm
Calcium	520
Magnesium	360
Sodium	610
Bicarbonate alkalinity	36
Chloride	790
Sulfate	640
Silica	16
Chromium (CrO_4^{-2})	13
Zinc (Zn)	2.1
pH (pH units)	6.9
Conductivity (μmho/cm)	3,000

Expressed as $CaCO_3$ except as noted

Test exchanger tube B was pretreated in two steps. First, the tube was soaked in a 20,000-ppm (2.0%) (as PO_4^{-2}) solution of an alkaline-polyphosphate-based cleaner at 130–140°F, with mild agitation for 2 h. Both tubes were free of rust, but were coated with oil to simulate an actual new exchanger tube. After the cleaning procedure, tube B was passivated in a 1,000-ppm (0.1%) (as CrO_4^{-2}) chromate-zinc solution at 130–140°F. Mild agitation was again done for 1 h.

The two pretreated tubes were placed in refinery cooling-water service for 54 d under normal operating conditions. A typical cooling-water analysis is given in Table II. Fig. 2 illustrates the improved corrosion protection attained following this two-step pretreatment procedure. From these runs, full-scale treatments were developed.

We now use two methods for offstream pretreatment of cooling-water exchangers:

- Recirculation.
- Steam heating and compressed-air agitation.

Pretreatment by recirculation

Pretreatment by recirculation requires a portable, heated solution-tank and a circulating pump (see Fig. 3). As an alternative, a tank truck supplied by a chemical cleaning company may be used.

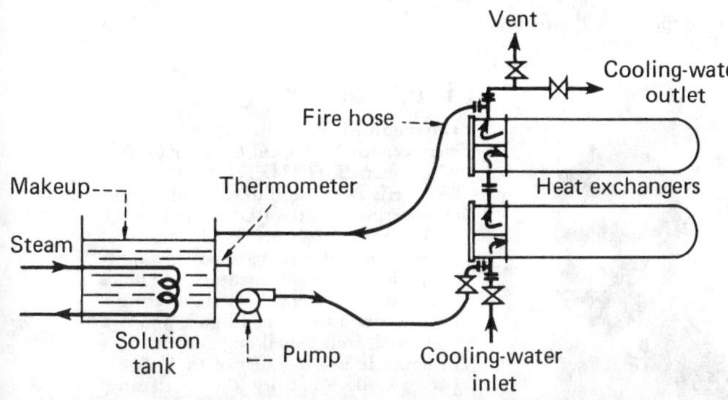

Pretreatment by recirculation requires a hot solution-tank and a pump **Fig. 3**

After installation of the new cooling-water exchanger, water-flush it for approximately 15 min. Cooling water or service water can be used.

Prepare the polyphosphate cleaning solution in the tank, and heat it to 130–170°F. Use the nomograph (see Fig. 4) to determine the amount of chemical required to attain a 10,000-ppm, or 1% (as PO_4^{-2}), cleaning solution. Adjust the solution pH to 6.0–7.0. In this range, the polyphosphate cleaner is most effective, and calcium phosphate scaling-potential is minimal.* Circulate the hot cleaning solution through the exchanger for a minimum of 4 h. A maximum recirculation flowrate along with proper exchanger venting ensures complete contact of the pretreatment solution with the metal.

After draining and flushing the cleaning solution from the exchanger and tank, prepare the chromate-zinc passivating solution and heat it to 130–170°F. Use the nomograph to determine the amount of chemical required to make a 1,000-ppm (as CrO_4^{-2}) passivating solution. Circulate this solution for a minimum of 2 h.

The exchanger is now ready for service. If it is not to be used for a long time, seal it with the passivating solution inside to ensure corrosion protection. When putting the unit in service, flush it and the solution tank, and add the effluent to the cooling-water system. This allows for partial recovery of the corrosion inhibitor, along with slower bleedoff to the facility's wastewater treatment unit. (Always follow proper safety and environmental procedures when mixing and draining any chemicals.)

Pretreatment by agitation

When a solution tank is not available, use steam heating and air agitation (see Fig. 5).

This procedure begins with a thorough 15-min flush of the exchanger, with cooling water or service water. Drain the exchanger. Next, prepare a slurry of the polyphosphate cleaner and pump it into the exchanger. A 55-gal drum is ideal for making up the slurry. Fill the exchanger using a pneumatic or manual drum-pump.

The exchanger should be vented and filled with serv-

*Some pretreatment products are pH-buffered, thus eliminating acid handling and pH control during application. Consult a water-treatment-chemical supplier.

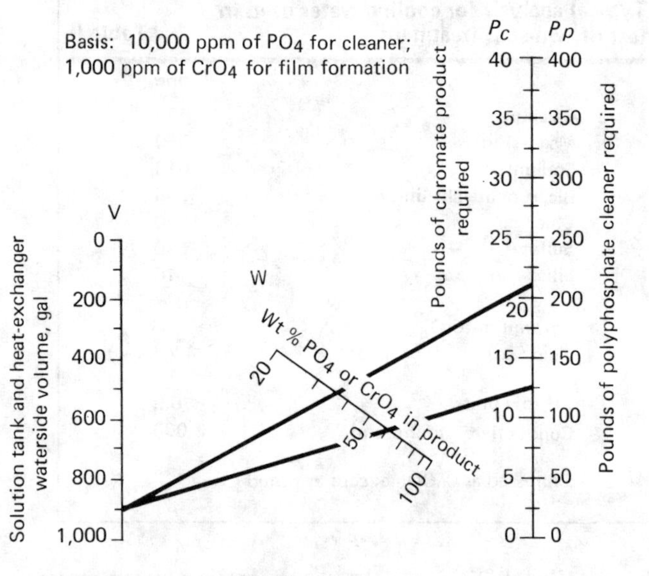

Polyphosphate-cleaner/chromate requirements for offstream pretreatment **Fig. 4**

Nomograph

Example: A bank of two exchanger bundles is to be pretreated by the circulation method. The capacity of the solution tank is 700 gal. The total waterside capacity of the two exchangers is 200 gal. Estimate the pounds of polyphosphate cleaner and chromate product required for pretreatment. The polyphosphate cleaner is 60% by weight PO_4^{-2} and the chromate product is 35% by weight CrO_4^{-2}.

Solution: Total-system water-capacity is 700 gal + 200 gal = 900 gal. On the figure:

1. Connect 900 gal on the *V*-scale with 60% on the *W*-scale. Read the answer as 125 lb of polyphosphate cleaner on the P_p-scale.
2. Connect 900 gal on the *V*-scale with 35% on the *W*-scale. Read the answer as 21 lb of chromate product on the P_c-scale.

Equations used for the graph:

Polyphosphate required:

$$P_p = 8.35 \frac{V}{W}$$

where:

P_p = Amount of polyphosphate product required to attain 1% PO_4^{-2} solution, lb
V = Volume of system, gal
W = Weight percent PO_4^{-2} in product, %

Chromate required

$$P_c = 0.835 \frac{V}{W}$$

where:

P_c = Amount of chromate product required to attain 1,000-ppm CrO_4^{-2} solution, lb
V = Volume of system, gal
W = Weight percent CrO_4^{-2} in product, %

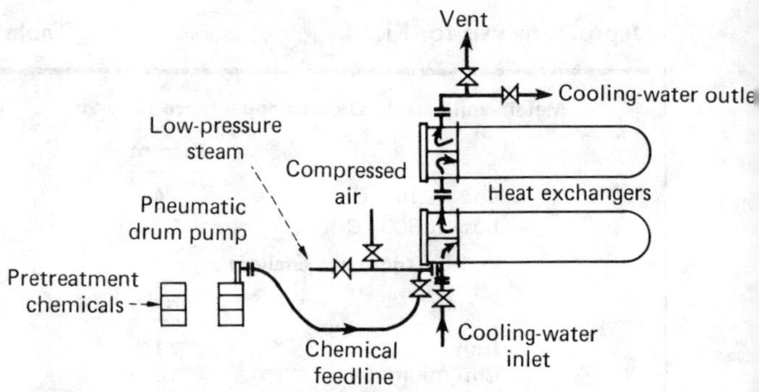

Pretreating may be done by employing steam heating and air agitation **Fig. 5**

ice water rapidly to promote internal mixing of the cleaner.

Begin injection of low-pressure steam and compressed plant air to heat and agitate the cleaning solution. Maintain the solution temperature at 130–170°F by intermittent steam injection for a minimum of 4 h. Water-flush and drain the exchanger to prepare it for passivation.

Pump the liquid chromate-zinc corrosion inhibitor into the exchanger. Again, rapidly fill the exchanger to promote mixing of the corrosion inhibitor within the unit. Begin injection of the low-pressure steam and agitate with plant air. Maintain the corrosion-inhibitor solution at 130–170°F for a minimum of 2 h.

Nonchromate programs

While the outlined procedures employ a chromate-zinc corrosion inhibitor, the basic two-step procedure can be adapted to a nonchromate corrosion-control program. Consultation with your water-treatment-chemicals supplier, along with implementation of a bench-scale study, will aid in establishing an effective pretreatment procedure for a nonchromate program.

Acknowledgements

Special thanks to Nalco Chemical Co. and P. R. Puckorius for help in establishing this program.

References

1. Puckorius, P. R., Proper Startup Protects Cooling-Tower Systems, *Chem. Eng.*, Vol. 85, No. 1, Jan. 2, 1978, p. 101.
2. Nalco Reprint 98, Pretreatment—The Key to Effective Protection of Cooling Water Systems.

The author

Raymond M. Pasteris is a senior utilities engineer for Mobil Oil Corp., P.O. Box 874, Joliet, IL 60434. Tel: (815) 423-5571. His responsibilities include refinery engineering studies of boiler and cooling-water treatments, and steam-generation and cooling-tower operation for performance optimization. He holds a B.S. degree in chemical engineering from the University of Illinois, and is a registered professional engineer in Illinois. He is a member of the Joliet Junior College Faculty, the American Water Works Assn. and AIChE.

Section II
MATERIALS SELECTION AND IDENTIFICATION

Materials selection tips for process plants

By purchasing to a comprehensive specification, less-expensive materials of construction can be used.

Lee Evans, Consultant, Plymouth, England

☐ Corrosion is the biggest single cause of plant and equipment breakdowns in the chemical process industries. Most of these breakdowns can be prevented by correct choice of materials from the very wide range available to the engineer. This includes not only metals but plastics (both thermosetting and thermoplastic), rubbers and paints.

In large integrated plants operated continuously, such as oil refineries and petrochemical installations, production losses from plant breakdowns are so costly that the most expensive materials may be justified—for example, tantalum bayonet heaters in acid recovery units. In smaller batch operations—for example, textile dyeing—it may be more economical to select less-costly materials and schedule frequent cleaning of equipment to prevent corrosion. Whatever the scale of the plant, adequate information on the range of materials available and their corrosion characteristics is essential at the design stage.

Establish corrosion levels

To ensure that the most economical solution to a corrosion problem is found, it is necessary to fully understand the type of corrosion likely to occur, and to be confident that the operating conditions of the equipment are fully known—not only the normal operating conditions but also any abnormal circumstances that may occur, for example, at startup and shutdown. It is not usually necessary or economical to design and build a plant so that it will be completely free from corrosion. Corrosion penetration rates of less than 0.005 in./yr (ipy) are usually acceptable, and rates between 0.005 and 0.025 ipy may be economical, depending on the cost of alternative materials and the degree of contamination acceptable in the process materials. Sometimes, corrosion rates are expressed in weight loss per unit area rather than in reduction of thickness—the correlation between these for materials of different densities is given in the Table.

Where metallic contamination is unacceptable—e.g., in the manufacture of chemicals for pharmaceuticals—carbon steel vessels with a resin coating, such as a stoved phenolic/epoxy lining, are often an economical solution. Coatings of this type can be used up to 150°C. For higher temperatures, coatings based on fluorinated resins such as fluorinated ethylene propylene (FEP) and polyvinylidene fluoride (PVDF) are necessary. These resins are more expensive and may increase the cost of a stainless steel vessel.

Most pressure vessel codes, such as ASME VIII, provide for a corrosion allowance to be added to the design thickness of the vessel wall to compensate for the predicted corrosion rate. To do this requires a thorough knowledge of the corrosion characteristics of the fluids to be handled and, in particular, confidence that the corrosion will be uniformly distributed. However, corrosion is often not uniform, and there are many localized forms that must be considered, such as: pitting, crevice, stress cracking, fatigue, erosion, intergranular, heat transfer, and bimetallic or galvanic.

Design against corrosion

Often, localized corrosion can be economically prevented by careful attention to design details rather than by selecting more-corrosion-resistant materials. For example, cold concentrated sulfuric acid can be handled in carbon steel piping, provided that the flowrate does not exceed 3 ft/s. Thus, it is less expensive to use carbon steel and increase the pipe diameter so as to keep within this limit, rather than use a smaller-diameter pipe made of stainless steel.

If heating coils for caustic soda storage tanks are placed outside the tanks, they can be made of carbon steel, whereas if placed inside the tanks they will need to be made of a nickel alloy because carbon steel is subject to caustic cracking under heat transfer conditions. The corrosion fatigue of sieve trays in distillation columns can usually be prevented more economically by increasing the stiffness of the trays than by changing to a higher-alloy stainless steel.

The pitting and stress corrosion cracking of equipment made of the 304 and 316 grades of stainless steels can often be avoided by careful attention to design details—for example, siting feedpipes below liquor levels so that splashing does not cause concentration of chloride, and eliminating dead zones where chloride-containing liquors will cause pitting.

Simple precautions in operational procedures also

Originally published May 4, 1981

Weight loss equivalent to a 0.005-ipy corrosion rate

Alloy-based metal	Mdd (mg/dm²/d)
Aluminum	10
Copper	32
Gold	75
Iron	25
Nickel	32
Tantalum	60
Titanium	18

help to prevent premature failure of equipment. Pitting of brine-cooled stainless steel plate heat-exchangers can be prevented by draining out the brine when the equipment is not in use. The stress corrosion cracking by polythionic acids of refinery units can be prevented by neutralizing the condensation that occurs when the units are shut down.

Specify accurately

After selecting materials, it is essential to ensure that they are properly and fully specified in respect to properties and fabrication procedures. If this is not done adequately, the equipment may not provide the durability and reliability expected of it. For metals and plastics, there are usually national and often international specification standards that can be cited.

For example, if a molybdenum-containing steel is required, it is not adequate to quote just AISI 316 or UNS 31600 (Unified Numbering System), as these designations cover only composition limits and do not necessarily ensure the delivery of material with the required mechanical properties, corrosion resistance, dimensional tolerances, etc. For one relevant specification, ASTM A167, the alloy required is either 316 (31600) or 316L (31603), depending on whether the material will be welded and on its thickness.

If pressure vessels designed to ASME VIII are involved, it will be necessary to order plate and sheet to ASTM A240. This specification has been approved by ASME for the construction of pressure vessels and will also be found as SA 240 in Section II of the ASME Boiler and Pressure Vessel Code.

In addition to composition limits for the 300 and 400 stainless-steel series, these specifications tabulate the minimum mechanical properties that are acceptable, and there are optional clauses that can be specified, such as an intergranular corrosion test, if the purchaser considers them necessary for the success of the fabrication procedures.

Welded pipe and tube are appreciably less costly than seamless, but for corrosive conditions it is advisable to specify material that has been heat-treated after welding. For example, in ASTM A312, which covers seamless and welded austenitic stainless steel pipe of the 300 series, heat treatment at 1,040°C after manufacture is mandatory. For tubing for heat exchangers and condensers where tube/tube-plate joints have to be made, A249 should be specified to ensure the supply of tube that has been tested to withstand expansion cracking.

If it has been decided to use a lead-lined vessel, then it is important to ensure that chemical lead, which is of higher purity than plumber's lead, is used by specifying material to ASTM B29. Furthermore, chemical lead contains trace quantities of silver and copper, which help to prevent premature failure of linings by fatigue and creep corrosion.

Plastic equipment also requires careful and comprehensive specification, although national standards for plastics are not as widely available as they are for metals. For example, there are several types of polyester resins that can be used for the production of glass-reinforced plastic vessels, the most common of which in ascending order of corrosion resistance and cost are: orthophthalic, isophthalic and bisphenol. Orthophthalic resins are used mainly for consumer goods, such as car bodies and boats, and should be excluded in specifications for chemical plant equipment.

The choice between isophthalic and bisphenol resins should be determined by corrosion tests in the particular environment or from previous experience, but the purchase specification should leave no doubt as to which is to be used. An example in the thermoplastic field concerns polypropylene piping. If it is to be used for handling chemicals at high temperatures, then, to ensure long life, a grade containing the best antioxidant additives is required. This grade would probably not be suitable for foodstuffs or potable water because the best antioxidants are often slightly toxic.

Careful attention should also be given to the specification of painting schemes for protecting the exteriors of equipment and structures. There are now available many paints based on synthetic resins, such as epoxy and vinyl, which will withstand splash or spillage from most chemicals and give long-term protection in heavily polluted environments. However, satisfactory performance from these paints requires good preparation of the steel surface before application. Therefore, the painting contract should specify shot blasting.

Guidance on specifications for paint schemes is given in the "Steel Structures Painting Manual" (Steel Structures Painting Council, Pittsburgh, 1973) and the "Paint Handbook" (McGraw-Hill, N.Y., 1981).

Acknowledgement

This paper was first presented at the Eurochem 80 Conference, June 23–27, 1980, Birmingham, England.

The author

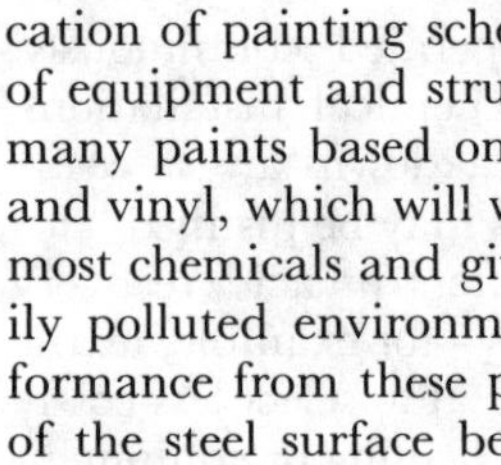

Lee Evans is an independent consultant with an office at 8 St. Paul Street, Stonehouse, Plymouth, England. Tel: Plymouth (0752) 265112. For the past 20 years, he has been a materials specialist in the chemical and synthetic-fiber industries. Mr. Evans is a Fellow of the Institution of Metallurgists and the Institution of Corrosion Science and Technology. He is the author of "Chemical and Process Plant: a Guide to the Selection of Engineering Materials," the second edition of which has recently been published by Hutchinsons in the U.K. and by John Wiley & Sons, N.Y., in the U.S.

Materials identification in the plant—I

Mixups involving materials of construction in processing plants can cause catastrophic failures. This three-part series will show you what techniques are available for identifying and differentiating among metals.

Bernard Ostrofsky, Standard Oil Co. (Indiana)

☐ In large processing plants, it is not uncommon for materials to be incorrectly marked, or for inadequate materials to be substituted by mistake for those specifically chosen to withstand particular process environments. Suppliers can add to this problem by delivering the wrong materials, as happens perhaps 10% of the time. The use of the wrong material in process equipment endangers personnel and can cause heavy financial loss.

In most cases, applying a reliable materials test can avoid the unexpected and unwanted. Many techniques are now available for identifying and distinguishing among various materials. They range from simple screening for broad categories of metals, to detailed identification of specific alloys.

The most widely used techniques include: chemical and acid spot tests, spark tests, fracture tests, hammer and chisel tests, file tests, thermoelectric characterizations, eddy-current sorting of alloys, optical emission spectroscopy, X-ray emission spectroscopy and microprocessors. The suitability of these techniques for specific identifications depends on how much information is required, where the test takes place, the time available, and inspector competence.

Chemical and acid spot tests

Spot tests [*1-3*] offer an easy way to identify alloys by composition. Some of the material to be determined is first put into solution. Appropriate chemical indicators are then added to detect the presence of specific elements. For example, a test to distinguish between Type 304 and Type 316 stainless steel would begin with the use of a magnet. Because both types are austenitic stainless steels, they are nonmagnetic. After this category verification, one would proceed to test for the presence of molybdenum. If the test result were positive, then the unknown would most probably be Type 316 SS.

The above test involves the electrographic extraction of a few milligrams of the alloy with nitric acid, aided by a small d.c. voltage to concentrate the metallic ions at the cathode (clean aluminum strip) on a piece of filter paper. The addition of hydrochloric acid and sodium thiocyanate resulting in a dark red spot verifies the existence of a ferrous alloy. Further, addition of sodium thiosulfate will result in a pink halo if molybdenum is present; if the concentration of molybdenum is greater than approximately 9%, a dark blue spot will develop.

There are numerous acid tests; however, none of them can be performed on metal that is hotter than about 100°C, because the acid solution will boil away. The following common metals can be readily identified by reliable acid tests: plain low-carbon steel; 2, 5, 7 and 9% chromium alloy steel; ½% molybdenum steel (carbon-moly); 3½% nickel steel; and pure nickel, Monel, magnesium and aluminum.

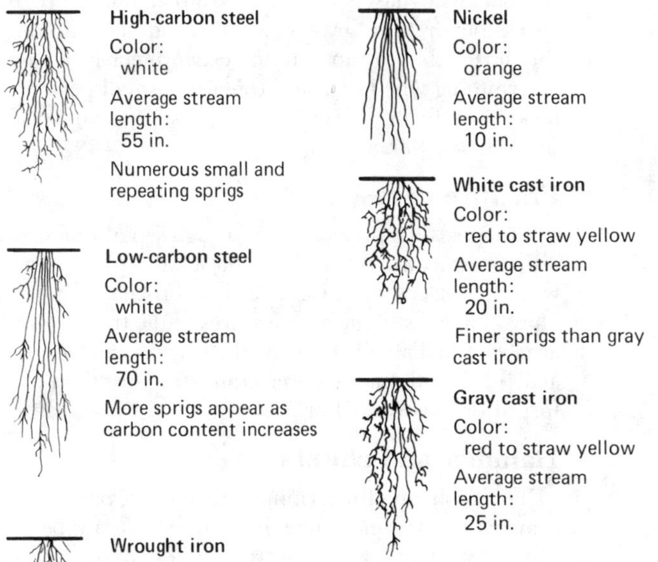

Spark characteristics of common ferrous alloys — Fig. 1

Originally published December 15, 1980

Fracture-test surfaces for common ferrous metals — **Table I**

Metal	Feature appearance
Gray cast iron	Coarse grain, gray
Malleable iron	Fine grain, black
Wrought iron	Fibrous (like hickory), light gray
Low-carbon steel	Fine grain, light gray
Tool steel	Very fine grain (silky), light gray

Care should be exercised to remove any potentially corrosive reagents from the surface after testing.

Spark testing

Spark tests are mainly useful in identifying different grades of steel and iron. A lightweight portable grinder is commonly used, and the operator notes the appearance of the sparks given off when the grinding wheel contacts the metal. Different compositions of steel produce different "spark pictures," as illustrated in Fig. 1. Pure iron held against a grinding wheel will throw off burning particles that leave a reddish-yellow streak as they speed through the air. When steel is spark tested, the carbon present in the red-hot particles turns to gas and explodes with "bursts" that resemble the forks of light given off by Fourth of July sparklers.

The best way to master this method of steel identification is to obtain a series of samples of known composition and become familiar with the spark picture of each. Anyone who has once seen the great difference in the spark pictures of gray cast iron and low-carbon steel should be able to separate these two metals; by contrast, determining whether a steel contains a small percentage of nickel is difficult. Of course, spark testing is not feasible in fire-restricted areas.

Fracture tests

Fracture tests provide only a rough separation of widely differing metals. A small sample of the piece is notched with a hacksaw or chisel and then broken by a hammer blow. Some distinguishing features of the fracture surfaces are listed in Table I. Usually, the fracture surface of a cast article will exhibit a coarser grain structure than that of a forged or hot-worked article.

Hammer and chisel tests

The problem of determining whether a part is made of gray cast iron, malleable iron, or steel can be readily solved by anyone who has access to a hammer and chisel. The chisel is placed on a sharp edge of the metal object and tapped lightly and steadily with the hammer. The chisel should slant at about a 45-deg angle from the edge of the object so that cutting can continue.

Fig. 2 shows the type of chips obtainable from the three metals. It is impossible to get a sliver of curled chip from cast iron because particles keep flying from the metal just ahead of the chisel edge. In the case of malleable iron, a short slightly curved chip can be obtained, due to the ductility of this metal.

Types of chips obtained by the chisel test — **Fig. 2**

If the tested piece is steel, a careful operator can obtain a well-curved chip over a half-inch in length.

Do not use this test in a fire-restricted zone.

File tests

File tests are valuable for detecting wrought iron, for differentiating chromium plating from other bright-surfaced metals, and for other simple identifications. A fine-toothed file is used to scrape a bright surface transversely (crossways) on the metal object to be identified. If the material is wrought iron, fine, parallel dark lines, or slap stringers, will be faintly visible, running in the direction of rolling.

It is usually easy to identify chromium plating just by its bright, shiny appearance. However, if there is any doubt, one can try to cut the plating with a file. Chromium plating is so hard that a file will scarcely mark it. By contrast, a stainless steel or nickel surface, although having the appearance of chromium plating, is readily cut by a file. In a similar manner, a file can be used for the separation of hardened from annealed steels.

Acknowledgements

This three-part series is an update of a paper presented at the 37th National Fall Conference of the American Society for Nondestructive Testing and subsequently published in *Materials Evaluation,* Vol. xx, August 1978.

References

1. Nondestructive Rapid Identification of Metals and Alloys by Spot Test, ASTM STP550, Philadelphia.
2. Feigel, F., "Spot Tests," Elsevier Publishing Co. 1958.
3. "Rapid Identification of Some Metals and Alloys," International Nickel Co., Inc., 1948.

The author

Bernard Ostrofsky is a senior research associate for the Materials Research Department of Standard Oil Co. (Indiana), Amoco Research Center, P.O. Box 400, Naperville, IL 60540. Tel: 312-420-5744. He conducts research and development in all areas of nondestructive testing. An associate technical editor for *Materials Evaluation,* he is a fellow of the American Society for Nondestructive Testing and a P.E. in California. He has a B.S. in physics from City College of New York and a Ph.D. from Polytechnic Institute of Brooklyn.

Materials identification in the plant—II

Applying a reliable materials test can avoid costly and dangerous mixups. This article, the second of three, discusses how to differentiate among metals by electrical characteristics.

Bernard Ostrofsky, *Standard Oil Co. (Indiana)* *

☐ Part I of this series presented chemical and physical techniques for identifying unknown metals. Part III will focus on optical emission and X-ray emission spectroscopy, and the use of microprocessors.

Thermoelectric characterization

The thermoelectric characterization of metals is based on the net voltage caused by a temperature difference at the junction of two dissimilar metals. Since the net voltage observed is a function of the metal couple, it is possible by reference to calibration tables to identify metals [*1,2*]. A copper lead is clipped to a section of an unknown metal and attached to one terminal of a zero-center-scale galvanometer. This contact is the cold or reference junction. Another copper lead connected to the opposite-polarity terminal of the galvanometer is attached to a steel file. The file is rubbed briskly on a clean spot on the sample until a steady voltage reading is obtained, this value being indicative of the unknown sample. It is assumed that the hot junction always reaches the same temperature.

Table I shows some typical readings. There is some overlap of voltage, which could result in uncertainty as to alloy identification. Checking against a known standard could help alleviate this problem. Use of standards to check calibration periodically is recommended.

Eddy-current sorting

Eddy-current techniques may be applied to sort and identify metals [*3*] by making use of the conductivity property of a metal. An eddy current is a circulating electrical current induced in a conducting material by an alternating magnetic field. The conductivity of standard alloys is reasonably uniform from sample to sample and identification of many common alloys and their heat-treatment condition is possible. This technique is based on the fact that, from an electrical viewpoint, an alloy may be considered as a pure metal with a high impurity content.

In eddy-current testing, all information about the material tested is obtained through the test coil that is applied to the metal surface. Coil characteristics are indicative of the test sample. The coil has an apparent impedance that will change as the sample it reacts with is changed. An impedance plot constructed for the change in test-coil impedance at three different frequencies for a series of nonferromagnetic metals with conductivity varying between 3% IACS—International Annealed Copper Standard—(titanium) and 100% IACS (copper) is shown in Fig. 1. All data were obtained with the same test coil. It can be seen that the lower frequencies allow better separation of the higher-conductivity materials, and vice versa.

When test samples are ferromagnetic, sample permeability, μ, becomes important. If the relative permeability, μ_r, is considered constant (low exciting field) and reasonably large (>50), and the filling factor (degree of coupling) is appreciable, then a graph similar to Fig. 2 derived for some common steels can be constructed and used for ferritic alloy sorting. It is also possible to use phase-discrimination techniques for alloy sorting.

There are many variables active in eddy-current testing that affect the results. In alloy sorting, we are interested in conductivity changes; however, the impedance values observed can be influenced by heat treatment, grain size, grain-boundary precipitation, grain orientation and geometrical factors, among others. Therefore, differentiation among materials may be difficult and may even not be possible. If all factors were equal so that only conductivity were the governing parameter, it would be difficult to separate strictly on the basis of eddy-current tests a 6061T6 aluminum with 42% IACS from a bronze 14 with 43% IACS, or 304 SS with 2.5% IACS from a high-alloy steel with 2.9% IACS.

An example of a problem that occurred in some field work involved a forged chrome-steel heat-exchanger baffle that appeared to be the incorrect alloy. It was suggested that the surface was decarburized, resulting in the wrong conductivity for the base material. After a thin surface layer was removed, the material identification was verified. Care must be taken in the application of this method, with extensive use of standards for comparison. If exact alloy composition is required, other methods should be used. Many commercial eddy-current instruments suitable for alloy sorting are available (see Part III).

*To meet the author, see p. 64.

Originally published January 12, 1981

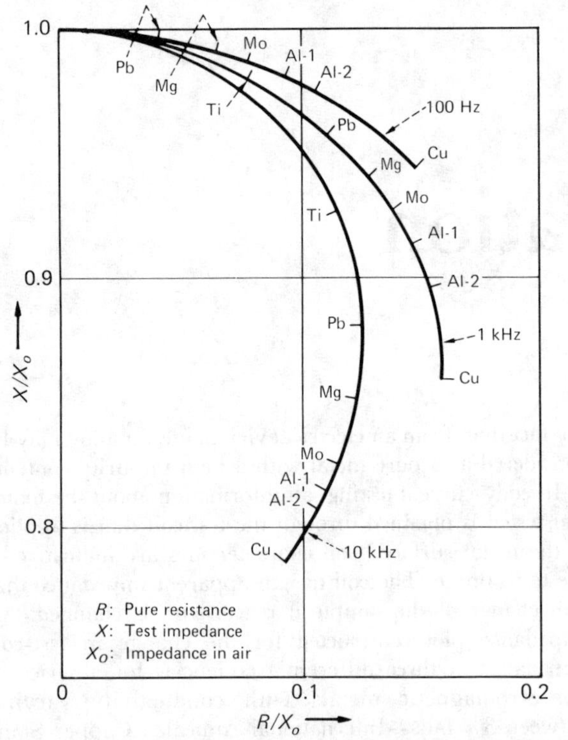

Impedance plot from eddy-current testing **Fig. 1**

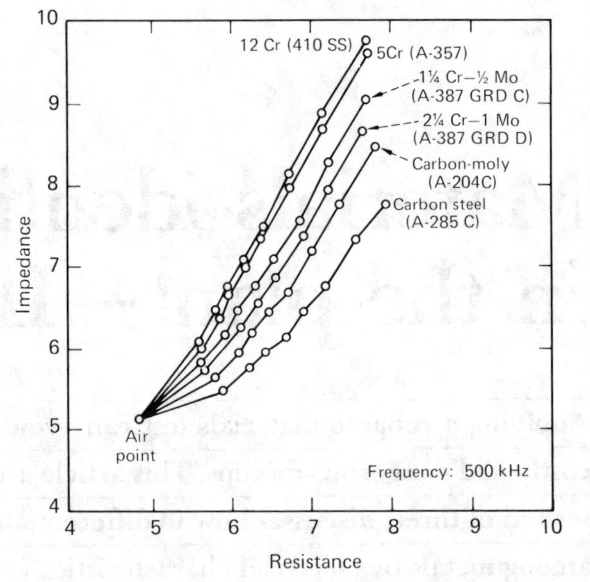

Impedance graph of refinery steels **Fig. 2**

Thermoelectric reading for common alloys **Table I**

Alloy	Reading*	Range
−10.0	Monel	−10.0 to off scale
−8.0	Hastelloy "D"	−8.0 to −8.5
−5.0	Nickel	−6.0 to −6.5
−3.0	Cast iron	−3.0 to −3.3
−2.1	3½% Nickel	−2.1 to −2.3
−2.0	18–8	−1.5 to −2.5
−2.0	25–20	−1.5 to −2.4
−1.7	Stellite	−1.7 to −2.0
−1.5	Malleable	−1.5 to −1.7
−1.5	Alloy 20	−1.5 to −1.6
−1.1	Incoloy	−1.1 to −1.5
−0.5	Hastelloy "F"	−0.5 to −0.7
−0.2	Brass	
−0.1	Aluminum	
0.0	Copper	
+1.5	Hastelloy "C"	1.5 to 1.9
+1.8	Carbon moly	1.8 to 2.0
+2.2	Carbon steel	2.2 to 2.5
+2.5	1% Cr–½ Mo	2.5 to 3.0
+3.5	Titanium	3.5 to 4.5
+4.0	Inconel	3.5 to 6.0
+5.0	12% Chrome	4.5 to 5.5
+6.0	9% Chrome	6.0 to 6.5
+7.0	5% Chrome	7.0 to 7.5
+8.0	Hastelloy "B"	8.0 to 10.0

*Note: These are approximations. For best results, compare with a known sample. Units are arbitrary, a function of galvanometer scale.

Electrical methods

Though the techniques listed below have only limited use [*4*], one might provide a valid test for a particular materials identification problem.

■ Triboelectric effect based on electrification by rubbing—A standard substance such as rubber is rubbed against a series of standards, and the electrification in the range of microvolts measured and tabulated is used for identification. There is difficulty in obtaining reproducible results by this method.

■ Fall-of-potential technique based on electrical conductivity—A direct-current supply is applied between two probes on the surface of an electrically conducting material. The voltage drop between the probes is measured by a galvanometer. Variations in conductivity are tabulated for identification. Unfavorable surface conditions tend to interfere with accuracy, and a large amount of calibration is required, as is a delicate detector and a high-current power supply.

■ Magnetic permeability to identify ferromagnetic materials—Alternating-current bridges and a commercial instrument, the Magnatest Q, are available for these measurements. This method is advantageous if many of the same items are to be identified and different items are to be rejected. It will not, however, specifically identify a material.

References

1. "Fundamentals of Temperature Pressure and Flow Measurements," John Wiley and Sons, New York, 1969.
2. Use of the Ferretmeter, private communications, R. P. Buhrow, Standard Oil Co. of Ohio, 1977.
3. McMaster, R. C. (ed.), "Nondestructive Testing Handbook," American Society for Nondestructive Testing, Columbus, Ohio, 1959.
4. Hinsley, J. F., "Nondestructive Testing," Macdonald & Evans Ltd., London, 1959.

Materials identification in the plant—III

This final installment of the series focuses on how to use optical and X-ray emission spectroscopy, and microprocessors, to differentiate among metals.

Bernard Ostrofsky, Standard Oil Co. (Indiana) *

☐ Part I of this series presented chemical and physical techniques for identifying metals. Part II discussed the application of materials tests based on the electrical characteristics of metals.

Optical emission spectroscopy

Under suitable excitation conditions, many metallic elements and alloys emit characteristic wavelengths in the optical wavelength region. These emissions may be used to identify and determine the concentration of the elements present [*1*]. The usual procedure for excitation of solid samples requires the passage of an electrical discharge between two portions of the sample or between the sample and counterelectrode that does not contain the elements being determined. (In visible and ultraviolet spectroscopy, the emission process is the result of an electronic rearrangement in atoms.) The counterelectrode may be pure carbon in graphite form, which has the advantage of being highly refractory and does not introduce any interfering lines of its own.

The d.c. arc (2,200 V, 5 A) is the most sensitive method of excitation. The chief disadvantage is that it is not adequately reproducible, and the discharge tends to localize, causing hot spots in the material surface, resulting in uneven sampling. This deficiency is overcome by the a.c. arc method, which requires high voltage (up to 5,000 V at 1 to 5 A) with 120 interruptions/s.

The method most suitable for application of a portable instrument for field use is the spark source. This requires high voltage (4,000 to 15,000 V) from a stepup or Tesla coil. Spark excitation is better suited than arc to precise timing of exposures, and only minute amounts of the sample are vaporized. A recent significant development is the application of a laser as a high-energy source. This beam may be focused on a specimen area as small as 50 μm dia. Such an analysis on a micro scale may not be representative of the bulk of the sample. With laser excitation, the sample does not have to be electrically conducting.

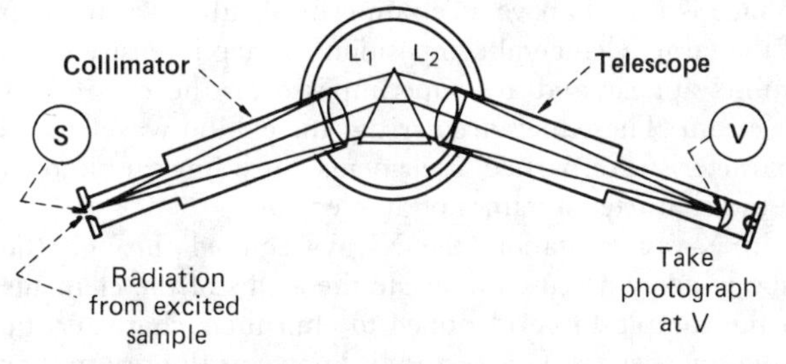

Prism spectroscope separates radiation by wavelength **Fig. 1**

The radiation emitted after the specimen has been excited consists of many different frequencies, which are resolved by wavelength and measured to determine the constituents of the specimen being analyzed. The means of dispersion may be either a prism or a grating. Fig. 1 is the schematic diagram of a prism spectroscope. Radiation from the excited sample enters the slit, S, and is made parallel by the collimating lens, L_1. The radiation is separated according to wavelength after passing through the prism and focused by lens L_2. If the emissions used are in the visible range, they may be viewed by eye or the record may be photographed. If the emission is in the ultraviolet range, the spectra must be photographed at point V.

A prism has a nonlinear dispersion and a lower spectral resolving power than a grating. Diffraction gratings are now produced with high quality and are preferred for high-resolution quantitative-analysis instruments.

X-ray emission spectroscopy (XES)

Characteristic X-ray emissions from materials may be used to identify and quantitatively determine elemental composition [*2,3,4*]. The method is nondestructive. It requires a form of excitation to produce the characteristic X-radiation and techniques for determining the wavelengths characteristic of the source of radiation.

X-rays are electromagnetic radiation of the same nature as light but with a much shorter wavelength. The wavelength of visible light is about 6,000 Å, whereas X-rays are on the order of several Å. X-rays occupy the region between gamma rays and ultraviolet rays.

X-rays are produced when high-speed electrons collide with a target material. This is normally accomplished in

*To meet the author, see p. 64.

Originally published February 9, 1981

an X-ray tube, which contains the filament and target in an evacuated enclosure. Electrons from the filament are accelerated to the target by high voltage (normally 30 to 50 kV) applied across the tube. The electrons collide with the metal target, producing X-rays at the point of impact, which radiate in all directions. Analysis of these X-rays shows that they are a mixture of different wavelengths and intensity depending upon the tube voltage. Up to what is called the short-wavelength limit, indicated SWL in Fig. 2, the intensity is zero. Intensity increases rapidly to a maximum, and then decreases to a minimum on the long-wavelength side. Up to 20 kV, the radiation is entirely continuous (white radiation). Like white light, it is a mixture of many X-ray wavelengths. When the tube voltage is raised above a certain critical value—a function of the target elemental composition—sharp intensity maximums appear and are superimposed on the continuous spectrum. These lines are narrow and exhibit wavelengths characteristic of particular elements. Such lines are known as the characteristic lines of an element.

In X-ray excitation, the X-rays emitted through the tube window are used to excite the atoms of the elements in the sample to be identified to emit their characteristic radiation spectra. This process is known as fluorescence or secondary X-ray emission. Primary emission results from the impact of electrons on the target material. In the application of γ-rays for the stimulation of characteristic X-rays from the material to be identified, a radioisotope, emitting γ rays, is substituted for the X-ray tube.

The irradiation of a specimen containing a number of elements by an exciting X-ray or γ-ray source results in the emission of many characteristic X-ray wavelengths, which have to be sorted in order to identify the source. Since all the characteristic X-ray wavelengths from the elements are known, identification is possible. In addition, since the intensity of the radiation is a function of the concentration, it is possible, after suitable calibration, to determine quantitatively how much of each element is present in an unknown sample.

The two primary methods of sorting out X-ray emissions are dispersive and nondispersive analysis technique whereby wavelengths are separated and measured by means other than crystal dispersion.

Electronic discrimination, also known as energy-dispersive discrimination, is the most useful nondispersive system. Proportional and scintillation X-ray detectors are energy-sensitive, emitting pulses of mean amplitude proportional to the X-ray energy being measured. The proportional counter is similar to the Geiger counter except for the operating potential. It consists of a metallic tube with a central anode and is filled with a gas. The tube may be sealed or of the thin-window gas-flow type, the latter being more useful for detection of longer wavelengths.

The scintillation counter currently is the most widely used because of high efficiency for detection of wavelengths in the useful analytical region. It consists of a thallium-activated sodium iodide phosphor that converts a fraction of the X-ray energy into visible light. The light is transformed into electrical pulses by means of a photomultiplier tube. It is of higher efficiency than the gas detectors but pulse distribution is broader, resulting in an overlap of neighboring wavelengths and poorer resolution.

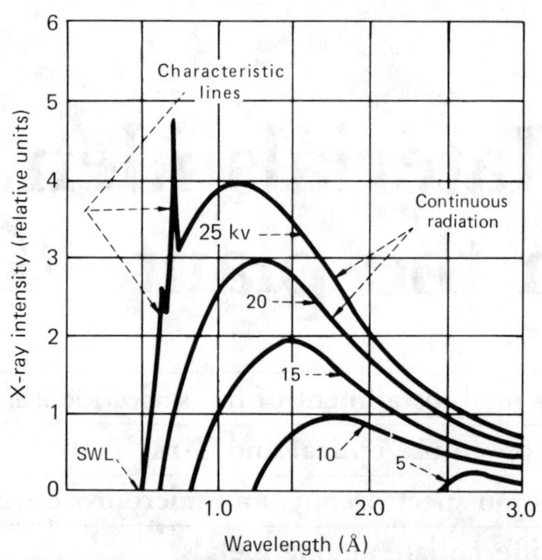

X-ray spectrum of molybdenum shows characteristic lines **Fig. 2**

Solid-state (lithium-drifted silicon) detectors are available for a broad energy-detection range with good resolution. This type of detector must be kept at liquid-nitrogen temperature for proper operation. Because of the low resolving power, some type of selective X-ray filter system, such as balanced Ross-type filters, is used. A narrow wavelength band can be established by using two filters with absorption edges close to each other. These edges are chosen to be just above and just below the Kα wavelengths of the element to be measured (see Fig. 2). The filter pair thus consists of a Kβ filter (one atomic number higher) and a second filter (two atomic numbers lower) adjusted so that its attenuation of the Kβ is the same as that of the Kβ filter.

Two counts of X-ray amplitude are made: first with the Kβ filter, then with the second filter. The difference in counts corresponds to the desired element's intensity. Obviously, a set of filters is required for each element to be analyzed.

Since the detector output pulses are of low amplitude, they must be amplified. Optimum settings of detector voltage and amplifier gains are required to yield maximum signal-to-noise ratio for the measured element.

Fig. 3 is a schematic of γ-ray, X-ray spectroscopy geometry for an energy-dispersive system. Pressing the sample to the top of the housing retracts a shield, exposing the specimen to the γ-ray source. Excited X-rays are detected by the X-ray detector and the output is fed to the amplifier pulse-height analyzer for measurement as previously described.

The principal advantages of energy-dispersive X-ray systems are simpler instrumentation and high intensities. The most serious disadvantage is the poor energy resolution resulting in difficulty in resolving neighboring elements with a two- or three-atomic-number separation. A consequence of poor resolution is difficulty in determining minor concentrations of elements.

Crystal or wavelength dispersion is accomplished by allowing the X-rays from the sample to be diffracted by a crystal or known atomic interplanar spacing and crystallo-

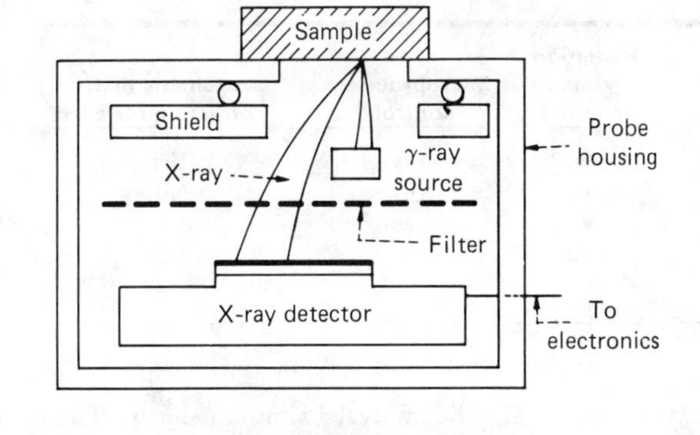

γ-ray, X-ray spectroscopy geometry for an energy-dispersive system **Fig. 3**

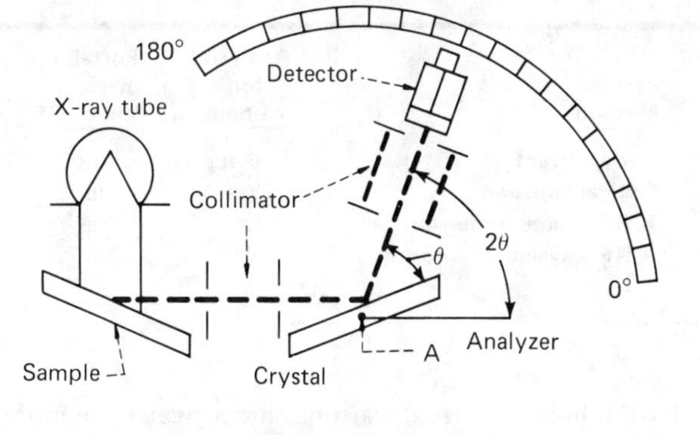

Flat-crystal X-ray spectrograph **Fig. 4**

graphic orientation. The X-rays are diffracted in accordance with Bragg's law:

$$n\lambda = 2d \sin \theta$$

where: n = order of diffraction, λ = wavelength (Å), d = interplanar spacing, and θ = angle between incident beam and crystal surface.

Fig. 4 is a schematic of a flat crystal spectrograph. The fluorescent X-radiation from the sample is diffracted by the analyzing crystal and received by the detector. The diffracted radiation leaves the crystal at an angle θ with respect to the crystal face and 2θ with respect to the incident beam. The goniometer is arranged so that the detector moves at twice the angular speed of the crystal and is always in position to detect diffracted radiation should it exist. A spectrum to cover the complete range of analytical wavelengths can be recorded as the crystal rotates from 0 to 90° with the detector moving through a 0-to-180° arc.

Curved crystals or focusing systems are used to obtain increased intensity. They behave in a manner analogous to concave diffraction gratings used in optical emission spectroscopy.

The crystal spectrometer using an X-ray tube source with high-voltage power supply is not as conducive to portable packaging or self-contained power supply. It does, however, provide better emission-line resolution and can be used to scan the entire range of analytical wavelengths, providing greater flexibility for diverse element analysis.

If a limited number of elements are to be checked, an energy-dispersive system that is lightweight and battery-powered is advantageous. If a broad spectrum of elements is to be determined and high resolution and sensitivity required, economics determine that a wavelength-dispersive system be employed.

Microprocessors

The current trend in instrumentation for rapid identification of materials in the field is automation under microprocessor control. The Table is a summary of instrument capability.

The Sensor Corp. portable Metal Analyzer III is a computerized eddy-current instrument that identifies a metal sample by comparing phase shift at resonance against a signature of a known material. In this instrument a total resonance scan of 256 bits is divided into four 64-bit scans. This is done to reduce the number of memory chips. Signatures may be stored temporarily in RAM or permanently in EPROMs. As many as 48 signatures may be stored. Resonance condition of the inductive sensor is changed by varying capacitance connected across the sensor. Phase-shift nonlinear response is related directly to the composition of the metal.

The resonance condition of the unknown is determined and stored in digital form and compared to the reference material's resonance curves. When a match is found, the 16-digit alphanumeric display yields the name or number identifying the unknown sample. It is also possible to sort materials simply and rapidly by storing a known material signature and comparing a group of components to determine whether it is the same material, with the display stating Yes or No. Other variations can readily be accomplished. This system is nondestructive and has the capability of distinguishing between grades of materials where carbon content is the only variable. No other really portable instrumentation is at present capable of accomplishing this type of analysis. In order to improve the resolution based on carbon content it will be necessary to store more points on the curve and thereby increase the memory bank.

Carbon may be determined with an optical emission spectrograph now available from Applied Research Laboratories, Div. Bausch & Lomb, Model 3600. This unit is mobile rather than portable and capable of being programmed to analyze for up to 10 elements. A spark gap is enclosed at the end of a hose. This is applied to the material being analyzed. The optical spectrum developed is transmitted via fiber optics to a grating where it is dispersed and analyzed in a manner similar to that of a laboratory instrument. The instrument may be truck-mounted to be transported to field locations, and powered from a generator or the mains. It requires a two-man crew for operation.

Another such spectrographic unit, the Spectrotest, is available from Technic, Inc. This has the electrode in a pistol assembly and transmits the optical spectrum along a 10-m-long quartz-fiber light pipe to the spectrometer for analysis. It is claimed the time spent for each analysis is

Comparison of capabilities of instrumented methods

Method	Analysis for carbon	Portable battery-operated	Multi-element analysis	Radiation source required	Microprocessor control	Automatic matrix correction for conc.
Eddy current	Indirect	Yes	No	No	Yes	Yes
Optical emission	Yes	No	Yes	No	Yes	No
X-RF energy dispersion	No	Yes	Yes	Yes	Yes	Yes
X-RF wavelength dispersion	No	No	Yes	Yes	No	No

2-4 s. Audible and visual warnings are activated when the material being analyzed deviates from the selected element tolerance value.

The Metascope (Cooperheat) also identifies alloys by optical emission spectroscopy. It is portable, the entire instrument being hand-held. Since it uses a prism for wavelength dispersion, the resolution is not as effective as large-grating dispersion; spectral lines close in wavelength are not easily resolved. Carbon cannot be analyzed, and spectral lines for some elements must be present in relatively high concentration to be detected by eye. However, where one-man operation and only limited elemental detection is required, it will do the job.

Laboratory X-ray fluorescence wavelength-dispersive instruments are now microprocessor-controlled. Currently the only portable such instrument, the Portaspec (Pitchford Scientific Instruments, Div. Hankison Corp.) is not microprocessor-controlled. It is expected that control of this type of instrument will be automated so that the goniometer will be set to the required element angles, X-ray peaks identified and the quantitative calculations, including matrix and fluorescence corrections, be performed. X-ray spectral data from standards will be stored in memory, and identification made by comparison with the standards—all functions being carried out by the microprocessor.

Microprocessor operation of energy-dispersive portable X-ray fluorescence analyzers has already been accomplished and instrumentation is commercially available. Systems are available from Texas Nuclear, Div. Ramsey Engineering Co. ("Alloy Analyzer"), Inax Instruments ("Inax Portable Spectrometer") and Columbia Scientific Industries, Inc. ("CSI Model 740").

The Texas Nuclear instrument, with the proper combination of source-detector-filter heads, is capable of analyzing for 11 elements—namely Ti, V, Cr, Mn, Ni, Co, Cu, Nb, W, Fe, Mo. Liquid-crystal readout gives the individual element concentration or alloy type by comparison with spectral data from 100 reference alloys stored in PROMs. The instrument weighs 15 lb and has a head design that permits analysis of difficult areas such as butt welds and nipplettes..Each head is capable of holding two isotope sources.

The Inax instrument uses a Si (Li-drifted) detector and is capable of determining any element from Al on up in the periodic table. Though the detector must be kept at liquid-nitrogen temperature, it may be allowed to read ambient temperature when it is not in use. The probe unit incorporates three isotope-collimator assemblies as well as the detector. No filters are required in this system and up to 36 elements may be analyzed simultaneously. The data are read out on twelve 4-digit liquid-crystal displays. The pulses from the detector are analyzed automatically by a pulse-height analyzer and the portable analyzer uses programmable calculator logic. The analyzer may also be used in the laboratory with a larger computer programmed in Basic.

The unit may be operated in the sort mode, where the unknown spectrum is compared with spectra from standards and identification is completed in 2 s. If no match is found the word NO is displayed. In the identify mode, about 2 min is required for identification. When the proportion of all elements is very close to that of the standard, within the limits of accuracy, the display will give the alloy number. In the analysis mode specific selected elements will be analyzed and the % concentration by weight will be displayed. Normal accuracy will be 0.01%, and as good as 0.001% for Mo. There are no controls on the probe. As soon as the probe is pushed against the sample the shutter opens and the complete analytic sequence begins. The results read out on the LCD display.

The Columbia Scientific Model 740 is another X-ray energy-dispersive analyzer. It makes use of a proportional counter/detector and pulse-height analyzer under microprocessor control. Three probes are available. The surface probe is for in-situ measurements, no sampling required. Solid, powdered or liquid samples are analyzed with the sample probe. The light-element probe is used for the analysis of elements aluminum through titanium. The complete elemental analytical capability is for aluminum through uranium in the 100-ppm and greater concentration range.

There are 8 stored calibration modes; 4 elements may be determined simultaneously in each mode, and all interelement-effect corrections are made by the microprocessor. The instrument weighs 19 lb and will function for 12 h with fully charged batteries.

Since X-ray fluorescence analysis is a surface method and only minute depth of penetration of the sample surface occurs, care must be taken to have a clean surface for analysis.

References

1. Boode, U. R., "Chemical Spectroscopy," Wiley and Sons, New York, 1943.
2. Van Olpen, H., and Pattis, W. (editors), "X-Ray and Electron Methods of Analysis," Plenum Press, New York, 1968.
3. Jenkins, R., "An Introduction to X-Ray Spectrometry," Hegdon and Sons, Ltd., 1974.
4. Kaelble, E. F., "Handbook of X-Rays," McGraw-Hill, New York, 1968.

Section III
PAINTS AND COATINGS

Paints and coatings for CPI plants and equipment

The first, and sometimes only, barrier against a corrosive environment is the surface coating applied to equipment and structural materials. Proper specification of coating, surface preparation, and application is critical to successful performance.

Guy E. Weismantel, Senior Regional Editor

☐ A basic knowledge of paints and coatings is essential for the chemical engineer. This should involve familiarity with: paint fundamentals, surface preparation methods, application and inspection techniques, coating-system selection criteria, specification guidelines, troubleshooting and economics. The engineer who can identify and avoid coating-related problems can prevent costly corrosive damage to equipment and structural steel.

Paints are commonly called "surface coatings," and the terms are used interchangeably. A paint is a decorative, protective or otherwise functional coating applied to a substrate, which may be another coat of paint. Paint should be considered an engineering material; it should be specified in accordance with the performance desired, and the conditions and techniques of its application should be specified and controlled. Most coating failures are due to a misunderstanding of the importance of careful surface preparation and application.

A paint normally consists of film formers, solvents, pigments and additives. The film former may be drying oil, varnish, resin solution, dry resin, plasticizer, or a combination of these. The solvent may be a free agent or a component of varnishes or resin solutions. The pigments and the additives are usually distinct compounds included for their special properties.

The percent concentration of the principal ingredients is the most important factor in determining the difference between paint types. A key factor is the pigment-volume-concentration—that is, the volume of the pigment in the dried film compared with the total volume of dried film, as shown in Fig. 1. Up to a certain volume ratio, the addition of pigment actually reinforces and improves film-forming properties. With continual pigment addition, a paint formula passes through a critical value called the critical-pigment-volume-concentration. When this critical volume relationship is exceeded, the resistance properties of the paint rapidly deteriorate. The film former no longer bridges all the voids between the pigment particles. Striking-in increases, the film becomes porous and, as a result, weathers faster, is less washable, and loses abrasion resistance, flexibility and other desirable properties.

Surface preparation

Approximately 90% of protective-coating failures occur because the surface to be painted is improperly prepared. Most surfaces require extensive cleaning to remove mill scale, rust, oil, grease, chemical deposits and other contaminants.

There are many surface-preparation methods. Acid or alkali chemical cleaning, solvent washing, grinding, scarifying, high-pressure water blasting, scraping, and wire-brushing all are commonly used. Choosing the right method is most important. Selection depends on the type of substrate being prepared (steel, galvanized metal, wood, concrete, etc.) and the coating used.

Generally, synthetic-resin coating systems such as vinyls, epoxies, chlorinated rubbers and phenolics require more-stringent surface preparation (for example, blast cleaning or pickling) than do oil-based coatings such as alkyds and epoxy esters.

Cost is also a factor. For new construction, surface preparation and painting in a fabricating shop is almost always less expensive than painting after field erection or installation. This is true because pickling, rotary-wheel blast cleaning, and other cost-effective surface-preparation methods cannot be carried out in the field.

Personnel safety, and hazards to adjacent equipment or operations, are also limiting factors. Many jobs do not allow open blast cleaning because of the danger of sparking and, ultimately, of explosion. Abrasive blast cleaning is also discouraged if electric motors, hydraulic equipment, and the like are in close proximity. Also, government regulations prohibit open sandblasting or

The pigment-volume concentration (PVC) is critical in determining a paint's properties **Fig. 1**

other surface-preparation methods that contribute atmospheric emissions or otherwise expose workers to an unhealthy environment (e.g., airborne contaminants, excessive noise).

Both mechanical and chemical methods of surface preparation are used in plants. Of the mechanical methods, water blast and abrasive blast cleaning are by far the most important and will be covered in detail. Specific recommendations for hard cleaning, power-tool cleaning, flame cleaning, and steam cleaning can be found in the literature [*1,2*].

Water blast cleaning

Water blast cleaning uses a high-pressure water stream to remove surface contaminants that are not tightly adherent, such as loose and blistered paint, chalking, grease, and other cumulative residues. In most cases, oil and grease are removed sufficiently. For heavy, well-adhered rust, scaling tools may be required before the water blaster is used.

A water blasting unit will pump approximately 10 gpm at pressures of up to 12,000 lb/in., thus accommodating a wide range of pressures for various applications. Units are rather small and highly mobile. Water blast cleaning is competitive with hand and power-tool cleaning, and in general produces similar results.

The advantages of water-blast cleaning are that it leaves no dust or other loose material on the cleaned surface and no abrasives are necessary. Thus, abrasive-contamination hazards as well as abrasive cleanup are avoided.

Water blasting is most efficient when used on surfaces that are irregular in shape and inaccessible to hand or power tools. For example, it is excellent for use in cleaning expanded metal and open grating on floors and catwalks. It can also be used readily on crevices, flanges and back-to-back angles. In addition, it can be employed effectively on structural steel, floor plates, piping, and storage tanks as well as in cleaning concrete and masonry surfaces. Since water blasting is often used for purposes other than surface preparation, e.g. descaling or defouling heat-exchanger bundles, maintenance personnel are often familiar with the operation, a big plus.

The disadvantages of water blast cleaning are that the entire work area becomes wet, the surface being cleaned must be dried thoroughly prior to painting, and tightly adhering contaminants such as tight paint and mill scale are not removed. Also, there have been cases where water blast cleaning of previously painted surfaces has caused pinholing and coating failures. In addition, since holding the high-pressure water nozzle within approximately 3 in. of the surface (as recommended) can be very tiring, operators tend to hold it farther from the surface, thus decreasing effectiveness.

For water blasting, effective rust inhibitors are 0.2% solutions by weight of chromic acid, sodium chromate, sodium dichromate, or potassium dichromate. In addition, a 2% solution of a mixture of four parts diammonium phosphate and one part sodium nitrite can be used. If experience shows that the concentration is not high enough to prevent rusting, the amount of inhibitor should be increased. It should be understood that the protection offered to the metal by any of these treatments is very limited.

Abrasive blast cleaning

In abrasive blast cleaning, an abrasive is propelled most commonly by air or rotary wheel against a surface. The abrasive impacts and abrades the metal, removing contaminants and roughening the surface. Preparation obtained by abrasive blast cleaning is the best of any mechanical method.

Air blast cleaning uses high-pressure compressed air that expands through an orifice to propel the abrasive against the surface. Nozzle discharge is approximately 100 lb/min of abrasive. Open air-propelled blast cleaning (commonly called sandblasting) may be the most thorough and economical means of cleaning scale and heavy contaminants from field structures. In many cases, air blasting in cabinets or rooms is also the best method of cleaning or polishing small parts.

Vacuum-blast cleaning is similar to air blasting, except that the abrasive is reclaimed from the immediate blast area by a vacuum. The blast nozzle is enclosed in a hollow cup with a rubber or brush seal around its perimeter. The abrasive flows through the nozzle, impacts on the metal being cleaned, and is withdrawn by the vacuum intake surrounding the nozzle. Abrasives with a low breakdown rate can be recycled, and those with a high breakdown rate, such as sand, are vacuumed into an enclosed hopper or bin and discarded.

The obvious advantage of this system is that abrasive cleanup and dusting are almost completely eliminated. Major disadvantages compared with conventional open air blasting are that the equipment is more expensive and not as portable, the cleaning rate is much slower, and the system works best on flat pieces and plates (irregular shapes can be cleaned but with difficulty).

While vacuum-blast cleaning should not be used for high-volume production work, it may often be the best method of surface preparation where a high degree of surface cleaning is required but abrasive contamination must be eliminated, such as areas where hydraulic equipment and electric motors are used.

In **rotary-wheel blast cleaning,** abrasive particles are propelled at high velocity by electrically driven wheels. The object being cleaned is placed within the blasting machine. A typical rotary-wheel machine has from four to sixteen or more wheels, although in steel-fabricating shops eight-wheel machines are most common. A single wheel can discharge up to about 1,600 lb/min of steel shot or grit.

The advantage of rotary-wheel blast cleaning is that the rate of cleaning is much faster than with any other method. All kinds of abrasives can be used, and all abrasives can be reclaimed and recycled. Dusting of the work area is eliminated, as is abrasive contamination. The degree of cleaning with rotary-wheel blasting conforms to requirements for white, near-white, commercial and brushoff blast cleaning (these are discussed later), depending on the speed at which the object being cleaned is moved past the abrasive-throwing wheels.

The disadvantages of rotary-wheel blast cleaning are that the initial equipment cost is considerably higher than for any other cleaning method, although operating costs may be lower, and that the units are not portable and work must be brought to them. Objects exceeding a certain size may not fit into the blasting machine.

Also, the interior of partially protected or enclosed surfaces, such as box trusses, cannot be adequately cleaned. In these cases, surface contaminants must either be ground off or removed by air blast cleaning. Depending on the size and capacity of the objects being cleaned, wheels can be added to make the blast cleaning pattern more effective.

Note: Never use shot alone in rotary wheel blasting. Shot alone will peen the surface, hardening it, and adhesion will become impossible due to an improper surface profile. In one plant, this mistake resulted in massive paint peeling and almost the entire facility had to be recleaned and repainted.

Use of wheel blasting by custom pipe-coaters has markedly reduced the painting costs of straight-run piping.

Blasting standards

Blast cleaning standards have been established by the Steel Structures Painting Council (SSPC) and the National Assn. of Corrosion Engineers (NACE), as shown in Table I. Both organizations describe the requirements for white blast cleaning, near-white blast cleaning, commercial blast cleaning, and brushoff blast cleaning. Excerpts from the SSPC Surface Preparation Specifications can be found in the accompanying box.

NACE standards [3] are similar to the SSPC definitions. NACE offers plastic-encapsulated steel panels that have been sand-, grit-, or shot-blasted to the specified standard. The SSPC also has photographic standards, showing various degrees of surface cleaning (hand and power-tool cleaning and the four degrees of blast cleaning) over four rust grades of steel (adherent mill scale, rusting mill scale, rusted steel, and pitted and rusted steel). Often, job standards are prepared at a worksite to avoid arguments about interpretations of the degree of workmanship required.

Surface profile

The surface profile, or roughness, of the blast-cleaned surface depends on the type of abrasive used and the force with which it impacts on the surface. Although there is some question as to the best surface profile for a given coating system, it is generally recognized that too deep an anchor pattern will result in peaks that

Surface preparation specifications **Table I**

SSPC*	NACE†	Description
SP 1-63	–	Solvent cleaning
SP 2-63	–	Hand-tool cleaning
SP 3-63	–	Power-tool cleaning
SP 4-63	–	Flame cleaning of new steel
SP 5-63	No. 1	White-metal blast cleaning
SP 6-63	No. 3	Commercial blast cleaning
SP 7-63	No. 4	Brush-off blast cleaning
SP 8-63	–	Acid pickling
SP 10-63	No. 2	Near-white blast cleaning

* Steel Structures Painting Council
† National Association of Corrosion Engineers

Steel Structures Painting Council Surface Preparation Specifications

White blast metal cleaning. "A White Metal Blast Cleaned Surface Finish is defined as a surface with a gray-white, uniform metallic color, slightly roughened to form a suitable anchor pattern for coatings. The surface, when viewed without magnification, shall be free of all oil, grease, dirt, visible mill scale, rust, corrosion products, oxides, paint, or any other foreign matter. The color of the clean surface may be affected by the particular abrasive medium used."

Near-white blast cleaning. "A Near-White Blast Cleaned Surface Finish is defined as one from which all oil, grease, dirt, mill scale, rust, corrosion products, oxides, paint or other foreign matter have been completely removed from the surface except for very light shadows, very slight streaks, or slight discolorations caused by rust stain, mill scale oxides, or slight, tight residues of paint or coating that may remain. At least 95% of each square inch of surface area shall be free of all visible residues, and the remainder shall be limited to the light discoloration mentioned above."

Commercial blast cleaning. "A Commercial Blast Cleaned Surface Finish is defined as one from which all oil, grease, dirt, rust scale, and foreign matter have been completely removed from the surface and all rust, mill scale, and old paint have been completely removed except slight shadows, streaks, or discolorations caused by rust stain, mill scale, oxides or slight, tight residues of paint or coating that may remain."

Brush-off blast cleaning. "A Brush-Off Blast Cleaned Surface Finish is defined as one from which all oil, grease, dirt, rust scale, loose mill scale, loose rust and loose paint or coatings are removed completely, but tight mill scale and tightly-adhered rust, paint and coatings are permitted to remain provided that all mill scale and rust have been exposed to the abrasive blast pattern sufficiently to expose numerous flecks of the underlying metal fairly uniformly distributed over the entire surface."

may not be covered sufficiently by the coating system, causing corrosion-initiation sites. Similarly, too shallow an anchor pattern will result in a smooth metal surface, to which paint will not adhere. Finally, good adhesion requires a clean surface and a coating with proper viscosity for wetting [*4*].

The greater the surface to which the coating adheres, the greater the total adhesive strength compared with the tensile strength of the system. A uniform anchor pattern (Fig. 2) is vitally important to adhesion. It is obtained by using a clean, well-graded, and sized abrasive. Generally, in a field operation, a hard, sharp and correctly sized silica sand will produce the desired results at the least possible cost. In comparison, a soft, round, ungraded bank sand will not cut steel fast, will cut it with uneven etch, and must be used in excessive amounts.

Abrasive selection implies knowledge of how abrasion size relates to anchor pattern, abrasion shape relates to surface finish, and abrasion hardness relates to cleaning speed.

Equipment

The key to a successful air blast operation is proper air pressure and abrasive selection. A difference in pressure from 60 psi to 100 psi at the blast nozzle can more than double the cost of the work. To ensure proper pressure, the following equipment is mandatory: adequately sized air compressor, large air-supply hose, portable high-production blast generators equipped with abrasive-metering valves, large abrasive hose, venturi nozzle, and automatic moisture separators of proper size.

In addition, the blasting operators must be protected and this requires special equipment, as discussed in the accompanying box.

Inspection

Details of inspection are beyond the scope of this article. However, chemical engineers should recognize that the best specification for surface preparation and application, and for a coating system, is worthless unless accompanied by detailed intermediate and final inspections. Shoddy workmanship is not uncommon, so inspection is an integral part of painting.

As a bare minimum, steel should be blast-cleaned to the specified surface condition and checked with a visual comparator, and final inspection should include a measured dry-film thickness. This is accomplished with the use of several types of dry film gages. Inspection test equipment should include: nozzle orifice gage, needle air-pressure gage, anchor pattern standards, visual comparators, surface magnifier, humidity gage, wet-film gage, dry-film gage (for example, the Mikrotest, Inspector, Elcometer, or Tinsley magnetic tape devices), and U.S. Standard Film Thickness cards, for verification of gage readings.

Two widely used commercial inspection devices are the surface-profile comparator (KTA-Tator Associates, Inc., Coraopolis, Pa., Fig. 3a) and the Clemtex anchor-pattern standards, CAPS (Clemtex, Ltd., Houston, Tex., Fig. 3b). These devices are designed for quick field inspection of a blast-cleaned surface. CAPS are made from stainless steel with the same hardness as mild steel.

Abrasive-blasting safety considerations

Protection of the blast-cleaner operator's eyes and respiratory system is a major safety consideration in any open blast-cleaning operation. Two types of helmets are commonly used in the blast-cleaning industry.

One is a slipover protective device against ricocheting abrasive, usually made of canvas with a plastic face mask. Because this type has no provisions for eliminating dust particles from the air breathed by the operator, it should be used in conjunction with a respiratory filter fitting over the nose and mouth.

The second type is an air-fed helmet made of metal or plastic, into which a separate supply of air is fed. Because the hood is under positive pressure, the operator does not inhale any dust resulting from the blast-cleaning operation.

A recent innovation, particularly in hot climates, has been the use of an air-conditioned hood or suit. Air to these units is not only filtered but cooled for greater comfort.

Helmet air purifiers take air from the compressor, regulate it to the lower pressures required for the operator's helmet, and remove dust, moisture, and oil fumes, thus providing dry breathing air to the wearer. However, the purifier does not remove carbon monoxide or other gaseous contaminants from the helmet air.

In the past, it was sufficient to ensure that the air-intake manifold of the compressor be located away from the exhaust of any adjacent machinery. It is now mandatory that an automatic shutoff device controlled by the blast cleaner be installed on the blast machine. The device, commonly called a "deadman," will shut off the air supply to the machine if the spring-loaded control lever is released by the operator. This will prevent the dangerous whipping of an operating blast hose if an operator becomes disabled. Note: pay particular attention to government standards for cleaning and handling respiratory air.

Chemical cleaning

With the exception of solvent cleaning, all chemical cleaning methods change the surface of the metal by forming a surface complex with the residual base metal. In many cases, this complex is beneficial; in others it is not and must be removed.

In solvent cleaning, the contaminants themselves are changed and removed from the surface, but the metal is not affected. In acid cleaning, the contaminants are dissolved, as is a portion of the metal. In alkali and emulsion cleaning, the contaminants are usually saponified or emulsified and then removed.

Coatings used in the CPI

Inorganic zinc

By far, the most widely used coatings are the inorganic zinc silicates. They have the best weather and solvent resistance of any protective coating. They contain no organic matter and can be used in virtually any solvent, including chlorinated materials. The life expectancy of these coatings in severe weathering service has not yet been established, but applications more than 20 years old still provide complete protection to the steel substrate.

The main disadvantages of inorganic zinc silicates are their limited chemical resistance and critical application properties. Since the films contain zinc, they are suitable for neither strong-acid nor strong-alkali environments. Successful application requires good surface preparation, and only minimal special equipment. Inorganic zinc silicate coatings have three outstanding properties:

- The best resistance to petroleum products and chemical solvents of any known coating.
- The best resistance to severe weathering environments of any known protective coating.
- Nonsusceptibility to underfilm corrosion, even after exposures of a decade or longer.

Cured inorganic-zinc-silicate films can quite properly be thought of as a cross between hot-dip galvanizing and a fused ceramic. From the standpoint of hardness and abrasion resistance, they are quite comparable to 2-oz American Petroleum Institute quality hot-dip galvanizing. While somewhat inferior to galvanizing in flexibility, they have no trouble following the normal contraction and expansion of steel surfaces. It is interesting to note that the hardness and abrasion resistance of inorganic films improve with aging, undoubtedly as a result of contact with atmospheric carbon dioxide.

Properly formulated inorganic zinc films have been used in freshwater immersion for periods exceeding

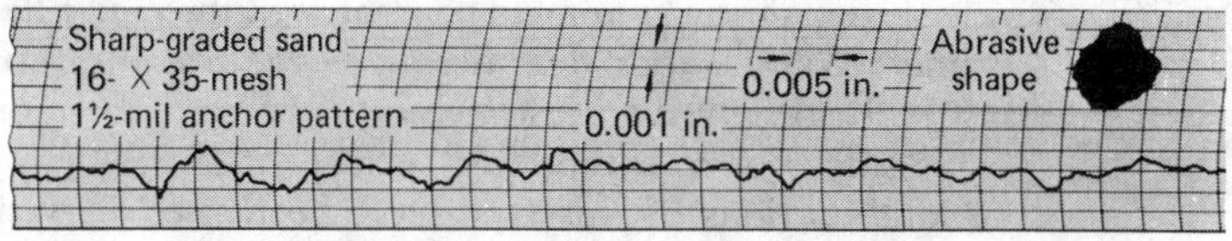

A uniform anchor pattern is vitally important for paint adhesion **Fig. 2**

eight years without displaying any signs of coating failure. Their performance in salt brine or seawater, however, is not outstanding, and in most cases the films are destroyed in approximately 18 mo.

Inorganic zinc-silicate coatings have an excellent history in hot, dry services, such as on the exterior of mufflers and stacks. Used alone, they can be successfully employed at temperatures to 800°F (427°C). When topcoated with a high-heat silicone coating, they can be used at temperatures up to 1,000°F (538°C). In the latter case, it is imperative that the silicone be applied to the inorganic coating prior to the inorganic film's being heated to any temperature above 300°F (149°C).

Inorganic coatings topcoated with chemically resistant organic films—such as epoxies, chlorinated rubbers,

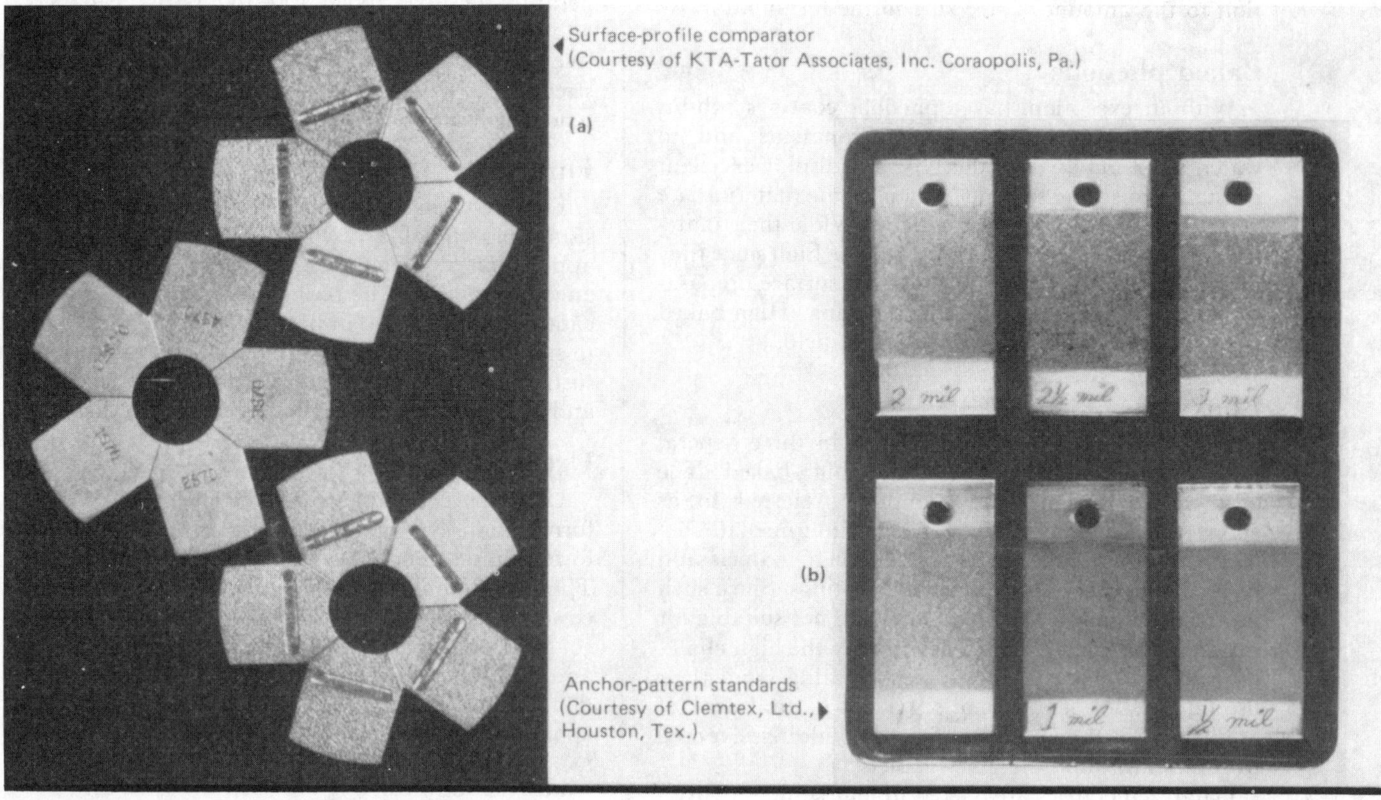

Two widely used inspection devices for assessing anchor pattern of surface preparation **Fig. 3**

and vinyls—provide the longest-lasting protection for steel in difficult chemical-plant environments. The organic topcoats, in effect, prevent the aggressive chemical solutions from coming in contact with the zinc-filled primer. Water vapor does permeate the organic film, but no rusting of the substrate occurs because the inorganic-silicate film cathodically protects the steel.

Combinations of this type have been used in very severe chemical plant environments, with no more than 0.5% failure after 10 years of service in coastal areas.

Some principal CPI applications for inorganic zinc coatings follow:

- Interior of petroleum-storage tanks.
- Interior of clean-oil tankers.
- Offshore platforms.
- Barges and small watercraft.
- Structural steel and piping for the petroleum, petrochemical and chemical industries.
- Decks of floating-roof tanks.
- Steel used in docks and wharves.
- Cooling-tower piping.
- Stacks and hot surfaces.

Organic zinc

Some zinc-rich coatings are made by combining zinc dust with an organic resin or elastomer. Typically, these are epoxies, chlorinated rubbers, vinyl, acrylics, epoxy esters, phenolics, and organic silicones. There is no reaction between the zinc and the resin; typical loadings of zinc pigment reach 95 percent by weight or up to 70% by volume. Organic zinc-rich primers have been used successfully, but organic binders can encapsulate zinc particles, thus limiting the amount of galvanic protection to the amount of free zinc in the formulation.

Baked phenolic

Without exception, baked-phenolic coatings exhibit excellent resistance to acidic environments and to water. Most materials of this type also display excellent resistance to strong solvents. As a class the materials are very weak in alkaline environments. While their material cost is low, their applied cost is quite high since they demand the ultimate in multiple-coat surface preparation as well as high-temperature baking. High-baked phenolics are difficult to repair in the field.

Epoxy

Epoxy-resin coatings are available in three general types: oil-modified, catalyzed, and high-baked. The oil-modified varieties are commonly referred to as epoxy esters. They have properties intermediate between those of high-quality conventional enamels and truly chemical-resistant protective coatings. Since such products contain a drying oil, they are not suitable for exposure to strong alkalis. They display the high chalking rate characteristic of all epoxy coatings, and their use on exterior steel is limited. Their natural area of application is the interior surfaces of buildings exposed to fumes and mild alkaline cleaners.

Catalyzed epoxies are offered in four common variations that differ in the catalyst employed. Amine-cured epoxies display the best solvent and acid resistance of coatings of this generic type. Polyamide-cured epoxies have less general chemical resistance, but are superior to amine-cured types in water resistance, weather resistance, and ability to adhere to difficult surfaces.

Amine-adduct-cured materials are based on the prereaction of a portion of the catalyst with the epoxy resin. Such materials are less sensitive to climatic conditions than amine-cured epoxies and are considered the equal of either amine or polyamide variations.

High-baked epoxies display the best chemical and solvent resistance of any type of epoxy but, as their name implies, require very-high-temperature baking to achieve complete polymerization. Use of this type of epoxy is normally limited to the interior of tanks handling concentrated caustics or solvents.

All epoxy-based materials are extremely weak in acid and water service, and they also show a marked tendency to discolor and chalk. Inasmuch as catalyzed and baked epoxies are polymerized to form high-molecular-weight, very solvent-resistant films, they present overcoating problems. Most authorities agree that mechanical roughening of some type is imperative to secure good intercoat adhesion.

Vinyl

Vinyl-based protective coatings display the greatest versatility of all the commonly used resin types. These coatings have good resistance across the pH scale, excellent water resistance, and perhaps the lowest chalking rate of any organic product. The films are thermoplastic, thus ensuring ease of recoating.

The main disadvantages of vinyl-resin coatings are their limited solvent and heat resistance (they are temperature-sensitive above 150–180°F, or 66–82°C). In addition, they display a marked tendency to be sensitive to intercoat contamination. The films have greater strength in tension than in adhesion, and there is a tendency to peel when surface preparation is improper.

Furan

Furan coatings are among the most versatile and resistant organic films discovered to date, but because of application problems they have not found wide acceptance in industry. The resins are set with a strong-acid catalyst that makes it impossible to apply them directly to steel surfaces and limits their adhesion to any prime surface. Once cured, the films become extremely hard, and it is virtually impossible to maintain them.

Urethane

Urethane coatings are offered in a wide variety of formulations, varying from varnishes for wood surfaces to multiple-component materials for industrial service. They are finding increased acceptance with industry as cosmetic finishes, and are used over epoxy protective coatings or primers because of their outstanding gloss and color retention. As corrosion barriers, they offer no demonstrable superiority to well-formulated activated epoxy or high-molecular-weight vinyl coatings.

Coal-tar epoxy

As their name implies, coal-tar-epoxy coatings are mixtures of coal-tar pitch and low-molecular-weight epoxy resins. In the simplest terms, they are a combination of thermoplastic and thermosetting resins. Common formulations contain up to 35% of the epoxy resin. The resulting films are very resistant to acids and water, and they also have good resistance to solvents of moderate strength. While these coatings are never recommended for immersion in strong caustics, they do display good resistance to spills of mild caustics.

Chlorinated rubber

The principal chlorinated-rubber resin is manufactured by chlorinating natural rubber; it is almost always modified with alkyd resins in manufacturing protective coatings. Since commercial formulations generally contain a high percentage of the alkyd resin, they are much less sensitive than, say, vinyl coatings to intercoat contamination.

Liquid-applied chlorinated-rubber coatings should not be used in highly corrosive environments such as the interior of acid-storage tanks. The reason is that even though the coating is resistant to the chemical, rapid attack of the underlying surface will occur at any pinhole or discontinuity in the coating film. For extreme services, thick membrane linings such as rubber or polyvinyl chloride sheets are required. For severe abrasion or uses in which temperature is also a factor, the membrane should be overlaid with acidproof brick.

Silicones

Silicone coatings show high heat resistance, excellent outdoor durability, excellent water resistance, excellent electrical resistance, and high resistance to corrosive atmospheres. General practice includes baking, because the material is rather soft when air-dried. Clear silicone coatings are used when high heat resistance is desired, since silicones undergo less decomposition under heat than do most polymers.

Silicone-acrylics have high gloss and extended protection against blistering, peeling, checking, and crazing. Pigmented, they will last from 10 to 15 years. As clear finishes, they are much less durable, but they have much better film properties than phenol-resin or alkyd spar varnishes.

Silicone-modified alkyds contribute markedly to exterior durability. Their gloss retention depends on the ratio of silicone to alkyd. Federal Specification TT-E-00490 describes a silicone-alkyd enamel containing 30% silicone-alkyd resin. As a clear coating, it is second only to the fluorocarbons in durability. Its durability is vastly increased when pigmented. Heat resistance is directly proportional to the amount of silicone resin.

Oil-based paints

Oil-based paints are unsatisfactory in most chemical-industry environments. Some of the reasons are given below:

- Slow drying time—Chemical dust and fumes would be deposited between coats because each has to dry 24 h or longer before a subsequent coat can be applied. This sandwiching of chemicals between layers causes the paint to deteriorate rapidly and permits chemical attack on the steel through thin prime coats.
- Lack of resistance to chemicals—Many chemicals quickly destroy pigments and vehicles in oil paints, leaving the surface vulnerable.
- Oil solubility—Ordinary cleaning fluids and lubricating oils dissolve oil-based paints, making equipment covered with these paints look unsightly and destroying the protective qualities of the coatings.

Other vehicles

The above list of coatings does not include standard alkyd enamels, water-based acrylics, polyvinyl acetate emulsion-based coatings and numerous other finishes generally considered to be architectural coatings. These do find widespread use in the CPI in general construction and services that are not located in extreme environments.

In addition, fluorocarbons, polyvinyl and polyvinylidine copolymers, bituminous materials, polyfluoroethylene, polyethylene, polysulfides, polyesters and nylon have specialized functional uses.

Coating system specification

For very mildly corrosive conditions, in which relative humidity is rarely above 50% and the corrosion rate is less than 25 mdd (milligrams per square decimeter per day, or 0.19 lb of steel per square foot per year), it would be difficult to justify painting from the standpoint of corrosion. However, such mildly corrosive conditions rarely are found in industrial plants. Severe corrosion rates of 300 mdd are not uncommon in industrial and marine situations. Carbon steel and many other metals should be protected from such exposures.

Many chemicals, among them chlorinated solvents, hydrocarbon solvents, ketones, sodium hydroxide (concentrations not over 50%) and phenol, are not considered corrosive to carbon steel at ambient temperatures. However, spills of these materials may promote corrosion by removing the protective coating from the steel, thus permitting the metal to be corroded by atmospheric exposure. Even if the atmosphere is not corrosive, the spills produce an unsightly appearance by stripping away some of the coating.

When spills do occur, it is good practice to recoat the stripped area as soon as possible—within 1 week if the atmospheric corrosion rate is 100 mdd or more, or within 4 weeks if the corrosion rate is less than 100 mdd.

Table II rates the relative resistance of typical generic coatings to sunlight and weather, stress and impact, abrasion, heat, water, salts, solvents, alkalies, acids, and oxidation, using a value of 10 as the best for each property.

Top-quality coatings should be compared generically, but identification by generic name per se is no guarantee of quality. Coatings should be purchased on specifications from reliable coatings manufacturers. It is false economy to purchase a protective coating without knowing its solids content and resin content.

Cost plays the major role in selecting a coating system. Specify the system that performs best for the least amount of money over a given period of time. This does not necessarily mean using the least expensive or the most expensive system. Because of repainting costs and corrosion, a "cheap" system may very well cost more

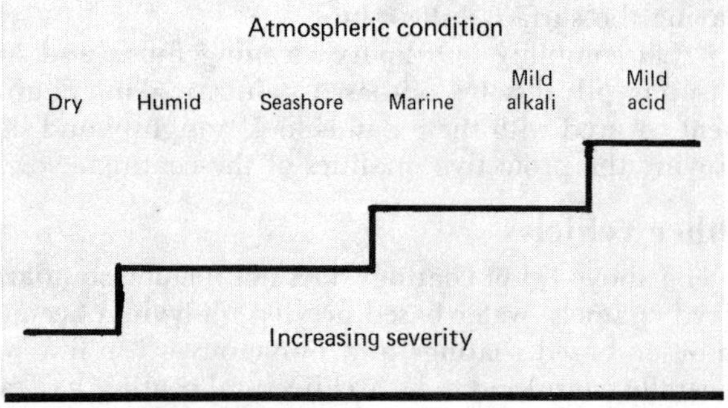

Corrosive severity of industrial environments **Fig. 4**

per year than a more expensive system. But don't gild the lily by using very expensive coatings. If, for example, a coating system performs adequately for 5 years and costs only $0.75/ft^2, a coating costing $1.50/ft^2 cannot be justified for the same service even if the more expensive coating performs indefinitely.

Selection criteria

There are numerous criteria for selecting coating systems. For coatings that will be exposed to the weather, climatic conditions are of prime importance. If the relative humidity does not exceed 50 percent and there are no chemical or other industrial exposures, steel need not be sandblasted. In such areas cleaning with power tools or a wire brush, followed by an oleoresinous primer and a topcoat, is ample.

Comparative resistance value of typical commercial coating formulations **Table II**

Condition	Generic type: Vinyl	Epoxy	Phenolic	Alkyd	Oil-based
Sunlight and water	10	9	9	10	10
Stress and impact	8	3	2	4	4
Abrasion	7	6	5	6	4
Heat	7	9	10	8	7
Water	10	10	10	8	7
Salts	10	10	10	8	6
Solvents	5	8	10	4	2
Alkalies	10	9	2	6	1
Acids	10	10	10	6	1
Oxidation	10	6	7	3	1
Total	**87**	**80**	**75**	**63**	**43**

Note: Based on a top rating of 10 per category [1].

Criteria for selecting coatings **Table III**

Abrasion resistance	Temperature resistance
Adhesion properties	Drying time
Impact resistance	Appearance
Flexural qualities	Wetting time
Resistance to specific chemicals	Applied cost
Resistance to sunlight	Antistick properties

Steel exposed to moderately severe environments should be sandblasted to white metal or nearly to white metal, primed with a zinc-rich coating, and topcoated with a coating resistant to the environment.

Steel exposed to severely corrosive environments, such as marine, humid, coastal, or chemical exposures, should be sandblasted to white metal, primed with a zinc-rich coating, and topcoated with a coating suitable for the exposure.

Table III lists several factors that the chemical engineer must consider when choosing paint. Table IV gives typical examples of some of the systems used.

Industrial environments are dynamic, i.e., ever-changing—especially in coastal areas. A coating or lining suitable for two different environments may not be suitable under combined conditions or separate environments in sequence. For example, an epoxy is excellent in alkaline or in solvent environments, but can fail quickly if exposed to the two alternately, because of shrinking (alkaline conditions) and swelling (solvent conditions).

Thus, the coating must be carefully chosen to fit the environmental exposure conditions (Fig. 4). This is true for all parts of the system—primer, intermediate and top coats. Here are the basic principles.

A primer should meet most if not all, of the following requirements:

- Good adhesion when the surface has been cleaned or prepared according to specifications.
- Appropriate flexibility-distensibility.
- Satisfactory intercoat bonding-surface for the next coat.
- Ability to stifle or retard the spread of corrosion from discontinuities such as pinholes and breaks in the coating film.
- Enough chemical and weather resistance (inertness) to protect the surface for a time period in excess of that anticipated before application of the next coat in the system.
- In certain conditions, notably tank lining, chemical resistance equivalent to the rest of the system.

Intermediate coats may be required in a system to provide one or more of the following properties:

- Adequate film thickness of the system (body coat).
- Uniform bond between the primer and the top coat (tie coat).
- Superior barrier with respect to aggressive chemicals in the environment.

The finish coat is the initial barrier to the environment as well as being the surface seen by management and the public. However, in some situations the barrier to the environment is primarily a function of the body or prime coat, while the finish coat serves to provide a pleasing appearance, a nonskid surface, a matrix for antifouling agents, or other specialized purposes. Obviously, the chemical resistance of the finish coat in such a situation must be sufficient to ensure its remaining intact in the environment.

While thousands of different environments exist within the CPI, Table V lists some common denominators for system selection for refineries, chemical plants and wastewater treatment plants. The table does not cover all of the coatings requirements for a facility (e.g.,

Linings for carbon steel tanks, pipes, and miscellaneous equipment (excerpted from "Paint Handbook") **Table IV**

	Purpose of lining			Lining systems*								
Product	**Product purity**	**Non-stick**	**Corrosion resistance**	**No. 1: Epoxy**	**No. 2: Epoxy phenolic**	**No. 3: Vinyl, odorless**	**No. 4: Vinyl**	**No. 5: Phenolic, baked**	**No. 6: Coal-tar epoxy**	**No. 7: Neoprene**	**No. 8: Inorganic**	**No. 9: Rubber, sheet**
Carbon tetrachloride	X										A	
Castor oil	X			A	B	B						
Caustic soda, 10 and 20%	X			A	B		B		B	B		B
Caustic soda, 50%	X			A	B		B		B	B		B
Caustic soda, 73%			X		B							
Cellosolve	X										A	
Chlorine (dry)												
Chlorine (wet)			X									A
Chlorobenzene (dry)	X										A	

Note: An "X" in one or more of the first three columns indicates the purpose of the lining. "A" in a lining-systems column indicates that this is the recommended lining. "B" indicates that the lining is suitable for the product but is not necessarily the recommended lining. For example, the lining may be more expensive than is necessary for a given product. A blank space does not necessarily indicate that a given lining is unsuitable; it means only that the system was not considered for the product or products in question. Unless otherwise noted, the lining performance is based on a product temperature not exceeding 100°F (38°C), although many linings will perform well up to 200°F (93°C) or higher. Each case should be checked or, preferably, tested for performance above 100°F.

*Descriptions of the lining systems follow:

No.	Description	Dry film thickness (mils)
1	Two-coat, 100% solids epoxy	20
2	Five-coat epoxy phenolic, low-baked	10
3	Five-coat odorless food-grade vinyl	10
4	Seven-coat chemical-resistant vinyl	6
5	Five-coat baked phenolic, high-baked (400°F)	4.3
6	Two-coat coal-tar epoxy	16
7	Four-coat neoprene	15.5
8	Two-coat system; one coat, lead-free inorganic zinc, and one coat, inorganic topcoat	4
9	Sheet rubber (various types; some pressure-cured, some exhaust-steam-cured; consult manufacturer for specific rubber for a given product).	3/16 in.

blowers, compressors, conveyors, cranes, extruders, fans, hoists, pumps and exchangers). Obviously, an engineer should not rely on a vendor's standard paint for an application that requires a special coating. For example, an operating engineer at a large fertilizer warehouse complained that the vendor's standard coating on forklifts and conveyors was deteriorating, causing the equipment to rust and corrode. To avoid this, special painting instructions and specifications should become part of the ordering specification for this kind of equipment. If the problem is very serious, inspectors should be onsite to inspect the equipment before it is painted (for surface cleanliness, etc.) and to assure the correct coating (as specified by the operating company) has been used. There are many cases where the painting contractor deliberately or inadvertently uses the wrong paint. In cases where repainting is necessary, the job can become expensive and subject to litigation.

Shop priming specifications

The specification for shop priming must establish the *minimum* requirements for surface preparation and shop priming of equipment, structural steel, steel tanks, and piping. It should note that "shop priming" may not limit the supplier's responsibility for priming to the shop location. Field surface preparation and touch-up or painting of abraded or damaged areas of surfaces left unprimed can be part of the bid requirements.

A shop-priming painting schedule must identify the extent of shop priming of both the insulated and the noninsulated portions of equipment. It will also identify surfaces that are to be galvanized (generally structural steel that is not fireproofed). Also included are surface-preparation procedures as well as the complete identification of all classes and types of material requiring painting. This is coupled with a complete shop-priming schedule (see Table VI).

Shop priming is proving to be a cost-effective means of reducing painting costs, especially for straight-run piping, as shown in Table VII. Detailed economics of shop vs. field painting is the subject of several NACE publications [7].

Field specifications

The specification for field painting follows that for shop painting, but the inspection requirement to assure

Part of coating selection guide for CPI facilities from "Paint Handbook" **Table V**

Item	Exposure	Surface preparation	Primer (Generic type)	Topcoats (temperature limits) (Generic type)	Dry film thickness (mils)	Remarks
Structural steel, including equipment supports	Fumes and spills of acids and acid salts	Near-white metal blast	One coat, inorganic zinc	Two coats, high-build vinyl (up to 150°F)	Primer 3.0 Topcoats 6.0 **Total 9.0**	The first coat of vinyl should be applied over the primer before exposure to acid fumes or spills.
		Commercial blast	One coat coal-tar epoxy	One coat, coal-tar epoxy (up to 200°F)	Primer 8.0 Topcoat 8.0 **Total 16.0**	This material is black. It is a good heavy-duty coating when looks are not important.
	Alkalies and alkaline salts spills	Near-white metal blast	One coat, inorganic zinc	Two coats, polyamide-cured epoxy (up to 200°F)	Primer 3.0 Topcoats 4.0 **Total 7.0**	The first coat of epoxy should be applied over the primer before exposure to alkaline spills.
		Commercial blast	One coat, coal-tar epoxy	One coat, coal-tar epoxy (up to 200°F)	Primer 8.0 Topcoat 8.0 **Total 16.0**	This material is black. It is a good heavy-duty coating when looks are not important.
	Solvent spills	Near-white metal blast	One coat, inorganic zinc	Two-coats, polyamide-cured epoxy (up to 200°F)	Primer 3.00 Topcoats 4.00 **Total 7.00**	This system performs well when exposed to spillage of aliphatic solvents, aromatic solvents, and some chlorinated solvents.
		Near-white metal blast	One coat, inorganic zinc	One coat, inorganic topcoat (up to 750°F)	Primer 3.0 Topcoat 6.0 **Total 9.0**	This system is resistant to spills of strong chlorinated solvents, provided they are acid-free. It is also resistant to phenol spills.
	Calcium and sodium hypochlorite spills	White metal blast	None	One coat, vinyl ester (210°F, all concentrations of calcium hypochlorite)	**Total 20.0**	This material has very good resistance to calcium and sodium hypochlorite (up to 15% sodium hypochlorite, up to 180°F).
Steel piping, aboveground, uninsulated	Fumes and spills of acids and acid salts	Near-white metal blast	One coat, inorganic zinc	Two coats, high-build vinyl (up to 150°F)	Primer 3.0 Topcoat 6.0 **Total 9.0**	The first coat of vinyl should be applied over the primer before exposure to acid fumes or spills.
		Commercial blast	One coat, coal-tar epoxy	One coat, coal-tar epoxy (up to 200°F)	Primer 8.0 Topcoat 8.0 **Total 16.0**	This material is black. It is a good heavy-duty coating when looks are not important.

quality control becomes more important. Use appendixes to cover details of paint systems and coating schedules.

For each paint system include such information as the substrate, service conditions, operating-temperature range, minimum surface preparation and profile, manufacturer, and prime coat and/or final coat or coats as necessary. For each coating schedule show a complete listing of all surfaces to which paint is to be applied, together with equipment or item number and designation, surface temperature, paint system to be used, and manufacturer's code for the final color indicated.

For example, carbon steel surfaces that have been primed with organic or inorganic zinc-rich materials should be hosed down with clean salt-free water to remove surface salts and other contaminants, and allowed to dry prior to topcoating. Sufficient dropcloths, shields, and other protective equipment should be used to prevent surface-preparation abrasives, paint overspray, or paint droppings from fouling adjacent surfaces.

There are conditions that limit paint application in the field. Paint should not be applied:

- When the ambient temperature is below 40°F (4°C) or above 110°F (43°C).
- When the ambient temperature is expected to drop to 32°F (0°C) before the paint has had time to dry.
- In rain, snow, fog, or mist.
- In extremely windy or dusty conditions.
- When the steel temperature is at or near the dew point.
- In the vicinity of sand- or grit-blasting operations.
- When the steel surface temperature exceeds limits

Part of coating selection guide for CPI facilities from "Paint Handbook" **Table V (continued)**

Item	Exposure	Surface preparation	Generic type: Primer	Generic type: Topcoats (temperature limits)	Dry film thickness (mils)	Remarks
Steel piping, aboveground, uninsulated (continued)	Alkalies and alkaline salts spills	Near-white metal blast	One coat, inorganic zinc	Two coats, polyamide-cured epoxy (up to 200° F)	Primer 3.0 Topcoats 4.0 **Total 7.0**	The first coat of epoxy should be applied over the primer before exposure to alkaline spills.
		Commercial blast	One coat coal-tar epoxy	One coat, coal-tar epoxy (up to 200° F)	Primer 8.0 Topcoat 8.0 **Total 16.0**	This material is black. It is a good heavy-duty coating when looks are not important.
	Solvent spills	Near-white metal blast	One coat, inorganic zinc	Two coats polyamide-cured epoxy (up to 200° F)	Primer 3.00 Topcoats 4.00 **Total 7.00**	This system performs well when exposed to spillage of aliphatic solvents, aromatic solvents, and some chlorinated solvents.
		Near-white metal blast	One coat, inorganic zinc	One coat, inorganic topcoat (up to 750° F)	Primer 3.0 Topcoat 6.0 **Total 9.0**	This system is resistant to spills of strong chlorinated solvents, provided they are acid-free. It is also resistant to phenol spills.
	Calcium and sodium hypochlorite spills	White metal blast	None	One coat, vinyl ester	**Total 20.0**	This material has very good resistance to calcium and sodium hypochlorite (up to 15% sodium hypochlorite, up to 180° F).
	Hot piping, 201° F to 750° F, all areas	Near-white metal blast	One coat, inorganic zinc	One coat, modified silicone	Primer 3.0 Topcoat 2.0 **Total 5.0**	Apply topcoat before exposing to acid or alkaline fumes or spills.
	Hot piping, 751° F to 1,200° F, all areas	White metal blast	One coat, silicone for hot surface	One coat, silicone for hot surfaces	Primer 1.0 Topcoat 1.0 **Total 2.0**	Do not apply thick coats. They will spall off.
Steel piping, aboveground, insulated	Fumes and spills of acids, acid salts, alkalies and alkaline salts; process temperature, 0° F to 200° F	Near-white metal blast	One coat, inorganic zinc	One coat, polyamide-cured epoxy	Primer 3.0 Topcoat 2.0 **Total 5.0**	No coating is needed on insulated surfaces with continuous operating temperatures below 0° F or above 200° F. Omit coat of epoxy when not subject to chemical spills.

Reference: [*1*]

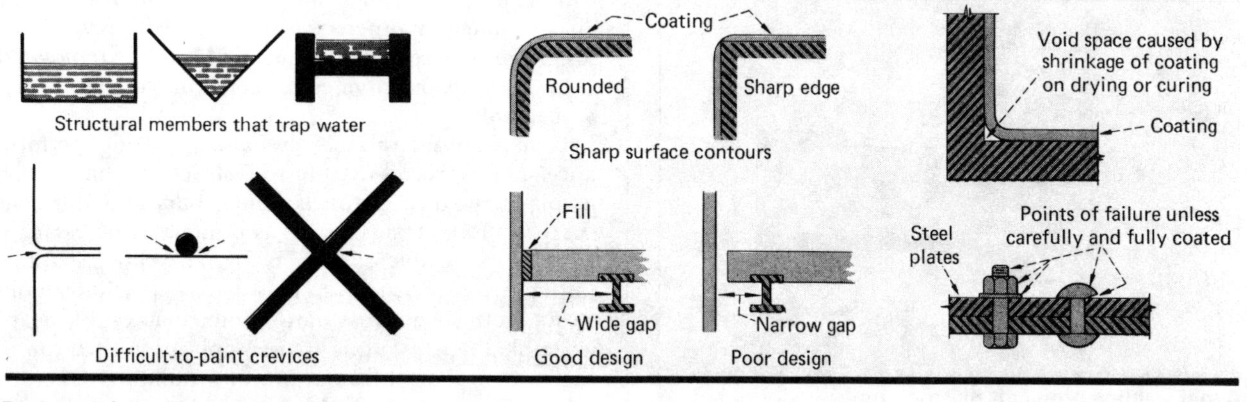

Design pitfalls that may cause paint failure **Fig. 5**

Typical page from a shop priming schedule — Table VI

Carbon steel surfaces	Operating temperature range (°F) normal environment not in acid areas [1]: 250 and below	250 to 700	700 to 1,000
Furnaces and exchangers			
a. Casing steel	P.S. 1	P.S. 2	P.S. 3
b. Buckstays	P.S. 1	P.S. 2	P.S. 3
c. Nozzles	P.S. 1	P.S. 2	P.S. 3
d. X-over piping	P.S. 1	P.S. 2	P.S. 3
e. Breeching	P.S. 1	P.S. 2	P.S. 3
f. Stacks	P.S. 1	P.S. 2	P.S. 3
g. Structural steel	P.S. 1	P.S. 2	P.S. 3
Structural steel			
a. All structural steel—except as otherwise specified	1	—	—
b. Grating and checker plate	1	—	—
c. Stairway hardware	1	—	—
d. Ladders and cages	1	—	—
e. Toe plates and stringers	1	—	—
f. Handrails	1	—	—
g. Bolts, nuts and washers	1	—	—
h. Anchor bolts	1	—	—
Piping and fittings			
a. Pipe	P.S. 1	P.S. 2	P.S. 3
b. Fittings	P.S. 1	P.S. 2	P.S. 3
c. Flanges	P.S. 1	P.S. 2	P.S. 3
d. Valves	2	2	2
e. Spring hangers	2	—	—
Pumps and motors			
a. Casing	2	2	2
b. Nozzles	2	2	2
c. Baseplates	P.S. 1	P.S. 1	P.S. 2
d. Motors	2	2	2

Notes:
1—These surfaces shall be galvanized.
2—Manufacturer's standard finish is acceptable.
P.S. 1—Paint Standard No. 1
P.S. 2—Paint Standard No. 2
P.S. 3—Paint Standard No. 3

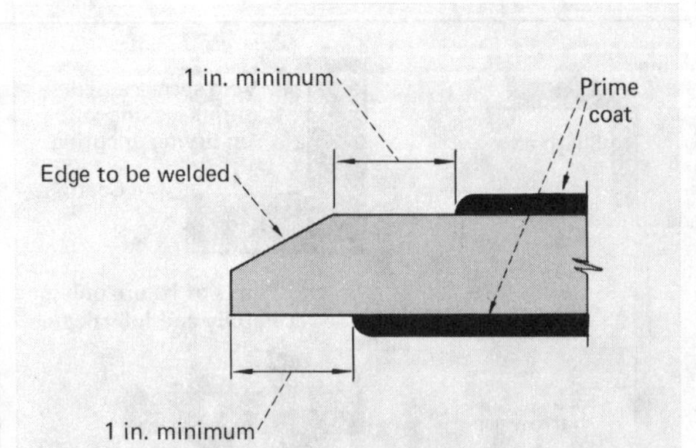

Avoid welding problems by stopping prime coat 1 in. from edge — Fig. 6

Cost comparison of field vs. shop blasting and priming of straight-run pipe — Table VII

Item [1]	Shop*	Field: Nonunion	Field: Union
1. Blast and prime	\$0.35†	\$0.40	\$0.45
2. Unload railcars	0.018	0.095	0.095
3. Freight from shop to jobsite	0.068		
4. Unload truck at jobsite	0.069	0.069	0.069
5. Movements of pipe at field			
Railcars to laydown		0.069	0.069
Laydown to blast area		0.069	0.069
Blast area to laydown		0.069	0.069
Laydown to erection	0.069	0.069	0.069
6. Clean up sand		0.003	0.003
7. Dunnage and yard preparation		0.028	0.028
8. Rain delay at 13 percent		0.0776	0.107
Total	**0.574**	**0.9486**	**1.028**
Difference		0.3746	0.454
Savings possible at 500,000 ft²		\$187,300	\$227,300

*Color coding and heat number retention carry an extra price. For colo coding add 1½ cents per square foot; for heat number retention, 3 cent per square foot.
†Unloading and loading trucks included in paint price.

specified by the paint manufacturer for the specific coating materials being used.

Design and troubleshooting

Design pitfalls can lead to paint failure (see Fig. 5). Particular care should be taken to get a continuous coating on all metal surfaces. If protrusions, such as weld spatter and laminations, can be power-ground flat and smooth during surface preparation, it will pay off in total paint performance. Sharp edges, such as the edges of structural members and rough flame- or saw-cut edges, should be rounded by grinding, preferably to a 1/8-in. minimum radius. Crevices, sharp corrosion pits and deep gouges in the metal should be filled with weld metal. Corrosion design problems, such as tack welds, back-to-back angles, and lap joints, should be eliminated in the design, if possible, or sealed by welding or caulking.

Sharp metal edges and pipe threads are difficult to cover adequately with primer. Frequently these are the first areas to rust after painting. Painting bolts and nuts with a zinc-rich primer followed by a Teflon topcoat is an economical maintenance aid because it prevents corrosion from "freezing" the nut in place. The removal of corroded nuts on large equipment can cost more than \$5 per bolt.

Design considerations also affect coating performance. This is true of most ferrous structural shapes, such as plates, beams, channels, pipes, bars and thin-gage sheets. For these surfaces, it is sometimes necessary to apply a protective coating before they are assembled into an overall structure. This is particularly true of a structure that has areas that become inaccessible as it is assembled. An example is the application of siding to the structural-steel framework of a building. Once the siding has been attached to the structure, it is impossible to reach the face of the steel in contact with the

Crevice corrosion

Two types of crevice corrosion are particularly difficult to combat. The first of these is commonly referred to as an oxygen concentration cell, and the second is usually called an ion concentration cell. These types of corrosion can be expected whenever crevices or pockets exist as a result of poor design or construction practices.

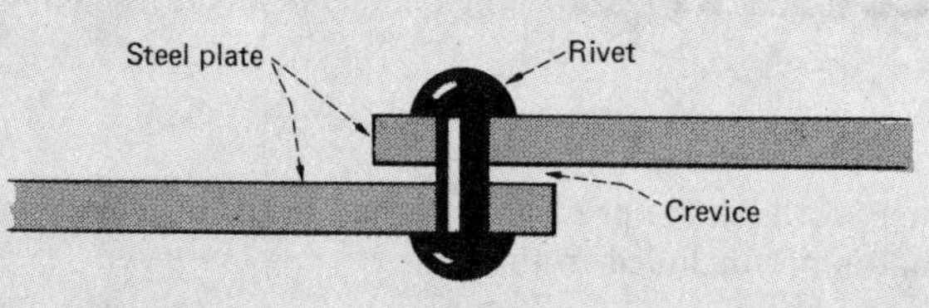

The figure shows two steel plates riveted together in such a manner as to create a crevice. On exposure to the atmosphere, the crevice will be filled with water containing ions or chemical particles. It is easy to understand that there will be a limited amount of oxygen on the inside of the crevice, but a very plentiful supply on the flat steel surfaces immediately outside of it. The area exposed to the most oxygen will become the cathode, thereby causing the oxygen-starved area to act as an anode and be corroded. The metal in the crevice is oxygen-starved and, therefore, subject to vicious pitting corrosion.

The most practical way to eliminate this type of crevice corrosion is to avoid structural design that creates crevice conditions. Before existing structures are coated, crevices should be closed by welding or filled with an appropriate mastic. Similar corrosion occurs where salt brine and similar materials are present.

siding. The face must therefore be painted well before the siding is applied.

Another problem arises from the use of skip welds instead of continuous welds for joining members. Skip welds are chosen for their strength, but they create a crevice (the "skip") where the two surfaces being joined are not welded. The protective coating being applied over the surfaces must then bridge the skip and provide protection in the unpainted crevice. Many steel structures have failed because of crevice corrosion in the area where a coating couldn't be applied (see box above).

For industrial plants, surface-preparation and coating standards are in effect before the plant goes into operation. Structural steel and all steel vessels usually are coated at the fabricating shop. Paint inspection must start at this point because the steel often is shipped with either a prime coat or prime and finish coats. Coatings are selected, among other reasons, for their resistance to damage in being handled by slings, wire ropes, etc. They must be capable of easy repair in the field with minimum surface preparation and with maximum longevity and protection.

In addition, many refinery and petrochemical plants face dangers of fire and explosion. No sparks of any kind are allowed, and the use of static-dissipating sandblasting hose and the electrical grounding of equipment and hoses are therefore required. Because of this requirement, some operators elect to compromise on allowable surface-cleaning techniques, and this in turn impacts on coating selection.

Welding is a problem before and after painting. If welding is required after shop coats have been applied, the paint can interfere with the weld. To avoid such problems, specify that the prime coat be held back a minimum of 1 in. on both sides from the edge prepared for welding, as shown in Fig. 6. The problem is exacerbated when steel is prepainted with a zinc-rich primer prior to fabrication. This prime coat can reduce welding and gas-torch cutting speeds, and, more important, reduce weld strengths and corrosion protection at and near the seam.

Field welds, used to assemble piping, equipment, and structural steel that have been shop-primed, require touch-up of the prime coat for a minimum distance of 1 in. on each side of the weld. Such painting should be done only *after* the removal of all weld splatter and proper surface preparation. Painting over welding flux is a serious cause of corrosion and structural failure. *Remove all weld splatter* prior to sandblasting.

Acknowledgements

The author specifically recognizes P. James Bennett, Clemtex Ltd., Houston, Tex., for his assistance in preparing portions of this article. He also thanks Kenneth B. Tator, KTA-Tator Associates, Coraopolis, Pa., Paul Weaver, Consulting Engineer, Houston, Tex., Jon Rodgers, Reliance-Universal, Houston, Tex., and Trevellyan V. Whittington, ATI Division of Parke Davis & Co., North Hollywood, Calif., all of whom contributed to the "Paint Handbook," which formed the basis for much of this article.

Finally, the author expresses his gratitude to Hal Crawford of McGraw-Hill Book Co., Professional and Reference Book Div., for permission to use excerpts and illustrations from the "Paint Handbook," published in January 1981.

References

1. Weismantel, Guy E., ed., "Paint Handbook," McGraw-Hill, New York, 1981.
2. Gross, William F., "Application Manual for Paint and Protective Coatings," McGraw-Hill, New York, 1970.
3. "Surface Preparation Handbook," National Association of Corrosion Engineers, June, 1977.
4. Doherty, C. B., Surfaces Produced by Abrasive Blasting of Steel, *Materials Performance*, November, 1974, p. 12.
5. Roebuck, A. H., Safe Chemical Cleaning—The Organic Way, *Chem. Eng.*, July 31, 1978, p. 107.
6. Holland, F. A., Watson, F. A., and Wilkinson, J. K., Engineering Economics for Chemical Engineers, *Chem. Eng.*, June 25, 1973 to October 28, 1974.
7. *Proc.,* NACE Annual Conf., Atlanta, Ga., March, 1979.

The author

Guy E. Weismantel is a Senior Regional Editor for CHEMICAL ENGINEERING, 375 Dresser Tower, 601 Jefferson Street, Houston, TX 77040. Tel: 713-659-4551. Before joining *CE* in 1967, Mr. Weismantel had ten years of experience in the manufacture and use of paints, varnishes and lacquers. He is Editor-in-Chief of the "Paint Handbook," McGraw-Hill, 1981. Mr. Weismantel holds a B.S. in chemical engineering from the University of Notre Dame. He is a member of the Houston Soc. for Coating Technology, the Houston Paint and Coatings Assn., AIChE, WPCF, and the Houston Chamber of Commerce Environmental Committee.

How to specify coatings

There are many ways that a paint job on the plant can break down. The surest route to success is to make comprehensive specifications and to check that they are being fulfilled at every stage of the work.

Dean Berger *and* ***Frank Border***
Gilbert/Commonwealth

☐ The engineer's choice of a coating system is based on past performance, appearance, cost, ease of repair, and ultimate satisfaction to the owner. However, the properties are evaluated in relation to the needs of the environment where the system is to be used.

The locks and dams in the upper Mississippi River, for example, can only be painted during winter when the gates may be removed. In this case, a vinyl system is used because it can be applied in cold weather and performs very well in water-immersion service. Other systems do not perform as well.

Physical properties such as adhesion; resistance to abrasion, chemicals, salt, fog, humidity, fire, and water-immersion; flexibility; gloss retention; and durability are important in the selection of any system. Structural design will also affect selection.

Choosing tank linings becomes even more involved. It is of paramount importance to use materials with a proven record in this application. If new materials are used, they have to be carefully screened prior to acceptance.

Coating specifications for chemical plants, power plants and similar projects should include:

- The scope of the work.
- A list of definitions.
- A list of applicable codes and standards.
- General requirements relating to storage and mixing of materials, and use of solvents and thinners.
- A list of coating systems and requirements.
- A list of acceptable coating manufacturers.
- Surface preparation and special requirements.
- Application procedures.
- Inspection of work.
- A quality-assurance program.

Scope of work

This section generally describes the project and materials to be used, surface preparation, application, inspection, and quality assurance in the performance of the work. In other words, an expression in broad terms of what is and what is not included in the job.

Definitions

It is necessary to define certain terms to preclude misinterpretation:

Owner	Quality Assurance
Engineer or Architect	Quality Control
Contractor	Items
Work	Services
Equal	Contract Documents

Reference codes and standards

It is important, particularly with regard to nuclear and other highly sophisticated projects, to make reference to all applicable codes and regulations, such as ASTM, ANSI and SSPC (Steel Structures Painting Council). By reference to these codes and regulations, the specification writer's work is lessened; but at the same time, of course, the reference standards assure proper and complete technical coverage of the subject.

Storage and mixing of materials

This spells out how, and under what conditions, materials are to be stored for protection from the elements, as well as for optimum preparation and safety. The use of solvents and thinners must be strictly regulated—hence it is important to specify precisely what will or will not be acceptable or tolerated. More importantly, the specification should describe precisely the action to be taken should any materials be mishandled or contaminated.

Coating formulation

This is a precise description of the type of material, wet or dry (or both) film thicknesses, sequence of application, number of coats, etc. There may be only one or there may be many systems on any given project. Should there ever be a need to add to, delete from, or change a system, it is obvious that the effort involved needs to be minimized. The specification sets forth the manner in which alternative materials and methods may be offered for consideration.

Surface preparation

No one will deny that surface preparation is the single most important item for a successful coating system. Here, we address the methods, materials, equipment, housing, protection and storage of items. If grit or sand blasting is to be used, for example, the material, air pressure and equipment must be specified. Results to be obtained (e.g.,

Originally published January 14, 1980

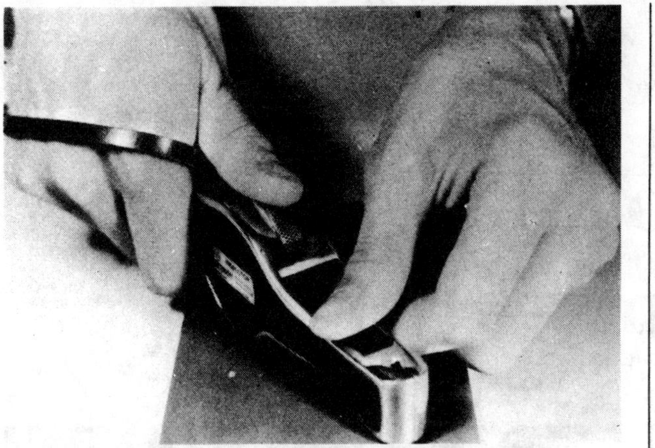

Thickness measurement with Mikrotest

minimum profiles) must be set forth. If other surface contaminants are to be removed, the method by which this is to be done must be spelled out.

Application procedures

In this section the specifier must spell out: (1) The system or systems that will effectively achieve the finish desired, whether it be by spray, brush, roller, squeegee, trowel, or some other method; (2) The mechanics—e.g., if by spray: air pressure, nozzle types, thinner, pattern, distance and so on; the coatings manufacturer should be consulted on this; (3) Periodic site visits by the coating manufacturer; (4) Qualifications of personnel; (5) Protection against drippings on floors, walls, personnel and equipment by the use of tarpaulins, dropcloths, masking or other protective covering.

Inspection of work

■ Each phase of the coating application should be inspected by personnel familiar with coating materials, application, and surface preparation techniques.

■ The inspection should conform to the applicable codes and regulations such as, for example, ANSI N15.12-1974, Sect. 10.

■ The coating inspector's qualifications should include prior training and experience for similar work of comparable scope. This inspector should receive training either onsite or at a school provided by a coatings supplier.

■ The coating inspector should be responsible for control of: the quality of air from the sandblast nozzle and compressor during surface preparation; surface profile or anchor pattern cleanliness; quality of sand or grit used; dry film thickness (DFT) of primer and top coats; the primed surface; epoxy top coats; the proper film thickness (this must be established for no less than one-tenth of the total surface).

Measurements of film thickness should be made at the rate of two per square foot on 10% of the total surface taken over a uniform surface area. For example, a 250-ft^2 surface-area plate will require 50 film-thickness measurements. The Inspector Gage or the Mikrotest (see fig.) are two suitable instruments. For fast readings on large surfaces, the Dermitron may be used.

After an epoxy top coat has been applied and allowed to cure 30 days, the inspector should run Elcometer adhesion tests. Such tests would normally be conducted on special surfaces or panels representative of the applied coatings and applied with the materials at the jobsite. The coating system should meet 200-psi adhesion tests of epoxy topcoat over inorganic zinc or bare substrate. The DFT of the topcoats should be specified, with a tolerance of -1 to $+3$ mil; also the total DFT (-1.5 to $+5$ mil).

■ The manufacturer of the coating should be accorded access to the work for the purpose of advising the owner, contractor, inspector, and quality-assurance engineer or quality-control manager on all aspects of the work.

■ The contractor and the quality-control manager for field work, or the contractor and his quality-control representative for shop work, should check the DFT at least twice a workshift for each application, to ensure that the coating is applied in accordance with specifications.

■ The accepted standard for determining film thickness for magnetic substrates should be the DFT measurement. The wet-film-thickness measurement should be used as a guide during application.

■ Procedures for handling uncorrected deviations from the specification should be detailed.

■ Major runs or sags are not acceptable; if any occur, they should be removed down to the bare substrate, and the area should be recoated.

■ Skips and damaged areas are not acceptable and should be corrected in the field.

■ A group of observers should periodically remove the coatings from the selected test areas, down to the base substrate (primer), to check coating-system thickness, coating adhesion to the substrate using an Elcometer adhesion tester, and condition of the substrate. This group may include the owner, the owner's or engineer's quality-assurance representative for *shop work*, and the quality-control manager for *field work*.

Each test area should be a patch no greater than 2 in^2 (12.9 cm^2). These tests should be used sparingly to demonstrate adequacy of the coating system.

If the results of a number of these tests indicate unsuitable substrate preparation, or lack of substrate or coating-system adhesion, the owner or the engineer may then extend the test areas as required.

Quality-assurance program

ANSI N101.4-1972 provides a list (and samples) of forms to control the entire coating process. On major tasks, the number and type of forms to be used should be determined by the owner, engineer, contractor, and quality-assurance representative. A quality-assurance program, describing the contractor's method of implementing the requirements of the specification, should be submitted for approval of the engineer and owner.

The authors

Dean M. Berger is a coating specialist in the chemical engineering dept. of Gilbert/Commonwealth, P. O. Box 1498, Reading, PA 19603, telephone 215-775-2600. He received his B.S. in chemistry from North Central College, and has done graduate work at the University of Wisconsin. His experience includes 20 years in research and development with PPG Industries and work with Union Carbide Corp. He has devoted many years to the Steel Structures Painting Council, National Assn. of Corrosion Engineers, American Soc. for Testing and Materials, and Federation of Societies for Coating Technology.

Frank Border is a project architect with Gilbert/Commonwealth. He attended Wyomissing Polytechnic Institute, Wyomissing, Pa., and is a registered architect in Pennsylvania. He is a member of the Construction Specifications Institute.

What you need to know about maintenance paints—Part I

How do you choose among the numerous kinds of paints and coatings? Here is a guide to proper selection.

Gary N. Kirby, Ciba-Geigy Corp.

☐ In 1980, the annual cost of corrosion in the U.S. was estimated at $70 billion [*1*]. The use of paints and industrial coatings is the most common way to fight this problem. Proper selection of such coatings can save maintenance money by extending repainting times. Industrial coatings can provide superior resistance to chemicals, water, heat and abrasion.

However, it is seldom simple to specify these coatings, mostly because there are so many of them. They range from simple oil-base paints through alkyds, to vinyls, chlorinated rubbers, epoxies, urethanes, and more. There are also variations of these basic types, and mixtures of these variations. The result is industrial-paint catalogs containing, for example, 70, 90, or even 160 coatings.

Here, we will present a brief guide that simplifies the often-confusing process of writing maintenance-coating specifications for protecting steel structures or equipment. We will assume that reliable paint suppliers have already been located. We will also assume that their prices are acceptable, their deliveries are fast and reliable, and their technical service is responsive and sound.

Performance characteristics

Table I lists the typical coatings used in chemical plants, and shows how they resist acids, alkalies, salts, organic solvents, water, oxidation, weather and abrasion [*2–4*].

Table II is a checklist that gives properties for the common maintenance coatings. Most paint specifications involve a primer plus at least one top coat.

Primers

Primers provide adhesion to the metal plus corrosion control to protect it. There are three main types of primers [*5*]:

1. Barrier primers
2. Inhibitive primers
3. Zinc-rich primers

Steel-supported skylight. Repaint steel when rust begins to break through old paint

1. *Barrier* primers are impermeable films, such as chlorinated rubber or aluminum-flake-loaded catalyzed epoxies. These primers reduce the access of water, chlorides or sulfates to the steel (or other metal) surface. They impede corrosion by artificially increasing the electrical resistance of the corrosion cell. Barrier primers are favored in environments where there is continuous exposure to corrosive electrolytes, such as water storage tanks or buried pipelines [*5*].

2. *Inhibitive* primers are commonly composed of oil-base, alkyd, or phenolic-alkyd vehicles that contain small additions of soluble inhibitors, such as chromates or molybdates. These inhibitor chemicals dissolve as moisture diffuses through the paint system to the steel surface. The inhibitors then retard the electrochemical corrosion reactions at the microscopic anodic or cathodic areas on the steel surface [*6,7*].

These inhibited anodes and cathodes, therefore, re-

Originally published July 26, 1982

The resistance of common maintenance coatings to chemicals, weather, abrasion and oxidation **Table I**

	Acids	Alkalies	Salts	Solvents	Water	Weather	Oxidation	Abrasion
Oil-base	1	1	6	2	7	10	1	4
Alkyd	6	6	8	4	8	10	3	6
Chlorinated rubber	10	10	10	4	10	8	6	6
Coal-tar epoxy	8	8	10	7	10	4	5	4
Catalyzed epoxy	9	10	10	9	10	8	6	6
Silicone aluminum	4	3	6	2	8	9	4	4
Vinyl	10	10	10	5	10	10	10	7
Urethane	9	10	10	9	10	8	9	10
Zinc (inorganic)	1	1	5	10	5	10	10	10

A value of 10 represents the best protection.

sult in lower corrosion currents, and thus lower corrosion rates for the steel. The inhibitive primers are most valuable where continuous access of corrosive electrolytes (such as chlorides or sulfates) is unlikely [*5*].

3. *Zinc-rich* primers are highly loaded with metallic zinc, e.g., 86% by weight. The zinc sacrificially corrodes to protect the steel, which is the noncorroding cathode. This type of primer protects against corrosive undercutting of the paint system and, in addition, offers good abrasion resistance.

There are organic (e.g., epoxy) and inorganic (e.g., ethyl silicate) zinc-rich primers. Organic primers can tolerate slightly poorer surface preparation and are easier to apply than the inorganic ones. Organic primers are also slightly more chemical-resistant. However, dry-heat resistance for organic primers is limited to 250°-300°F (121° to 149°C), compared with up to 1,000°F (537°C) for the inorganic formulations.

Zinc-rich primers are generally applied by a conventional sprayer. Airless spray can be used, but tips or packing will wear excessively. Film builds must be kept low (usually 2.0 to 5.0 mils) to avoid "mudcracking" in angles, webs, etc. There are brands of modified inorganic-zinc primers that allow up to 12 mils dry-film thickness, and are rollable and brushable. The author is currently testing one such primer outdoors. It is rollable, apparently, because the zinc powder is extremely fine. For large jobs, it would still be sprayed.

Inorganic zinc-rich primers have better organic-solvent resistance than does any organic coating. However, some are very porous and tend to create bubbles when topcoated. A thick "mist coat" of the topcoat is therefore helpful before application of the final full top coat. Inorganic primers will not cure at relative humidities below about 40%.

In outdoor exposures or in other environments at an essentially neutral pH, the inorganic zinc primers can be left untop-coated (ignoring aesthetics). However, if the pH of the environment is below 5 or above 10, untop-coated zinc-rich primers will corrode quickly. Zinc is an active amphoteric metal, and will react with either strong acids or strong bases. For such corrosive environments, top-coat with epoxy, chlorinated-rubber, vinyl or urethane paints. Alkyd top coats saponify with zinc, and so are not recommended.

Zinc-rich primers are favored for shop-priming steel that is slated for severely corrosive environments, such as continuous or intermittent immersion in seawater or for long life outdoors. The zinc-rich primer has been called the "ultimate contribution of the paint technologist to the war on corrosion" [*5*].

Top coats

The role of the top coat(s) is to:

1. Protect the primer (as it protects the steel).
2. Provide the protection listed in Table I.
3. Provide aesthetics, including long-lasting color and gloss, preferably without excessive chalking, fading or yellowing.

Enough paint must be applied—for corrosive environments, this is usually a minimum of three coats with a total film thickness of at least 5 mils. In one comparison [*5*], oil-based coatings 4 to 5 mils thick gave 8 to 10 years' protection in rural Fargo, North Dakota; at least a 9-mil dry-film thickness was required to give the same protection in oceanside Kure Beach, North Carolina. In the moderately severe environment of the Kure Beach ocean lot, the minimum paint-film thickness for long-term protection was 9.7 mils for oil-base paints, 5 mils for alkyds, 4.5 mils for epoxies, 4.2 mils for chlorinated rubber, and 3.3 mils for vinyls [*8*].

High-build coatings provide an economical means of adding more mils of coating per application, but are not as protective as regular-build coatings applied in several coats to give an equivalent film thickness.

Keep in mind the strong points of the various coatings. For resistance to a strong organic solvent, specify an inorganic zinc primer, without a top coat (if the solvent doesn't hydrolyze to create acidic conditions) [*6,7*]. For lighter organic-solvent exposures, specify a catalyzed-epoxy system or a urethane top coat.

For temperature resistance, consider a catalyzed epoxy (300°F), silicone alkyd (500°F), silicone aluminum (1,000°F), or inorganic zinc primer (1,000°F). Some silicone-aluminum coatings require a minimum 500°F service temperature to cure properly.

For broad chemical resistance, specify an epoxy amine, a ketimine epoxy, or a two-component urethane.

Always look at the recoatability of the finish coat. Will it require roughening or a solvent wipe when it is eventually recoated? Will the maintenance coating itself become a maintenance problem?

Here are important points to know about the various coatings you can select **Table II**

Coating	Advantages	Remember
Oil-base	Good rust penetration. Good substrate wetting.	Poor acid, alkali, solvent and oxidation resistance. Slow-drying, embrittles and yellows with aging.
Alkyd		
Short-oil	Inexpensive, fast-drying, good adhesion, easy recoating.	Do not use over zinc-rich primers. Fair acid, alkali, solvent and oxidation resistance. Fair impact resistance.
Long-oil	Good durablity, good weathering, flexible, easy recoating.	Poor acid, alkali, solvent and oxidation resistance.
Chlorinated rubber	Good water, acid, alkali resistance. Easy recoating.	Poor aromatic solvent resistance; "strings" when brushed. Moderate temperature resistance.
Epoxy	Versatile.	May surface-chalk in sunlight. Use light colors outdoors.
Coal-tar	Excellent resistance to water, seawater and soil immersion.	Check touch-up and recoating limits. Check flexibility and abrasion resistance. Dark colors only.
Amine	Good water, acid, alkali, solvent resistance. Hard, good temperature resistance.	Skin sensitizer. Check recoating limits.
Polyamide	Good water, alkali resistance. Tough, good temperature resistance.	Check recoating limits.
Ketimine	Indefinite recoating time. Wide range of chemical and solvent resistance.	Ketone fumes while drying.
Urethane (2-component, aliphatic)	Good water, acid, alkali, solvent resistance. Good abrasion resistance. Good gloss and color retention. Low-temperature application.	Check temperature resistance. Check solvent fumes. Expensive. Check recoatability.
Vinyl	Good water, acid, alkali resistance. Tough, good recoatability, rapid drying. Low-temperature (5°F) application.	Low temperature resistance (150°F). Poor aromatic ketone solvent-resistance. Low percent solids. Needs good surface. Mostly sprayed. Ketone fumes.
Zinc-rich primers (untop-coated)		
Inorganic	The best organic-solvent resistance. Good abrasion resistance. Temperature resistance to 1,000°F.	Requires good surface preparation. Limited (untop-coated), to pH of 5 to 10. Generally sprayed (conventional). Film thickness limited.

The factors to consider in specifying a maintenance coating are:

- Structure or equipment to be painted.
- Location of the work.
- Service environment.
- Level of protection desired.
- Costs.
- Workforce.
- Writing a specification.

The first two of these factors will be discussed here. Part II will cover the last five.

Structure/equipment

Is the steel already painted? Always remove loose paint. If you are painting over an oil-based or alkyd paint, remember that the stronger organic solvents in industrial coatings—solvents like xylene or methyl isobutyl ketone—will readily lift these paints. Apply a "universal" primer, such as a phenolic alkyd, as a tie coat first, then apply the strong-solvent coating. Or try a water-based epoxy.

Caution: Whenever specifying a primer, check the minimum and maximum recoat times. A slow-drying primer may hold up the job. A fast-drying primer may not penetrate rust for adhesion, or it may not be recoatable if not top-coated within a few days.

Has the steel started to rust? The best surface preparation for rusting steel is sandblasting, at least to a commercial-blast finish (Steel Structures Painting Council SP-6) [*2*]. In one outdoor test in Sheffield, England, the durability of two paint systems was checked comparing weathered and wire-brushed steel with sandblasted steel for each [*9*]. After painting, the sandblasted steel gave 4.5 and 5.3 times as long service as the wire-brushed steel.

The resulting blast profile should be about one-third the coatings' expected dry-film thickness (DFT) [*10*]. The blasted steel has to be primed on the day of blasting, before rusting resumes.

The most important single factor influencing the life of a paint is the proper preparation of the metal surface. An average paint system on a properly prepared metal surface usually outperforms a better paint system on a poorly prepared surface. Up to 70% of coating failures have been blamed on poor surface preparation [*11*].

However, the location of the work can rule out sandblasting (see below). If blasting is not feasible, specify power-tool cleaning (SSPC-SP-3) or hand cleaning (SSPC-SP-2). The goal is to remove all loose rust, paint, dirt, grease, chemicals and other contaminants. Grease can be removed by solvent cleaning (SSPC-SP-1). Acid can be neutralized by applying a soda ash solution, followed by a water rinse.

If paint must be applied over adherent rust, then use a primer that penetrates the rust. There are several approaches. Fish-oil or linseed-oil primers penetrate rust very well, but few can withstand the strong solvents found in organic coatings without being lifted.

One paint company recommends a 30-day drying time for its fish-oil primer before covering with an epoxy. This is hardly practical, if only because of the contamination that would coat the primer during this time. There is one fish-oil primer that reportedly will accommodate a chlorinated rubber top coat after drying overnight.

Another option is to prime with an oil-base primer, and then cover it with a universal primer that is compatible with the industrial top coat. Finally, there are various aluminum-epoxy universal-maintenance primers that are rust-penetrating.

When should you paint? The minimum surface preparation is required when you paint just as the existing paint is starting to fail. For a horizontal I-beam, the moisture or condensate in a humid environment collects along the edges of the flanges. When the liquid evaporates, it leaves a dark ridge of dust or dirt along the edges. These edges are relatively sharp and have a thinner paint film than do the flat areas of the beam. As the moisture penetrates with time, these edges will start to blossom with small patches of rust.

The optimum time to paint is when these patches multiply and start to cover the flange (see photo). If you wait until the rust grows heavy and the flange surfaces roughen, about three times as much hand cleaning is required, and this will cost about three times as much. Waiting beyond this point will allow the steel to pit or form rust pustules that can be very difficult to hand-clean. Sandblasting may then be required to clean the steel adequately.

Location

Outdoor painting offers the advantage of better ventilation; spray painting, even with strong solvents, is easier. This makes it easier to use zinc-rich primers or vinyl paints, both of which are mostly sprayed. For heavy coatings, air (conventional) spray can apply paint at four times the rate of brushing [*3*]. Of course, the wind may blow paint-spray onto other areas, including the plant parking lot.

Outdoor painting can be limited by the temperature. Many paints cannot cure in a reasonable time below 50°F, although some urethanes do cure adequately at lower temperatures. At the other extreme, many paints start to thicken and become hard to brush or roll above 90°F. Thinning helps in such cases.

Excessive humidity or moisture can also cause difficulties. For example, paint on one water tower peeled off in sheets because it was applied during a humid day when moisture was running down the tower.

Oil-base, alkyd, vinyl or urethane top coats are favored for outdoor service due to their excellent weathering characteristics.

Indoor painting involves different concerns. Painters have to have adequate ventilation. They cannot be expected to paint with strong solvents all day up among pipe racks, for example. Sandblasting is not advisable if there is nearby glass or plastic piping, exposed electrical equipment, or open product-containers. Containing the grit by hanging sheet plastic only works if it is done carefully. Otherwise, the grit will escape through the overlapped areas, even contaminating adjacent freshly primed surfaces. There are portable vacuum-grit blasting units that contain and confine their aluminum oxide grit, but some work slowly and are mostly suitable for painting floor-level equipment of limited size.

Spray-painting indoors can also cause problems through overspray. One inexperienced sales representative ran a plant spray-painting demonstration in a production bay and promptly supervised the spraying of his own jacket, which he had carefully left folded in an adjacent bay.

Indoor areas can also be too moist to paint. The paint on the structural steel in one production bay failed after only one year because it had been applied under highly humid, moist conditions.

References

1. Payer, J. H., et al., NBS-Battelle Cost of Corrosion Study ($70 Billion!), *Mater. Perform.*, Sept. 1980, p. 51.
2. Weismantel, G. E., Paints and Coatings for CPI Plants and Equipment, *Chem. Eng.*, April 20, 1981, p. 130.
3. Helms, F. P., Paints and Corrosion Control, "Process Industries Corrosion," Natl. Assn. of Corrosion Engineers, Houston, 1975, p. 346.
4. Uhlig, H. H., Organic Coatings, "Corrosion and Corrosion Control," 2nd ed., John Wiley & Sons, New York, 1971, Chap. 15, p. 245.
5. Hare, C. H., Anti-Corrosive Barrier and Inhibitive Primers, "Federation Series on Coatings Technology," Unit 27, Federation of Societies for Coatings Technology, Philadelphia, 1979.
6. Kirby, G. N., Corrosion of Carbon Steel in Water (How to Select Materials), *Chem. Eng.*, Nov. 3, 1980, p. 98.
7. Kirby, G. N., Corrosion Performance of Carbon Steel, *Chem. Eng.*, March 12, 1979, p. 72.
8. Tator, K. B., How Coatings Protect and Why They Fail, *Mater. Perform.*, Nov. 1978, p. 41.
9. Uhlig, H. H., Organic Coatings, "Corrosion and Corrosion Control," 2nd ed., John Wiley & Sons, New York, 1971, Chap. 15, p. 251.
10. Weaver, P. E., Surface Preparation, "Industrial Maintenance Painting," 4th ed., Natl. Assn. of Corrosion Engineers, Houston, 1973, p. 24.
11. Causes and Prevention of Coatings Failures, NACE Publication 6D170, Natl. Assn of Corrosion Engineers, Houston.

The author

Gary N. Kirby is materials and metallurgical engineer for the Dyestuffs and Chemicals Div. of Ciba-Geigy Corp. at its Toms River Plant, P.O. Box 71, Toms River, N.J. 08753; telephone: 201-349-5200. He consults regularly for several other Ciba-Geigy plantsites. Previously, Kirby was with Crawford and Russell, Inc. in Stamford, Conn. He is a member of the Natl. Assn. of Corrosion Engineers, the Amer. Soc. for Metals, and the Amer. Welding Soc. His Ph.D. in metallurgical engineering is from the University of Michigan. He has an M.S. in metallurgical engineering from Lehigh University and a B.S. from Cornell University.

What you need to know about maintenance paints—Part II

Selecting a particular coating depends upon factors including the service environment, the amount of protection desired and the costs of labor and paint.

Gary N. Kirby, Ciba-Geigy Corp.*

☐ So far, we have discussed the types of paints and coatings available, and their performance characteristics. Also, we have covered two concerns to consider in writing a coating specification: the structure or equipment to be painted, and the location of the work. Now, we will look at other considerations.

Service environment

Factors to consider here are corrosives, abrasion and temperature:

1. Corrosives—Table I shows the corrosion rates for steel in some chemical-plant exposures, including water, both cold and hot. These are common, but often-disregarded corrosives. Corrosion by hot water can be particularly severe.

In Table I of Part I , the industrial coatings—chlorinated rubbers, catalyzed epoxies, vinyls and urethanes—are rated at 10 for water exposure. Oil-based paints are rated at 7. This implies a 42% advantage in using the industrial coatings. In services involving high moisture, condensation or humidity, this could mean that one of these coatings might have a recoat time of 3 yr, compared with 2 yr for an oil-based paint. One author [3] recommends heavy-duty coatings such as epoxies, chlorinated rubbers, or vinyls when the corrosion rate of the steel exceeds 9 mils/yr.

Regular maintenance painting will protect steel in corrosive-fume environments. The author has measured the wall thickness of steel conduit situated above filter presses that give out hot, moist acidic fumes. The conduit had been protected by periodic maintenance painting, including using a chlorinated rubber for the past 5 yr. The wall thickness of the conduit was virtually unchanged after about 20-yr service.

The only effect of the most severe acid fumes was to produce a layer of tightly adherent oxide just under the paint film. The amount of protection provided by the paint is significant, because steel can corrode at up to 0.3 in./yr in hot water at 120°F (49°C). This condition is present as the condensate drops on the steel.

Common water systems can be highly corrosive to structural steel in the plant — Table I

Corrosive	Corrosion of steel: Rate, in./yr	Corrosion of steel: Years to corrode 1/8 in.
Plant atmosphere	0.0005–0.002	62.5–250
Cold water (22°C) pH = 7	0.010–0.012	10.4–12.5
Hot, corrosive tapwater (49°C)	0.3	0.4
Hot steam condensate (60°C) with CO_2/O_2 at pH 5 to 7	0.030–0.050	2.5–4.2
Acidic water (HCl) (22°C) pH = 2.67	0.076	1.6

Source: Refs. [*1*] and [*2*].

The choice between simple oil-base paints and industrial coatings becomes even more clear-cut when acid and alkali exposures are considered. Oil-base paints have very poor acid and alkali resistance. Alkyds have a 6 rating, which is better, but chlorinated rubbers, epoxies, vinyls and urethanes have 9-to-10 ratings.

Resistance to organic solvents varies quite a bit.

Dry-heat resistance of maintenance coatings — Table II

Coating	Service temperatures °F	Service temperatures °C
Chlorinated rubber	140–160	60–70
Vinyl	140–160	60–70
Oil-base	150–175	65–80
Alkyd	250	120
Coal-tar epoxy	160–250	70–120
Urethane	180–250	80–120
Catalyzed epoxy	250–300	120–150
Silicone alkyd	400–500	205–260
Zinc (inorganic)	750–1,000	400–538
Silicone-aluminum	1,000	538

*To meet the author, see p. 91.

Originally published August 23, 1982

Estimating costs for surface preparation and paint application for various conditions — Table III

Surface preparation

Sandblasting	Blasting rate, 100 psi 5/16-in. nozzle		Cost, $/ft^2
SSPC	Ideal	Condition 6	
SP-5 White metal	100 ft^2/h	83 ft^2/h	0.75–1.50
SP-6 Commercial blast	150	125	0.50–1.00

Hand cleaning — Rate of cleaning, ft^2/h or ft^2/8h

Condition of steel	Structural steel, h	Structural steel, 8-h	Tanks, h	Tanks, 8-h	Cost, $/ft^2
1	150	1,200	187.5	1,500	–
2	75	600	125	1,000	0.25–0.50
3	50	400	62.5	500	0.75
4	125	1,000	500	4,000	Do not paint
5	87.5	700	137.5	1,100	0.25
6	25	200	62.5	500	0.50–0.75
7	12.5	100	31.25	250	1.00

Note: See text for definition of conditions.

Application — Application rate, ft^2/coat/man-h

	Flat surfaces	Medium structural steel (250 ft^2/ton): Oil-base	Chlorinated rubber	Epoxy	Vinyl
Spray	300–600	210–240	180–210	160–195	145–180
Roller	200–400	150–180	125–150	100–125	100–125
Brush	100–200	70–80	65–70	60–65	60–65

Source: Some information is from Ref. [1].

Chlorinated rubbers and vinyls have poor solvent resistance, just a little better than alkyds or oil-base paints. The vinyls and chlorinated rubbers are thermoplastics, and are deposited by organic-solvent evaporation. They readily redissolve in the same solvents. This does have the advantage of making them easily top-coated later as the solvent in the new paint softens the old paint film. Catalyzed epoxies and urethanes polymerize by chemical reaction and have superior solvent resistance.

Oxidation resistance is important for exposure to such chemicals as chlorine, sodium hypochlorite and nitric acid. The vinyls rank highly here. One chlor-alkali plant uses them for maintenance painting because of such resistance. Urethanes and acrylics also have high ratings for oxidation resistance.

2. *Abrasion*—Abrasion resistance is important for applications such as stair treads, floors and steel decks. Urethanes are an obvious choice here, but epoxies also are used.

The author has successfully used ketimine-cured epoxies on kettle tops subjected to abrasion by workers' feet. A ketimine curing agent is made by reacting a primary amine with a ketone. After the film is applied, the moisture in the air reacts with the ketimine to regenerate the original ketone and amine. The amine then cures the resin.

3. *Temperature*—Table II shows the dry-temperature limitations of the common coatings. For high-temperature applications, zinc-rich primers and silicones are widely used. In applications such as chimney stacks, any corrosion of the outside steel occurs only during downtime. During high-temperature operation, there is no condensed water on the stack to act as a solvent for electrochemical corrosion.

Level of protection desired

Once you have considered the structure or equipment to be painted, plus its location and service environment, you must look at how much protection you want. The question comes down to: How much money is available? How soon do you want to repaint? How long do you want the structure or equipment to last?

One job concerned the structural steel supporting an outdoor wet scrubber that emits an occasional localized acidic spray. Selecting hand-cleaning plus an alkyd paint job might have necessitated repainting within a year or two. Instead, sandblasting plus two epoxy coats and a urethane top coat were specified. The steel looks fine after 2 yr, and will possibly last another 5 yr.

Costs

The bottom-line cost for the user is the cost per square foot per year. Table III shows the major cost components—surface preparation, paint application, and typical costs for the paint itself.

The material cost is not just the cost per gallon for the paint. The working paint film is formed from the solid component of the paint, as described by the percent solids. The overall cost per gallon also includes the cost for the solvent, which essentially evaporates. The paint cost can be calculated from the percent solids by means of a nomograph in Ref. [4]. Space does not permit us to show the figure here.

Table III does not include application costs, but gives paint-application rates. From these and other figures—percent solids, paint cost in $/gal, etc.—the paint and application costs can be found.

Surface preparation costs depend upon the condition of the steel. In Table III, there are seven conditions:

New construction:

1. Tight mill scale with little or no rust.
2. Loose mill scale and fine, amorphous rust, no pits.
3. Little mill scale, overall rust, some pitting.

Maintenance:

4. Thin finish coat, some primer showing, negligible rust.
5. Thin finish coat, primer showing, 10% of surface with loose scale and loose paint.
6. Finish coat thoroughly weathered, and badly blistered; approximately 30% of surface with pitting and hard scale.
7. Badly pitted with rust nodules.

Workforce

You must consider your workers' welfare, their skills, and their attitude towards using an industrial coating.

The workers' skills determine what application methods you can use. Spraying a zinc-rich primer requires care. Too heavy a coating (usually over about 5 mils) can cause "mud cracking." And remember that zinc-rich coatings are *generally* sprayed, in order the keep the zinc properly dispersed and to control the final film thickness.

Sample specification

A. *Surface preparation*

Sandblast to a commercial blast finish, according to the Steel Structures Painting Council Specification SSPC-SP-6, Commercial Blast Cleaning (add recommended blast profile, if given).

B. *Primer*

Within 4 h after blasting, brush or spray (airless preferred) on 3 to 5 mils dry-film thickness of:

Brand A "Product Name" aluminum polyamide epoxy primer

The dry-film thickness to be measured by the:

Brand B dry-film thickness gage [*2*]

Allow to dry at least 16 h, but not more than 5 d.

C. *Intermediate coat*

Airless spray (only) 2 to 3 mils dry-film thickness of:

Brand A "Product Name" vinyl coating, gray color

The dry-film thickness to be measured by the:

Brand B dry-film thickness gage. Allow to dry 2 h.

D. *Finish coat*

Airless spray (only) 2 to 3 mils dry-film thickness of:

Brand A "Product Name" vinyl coating, white

The dry-film thickness to be measured by the:

Brand B dry-film thickness gage

E. *Application*

1. Do not apply paint at temperatures under 50°F.
2. All surfaces to be painted shall be dry and free of dirt, rust, grease, and other contaminants.

F. *Manufacturer's directions*

All mixing, thinning and application shall be in accordance with the manufacturer's recommendations.

G. *Safety precautions*

1. Follow manufacturer's safety recommendations.
2. Keep all flammable materials away from heat, sparks and open flames.
3. Keep containers tightly closed and upright.

H. *Cleanup*

After completion of the work, remove all painting materials, tools, scaffolds and ladders.

Attitude is also important. Some workers prefer certain coatings over others. For example, many chlorinated rubber coatings "string" or "cob-web" heavily when they are brushed, adding greatly to worker fatigue. Thinning the paint improves brushability, but it also decreases the dry-film thickness.

One answer is to work with the painters. For trials of a new paint, apply some of it yourself. When one coating containing especially strong solvents was tried in an indoor test evaluation, both the author and the painter suffered respiratory problems and a sore throat that evening. Always inquire about brushability, coverage and paint fumes when evaluating a new coating.

The coatings manufacturer's representative can help here by suggesting easily-applied coatings and offering application tips.

The workers' welfare is connected with the health concerns attached to some of the materials in some of the coatings. Ask the manufacturer for a material-safety data sheet for each coating. The publication *Threshold Limit Values for Chemical Substances in Workroom Air Adopted by ACGIH for 1981* [*5*], from the American Conference of Governmental Industrial Hygienists (Cincinnati, Ohio), is helpful in evaluating the relative health concerns of various ingredients. Exposure limitations can be dealt with by providing adequate ventilation and by issuing the proper protective equipment for appropriate situations. This includes rubber gloves, cartridge breathing-masks, and industrial paper-suits. In tight enclosures, outside air can be supplied.

Another approach is to try new coatings that have lowered solvent concentrations. The paint manufacturer's concern here is not only with people, but also with economics. Some solvents are expensive, and their principal function is to evaporate. One such product is a high-solids epoxy (e.g., 90% solids) that contains much less solvent than do typical formulations. There are also water-base industrial coatings, including two-component epoxies and single-component urethanes. The water-base coatings the author has tried still contain some organic solvents and still emit some solvent odors.

A third approach is to find alternative coatings to do the job. One heavy black coating gave off particularly noxious and possibly carcinogenic fumes. It also had a very short recoating time—once the coating cured it could not even be touched up without a solvent wipe. An alternative catalyzed-epoxy coating was found that had less-objectionable fumes, plus the advantage of offering an indefinite recoat time.

As a further safety concern, keep solvent-containing materials away from heat, sparks and open flames. Keep containers tightly closed and upright to prevent leakage when not in use. Refer to application instructions for each material for further safety precautions.

Writing a specification

The simplest effective specification covers:

1. Surface preparation.
2. Method of application.
3. Dry-film thickness, and measuring methods.
4. The coating(s), including drying times.
5. Crucial application limitations.
6. Strict observance of the maker's directions.
7. Safety precautions.
8. Cleanup.

Acknowledgements

The author wishes to express his appreciation to F. A. Smiles of Con-Lux Coatings for his suggestions. Thanks also to Gail P. Jenkins for manuscript preparation.

References

1. Kirby, G. N., Corrosion of Carbon Steel in Water (How to Select Materials), *Chem. Eng.*, Nov. 3, 1980, p. 98.
2. Kirby, G. N., Corrosion Performance of Carbon Steel, *Chem. Eng.*, March 12, 1979, p. 72.
3. Weaver, P. E., Surface Preparation, "Industrial Maintenance Painting." 4th ed., Natl. Assn. of Corrosion Engineers, Houston, 1973, p. 24.
4. International Paint Co., Coatings for Industrial and Offshore Structures, brochure, Union, N.J.
5. "Threshold Limit Values for Chemical Substances in Workroom Air Adopted by ACGIH for 1981," American Conference of Governmental Industrial Hygienists, Cincinnati, 1981.

Specifying zinc-rich primers

These coatings protect steel surfaces from corrosion. Several types are available for different applications.

Dean M. Berger, Gilbert/Commonwealth

☐ Zinc-rich primers (ZRPs) are highly effective corrosion-resistant coatings for protecting steel surfaces. They are used to protect a broad range of products, ranging from small, mass-produced parts to large steel structures. Fossil-fuel and nuclear power plants, chemical processing plants, paper mills, transmission towers, offshore platforms, ships, bridges and many other large structures are coated with ZRPs. This discussion will concern large structures.

Selecting the primer

The engineer's choice of any given coating system is based on performance, appearance, cost, ease of repair, and ultimate satisfaction to the owner. Coatings are usually selected on the basis of past performance; however, properties of each generic system are evaluated with reference to the needs of the environment.

Physical properties such as adhesion, resistance to abrasion, chemicals, salt-fog, humidity, fire and water-immersion, and flexibility, gloss retention and durability are important in the selection of any system. The structural design of the object to be coated often affects the choice of any generic coating system. For example, the locks and dams in the upper Mississippi River can only be painted during winter when the gates may be removed. In this case, a vinyl-based ZRP is used, because it can be applied in cold weather, and it performs well immersed in water.

Picking primers for the inside of tanks is more complicated. Great care must be taken to use materials with a proven record. New materials must be carefully screened prior to acceptance. Specifications written for these systems usually require strict adherence to manufacturers' recommendations.

Zinc-rich primers have been successfully used to line tanks that hold chemicals, including organics such as styrene and xylene. In such cases, inorganic ZRPs are used because of their resistance to the media in the tank.

Writing the specifications

In preparing coating specifications for large steel structures, chemical plants, power plants and similar projects, it is important to detail the many related factors that the coating contractor will have to take into account. The specification should include [*1*];

1. Scope of work.
2. List of definitions.
3. List of applicable codes and standards.
4. General requirements relating to storage and mixing of materials and use of solvents and thinners.
5. List of coating systems and any special requirements for each.
6. List of acceptable coating manufacturers.
7. Special requirements for surface preparation.
8. Application procedures.
9. Inspection-of-work procedures.
10. Quality-assurance program.

Zinc-rich primers are classified according to the Steel Structures Painting Council SSPC 20X80P as inorganic or organic. They must meet specific test criteria and have a certain zinc content. Therefore, if one purchases a ZRP that meets this specification, then one can expect a high level of performance—provided the primer is properly applied. Now, the types of vehicles mentioned by the above specification will be discussed. Properties for the Council's SSPC 20X80P specification are found in Table I.

Types

Type I-A—Inorganic post-curing vehicles. These are water-soluble, and include alkali-metal silicates, phosphates and modifications of them. These must be cured subsequently by heat or the application of a solution of a curing compound.

Type I-B—Inorganic self-curing vehicles. These are water-reducible, and include water-soluble alkali-metal silicates, quaternary ammonium silicates, phosphates

Characteristics of various zinc-rich primers as set by Steel Structures Painting Council — **Table I**

Specification SSPC 20X80P Characteristics	Minimum requirements: Types I-A, I-B, I-C Inorganic	Minimum requirements: Type II Organic
Total solids, % by weight of paint (ASTM D-2369)		70
Pigment, % by weight of total solids (ASTM D-2371)		83
Total zinc dust, % by weight of pigment (ASTM D-521)	87	93
Total zinc dust, % by weight of total solids	74	77
Flash point (Tag closed-cup ASTM D-56)	I-A None I-B None I-C 55°F (12.8°C)	50°F (10°C)

Originally published December 14, 1981

U.S. government specifications serve as a guide in ordering various primers — **Table II**

Date of last spec amendment	Federal specification	Suggested vehicle	Solvent	Number of components	Government agency	End use
June 1977	TTP-641G					
	Type I	Linseed oil	Mineral spirits	2	GSA (General Services Admin.)	Repair galvanized surfaces
	Type II	Alkyd	Mineral spirits	2		
	Type III	Phenolic varnish	Mineral spirits	2		
May 1975	TTP-1046A	Chlorinated rubber	Xylene	1	GSA	Primer for steel and galvanized surfaces
June 1977	TTP-1561A	Petroleum/hydrocarbon or butadiene	Ethyl benzene Mineral spirits	1	GSA	Chain-link fence repair (aluminum modified)
Nov. 1979	CW.099440 VZ108d	Vinyl	Ketones	3	U.S. Army Corps Eng.	Primer--locks and dams
Nov. 1979	CW.099440 E303b	Epoxy-polyamide	Alcohols	3	U.S. Army Corps Eng.	Primer--steel, potable-water systems
May 1961	MIL-P-15145B	Phenolic varnish	Mineral spirits	2	U.S. Navy	Interior potable-water tanks
Nov. 1977	DOD-P-21035A	Phenoxy	Cellosolve acetate	2	U.S. Navy	Regalvanizing welds in galvanized steel
Nov. 1979	DOD-P-23236A	Na-Li silicate	Water	2	U.S. Navy	QPL,* fuel tanks, misc. storage tanks
	Types I and II	Ethyl silicate	Alcohol	2		
	Class 3	Phosphates	Water	2		
		Silicones	Solvent (xylene)	2		
July 1979	MIL-P-26915B					
	Type I Classes A & B	Phenolic varnish or chlorinated rubber	Xylol	2	U.S. Air Force	Steel ground-support equipment; galvanic protection
	Type II					
	Class A	Phenolic varnish	Xylol	2		Primer for other finish coats
	Class B	Styrene-butadiene copolymer	Xylol	2		
Mar. 1966	MIL-P-38336	Alkyl silicates	Alcohols or glycol ethers	2	U.S. Air Force •	Corrosive environments Weld-through primer
Mar. 1966	MIL-P-46105	Acrylic epoxy ester, urea formaldehyde	Xylol	2	U.S. Army	Automotive truck equipment
Jan. 1969	KSC Spec F0008				NASA	Steel protection
	Type I Class I	Alkyl silicate	Alcohols	2		
	Type I Class II	Na-Li silicate	Water	2		
	Type II Class I	Phenoxy or chlorinated rubber	Ketones, xylol	1		
	Type II Class II	Epoxy-polyamide	Ketones, xylol	2		

*QPL (Qualified Products Lost)

and their modifications. Such coatings cure by crystallization due to evaporation of water.

Type I-C—Inorganic self-curing vehicles. These are solvent-reducible, and include titanates, organic silicates, and polymeric modifications of these silicates. These systems are dependent upon moisture in the atmosphere to complete hydrolysis, thereby forming the polysilicate.

Type II—Organic vehicles. These are solvent-reducible, and include phenoxies, catalyzed epoxies, urethanes, chlorinated rubbers, styrenes, silicones, vinyls and other suitable resinous binders. The organic vehicles covered by this specification may be cured chemically or may dry by solvent evaporation. Under certain conditions heat may be used to facilitate or accelerate hardening.

Note that alkyds generally do not make good zinc-rich vehicles, although they have been used successfully.

This specification also provides many other criteria, the most stringent being passing a 1,000-h salt-fog test without a blemish. There are, however, a number of U.S. government specifications (see Table II). Most of these specifications set rigorous test criteria for specific end-uses. These specifications can be used to order a ZRP for such applications.

Since the main objective in using zinc-rich primers is to protect steel surfaces in a given environment, it is imperative that the coating engineer consider all facets of construction and coating work. Zinc-rich primers are used as:

1. Preconstruction and weld-through primers.
2. Single-coat systems.
3. Double-coat systems.
4. Delayed top-coat systems.
5. Multiple top-coat systems.
6. Repainting systems.

Preconstruction weld-through primers

Most zinc-dust-containing primers used in this manner are applied over near-white-metal blast-cleaned steel surfaces in rather thin coats of approximately 1.0 mil. These primers are used mainly in ship construction and protect the steel while it is being fabricated. Generally, the single-component inorganics are used with or without enhanced pigmentation. Often, the steel will be reblasted and the final selected coating system applied after fabrication. Epoxy-polyamide ZRPs are also used in this manner [*3,6*].

Single-coat systems

Zinc-rich primers will work equally well in protecting galvanized steel. Particularly suited are the post-curing water-base types, the self-curing water-based inorganics, and the alkyl silicate inorganics that have been used as a single coat over blast-cleaned steel. Many case histories have demonstrated the outstanding performance of these primers in marine environments. Generally, they are applied at 2½ to 3 mils in one coat. In cold climates, it is better to recommend the alkyl silicate types. Where warmer temperatures are expected, the water-base types will cure and perform very well.

Organic zinc-rich primers have been used for touch-up and repair of galvanized steel. They require less surface preparation and better tolerate contamination.

Often, storage tanks for chemicals such as styrene are lined with one coat of an inorganic ZRP.

Double-coat systems

The old adage that two coats are better than one has paid off in many areas. Although one coat of 2½ to 3 mils will last many years, it has been found that two coats of 4 to 6 mils will last much longer. Proper application and product selection must be evaluated before work is scheduled.

Delayed top coating

A basic concept in construction is to fabricate and erect with minimum cost and delay. Steel is primed in the shop, sent to the jobsite, erected, and later it is top-coated. Sometimes this might be one or two years later [*4*]. In many cases, 12 to 18 years go by before top-coating is done.

Usually, these surfaces must be spot-blasted, touched up with a selected primer, and covered with two more finish coats. The need for scaffolding and the accessibility of the structure are key factors in the cost of applying the coating. Such costs are so high that many engineers have reverted to painting the entire facility as soon as possible. In fact, in several jobs, the primer and two top coats are applied in the shop. The steel is shipped to the jobsite, erected, and touched up after erection. The control of the coating operation, inspection and total protection is excellent. This can be done where most of the steel is structural steel.

One of the most useful top coats for delayed top-coating is chlorinated rubber. It is chemically resistant, and easy to apply. Where spray application is permitted, vinyl top coats can be used. Vinyl systems are thermoplastic and may readily be re-topcoated years later. Epoxies and polyurethanes have been used. These, however, are difficult to recoat after weathering. Coal-tar epoxies are also used, but they, too, are difficult to recoat, even after one week's aging.

Multiple top coat systems

The broadest use of ZRPs is as the primer for other top coats. Many combinations have been used. A good one is a two-component alkyl silicate zinc-rich primer, followed by a vinyl intermediate and a vinyl acrylic top coat. Other popular top coats are two coats of epoxy-polyamide; one coat of an epoxy-polyamide intermediate followed by one of an aliphatic polyurethane; two coats of chlorinated rubber; two coats of coal-tar epoxy; one coat of wash primer that meets federal specification MIL-P-15328D, followed by two coats of latex, alkyds, or silicone acrylics.

Generally, these systems should be specified at a total dry-film thickness of about 8 to 10 mils. The top coat selected will depend upon the environment [*5*].

Repainting

If a structure has been coated with a primer other than a ZRP, it may be necessary to totally remove the existing paint. In such a case, it is nearly imperative that an organic zinc-rich coating be specified. The reason for this is obvious. It is nearly impossible to remove all the existing coating; crevices, latticework, abatements, joints, corners, and many hard-to-reach areas make the painter's job difficult. Inorganic zinc-rich primers simply will not adhere to previously-painted surfaces. Organic zinc-rich primers will at least adhere to such surfaces. Besides, they are much more forgiving when the steel may be otherwise contaminated.

Of all the organic vehicles, chlorinated rubber is the easiest to work with, since xylol is the only required solvent and the primer can be brush- or spray-applied. Epoxy polyamides are also easy to apply; however, they are multiple-component systems requiring two containers for the vehicle plus an additional one for the zinc dust. They offer the highest degree of performance of all the organic vehicles. Phenoxy has been used as a vehicle for zinc-rich primers. It works well with epoxy or epoxy and urethane finish-coats [*7*].

References

1. Berger, Dean, and Border, Frank, How to Specify Coatings, *Chem. Eng.*, Jan. 1980, p. 123.
2. Steel Structures Painting Council; Zinc Rich Primers Specification 20X80P.
3. Berger, D. M., Zinc Rich Coatings, *Modern Paint & Coatings,* June 1975, p. 19.
4. Berger, D. M., and Montz, E., How to Save $500,000 on Power Plant Coatings, *Electric Light & Power*, Nov. 1975, p. 22.
5. Keane, J. D., Bruno, J. A., and Weaver, R. E. F., Topcoats for Zinc Coatings, Five-Year Report, June 1981, Steel Structures Painting Council.
6. Rudlowsky, G., and Moffatt, W., Improved Fabrication Primer for Protection of Steel, MA-RD 930-77030, U.S. Dept. of Commerce, publication PB 262013, Oct. 1975.
7. Patton, W., and Pinney, S., Performance Characteristics of Zinc Rich Coatings applied to Carbon Steel, NASA publication TND 7336, Jul. 1973.

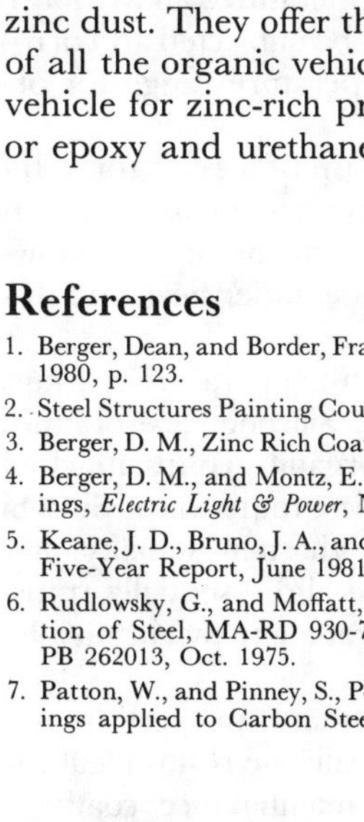

The author

Dean M. Berger is a coating specialist in the chemical engineering dept. of Gilbert/Commonwealth, P.O. Box 1498, Reading, PA 19603, telephone 215-775-2600. He received his B.S. in chemistry from North Central College, and has done graduate work at the University of Wisconsin. His experience includes 20 years in research and development with PPG Industries and work with Union Carbide Corp. He has devoted many years to the Steel Structures Painting Council, National Assn. of Corrosion Engineers, American Soc. for Testing and Materials, and Federation of Societies for Coating Technology.

High-temperature silicone-based coatings

These coatings are designed for temperatures of 300–1,400°F. Selection depends upon the temperature profile and the type of alloy of the substrate.

Paul E. France, The Dampney Co.

☐ High-temperature silicone-based coatings are specialized materials. Understanding how they work and how to specify and apply them will help to ensure proper service and eliminate such problems as disbondment, discoloration and early failure.

Basics of formulation

For high-temperature applications, the coating system is expected to retain its appearance and integrity while protecting metal substrates at temperatures above 300°F (150°C)*. Also, the coating may be subjected to corrosives. The figure shows the temperature range for organic and heat-resistant coatings.

In general, coatings are made up of a resin (or vehicle), pigments and solvents. Conventional coatings, such as alkyds, use organic vehicles as pigment binders. However, these vehicles may decompose under heat, and this can cause premature failure.

To overcome this problem, high-temperature coatings use heat-resistant silicone resins. Silicones are organosilicon compounds that contain organic groups attached to inorganic ones. The inorganic groups have Si-O-Si bonds. Silicones have excellent thermal stability and resistance to oxidation. They are also essentially transparent to, and resistant to degradation by, ultraviolet radiation.

The combination of heat-resistant properties and weathering characteristics makes silicone resins ideal for formulation into heat-resistant maintenance coatings. Unmodified silicone resins can be combined with organic resins to reduce costs or improve properties such as adhesion, abrasion resistance and curing time.

The pigments used must be compatible with the silicone and should not decompose at high temperatures. Pigments must also be color-stable over the entire working-temperature range of the coating.

Thermally stable inorganic pigments keep their color over time, unlike organic pigments, and so are used in silicone coatings. Traditionally, only black and aluminum-colored heat-stable pigments were available. Now, there is a wide range of colors, including earth tones.

*Conversions from °F to °C are rounded off to two significant figures.

Analyzing operating conditions

In specifying a high-temperature coating, the factors affecting performance must first be assessed. In addition to temperature, these include the nature of the substrate, its structure, stress due to thermal cycling, weathering, application limitations, corrosives, and life expectancy of the coating.

Two common pitfalls are made in specifying: (1) assuming that a single high-temperature coating will be right for all applications; and (2) "overspecifying" the coating. Too often, the substrate skin temperature is guessed at, and the guess is made on the high side for safety. Thus, the coating system specified may be suit-

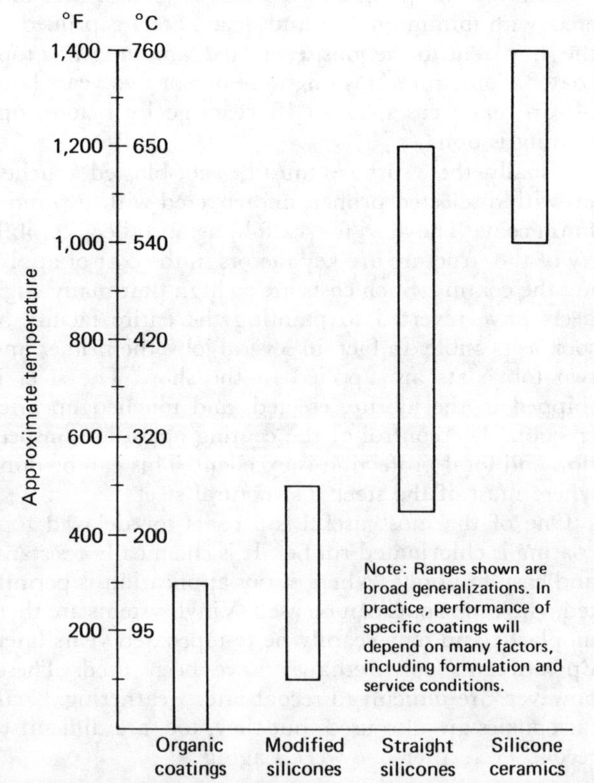

Here is a general guide to selecting a type of coating, based on temperature

Originally published June 27, 1983

able for operating temperatures much higher than those that will be encountered.

Also, in overspecifying, the coating may not cure properly. Coatings based on unmodified resins require curing at elevated temperatures to achieve optimum film properties—i.e., a threshold temperature must be reached before the silicone resin reacts with pigments and extenders to crosslink or polymerize. For this reason, a coating rated for 1,000°F (540°C) will not perform satisfactorily at a temperature below 450°F (230°C). Curing will never take place.

Measuring the temperature

Correct temperature determination is critical to writing a specification. Both the temperature range and the maximum temperature need to be found. So should thermal cycling and the rate of heat rise.

Surface thermometers and pyrometers are the most accurate ways to take temperature measurements. However, they must contact the surface. Equipment may not be easily accessible—or it may be so hot that it cannot be approached safely.

Also, temperature readings taken at the most accessible locations can be misleading. For example, at ground level, a stack may be heavily lined with refractories. It will have skin temperatures much lower than at its upper reaches where the lining is thinner.

When contact measurements cannot be made, other methods must be used. One is infrared-emissivity measurement. An infrared scan provides accurate temperature profiles of such equipment as smelters, blast furnaces and kilns.

For stacks, the stack-gas inlet temperature can be determined from the process-control temperature recorder. Once this temperature is known, the exit gas temperature can be found for an unlined stack of known height and diameter. This determination is done by using Fig. 9-40, in the "Chemical Engineers' Handbook," 5th ed. (Robert H. Perry and Cecil H. Chilton, eds., McGraw-Hill Book Co., New York, 1973).

Range of applications

There are two broad categories of high-temperature silicone-based coatings: those for service below 500°F (260°C), and those for service above 500°F to 1,200°F (650°C). In the former case, coatings are usually based on silicone-modified resins, which contain about 25-30% silicone. In the latter case, a higher percentage of silicone is required. As temperature is increased, more silicone is needed.

(Silicone-ceramics withstand temperatures to 1,400°F (760°C), but these are ceramic coatings; they will be briefly mentioned later on.)

If the maximum skin temperature is under 500°F, then a single coating system is needed. However, if a broader range of temperatures exist, two different systems are required. This is typically true for complex equipment such as units in chemical-processing plants and kilns.

In cases where there is an extremely rapid temperature rise, it is unlikely that any coating will work. This is because of the thermal stresses caused by the difference in coefficients of expansion between the substrate and the coating.

Design and maintenance factors

In writing a specification for a high-temperature coating, the equipment design and its condition must be considered. Usually, design changes can be made only on new construction, and only when a coatings specialist is consulted before fabrication begins.

If proper measures are not taken, premature coating failure can be caused by items such as bolts, rivets, corners, edges, inverted channels and poorly treated weldments. Sharp protrusions should be ground off, and welds power-brushed. Such areas should be spot-primed with a suitable silicone-zinc-dust heat-resistant primer.

The metallurgical makeup of the substrate must be considered, since not all equipment is made of carbon steel. Stainless steel that is to be insulated should be coated to prevent externally induced chloride stress-corrosion-cracking. The coating system must be chloride-free.

Any type of zinc-containing coating should be kept away from stainless steel, because subsequent welding might result in destructive alloying of the steel. Here, it is necessary to specify a coating that is free from chlorides, metallic zinc and any other harmful contaminants.

Prerusted or "weathering" steel may need painting. All products of oxidation must be removed from it before coatings are applied.

Mill scale must also be removed from any metal or alloy. Upon heating, the scale eventually shatters, disbonds and separates from the parent metal.

When refractories are used, their condition must be considered. Obviously, a gross failure of a refractory lining will result in overheating of the equipment's surface, destruction of the coating, and rapid fatigue-type failure of the steel casing.

Lesser refractory failures—e.g., thinning or cracking—may cause hot-spot failures of the coating. Local discolorations result, and are followed by disbonding, peeling and flaking.

Surface preparation

Once the conditions of application are known, the coating can be specified. However, no coating—no matter how well specified—will perform properly if it is not applied properly.

First, the surface must be correctly prepared. All organic contaminants must be removed completely, since they can cause poor adhesion and premature film failure. The procedures of the Steel Structures Painting Council should be followed for each type of substrate.

For carbon steel, abrasive blasting is the preferred method. It removes contaminants and creates a mechanical anchor pattern to hold the coating. The profile should not exceed 1 to 1.5 mils, since high-temperature coatings are applied in thin films to reduce internal thermal stresses.

For stainless steel, the removal of oil, grease, and cutting and drawing compounds can be done by either steam-detergent cleaning or wiping down with a high-flashpoint nonchlorinated solvent.

Priming

To avoid recontamination, priming should be done as soon as possible after surface preparation is finished. For carbon steel subjected to weathering and chemicals, use a silicone-zinc-dust primer. Topcoats are not always able to prevent under-film corrosion of the substrate. For indoor exposure in nonaggressive environments, a two-topcoat application offers a viable option.

When high-temperature equipment is to be painted, the nature of the previously applied coatings must be considered. Silicone-based topcoats should be applied only over either clean, dry, silicone-based primers or coatings of known composition. Inexpensive shop primers are not suitable.

Such primers and other coatings that are not heat-resistant must be removed. If this is not done, the previously applied coating will fail, and so will the heat-resistant coating. If the composition of the existing coatings cannot be determined, remove all of them.

During priming, the dry-film thickness of the primer should not exceed 1.5 mils for temperatures to 300°F (150°C), and less for higher-temperature service. Primers should be allowed at least a 24-hour wait before topcoating to ensure complete drying and "flashoff" of entrapped solvents (for curing, it is necessary for the coating to reach a threshold temperature). Brushing and rolling topcoats over an uncured primer can cause lifting of it.

Field-application methods

Equipment should be allowed to cool to ambient temperature before it is painted. (The only exception is coatings that are made to be applied to hot surfaces.) If equipment is warm–say 120°F (50°C)–rollers and brushes will produce excessively thick films, which can fail prematurely via disbonding, cracking and flaking caused by thermal stresses in the film.

Spray applications on hot surfaces can result in a condition similar to "dry spray." The film will not adhere properly, and will be extremely porous due to bubbling that results from rapid solvent evaporation.

Surface contamination is often a problem, especially for primers. Zinc dust is reactive and can give rise to surface contaminants that will produce poor adhesion. Any industrial "fallout dust" may be problematical; apply the topcoat as soon as possible after the primer is cured. If too much time passes after the primer is applied, remove any contaminants from its surface before putting on the topcoat.

Avoid prolonged exposure to wet weather, salt fog, or other corrosive conditions before a high-temperature coating is cured. Work should be scheduled so that equipment exposed to such environments can be put back into service as quickly as possible.

Poor control of film thickness can be a problem. If the film is too thick, it can check ("alligator"), crack and lift. The total system dry-film thickness should be about 2.5-3 mils maximum for temperatures to 500°F (260°C), and 1.5-2 mils for higher temperatures.

Applying coatings to hot surfaces

As already noted, most high-temperature coatings are made to be applied to surfaces at ambient temperature. What about equipment that either is rarely shut down or cannot be scheduled for painting due to short turnaround times?

For such problems, special coatings can be used. These materials—i.e., Dampney Co.'s Thurmalox 260 series—are made for in-service painting of equipment as hot as 400°F (200°C), and are rated for continuous temperatures to 500°F (260°C). Uses include coating of process vessels, piping and heat exchangers.

These coatings are based on a copolymerized silicone resin that contains a heat-activated catalyst. Heat triggers the catalyst, crosslinking the coating and producing a chemically resistant finish similar to one created by an oven cure in a factory. Thurmalox 260 coatings are self-priming and resist corrosion.

Their use removes abrasive blasting and painting from the turnaround schedule. One major benefit of this is elimination of the threat of blast particles contaminating equipment that has been opened up for inspection or repair.

For application, surfaces are blasted to a commercial grade. Usually, one coat of the material is applied by airless spray to a mimimum dry-film thickness of 1.5 mils. For surfaces previously coated with a silicate-based inorganic zinc-rich primer, one coat is sufficient.

Performance limitations

Although typical silicone-based heat-resistant coatings work well, they do have some limitations.

They are not intended for immersion service. Splash zones, mists and fumes can be a problem, too. Even when cured properly, these coatings are softer than most organic finishes. Adhesion is good, but the substrate must be as free as possible from grease and dirt.

When silicone-based coatings are modified with organic resins, heat resistance is reduced at the high-temperature end, and gloss and color-retention suffer somewhat. However, film properties (such as flexibility, chemical resistance and toughness) and curing are improved.

Colored enamels based on unmodified silicone resins withstand temperatures of 500-800°F (260-430°C). Black formulations perform up to 1,000°F (540°C), and aluminum combinations up to 1,200°F (650°C).

Heat-resistant paints containing ceramic frits operate successfully to 1,400°F (760°C). Here, the silicone resin thermally decomposes, and the frits fuse into a substrate-Si-O-frit bond. The resultant finish is very durable and heat-stable.

Finally, heat-resistant coatings require maintenance. They must be inspected, and repaired if damaged.

The author

Paul E. France is manager of field technical service for The Dampney Co., 85 Paris St., Everett, MA 02149. Tel: (617) 389-2805. He has devoted over 30 years to the development, field testing, marketing and specifying of industrial protective coatings. He joined the coatings division of Dearborn Chemical Co. in 1950. In 1963, he became responsible for coatings sales for Dampney, and then served as the firm's technical director for ten years before assuming his current position. He holds a B.S. degree in chemical engineering from Rose-Hulman Inst. of Technology, and is a member of the Natl. Assn. of Corrosion Engineers.

New maintenance coatings: A time for proving

A flood of radically reformulated products and entirely new materials are set for service in demanding CPI applications. To win a place in the market of the 80s, the key will be to demonstrate good on-the-job performance.

☐ Look forward to some new choices the next time you must select or specify a coating system for process plants or equipment. Be it an epoxy or a urethane or a vinyl, there are firsts in every category—but the trend is definitely toward products that are water-based or have higher solids content.

As many of the new products go to market this month (to coincide with the National Assn. of Corrosion Engineers meeting just held in Chicago), producers will be trying hard to get these first-generation materials off on the right foot. Only after long-term service will the products be able to prove their worth to CPI users.

Since California Air Resources Board (CARB) Rule 66* went into effect in 1967, considerable new coatings development work has been going on, for both the high-volume, mild-environment market and the industrial maintenance field. Finding reliable products for this latter market has presented the toughest technical problems, and the final word is not yet in. For example, there's a lot of hesitation over introducing water-borne coatings to replace products based on volatile organic solvents in environments that are more than moderately corrosive.

So far, only California is regulating the use of volatile organics in coatings. Even so, it has conceded that formulating new pollution-free, yet reliable, industrial maintenance coatings is a tall order. In fact, in new coatings rules put into effect last September, CARB has exempted maintenance coatings until September 1982, and could extend the reprieve if no adequate substitutes become available by then.

Meanwhile, the U.S. Environmental Protection Agency has not yet cracked down severely on coatings-related pollution. It is now in the process of writing an informational bulletin that recommends—but does not insist on—limits slightly less stringent than those set by CARB. But even there, EPA will specifically recommend that industrial maintenance coatings be exempted.

Nevertheless, companies feel there will be ever-tightening government scrutiny on coatings solvents, and manufacturers have also had extra impetus to develop new products due to the skyrocketing costs of hydrocarbon raw materials. The quest is to come up with environmentally acceptable products that maintain the corrosion-prevention performance levels of older, conventional products. Secondary goals include efforts to make these high-performance materials more versatile and more tolerant of varied application conditions.

This will lead to major rearrangements in the marketplace. A recently-released report by Chem Systems Inc. (New York) notes: "The industrial coating industry currently faces what could be the most traumatic transition in its history—the restricted use or elimination of traditional low-resin-content organic-solvent-based alkyd, acrylic, cellulosic, vinyl, epoxy and polyurethane systems in most industrial coatings applications . . . by 1990, the overall market share of conventional solvent-borne systems in industrial coating applications could reach five percent, a significant drop from the current level of approximately 50%."

Over the same period, Chem Systems projects waterborne industrial coating systems to grow from the current 15% market share to 25% in 1990, and high-solids systems (including reactives) from 15% to 30%.

A quick rundown of some of the myriad new products coming on the market illustrates what manufacturers are doing to meet the challenges:

JUST FOR TANKS—Responding to a need for a bottom coating for crude-oil storage tanks that would offer good corrosion resistance and improve structural strength as well, Carboline Co. (St. Louis, Mo.) has developed a new vinyl ester formulation (a polyester) expected to be released shortly. The product is more flexible than most polyesters and offers better solvent resistance than most epoxies. Paul Lodewick, laboratory project manager, says customers are interested in upgrading existing crude storage tanks so that finished product (especially gasoline) can be held in them. But, since no-lead gasoline has high toluene levels (for boosting octane), a stronger solvent than gasoline, better resistance is required of an interior tank-coating. In the past, epoxies had served well, but toluene affects them adversely.

For storage tank interiors, Glidden Coatings & Resins (Cleveland), a division of SCM Corp. reports it has a promising vinyl ester thick-film coating under development. It is being designed to provide a 20-mil-thick layer per coat to be used primarily to protect against hot strong acids.

COAL-TAR COATINGS IMPROVING—A number of firms are hard at work developing products that will have applications broader than the coal-tar paints' traditional use as a marine coating. The materials are composites of the cementitious coal tar plus a resin such as vinyl or epoxy. They are particularly suited to situations where proper surface preparation is difficult.

Ameron, Inc.'s Protective Coatings Div. (Brea, Calif.) is now producing a new vinyl resin coal-tar composition (Amercoat 245) that combines the chemical resistance of the vinyl with the water-immersion resistance of the coal tar. Its advantages, says Ameron, are that it is a one-can product, can be sprayed on, and will cure even in freezing weather. By contrast, the company notes, epoxy coal-tars are generally available only in two parts (requiring mixing) and need temperatures of 50-60°F for spraying and curing. Also, the new formulation may be easily repaired, the company says, whereas epoxy-tar coatings tend to get hard and brittle and are not compati-

* This rule limits the emissions of photochemically reactive hydrocarbons and oxidants. It was amended in 1971.

New coatings to protect equipment must prove their effectiveness

A few common coating terms

Efforts to cut down on solvents have taken two routes—the development of higher-solids coatings that simply use less solvent, and a switch to water-borne systems. A wide array of resins are employed in industrial coatings systems (see *Chem. Eng.,* Oct. 22, 1979, pp. 153-156). Common types include vinyl, epoxy, aliphatic urethane, chlorinated rubber, acrylic and alkyd. Alkyd enamels still serve as the workhorse industrial architectural paint.

High-solids—This is a rather vague term for coatings that contain relatively low amounts of solvent. Some people may consider the level to be 60% and others, 100%, depending on the specific resin used, the technology involved, and how the product compares with older, conventional products in solids volume.

Two-part reactives—These are systems generally considered to be high-solids, that are formed by reacting resin with a catalyst. Urethanes and epoxies are often formulated as two-part reactive systems for high-performance coatings.

Water-bornes—These are paints in which resin is water-dispersed, water-soluble or emulsified. The water-bornes do solve the organic-solvents problem of pollution and health hazards, but present their own difficulties—namely restricted application temperature (not lower than 50°F), poor penetration into the surface, and restricted modes of application.

Water-borne coatings on steel also suffer surface-staining phenomena known as flash rusting and early rusting (which occurs when a thin paint film is applied on a cool substrate and allowed to dry under high-humidity conditions).

ble with fresh repair coats.

Union Carbide is in the midst of developing a new water-based topcoat for a coal-tar-resin primer. The system would serve well in a plant where sandblasting cannot be done and oil-paint adhesion would be poor. Coal-tar primers are black and extremely difficult to topcoat with another color, so their use has been somewhat limited to date. Parker Helms, paint and coatings specialist for Union Carbide, views the new vinyl coal-tar coatings, such as Ameron's, as an important advancement in high-performance materials.

The Napko Industrial Marine Div. (Houston) of O'Brien Corp. is just beginning development work on coal-tar epoxies. L. D. Vincent, vice-president of the division, predicts that fillers such as glass and mica will be incorporated for greater internal strength, and that urethanes will be formulated in to increase solids. Look for the new products in about two years, he says.

And Carboline is working on a 100%-solids, solvent-free epoxy coal tar, expected to be released sometime this year. The firm's first coal-tar product, it will be designed to reduce costs of mixing and waste associated with present systems, by simultaneous application of two parts through dual-component spray equipment.

REACTIVE DILUENTS—Several companies have been working on creative ways to raise solids volume of paint high enough to meet CARB regulations. (A vinyl maintenance paint has spray solids normally in the range of 25-35%; viscosity restraints don't allow higher amounts when active solvents such as ketones and aromatics are used.) Rohm and Haas Co. (Philadelphia) has developed a reactive diluent (a methacrylate monomer) called QM 657 that allows replacement of solvent by the monomer. On application a dryer/catalyst is added so that the monomer reacts to become a part of the product. Effectively, solvent is replaced by diluent which, on application, polymerizes and becomes part of the system. The system is compatible with alkyds, chlorinated rubbers, and vinyls. An industry source reports that at least two companies are in the midst of developing new resin systems to go with the reactive diluent. A Rohm and Haas spokesman says the firm is committed to build a plant for manufacturing this material.

IMPROVING EPOXIES—New epoxies make one of the big stories in industrial maintenance coatings, and resin manufacturers particularly have contributed in a major way. According to consultants C. H. Kline & Co. (Fairfield, N.J.), epoxies are becoming increasingly important in industrial coatings because of their excellent adhesion, flexibility, overall chemical resistance, and toughness. Growing interest has developed in epoxies not only for high-solids products but also for water-based and powder coatings (powder coatings use is restricted to factory application).

Shell Development Co. (Houston) has developed a new family of low-viscosity epoxy resins that have improved exterior performance over conventional types. A conventional industrial-coating epoxy will, within 2 to 6 months, lose its gloss and begin to chalk and show a perceptible degree of yellowness. The new finishes, according to Ronald S. Bauer, project leader of Eponex resins, "still retain their initial gloss after 21 months and show no evidence of chalking and no significant increase in yellowness." He notes that encouraging accelerated-weathering results have been obtained with several solvent-borne, ambient-cured, two-package systems as well as with a two-package water-borne system.

Another resin manufacturer, Rohm and Haas, has introduced a new resin—an amine-functional acrylic (QR 765), to be used with a water emulsion of an epoxy resin. Rohm and Haas reports the CARB-complying two-component system (primer and topcoat) is showing performance, in large-scale trials, approaching that of current commercial solvent-based epoxies.

Both Sherwin-Williams and Glidden Coatings & Resins have active R & D efforts in epoxies. Sherwin-Williams is field-testing a new high-performance epoxy expected to go to nationwide marketing in the spring of 1981. And Glidden has targeted a 1982 deadline for the introduction of some new high-solids epoxies (which will be about 90% solids).

O'Brien Corp.'s Napko Div. is using new technology from Shell Chemical and Dow Chemical to produce epoxy-acrylic coatings. The high-solids product meets CARB requirements and offers color and gloss retention comparable to or better than urethane systems, says Vincent. The new product, Epoxacryl, has been field-tested and is about to be introduced to the market. Applications include structural steel, tanks, offshore structures and any steel subjected to a harsh, corrosive environment where high color and gloss retention are necessary.

WATER IN, ORGANICS OUT—Napko's Epoxacryl is a case-in-point of an epoxy-resin-containing product that can be considered high-solids but can also be formulated as a water-borne coating. The material can be made either with water or with solvent but the water-borne product is the one presently being introduced. The company is in the final stages of developing an organic solvents product which, in order to meet CARB regulations, will be a higher-solids product than the water-borne counterpart.

Ameron is also selling small quantities of water-borne coatings, mostly for external protection. The key, according to Daniel Gelfer, director of commercial development, is the use of stabilizers that make the epoxy and curing agent stable in water. As part of this work, Ameron is testing new Shell epoxies designed to resist ultraviolet exposure. "I think we as an industry need more time to evaluate water-borne epoxies," says Gelfer. "The indications are they will be alright, but they have not been on the market long enough to determine that yet."

Similar sentiments are being expressed about new water-borne maintenance coatings of all types. Most companies are developing water-based products but nothing suitable for heavy-duty use has made it to the market yet. Resin suppliers have reportedly made great strides on emulsions and latexes for the high-performance market in the last year, so much so that several coatings manufacturers indicate a strong optimism for the potential of these materials in chemical process uses.

Parker Helms of Union Carbide sees acrylic latex coatings making inroads right now into the chemical process industry. Carbide recently adopted an acrylic latex system for tank-farm use. The materials serve well in environments not subject to severely acidic conditions. The acrylic latexes have a natural affinity for alkaline surfaces, notes Helms, and for this reason work well on top of inorganic zinc primers (the workhorse primer used on blasted surfaces), which are somewhat alkaline themselves.

Addressing the concerns over uncertainties of new-product performance, the Steel Structures Painting Council (Pittsburgh) has undertaken a project to determine to what extent the new or proposed coating systems fulfill their primary function of protecting structural steel without the need for excessive maintenance. A broad range of products was evaluated and a report was issued last August, although tests are continuing on many of the materials. Called Performance of Alternate Coatings in the Environment (PACE), the study found that as a group, the water-based coatings did not perform as well as the solvent-thinned coatings. Certain formulations, however, approached the conventional alkyds in performance in accelerated tests, including water-soluble alkyds, an acrylic emulsion, a styrene-acrylic emulsion and an epoxy/acrylic emulsion. (Details on proprietary aspects of the most effective materials will be made available when approved by the suppliers.) Early outdoor-exposure tests tend to verify these promising results, but are not yet sufficiently definitive for pre-failure analysis. Water-thinned paints required better surface preparation than solvent-thinned ones, PACE workers reported. And several had application problems due to poor wetting, flash rusting instability, and insufficient thickness.

How to inspect paints and coatings—Part I

Proper inspection helps to ensure long life, thus cutting costs. The author outlines the inspector's job, and lists some instruments used in measuring coatings.

***Dean M. Berger**, Gilbert/Commonwealth*

☐ When coating dependability is essential, careful inspection of surface preparation and applied-coating thickness are important.

Inspection of shop- or field-applied coatings is not considered a profession. This is unfortunate, for if an inspector is to gain competence, he or she must understand the technical factors that dictate the specifications, and which instruments to use.

ASTM Committee D-33 (formerly D01.43), Coatings for Power Generation Facilities, concerns itself with inspection. This committee has prepared a manual of coating work for nuclear-power-plant primary containment facilities [*1*]. Chap. 5 deals with inspection. Also, the coatings industry has established a few schools for training inspectors. Such courses often take only one week.

It is not realistic to set up a list of specifications for a job and then fail to inspect the job. Inspection means checking not only the coating but also the equipment, surface preparation, and conditions of application.

The coatings inspector

The inspector must know enough about coatings to be confident in doing the job. On-the-job training cannot be overvalued. The inspector should participate in the following: prior work conference, prejob inspection, surface-preparation inspection, coating-application inspection, daily inspection reports, and final report.

Inspector qualification-requirements (for nuclear power plants) are listed in ANSI N45.2.6 [*2*] and the D01.43 manual [*1*].

Finally, the inspector is not a coatings consultant; and although he or she should be an expert inspector, rarely is an inspector an expert coatings formulator or corrosion engineer. The inspector is not expected to know all the answers; the best inspector is the one who knows when and how to get outside help.

There are numerous standards for coatings, as well as for inspections. For example, the design of storage tanks that require internal linings, as well as their fabrication, and surface-finish methods, are all covered in NACE Recommended Practice T-6A-29 [*3*]. Inspection of such linings is also described in NACE's "Coatings and Linings for Immersion Service," Chap. 4 [*4*]. ASTM D3276-80 is a standard recommended guide for use by paint inspectors.

Inspection procedures

Inspection of coatings applied to structural-steel or concrete surfaces is accomplished either visually or with instruments. Instruments are used to determine whether specified dimensional requirements have been met. Visual inspections determine the quality of the work.

Visual inspections are done with the unaided eye or with a magnifying lens. Sometimes, telescopic observation or low-power magnification is needed. A magnifier is shown in the figure.

By using standard visual techniques, the inspector identifies missed areas, thin areas, or damaged sections of a protective coating. White primers have been used

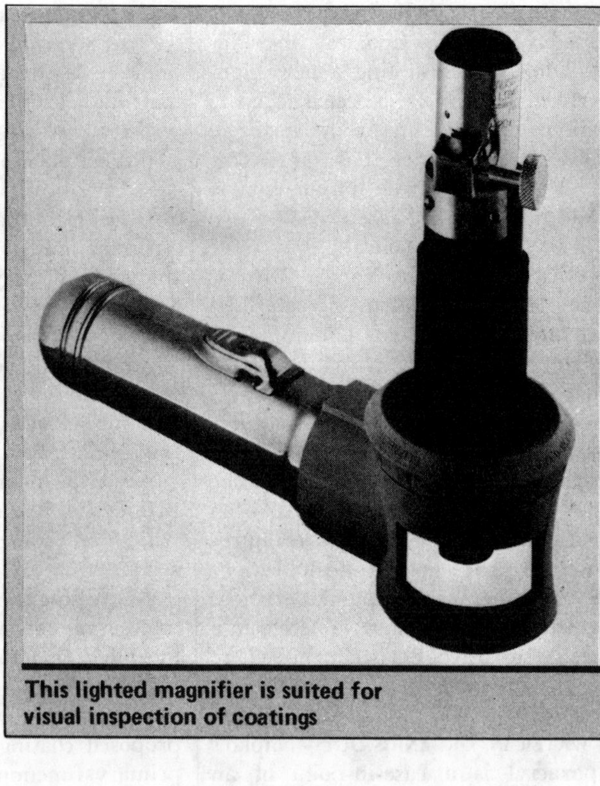

This lighted magnifier is suited for visual inspection of coatings

Gardner Laboratory, Inc.

Originally published March 7, 1983

A coating inspector must check the following items to determine the quality of the job

Item	Tank lining	Concrete surfacer	Concrete topcoat	Inorganic zinc primer	Organic primer, steel	Organic topcoat, steel
Pinholes	X	X	X			X
Blisters	X	X	X			X
Color and gloss uniformity			X			X
Bubbling	X	X	X		X	X
Fisheyes	X	X	X		X	X
Orange-peel	X	X	X		X	X
Mud cracking	X	X		X		
Curing properties	X	X	X	X	X	X
Runs and sags	X	X	X	X	X	X
Film thickness, dry	X			X	X	X
Film thickness, wet		X	X			
Holidays, missed areas	X	X	X	X	X	X
Dry spray	X	X	X	X	X	X
Foreign contaminants	X	X	X	X	X	X
Mechanical damage	X	X	X	X	X	X
Uniformity	X	X	X	X	X	X

effectively to easily spot areas of low film-thickness or inadequate coverage.

It takes years of nurturing and education to cultivate the eye of an inspector. Casual inspection by a trained inspector can spot trouble even at a good distance from a coated object. The inspector can visually define the general quality of the workmanship within minutes of arriving at a jobsite. The inspector should check over the items indicated in the table.

Inspection of surface preparation of steel

For a better understanding of the requirements of surface preparation, see the specifications found in the Steel Structures Painting Council's manuals, [*5*], and those of NACE [*4*]. The Soc. of Naval Architects and Marine Engineers has published a technical report [*6*] that describes various degrees of blast-cleaning of old, painted and rusted steel. NACE provides blasted-steel encapsulated panels that can be used to determine the cleanliness of the blasted-steel surface.

Before application of a paint or coating system, the surface must be properly prepared. This can be done in a number of ways. The most common one is by abrasive blast-cleaning. Prior to blast-cleaning, an inspector should look for the removal of all visible oil and grease.

Welds must be ground smooth, although not necessarily flush. Sharp protrusions should be rounded, and weld crevices should be manually opened enough for the coating to penetrate. If this procedure is not done, projections should be removed by grinding them down.

Back-to-back angles, tack- or stitch-welding, etc., cannot be properly cleaned and coated. Thus, these should be sealed with caulking to prevent crevice corrosion. Also, electric welding-flux should be neutralized or removed prior to painting.

Blast cleaning

When the substrate is to be abrasive blast-cleaned, inspection may include the blast-cleaning equipment:

The compressor—Can it deliver a sufficient volume of air at the required pressure (at least 90 to 100 psi at the nozzle)? Are the air receiver on the compressor and all the moisture and oil traps bled continuously? (Use a white cloth held in the airstream to check the air.)

The blasting gear—Does it look to be well-maintained? Can the machine be readily adjusted to reduce the volume of abrasives in the airstream? Is there excessive dusting during the blasting operation? Are there deadman controls on the equipment?

Are the proper hoses and hose lengths being employed? Are all external couplings wired? Will the hose dissipate electric charge? Finally, how worn are the nozzles? Normally, this inspection is the job of the applicator, and not that of the coating inspector [*7*].

Proper blast-cleaning technique

1. Regulate the blast cleaner so that the propelled stream of abrasive discharge (coming off the steel) is bluish and slightly discolored from the abrasive. Dusting should be minimal.
2. Make sure that the nozzle is held 2 to 3 ft from the surface at an 80–90-deg. angle. (This angle is for mill scale, old paint, rust, etc.) For lighter contaminants, an angle of 60–70 deg. is faster and quite suitable.
3. See that the operator cuts a pattern, rather than cleaning in nonuniform short jerks of the nozzle. Smooth actions produce better, more-uniform results.
4. Ensure that proper respiratory equipment is used.
5. Be sure that nearby equipment is properly shrouded and sealed off from blasting residues.
6. Finally, check that the area meets U.S. Occupational Safety and Health Admin. (OSHA) safety standards for rigging and air supply.

Checking the abrasives used

1. If sand is to be used for cleaning, it must be clean and free of friable material (soft particles that crush upon impact to form dust clouds). Heavy, coarse sands result in deeper profiles, and will also hit harder on the surface being blasted. Finer sands usually clean a surface faster. If a specific grade of sand is used, see whether it produces the blast profile specified in the

contract documents. A sieve analysis is available from the sand or grit manufacturer. Get a certified analysis from the supplier. If in doubt, have a sieve analysis run by an independent laboratory.

2. If a slag abrasive is used to increase blasting production, find out its grade and note that in an inspection book. It should not be reclaimed. If the aggregate is black, there may be problems obtaining a "picture-book" Steel Structures Painting Council SP-10 near-white metal-blast-cleaned or SP-5 white-metal blast-cleaned surface, due to residuals in the blast profile.

3. Metallic abrasives are normally used in automatic, rotary-wheel blasting machines. These aggregates also have specific grades; ask the production manager for the shot or grit number being used. Check the blast profile to ensure conformance with the specification. Check the amount of reused and new abrasive. No more than 80% should be reused.

4. Wet abrasives should not be allowed in dry-abrasive-blasting work. Abrasives should be stored in undamaged bags on pallets or on a false floor. Abrasives shipped as bag goods should be covered with a minimum 8-mil polyethylene film during transportation to the jobsite.

Checking the steel surface to be painted

There are two major reasons why coatings fail prematurely: The surface is not clean and the coating is not thick enough. Inspection of surface cleanliness of steel (or concrete) is most important.

Looking for contaminants—Look for mill scale. It is evident wherever there is a bluish-black oxide layer on steel, resulting from the hot-rolling process. If all intact mill scale is not removed, it will surely spall off later on, taking the coating with it. Rusty metal will appear at an early date.

To check for mill scale:

1. Look for the telltale blue-black surface.
2. Use the copper sulfate test—5% $CuSO_4$—in distilled water. Drop or brush on the solution. The copper will plate out on clean metal, and will not on mill scale.
3. Use the knife test—if the steel being inspected has already been overcoated, probe the surface with a knife. If mill scale is present, it will break into small black (coal-like) granules, or powder. If the metal is clean, there will not be any deposits.

To check for oil and grease:

1. Look for a thin layer of oil or grease. It will act as a bond breaker when the coating is applied. "Solvent wipe" and always check adhesion before giving the go-ahead, especially in dirty fabrication shops. Look carefully at surfaces in oblique light.
2. Use the blotter test on compressors: Hold a white blotter or cloth in the airstream at the nozzle, or near the air takeoff to the nozzle. Check for contamination, dirt, moisture and oil spatter.
3. Use an ultraviolet inspection light. Under these rays, certain oils fluoresce white.
4. Use the water-break test: Sprinkle water on a surface. If oil or grease is covering the steel, the water will not spread out; it will form beads. A dry white cloth will easily pick up any oil, if rubbed on the steel surface.
5. Check for oil or grease contamination of the shot in abrasive blasting equipment. Take a sample of the abrasive from the hopper in the machine. Place it in a vial of water, and shake. Oil and other contaminants will separate and rise to the surface.

Note: The inspector should work quickly so as not to delay the application of the primer. Steel can normally stand only a few hours before coating.

Dewpoint and water problems

Moisture interferes with coating adhesion and causes premature failure. Dewpoint condensation is a problem early in the morning, during spring and fall. Although not visible as a thin film (e.g., condensation on the sides of a tank), water can play havoc with a coating job. The inspection techniques used to discover water are:

1. Look for moisture in inaccessible areas. Look for dark spots and early rust bloom during abrasive-blasting operations. Inspect abrasives for moisture content.
2. Use a sling psychrometer to determine wet- and dry-bulb readings. Psychrometric tables will give the percent of relative humidity and the dewpoint. Record the surface temperature of the substrate to be coated. The surface must be 5°F above the dewpoint, and its temperature must be rising, before painting begins.
3. Use the blotter test described above to check the compressor.
4. For concrete, lay a rubber mat on the surface, overnight. If moisture is found under the mat, the concrete is not dry enough to apply a coating over it.

Also, have the surface cleaned before the coating is applied. Fallout is all around—brush off, blow off or vacuum any dust collecting on the surface to be coated.

References

1. Amer. Soc. for Testing and Materials (ASTM), ASTM D01.43, Manual of Coating Work for Light Water Nuclear Power Plant Primary Containment and Other Safety-related Facilities, 1979, ASTM, Philadelphia, Pa.
2. Amer. Natl. Standards Inst. (ANSI), ANSI N45.2.6, 1973, Qualification of Inspection, Examination, and Testing Personnel for the Construction Phase of Nuclear Power Plants, Amer. Soc. of Mech. Engineers (ASME), New York.
3. Natl. Assn. of Corrosion Engineers (NACE), Design, Fabrication and Surface Finish of Metal Tanks To Be Lined for Chemical Immersion Service, T-6A-29, March 1977, Proposed Standard Recommended Practice, NACE, Houston, Tex.
4. NACE, Coatings and Linings for Immersion Service and NACE Visual Surface Preparation Standards, TPC #2, Houston, Tex.
5. Steel Structures Painting Council, Vol I: Good Painting Practice; Vol II: "Systems, and Specifications," Vol. I & Vol. II, 'Surface Preparation Standards,' Pittsburgh, Pa.
6. Soc. of Naval Architects and Marine Engineers, Report 4-9, Abrasive Blasting Guide for Aged or Coated Steel Surfaces, 1969, New York.
7. Bigos, Joseph, Anchor Pattern Profile and Its Effects on Paint Performance, *Corrosion*, Vol. 15, No. 8, pp 428–432, Aug. 1959.

The author

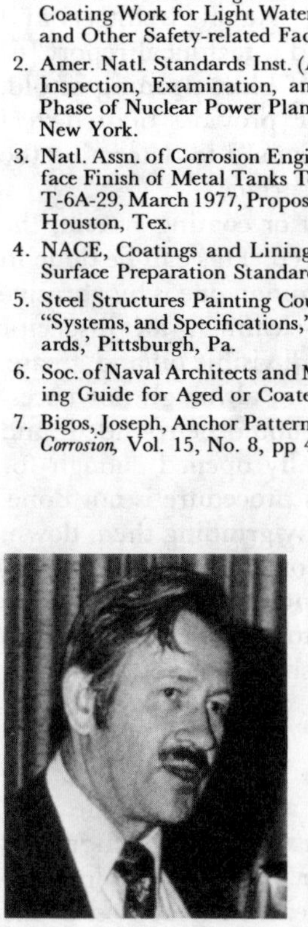

Dean M. Berger is a coating specialist in the chemical engineering department of Gilbert/Commonwealth, P.O. Box 1498, Reading, PA 19603, telephone 215-775-2600. He received his B.S. in chemistry from North Central College, and has done graduate work at the University of Wisconsin. His experience includes 20 years in research and development with PPG Industries and work with Union Carbide Corp. He has devoted many years to the Steel Structures Painting Council, National Assn. of Corrosion Engineers, American Soc. for Testing and Materials, and Federation of Societies for Coating Technology.

How to inspect paints and coatings—Part II

To help ensure a proper job, one must check the blast-cleaned surface profile, and how the coating is applied and if it meets thickness specifications.

Dean M. Berger, *Gilbert/Commonwealth**

☐ Here, we will conclude the discussion that began in Part I, which covered the role of the coatings inspector, blast cleaning, checking the surface to be painted, and dewpoint and water problems.

Surface profile

Without adequate preparation provided by abrasive blasting or mechanical grinding, many coating systems will not offer long-term performance. An insufficient anchor pattern (surface profile) will result in too smooth a surface and, thus, poor adhesion. On the other hand, a deep profile will require additional paint to cover the tops of the peaks. The blast profile for steel is about 25% of the total paint thickness. Therefore, if a 6-mil coating is specified, the profile should be 1.5 mils. The profile varies with the abrasive employed. There are many instruments to determine the actual surface profile.

(In discussing instruments used in evaluating surface-preparation methods and coatings, specific models and manufacturers will not be mentioned. Rather, the subject will be kept generic. Also, no attempt is made to cover all the various types of instruments and devices offered that can accomplish a certain job. Several examples will be given in each case to show some commonly used equipment.)

One device (not shown) to measure the surface profile contains metal disks with nominal profiles of ½, 1, 2, 3, and 4 mils. A flashlight magnifier is used with the disks to check sandblast-cleaned steel surfaces. Also available is a metal disk for comparing anchor patterns prepared with shot or grit abrasives. A similar method is to use a series of steel coupons, ranging in profile from 1 to 4 mils. These are handy pocket-size plates.

Several papers have been presented on surface profile and its measurement. An excellent one is by Bigos [*1*]. One technique advocated by him is to grind a small area of the blasted surface until only roughened surface is removed. The overall profile height is then measured with a depth micrometer, which is a dial micrometer with a pointed probe.

Although effective, this method is destructive. Nondestructive ways [*2*] include measuring surface reflectance, which is directly proportional to surface cleanliness. Also, a portable surface-profile microscope can be used. It is ideally suited to nondestructively measure surface profile.

Surface profile has also been measured using magnetic pulloff gages. Fig. 1 shows how one works. The device is placed on a coated steel surface and then is slowly pulled away. The coating will shield the effect of the magnet on the metal to a degree depending upon the paint's thickness. The scale on the gage can be calibrated to indicate thickness. The instrument is calibrated on a smooth steel surface. The base reading is then multiplied by 2.5 to account for the estimated void area (air space).

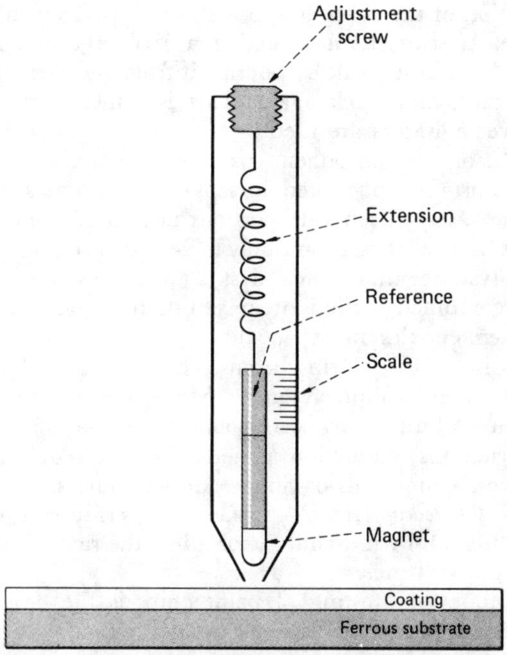

Magnetic pulloff gage is used to measure coating thickness on steel surfaces **Fig. 1**

*To meet the author, see p. 106.

Originally published April 4, 1983

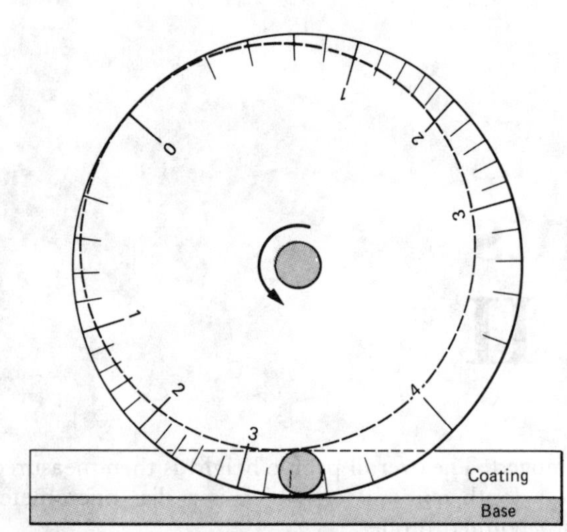

Eccentric-wheel-type wet-film gage is rolled in wet coating to show thickness **Fig. 2**

Numerous, more-sophisticated techniques are described in the recent Steel Structures Painting Council report, Surface Profile for Anti-Corrosion Paints [3].

Inspecting the coating application

Check out the mixing procedure—It is obvious that paints and coatings must be suitably mixed prior to application. Proper mixing redisperses the heavier pigments to ensure homogeneity. Follow the coating manufacturer's mixing instructions!

■ For one-component coatings—These are packaged to be mixed in their original containers. A paddle, mechanical agitator or power mixer is used. Thinners, if required, are added slowly after initial mixing. The painter should mix the paint long enough so that no portion of the coating appears as a "swirl" on the surface. After mechanical mixing, a "skilled journeyman" will "box" the paint by pouring it from one clean can to another, and back again. This is unnecessary when power agitators are used.

■ For two-component coatings—Catalyzed coatings are normally packaged in separate containers, or, in some cases, in separate portions of a larger container. The base must be thoroughly mixed prior to adding the catalyst. See that the catalyst is added slowly and that the combined portions are mixed thoroughly. Some catalyzed epoxies must be allowed to stand for about 30 min before painting begins. This "induction" period ensures more-uniform curing. Make sure that the temperature limitations correspond to the manufacturer's instructions. Induction periods are not required for polyester materials or amine adduct epoxies.

Is the coating cured?—Surface temperature, type of coating, and wind-chill factor affect the rate of cure in organic coatings.

Laboratory testing of paint chips is the only true means by which an inspector can verify that a coating is cured properly. Here are some other techniques:

1. Solvent rub—After curing of an epoxy paint or an inorganic coating, use a color-contrasting cloth dipped in a strong solvent (i.e., methyl ethyl ketone) to rub the surface. If the material was properly mixed and cured, no color will be transferred by the rubbing process. If the material was improperly mixed or cured, color will be seen on the cloth.

2. Sandpaper test—A number of paints and protective coatings will remain slightly tacky when they have not cured properly. When abraded with fine sandpaper, no coating material should be seen on the face of the paper. It should be removed as a fine powdery residue.

3. Hardness test—Mere use of a fingernail can go far in determining hardness. Also, special instruments or pencil-hardness tests can be used. For inorganic zincs, a coin scratched against the surface should not remove the film if the latter has been properly applied.

Checking paint application

Too much or too little on the surface?—A good painter who has "a handle" on spray application of coatings will always keep the gun parallel to the work surface. He or she will release the gun trigger at the end of each pass to avoid heavy deposits of paint, and will frequently use the cross-hatch method of laying up high-solids materials to prevent sagging.

When a painter applies paints or coatings too heavily for the temperature of the substrate, and for the viscosity of the material being applied, the results are runs, sags, drips and curtains. When these are found, the work should be stopped and the runs, and so on, should be brushed out before they start to cure. Careful adjustment of the gun and a good application technique will immediately reduce these problems.

Heavily applied paint may never cure properly, due to solvent entrapment. Or, if it does cure, film buildup can weaken the entire paint film.

1. Too little paint on the surface—When there are significant areas where the coating does not meet the specified dry-film thickness, tell the painter that the minimum specified thickness must be met.

Review the manufacturer's literature with the painter to determine what, if anything, must be done to prepare the already painted surface for an additional coat. Special considerations are required for a wide range of materials; check with the coating manufacturer *before* allowing the work to continue.

2. Too much paint on the surface—An inorganic zinc primer may "mud-crack" if applied too heavily. Follow the manufacturer's instructions for best results.

3. Overspray problem—This occurs whenever winds, temperature and humidity are rapidly changing. A partially dried coating that does not wet and flow into the previously applied material results. Overspray produces a loose, powdery coating that must be cleaned prior to subsequent topcoating. Inorganic zinc primers are susceptible to overspray; never topcoat unless the overspray is removed by abrading with a wire screen.

Overspray can be controlled, especially on hot, windy days, by changing the solvent used, to provide a "wetter" edge as the painter moves down the surface.

Film thickness gages

These tools measure both wet and dry film thicknesses. Wet-film measurements will spot errors in appli-

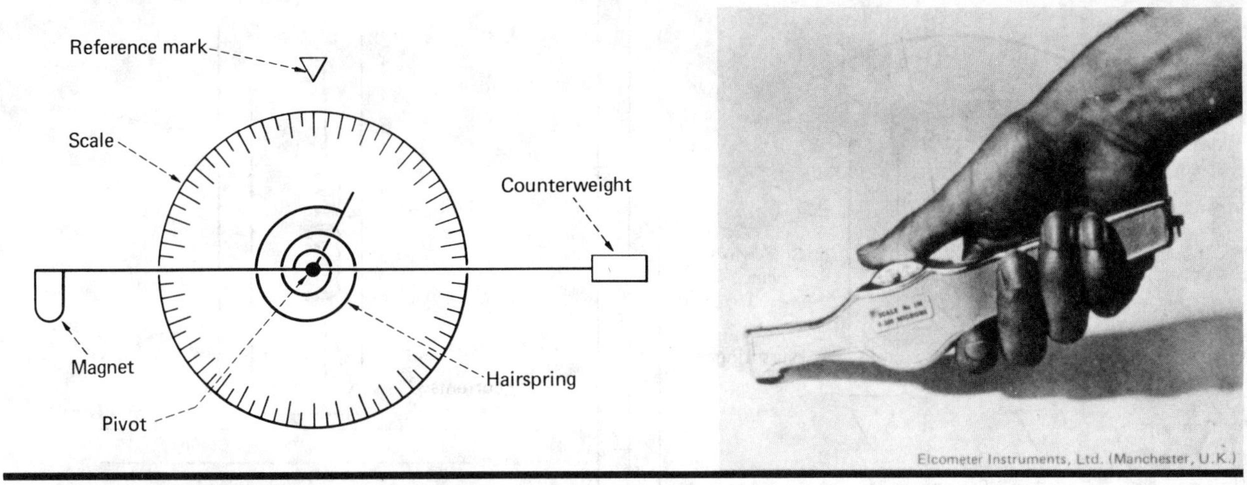

"Banana" pulloff gage measures thickness in any position without recalibration **Fig. 3**

cation before a large amount of work has been done improperly.

Estimates of the dry-film thickness can be obtained by measuring the wet film and multiplying this by the volume percent of solids of the paint.

The device in Fig. 2 is an eccentric-wheel-type gage that is rolled in the wet coating. Such devices can indicate thickness from 0 to 60 mils or 0 to 700 μm [*4*].

A variety of commercial instruments are available that can be used to see whether specifications are met. Selection depends upon requirements and budget. Details on measuring of paint film thickness are presented by various authors [*5–9*].

For dry films, nondestructive test instruments for steel substrates fall into two main categories: magnetic and eddy-current. The magnetic devices are more common. In their simplest form, these measure magnetic attraction, which is inversely proportional to coating thickness. There are several pulloff-type gages available. Fig. 1 shows one type; Fig. 3 shows another.

The pulloff gage serves as a rough guide to determine whether the coating is within specification. The manufacturer's stated accuracy is on the order of ±15%, provided that the gage is used in a vertical plane. If used in a horizontal or overhead position, more error will result. If the gage must be used in other positions than vertical, plot a correction curve for each different position.

Care should be exercised to inspect the hemispherically tipped magnet for dirt, small steel particles, tacky paint film, and tip wear, before use. Magnet wear alters the calibration of the gage. Never expose any magnetic-type gage to strong a.c. or d.c. fields, otherwise the magnet will change, affecting the tracking and calibration of the gage.

Besides accuracy limitations, the pulloff-type gage has other disadvantages:

1. The operator must record the coating thickness as the magnet breaks away from the coating.

2. Erroneous readings will result if the magnet is allowed to slide over the coating prior to breakaway and/or liftoff.

A more sophisticated version of the magnetic pulloff principle is incorporated into the "banana-type" thickness gage (see Fig. 3). This device is called a "banana" gage because of the shape of its case. A permanent magnet is mounted at one end of a balanced pivoted-arm assembly, and a coil spring is attached to the pivot and to a calibrated rotatable dial. The operator moves the rotatable dial forward until the magnet sticks to the coating. Variations in film thickness alter the attractive force of the magnet. This unknown force is determined by turning the rotatable dial backward, applying tension to the spring. When the spring tension exceeds the unknown attractive force, the magnet breaks contact. An audible click will be heard and the coating thickness will be shown on the graduated rotatable dial.

This type of gage will measure thickness in any position without need to recalibrate, because the pivot arm is balanced. External calibration is provided, and care should be exercised to follow the manufacturer's instructions when making adjustments.

Another version of the magnetic-principle device uses magnetic reluctance. Reluctance is that characteristic of a material that resists the creation of a magnetic flux in it (e.g., iron has less reluctance than air).

The gage in Fig. 4 contains a permanent magnet that is located between two soft-iron poles, resembling a horseshoe magnet. (The magnet is adjustable to produce an air gap.) A meter point-assembly with a soft iron vane is placed in the center of this horseshoe configuration, thus making a magnetic circuit with an indicating device. No power supply (or battery) is needed.

When the gage is placed on a dry coating (applied on steel), the magnetic flux changes in strength across the air gap in the magnetic circuit, moving the meter pointer across a calibrated scale. The scale gives coating thickness in either mils or μm.

The gage should be held at right angles to the surface to be measured. Tilting gives erroneous measurements. Always recalibrate the gage when changing from the vertical to the horizontal position.

In making film thickness measurements of any kind, do not take measurements close to the edge of a steel surface, to avoid distorted results. This is also true for

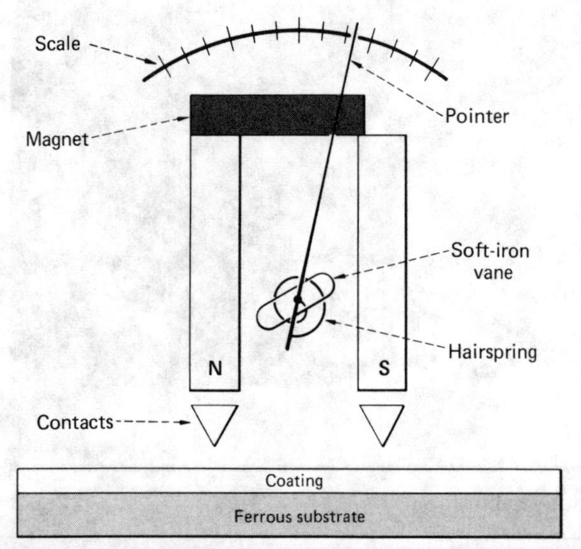

The principle of magnetic reluctance can be used to measure a coating put on steel **Fig. 4**

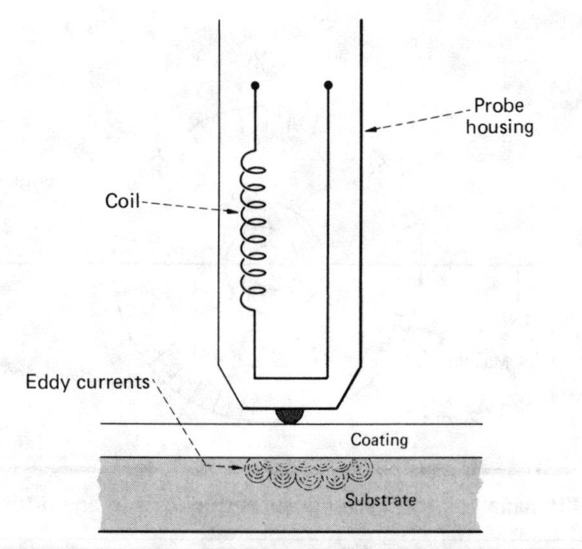

Eddy currents induced in a ferrous substrate are used to determine a coating's thickness **Fig. 5**

corners, angles, crevices, welds and joints. Recommended practice is to keep at least one inch away from the edge. Always measure on a clean surface, never on an oily or dirty one. Many readings are required.

Electronic gages

Some of these are more accurate than the previously mentioned mechanical gages. One such electronic gage uses the magnetic induction principle. The coating between the probe and steel acts as an air gap.

Another similar instrument uses a permanent magnet in its probe to produce flux, creating a magnetic circuit that measures coating reluctance. The thicker the coating, the higher the reluctance. When the probe is placed on a dry coating over steel, the magnetic flux adjusts to the particular thickness of coating. The change in flux is measured and displayed in units of thickness.

The device in Fig. 5 works on the eddy current principle. A coil of wire in the probe tip is energized with a high-frequency alternating current, creating a magnetic field. When the probe tip is brought close to a coated metal surface, eddy currents are induced in the steel substrate, altering the electrical characteristics of the probe coil. This electrical characteristic change in the coil is measured and displayed as coating thickness.

Destructive test instruments

Destructive test methods are normally employed when repairs can be made and when a controversy arises over the accuracy of dry-film-thickness measurements. The following are acceptable:

1. A small area of the coating is mechanically removed down to the bare substrate. Then the film thickness is measured with either a depth micrometer or a dial thickness gage.

2. A needle thickness gage can be used. A typical penetrating-needle-type instrument is battery-operated. A sharp needle probe is mechanically pushed through the coating until it touches the steel substrate, causing an electrical circuit to close and illuminate a lamp.

The inspector must be aware that the instruments used to measure film thickness are delicate and should be handled with care. Frequent calibration and good housekeeping practices will result in satisfactory data. In some instances, it may be necessary to remove the coating in order to calibrate the instrument on the exact substrate being measured. Here, the inspector will have to provide proper repair procedures.

The examples cited here are not exhaustive; indeed, they represent only a small cross-section of possible applications. Overall, it is hoped that this article will provide the inspector with a basic understanding of how to achieve a higher level of inspection.

Acknowledgements

The author appreciates the cooperation of those who supplied pictures and technical assistance, particularly Stan Mroz, Zormco Electronics; Ken Tator, KTA Associates; Frank Underhill, Gardner Laboratory, Inc.; and Ray Tooke, Micrometrics Co.

References

1. Bigos, Joseph, Anchor Pattern Profile and Its Effects on Paint Performance, *Corrosion*, Vol. 15, No. 8, pp 428–432, Aug. 1959.
2. McKelvie, A. N., Instrumentation as an Aid to Grit Control of Blast Cleaning Structural Steel Work, Ciria 38, Construction Industry Research and Information Assn., London.
3. Keane, J. D., et al., Surface Profile for Anti-Corrosion Paints, Steel Structures Painting Council, Oct. 1976.
4. Measurement of Wet Film Thickness of Paint, Varnish Lacquer and Related Products, Method A, Method B, ASTM D-1212-54 (1965) ASTM Part 27, 1975.
5. Keane, J. D., and Shoemaker, T. L., Development of Specifications for Measurement of Paint Thickness on Structural Steel, *J. of Paint Technology*, Oct. 1973, pp 46–67.
6. Hall, Charles H., Inspection and Inspection Instruments, NACE 12th Annual Liberty Bell Course, 1974.
7. Measurement of Dry Film Thickness of Organic Coatings, ASTM D-1005-51 (1972), Part 27, 1975.
8. Measurement of Dry Film Thickness of Non-magnetic Organic Coatings Applied on a Magnetic Base, ASTM D-1186-53 (1973), Part 27, 1975.
9. Plog, H., and Crosby, C. E., "A Guide to Methods and Instruments for Coating Thickness Measurement," Draper Ltd., Middlesex, U.K.

Electrochemical and galvanic corrosion of coated steel surfaces

Corrosion takes place when steel becomes the anode in an electrolytic cell. Here is how this process occurs.

Dean M. Berger, *Gilbert/Commonwealth*

☐ All metals assume their most stable state in nature. Unfortunately, the metals used by engineers—such as steel, zinc and copper—assume their most stable form as oxides, sulfides and similar compounds. By introducing sufficient energy, these compounds can be converted to pure metals. Then, these metals may be processed further into an unending series of items, the most common of which are steel structures. Steel offers the engineer the strength required for many applications.

Corrosion of structural grades of iron and steel, however, proceeds rapidly unless the metal is amply protected. This susceptibility to corrosion of iron and steel is of great concern because annual U.S. losses have been estimated at nearly $70 billion.

Metal corrosion

Corrosion, whether in the atmosphere, underwater or underground, is caused by a flow of electricity from one metal to another or to a recipient (i.e., soil) of some kind; or from one part of the surface of a piece of metal to another part.

An electrolyte is needed for this flow to occur. Water, especially salt water, is, of course, an excellent electrolyte. Simply stated, energy (electricity) passes from a negative area to a positive one via the electrolyte.

So, to have corrosion take place in metals, there must be an electrolyte, plus a metallic area or region with a negative charge in relation to a second area, and a second area positive in opposition to the first [*1*].

The recipient may be soil. This may happen because of the various compositions within a given soil: Soil frequently contains dispersed metallic particles or bacteria pockets that provide a natural electrical pathway to buried metal. If an electrolyte is present and the soil is negative in relation to the metal, the electric path will occur from the metal to the soil. And corrosion will result.

Water readily dissolves a small amount of oxygen from the atmosphere, and may become highly corrosive. When the free oxygen dissolved in water is removed, the water is practically noncorrosive, unless it becomes acidic or unless anaerobic bacteria incite corrosion. If oxygen-free water is kept neutral or slightly alkaline, it will be practically noncorrosive to steel. Thus, steam boilers and water-supply systems are effectively protected by deaeration of water.

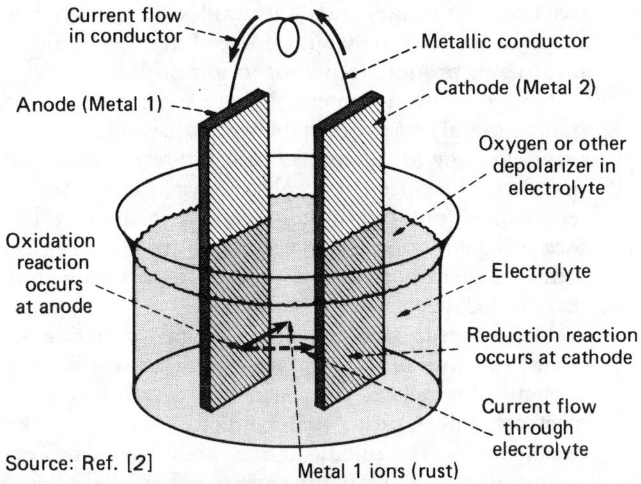

This simple corrosion cell results in billions of dollars of damage annually **Fig. 1**

Electrochemical corrosion

The cell shown in Fig. 1 illustrates the corrosion process in its simplest form. Oxygen is usually present as a depolarizing agent. Hydrogen gas is evolved when a metal corrodes in acid and when the corrosion rate is relatively rapid. A cathode having a layer of adsorbed gas bubbles as a consequence of the corrosion-cell reaction is said to be polarized. This reduces the consumption of metal by corrosion.

As can be seen in Fig. 1, the components form a closed electrical circuit. In the simplest case, the anode would be one metal, perhaps iron, the cathode another, say copper, and the electrolyte might or might not have the same composition at both electrodes. Alternatively, the electrodes could be of the same metal if the electrolyte composition varied.

For the cell shown, an electrical current would flow

Originally published June 28, 1982

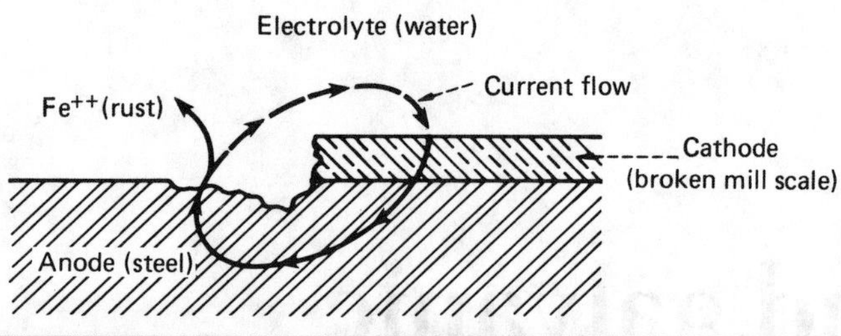

Mill scale is cathodic to steel, and establishes a corrosion cell with it **Fig. 2**

through the metallic conductor and the electrolyte. The anode would corrode (rust, if iron); this is an oxidation reaction. Simultaneously, a nondestructive chemical reaction (reduction) would proceed at the cathode, in most cases producing hydrogen gas on it.

The difference in potential that causes these currents is due mainly to contact between dissimilar metallic conductors or to differences in solution concentration, mainly with respect to dissolved oxygen in natural waters. Almost any lack of homogeneity of the metal surface or its environment may initiate attack, by causing a difference in potential. The result is corrosion that is usually localized.

Atmospheric corrosion differs from the action that occurs in water or underground, in that there is always a plentiful supply of oxygen. Such corrosion is mainly electrochemical, rather than being a chemical attack by the elements. The anodic and cathodic areas, however, are usually quite small and close together, so that corrosion is apparently uniform, rather than occurring as severe pitting, as is true for water or soil.

The larger the anodic area is in relation to the cathode, the faster the rate of corrosion. Anodes and cathodes exist on all iron and steel surfaces. They are caused by surface imperfections, grain orientation, lack of homogeneity of the metal, variation in the environment, localized shear and torque during manufacture, mill scale, or existing red iron oxide rust.

Rust equation

The formation of rust may be expressed as:

$$4Fe + 3O_2 + H_2O \rightarrow 2\,Fe_2O_3 \cdot H_2O.$$

The most stable form of rust is Fe_2O_3. At higher temperatures (900 to 1,300°F), Fe_2O_3 reverts to Fe_3O_4.

In an acidic environment, even without oxygen, the anodic metal is attacked rapidly. When acid corrosion results in salt formation, the reaction is slowed because of salt deposition on the surface being attacked.

Galvanic corrosion

Better known simply as dissimilar-metal corrosion, this occurs in the most unusual places and often causes the most considerable of professional problems.

The galvanic series of metals details how the galvanic current will flow between two metals, and which will corrode when they are in contact or near each other in the ground (see box). Metals near each other in the series do not have a strong effect on each other. The farther apart any two metals are, the stronger the corroding effect on the higher one in the galvanic series.

It is possible for certain metals to reverse their positions in some environments; but the galvanic series will generally hold in natural waters and in the atmosphere [*2*]. (The galvanic series should not be confused with the similar electromotive-force series, which shows exact potentials based on highly standardized conditions that rarely exist in nature.)

While the preceding galvanic series generally defines the available driving force to promote corrosion, the actual rate may vary. Electrolytes may be poor conductors, or long distances may introduce a large resistance into the corrosion-cell circuit. More frequently, scale forms a partially insulating layer over the anode.

The passivity of stainless steels or other metals or alloys is due to the presence of a corrosion-resistant oxide film on their surfaces. In most natural environments, such metals remain passive and thus tend to be cathodic to ordinary iron and steel. Change to an active state usually occurs only when chloride concentrations are

Galvanic series

Corroded end (anodic)

Magnesium
Magnesium alloys
Zinc
Aluminum 2S
Cadmium
Aluminum 17ST
Steel or iron
Cast iron
Chromium-iron (active)
Ni-resist
18-8 Chromium-nickel-iron (active)
18-8-3 Chromium-nickel-molybdenum-iron (active)
Lead-tin solder
Lead
Tin
Nickel (active)
Inconel (active)
Hastelloy C (active)
Brass
Copper
Bronzes
Copper-nickel alloys
Monel
Silver solder
Nickel (passive)
Inconel (passive)
Chromium-iron (passive)
18-8 Chromium-nickel-iron (passive)
18-8-3 Chromium-nickel-molybdenum-iron (passive)
Hastelloy C (passive)
Silver
Graphite
Gold
Platinum

Protected end (cathodic)

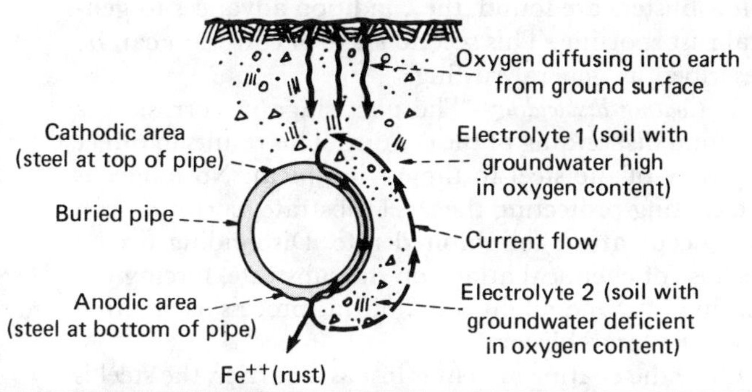

A steel pipe buried in moist soil often will corrode along its lower surface **Fig. 3**

high for stainless steel, as, for example, in seawater.

Accelerated corrosion of steel and iron can be produced by stray currents. Stray d.c. currents in the soil or water associated with nearby cathodic protection systems, industrial activities, or electric railways can be intercepted and carried for considerable distances by buried steel structures. Corrosion takes place when stray currents are discharged from the steel to the environment, and structural damage occurs rapidly [*3*].

Mill scale

If rust or mill scale is present on the surface of the steel, galvanic corrosion will occur. This is due to a dissimilarity with the base metal. The metal is the anode.

The difference in potential generated between steel and mill scale often amounts to 0.2 to 0.3 V; this couple is nearly as powerful a generator of corrosion currents as is the copper-steel couple. Fig. 2 shows how a pit forms where a break occurs in the scale. When contact between dissimilar materials is unavoidable and their surfaces are painted, it is important to paint both materials— especially the cathode. If only the anode is coated, any weak points, such as pinholes in the coating, will probably result in intense pitting [*3*].

In general, mill scale is magnetic and contains three layers of iron oxide, but the boundaries between the oxides are not sharp. The outer layer is essentially ferric oxide, Fe_2O_3, which is relatively stable and does not react easily. The layer closest to the steel surface and sometimes intermingled with the surface's crystalline structure is ferrous oxide, FeO. This substance is unstable, and the iron in it is easily oxidized to the ferric state. This process, accompanied by an increase in volume, may result in loosening of the scale.

The intermediate layer of magnetic oxide is best represented by Fe_3O_4. The actual thickness of mill scale on structural steel, which depends upon rolling conditions, varies from about 0.002 to about 0.020 in., and consists mainly of Fe_3O_4 and FeO. Much of the mill scale formed at high initial rolling temperature is knocked off in subsequent rolling [*3*].

Soil conditions

Differences in soil conditions, such as moisture content and resistivity, commonly are responsible for creating anodic and cathodic areas. Cathodes develop at points of relatively high oxygen concentration and anodes at points of low concentration.

Strained portions of metal tend to be anodic and unstrained portions cathodic. Thus, under all ordinary circumstances where iron and steel are exposed to natural environments, the basic conditions essential to corrosion are present to a greater or lesser degree.

A metal pipe buried in moist soil may corrode on the bottom (see Fig. 3). A variation in oxygen content at different levels in the electrolyte causes this. Thus, anodic and cathodic areas will develop, and a corrosion cell, called concentration cell, will form.

Concentration cells

Severe corrosion, leading to pitting, is often caused by concentration cells, particularly where differences in dissolved-oxygen concentration occur. When a part of the metal is in contact with water relatively low in dissolved oxygen, it is, of course, anodic to adjoining areas in contact with water higher in dissolved oxygen.

This lack of oxygen may be caused by exhaustion of dissolved oxygen in a crevice (see Fig. 4). The low-oxygen area is always anodic. Fig. 4 also illustrates another type of concentration cell; this cell, at the mouth of a crevice, is created by differences in concentration of the metal in solution. These two effects sometimes work together, as in a re-entrant angle in a riveted seam.

As a pit (perhaps caused by broken mill scale) becomes deeper, an oxygen concentration cell is started by the depletion of oxygen in the pit, and the rate of penetration is accelerated.

Coating systems

Using these systems is vital to protect steel. Coatings help prevent corrosion by providing:

1. Sacrificial or galvanic protection.
2. Passivation of the steel (inhibitive pigments).
3. A barrier against the environment.

Sacrificial coatings—Zinc-rich primers are applied at 3.0-mil dry-film thickness to provide galvanic protection. These primers are very effective, even in chemical environments, since the zinc will dissipate before the steel is attacked. Adequate high-performance topcoats are recommended to prolong the life of the coating.

Corrosion inhibitors—Most paint primers contain a partially soluble inhibitor pigment, e.g., zinc chromate, that reacts with the steel substrate to form the iron salt. Such salt slows down corrosion. Chromates, phosphates, molybdates, borates, silicates and plumbates are commonly used. Some pigments passivate by contributing alkalinity, thereby slowing down attack on steel.

Barrier coatings—Protective coatings are the most widely used and recognized forms of barrier material. These barriers may vary in thickness from thin paint films of only a few mils, to heavy mastic coatings applied at about ¼ to ½ in., to acidproof brick linings several inches thick. Barrier coatings are effective because they keep moisture and oxygen away from the steel substrate.

Coating breakdown

Most coating films permit chemicals, moisture and oxygen to permeate them and attack the steel. This

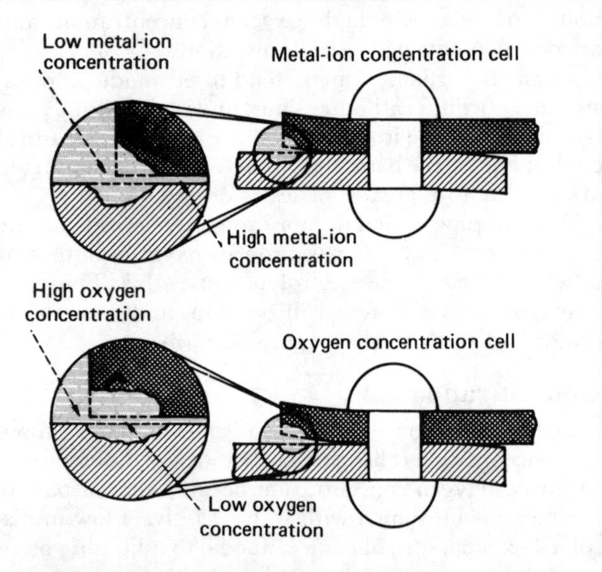

Concentration cells result in corrosion at crevices. These cells lead to pit formation **Fig. 4**

phenomenon is accentuated at temperatures between 150 and 210°F. Gaseous penetration occurs not only through pinholes and other micropores, but also through the coating film itself. The movement of penetrants through the film is fostered principally by osmotic and electroendosmotic pressures and the constant thermally induced movements and vibrations of the coating-film molecules [*4*].

The breakdown of protective coatings over steel substrates may be analyzed in a stepwise procedure. These steps usually are recognized visually and are a part of corrosion analysis:

1. Blistering—Early stages of corrosion, recognized as blistering, are often neglected. They have been described many times as "rust spotting." Standards for determining and evaluating the degree of rust spotting are found in the Steel Structures Painting Council standard Vis-2, or American Soc. for Testing and Materials (ASTM) D 610-68.

Frequently, blistering occurs without external evidence of rusting. The mechanism of blistering is attributed to osmotic attack or a dilation of the coating film at the interface with the steel, under the influence of moisture. Water and chemical gases pass through the film, dissolve ionic material from either the film or the substrate, and cause an osmotic pressure. This establishes a solute concentration gradient, with water building up at these sites until the film eventually blisters. Visual blistering standards are found in the standard, ASTM D 714-56.

Blistering is also dependent upon electrochemical reactions. Water diffuses through a coating also by an electroendosmotic gradient. After corrosion has started, moisture is pulled through the coating by an electrical potential gradient between the corroding area and the protected areas in electrical contact. Therefore, osmosis starts the blistering and, as soon as corrosion begins, electroendosmotic reactions accelerate the corrosion process greatly.

2. Rusting—After one rust spot has been observed or a few blisters are found, the condition advances to general rust spotting. This second stage of corrosion can be described as general rusting.

3. Coating disbonding—The next stage of corrosion is the total disbonding of the coating. This results in direct exposure of the steel to the environment. No longer is the coating protecting the steel substrate; corrosion can now occur at an uninhibited rate. Disbonding occurs because of chemical attack on the substrate, forcing the coating off. One might think of this process as creating one very large blister.

Once the coating system is lost as a barrier, the steel is left to be attacked directly. This attack most frequently is not uniform but rather localized, as many electrolytic cells.

4. Pitting and flaking—Pitting develops when the anodic (corroding) area is small in relation to the cathodic area (a corrosion cell). In a short time, the pitting undercuts mill scale, and flaking occurs. Pitting causes structural failure from localized weakening effects, even while there is still considerable sound metal remaining.

As the corrosion cell becomes more active, the rusting and pitting become more advanced. Deep pits in steel may eventually penetrate completely to create holes. Such penetrations generally result in structural loss. Within the corrosion cell, pitting occurs to such a degree that undercutting, flaking and delamination of the steel is seen. Most of these pits have a conical configuration. After a small hole develops, the electrolyte now can seek other fresh surfaces on the reverse side. This enables corrosion to occur on both surfaces, front and back.

5. Structural loss—As the chemicals attack the unprotected substrate, corrosion occurs at its most rapid and aggressive rate. Large gaping holes are found, causing considerable structural damage. These holes are rapidly enlarged because the electrolyte is ever present on both front and back surfaces of the steel. Catastrophic failure can result.

References

1. Introduction to Corrosion, Carboline Co., St. Louis, Mo., 1968.
2. "Paint Manual," 3rd ed., U.S. Bureau of Reclamation, U.S. Dept. of Interior, Denver, Colo., Chap III, 1976, Stock No. 024-003-00104-0.
3. Speller, F. N., "Corrosion: Causes and Prevention," McGraw-Hill Book Co., New York, 1951.
4. Hare, C. H., Anti-Corrosion Barrier and Inhibitive Pigments, Unit 27, Federation of Societies for Coating Technology, Philadelphia, Pa., 1979 (pamphlet).

The author

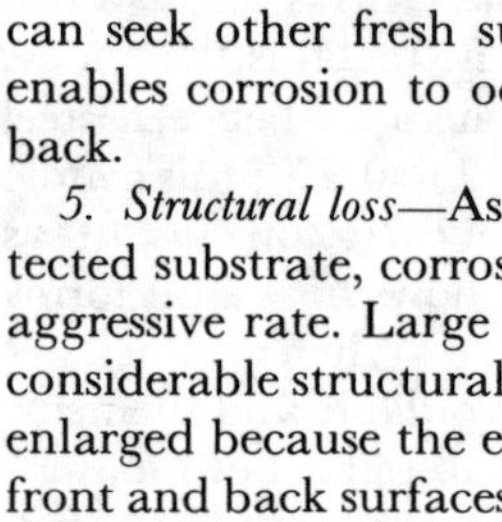

Dean M. Berger is a coating specialist in the chemical engineering department of Gilbert/Commonwealth, P.O. Box 1498, Reading, PA 19603, telephone 215-775-2600. He received his B.S. in chemistry from North Central College, and has done graduate work at the University of Wisconsin. His experience includes 20 years in research and development with PPG Industries and work with Union Carbide Corp. He has devoted many years to the Steel Structures Painting Council, National Assn. of Corrosion Engineers, American Soc. for Testing and Materials, and Federation of Societies for Coating Technology.

Structural steel: Galvanizing vs. painting?

Except in certain corrosive environments, galvanizing is generally the less expensive of the two methods of protection when lifetime project costs are considered.

Steven M. Seelinger, *BASF Wyandotte Corp.*

☐ In choosing among alternatives to suit an engineering need, the first question to ask is: Will each alternative perform the required job? Once this is answered, another is posed: What are the relative costs of the choices?

Often, the answer is easy—one system quickly proves to be much less expensive. Sometimes, however, an analysis of lifetime costs is needed. Such is the case when the choice is between painting and galvanizing structural steel for new-plant construction. Here, we will look at the durability of galvanizing and painting, and then compare their costs over a lifetime cost-cycle.

Durability of galvanizing

The corrosion resistance of galvanizing is strongest in neutral to moderately alkaline environments. In these, zinc develops a patina of stable, insoluble salts, mostly oxides, hydroxides or carbonates. The life of a galvanized component in these atmospheres is well beyond that required for a manufacturing plant.

However, acidic or highly basic atmospheres can shorten the life of galvanized surfaces. Soluble zinc salts are formed that offer no resistance to further attack. Acid gases present such a problem. But in these cases, corrosion takes place only if moisture is present. Where the climate leaves a structure alternately wet and dry, the cumulative corrosion will, of course, be less than would be shown, for example, in immersion-corrosion tests. No doubt this accounts for the long life of many exposed galvanized structures in chemical-process-industries environments.

Tests performed by the British Iron and Steel Research Assn. [*1*] on galvanized structures in an extremely corrosive atmosphere show that coating life varies linearly with coating thickness. The Galvanizers' Assn. of Australia [*1*] has shown that a 100-μm (4-mil) coating has a life ranging from 25 yr in a "severe marine/industrial" environment to 100 yr in a "rural" one.

The ASTM (Amer. Soc. for Testing and Materials) 20-yr-exposure tests [*2*] found that it takes 3.8 yr to corrode a galvanized coating 1 mil in a heavy industrial environment. Thus, a 4-mil coating should have a service life of about 15 yr. This is thin; typical coatings on structural steel are 7 mils.

On the other hand, direct spills of chemicals, and severe, localized exposure of fumes can severely shorten the life of galvanized structures. Zinc is ineffective at pHs below 6 or above 12 [*3*], conditions found in plants that handle acids or strong bases.

Thus, galvanizing will perform well for up to 25 yr in heavy industrial environments as long as there is no direct contact with acidic or basic spills, or with concentrated fumes of similar composition. For highly corrosive environments, high-performance paints outperform galvanizing.

Durability of paint

The life of a paint system depends upon the quality of its application (mostly on surface preparation) and the environment. A 7-yr life has been determined for a three-coat paint system (i.e., an inorganic zinc primer and epoxy intermediate and top coats) in a harsh industrial environment [*4*]. Life is defined as the time to

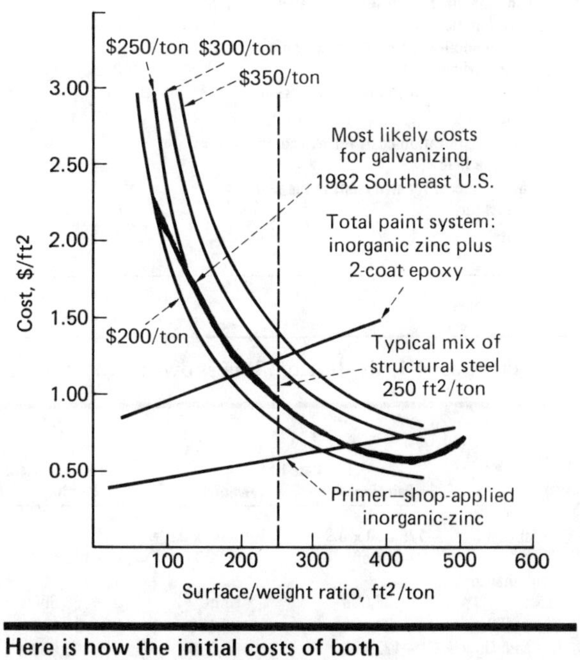

Here is how the initial costs of both methods compare as factors vary

Editor's Note: This article was originally presented at Corrosion/83, Anaheim, Calif., Apr. 18-22, 1983. Publication rights have been waived by the Natl. Assn. of Corrosion Engineers.

10 – 15% coating failure. Compared with a 25-yr life for bare galvanizing, painting is not effective.

While such a comparison appears at first to overstate the case for galvanizing, it is reasonable when the nature of galvanizing is compared with that of inorganic zinc primers. Galvanizing results in a diffusion coating with a metal-to-metal bond that is much stronger than that between the paint film and metal. Substrate bonding problems are rarely seen with galvanizing, but are common with inorganic zinc primers.

Also, while galvanizing is typically 7 mils thick, inorganic zinc primers are usually 2 – 3 mils. Unlike zinc primers, galvanizing is not porous.

Finally, inorganic zinc primers require labor-intensive abrasive blasting. More errors can occur during blasting than during galvanizing. Thus, galvanizing has a longer life and lower maintenance costs than does a paint system based on an inorganic zinc primer.

(This is not meant as a condemnation of these primers, since all paint films have some of the same shortcomings.)

Initial costs

Galvanizing of structural steel is usually quoted on a cost-per-weight basis ($/ton or $/hundredweight). A weight basis is used because galvanizing costs are more a function of weight than of area, since the steel is a major heat sink. Much of the cost comes from the energy needed to keep the zinc bath molten (840°F). Thus, for galvanizing, labor costs are not as much a function of area as they are for painting.

The galvanizing cost/ton is fairly constant over a wide range of medium-weight structural-steel pieces, but will vary for very light or heavy parts. A first-quarter 1982 cost survey of galvanizing shops was made by the author. For Texas, costs were about $175 – $200/ton for light and medium steel. A base cost of $200/ton is assumed. Total cost of galvanizing the steel also includes shipping and handling, straightening thermally distorted steel, and field touchup.

The author's estimates for these are $20/ton for straightening and $30/ton for extra handling and touchup. Thus, the final, installed cost would be $250/ton. This figure will be used here. It is not applicable to handrails, fasteners or floor gratings.

Often, the initial cost of painting is higher than that of galvanizing — Table I

Primer	Cost, $/ft²
Commercial blast using shop-based automatic equipment	0.24
Spray-application cost for inorganic-zinc primer	0.20
Material cost for inorganic-zinc primer	0.11
Top coats	
Spray-application costs for two coats in the field	0.34
Material cost for two coats in the field	0.18
Touchup	0.12
Total	1.19

Painting costs

These include such factors as paint cost, surface preparation, and labor charges. A comprehensive cost summary that accounts for such factors is published and periodically updated by Brevoort and and Roebuck [4]. Table I shows 1981 Gulf Coast figures for a three-coat system of an inorganic zinc primer with two epoxy coats. To compare galvanizing with painting, let us use $1.25/ft² for 1982 costs for the former.

Cost comparison

The figure shows the costs developed so far. The galvanizing cost of $250/ton is equivalent to $1.00/ft². Painting costs per square foot will increase as the surface/weight ratio increases. The opposite is true for galvanizing.

However, more than just initial cost is involved in comparing the two methods. A cash-flow analysis is needed to see which system is actually cheaper. Painting vs. galvanizing should be looked at as an investment decision, and this matter will now be covered.

Cash-flow analysis

The various factors that make up a cash-flow analysis will be mentioned briefly. Space does not permit either a detailed discussion of this topic or presentation of the formulas used to calculate net present value. These methods and formulas are standard, and are found in dozens of references. We will use the corporate tax rate of 46%, and a 10-yr life of the investment. The economic factors involved in performing a cash-flow analysis are:

Initial cost—The initial cost is offset by a 10% tax credit, which is allowed in the U.S. to stimulate economic expansion. So, the initial cash outflow equals the initial cost × (1 − investment tax credit), or initial cost × 0.90.

Depreciation—This results in a positive cash flow. Struc-

In all but Case III, the bottom line shows galvanizing to be the economical choice — Table II

All numbers are in dollars

	Case I		Case II		Case III		Case IV	
	Painting	Galvanizing	Painting	Galvanizing	Painting	Galvanizing	Painting	Galvanizing
Initial cost	−125,000 x 0.9 = −112,500	−100,000 x 0.9 = −90,000	−112,500	−90,000	−112,500	−90,000	−180,000 x 0.9 = −162,000	−145,000 x 0.9 = −130,500
Maintenance cost	−85,000	None	−60,500	None	−135,400	−182,800	−70,600	None
Depreciation	+24,100	+22,400	+24,100	+22,400	+24,100	+22,400	+34,700	+27,900
Net cash flow	−173,400	−67,600	−148,900	−67,600	−233,800	−250,400	−197,900	−102,600

tural steel, according to accountants, has a life of 10 yr; straight-line depreciation is used.

Maintenance costs—Maintenance involves either touch-up or major replacement painting. These are, of course, tax-deductible expenses.

Inflation—The cost of maintenance painting increases with the inflation rate. A figure of 10%/yr is used here.

Interest rate—The rate of return on an investment must be compared with some standard—some rate of return or interest rate. A value of 20%/yr is used.

The present value is then equal to the tax saving from depreciation less maintenance costs less initial costs.

Applying the economic analysis

Now, we will apply the above principles to determine whether galvanizing or painting is the better choice. (The reader will, of course, want to use his or her local figures and current prices). Four cases are considered. Only the results will be given.

Case I—An ideal application for galvanizing. It survives for 25 yr without maintenance. The painted steel is not touched up, but fully repainted periodically. Assume a heavy industrial environment, but without exposure to harmful chemical spills and fumes. Maintenance costs are assumed to be negligible. The steel will have 4 oz of zinc/ft² of area, resulting in the 25-yr life. The three-coat paint system is assumed to have a 7-yr life [*4*]. Assume that the plant will be completely painted after 7 yr, with no maintenance before that. Further, assume that maintenance-painting costs of erected steel are 50% greater than initial costs of painting the steel on the ground.

The plant has 100,000 ft² of steel or 400 tons at 250 ft²/ton. The lifespan is 20 yr. The cash-flow analysis for all cases is shown in Table III.

Case II—An ideal application for galvanizing. This is identical to Case I, except that the paint will be touched up, starting at the end of its "life" rather than be replaced totally at such time. Such an assumption is more in line with the author's intent that life is the time to 10 – 15% failure, not the time for total or large-scale failure. Assume that touchup costs are 10% of the initial installed costs, and rise with inflation each year.

Case III—A poor application for galvanizing. Here are problems of harmful spills and fumes. Assume that certain parts of the plant have a more corrosive environment than others, such that galvanizing on half of the steel will be lost after 2 yr, and on the other half after 4 yr. The size of plant is the same as before. All repainting is assumed as an expense.

Case IV—An actual case for an organics plant in Louisiana. The painted structural steel was not maintained for 3 yr, and then a touchup program was started that delayed repainting indefinitely. The yearly cost was $20,000. The plant has 575 tons of steel. Again, a weight ratio of 250 ft²/ton is assumed.

The initial cash flow, maintenance, depreciation and present value for each case appear in Table III. In all but Case III, painting will require about twice the cash outlay of galvanizing.

For cases above, the initial cost of galvanizing is lower than that of painting. What if this were reversed? If so, cash-flow analysis could be used to find the maximum justifiable ratio of initial galvanizing to painting costs.

This ratio is found by equating the total-lifetime painting and galvanizing costs, and solving the resulting equation for C_g/C_p, the initial cost of galvanizing/painting. A more detailed analysis of this is given by Kinstler [*5*]. Values of C_g/C_p appear in Table III. In some cases galvanizing is more economical even if it requires an initial outlay more than twice that of painting.

Acknowledgments

The author wishes to thank Gordon Brevoort of Sigma Coatings, Inc. and Thomas Kinstler of Metal Plate Galvanizing for help in developing this paper.

Sometimes, galvanizing can initially cost over twice that of painting, yet end up cheaper — **Table III**

Assumes ideal application for galvanizing. No maintenance costs; a 20-yr life.

	C_g/C_p
r = 0.10 R = 0.20 7-yr repaint cycle (repaint 2 times in 20-yr life)	1.94
r = 0.10 R = 0.15 7-yr repaint cycle (repaint 2 times in 20-yr life)	2.54
r = 0.10 R = 0.20 12-yr repaint cycle (repaint only once in 20-yr life)	1.40
r = 0.10 R = 0.20 5-yr repaint cycle (repaint 3 times in 20-yr life)	2.52

where r = inflation rate, %/yr
R = interest rate, %/yr

References

1. Galvanizers' Assn. of Australia, Melbourne, "Galvanizing," pp. 32-33.
2. Zinc Institute, New York, Zinc Coatings for Corrosion Protection, pamphlet, p. 28.
3. Shreir, L. L., "Corrosion," 1st ed., Vol. I, John Wiley & Sons, New York, 1963, p. 4.129.
4. Brevoort, G. H., and Roebuck, A. H., Coating Costs and Estimating, paper presented at Natl. Assn. of Corrosion Engineers' Corrosion/82, Houston, March 22-26, 1982.
5. Kinstler, T. J., Probability Functions in Corrosion Economics—Or a Corrosion Engineer Goes to Monte Carlo, paper presented at Natl. Assn. of Corrosion Engineers' Corrosion/82, Houston, March 1982.

The author

Steven M. Seelinger is materials engineering associate for BASF Wyandotte Corp., 100 Cherry Hill Rd., Parsippany, NJ 07054. Tel: (201) 263-5328. He has spent more than ten years in specifying materials of construction, corrosion-control systems, fabrication procedures and quality-control testing procedures for equipment in the chemical process industries. He holds B.S. and M.S. degrees in materials engineering and an M.S. degree in management science from Rensselaer Polytechnic Institute. Mr. Seelinger is active in the local chapter of the Natl. Assn. of Corrosion Engineers and is a member of the Amer. Soc. for Metals.

Section IV
MATERIALS TO RESIST CHLORIDES AND CHLORINE

Alloys to resist chlorine, hydrogen chloride and hydrochloric acid

High-nickel alloys are very useful in handling these corrosive chemicals. Handy reference charts show when carbon steel will suffice and when greater corrosion resistance is required.

C. M. Schillmoller,
VDM Div., The Ore and Chemical Corp.

☐ Gaseous chlorine at low temperatures and in the absence of moisture is not severely corrosive and is commonly handled by carbon steel. Dry hydrogen chloride behaves in a similar way. However, the strongly acidic HCl is harmful to steel.

Each of these three substances is discussed under various conditions. Materials considered include the high-nickel alloys, stainless steel, high-molybdenum alloys, titanium, zirconium and tantalum.

Corrosion charts

Corrosion is a very complex process. Seemingly unimportant variables, such as small amounts of moisture, impurities, or the presence of metal chlorides, can change the corrosion picture completely.

To present corrosion data in concise form, a variety of methods have been proposed. Basically, the author is opposed to the presentation of information via simplified charts if this alone is used for selecting materials of construction. However, concise and condensed information *is* valuable in that it presents a bird's-eye view of the situation and can be used for screening purposes, thus minimizing the number of materials to be tested or considered.

How far can one go in condensing information and still have it be of substantial value? Fig. 1, 2 and 3 attempt to condense corrosion data so the general picture can be obtained at a glance. Nickel and high-nickel alloys are among the few proven metallic materials that show good corrosion resistance in chlorine, hydrogen chloride and hydrochloric acid. These alloys are, of course, not new to the chlor-alkali industry, being more or less standard selections in caustic, brine and salt processing.

Table 1 provides a brief description of the alloys commonly in use, their ASTM specifications and the tradenames under which they are known. References in the text to Alloy 200, Alloy 400 and so on correspond to the VDM tradenames (shown in Table I) with which the author has the most recent experience.

In the chemical process industries, the common design parameter for tubing, valve trim and internals is 0.075 mm/yr (0.003 in./yr) maximum corrosion rate, while for vessels and pipe, an upper corrosion rate of 0.50 mm/yr (0.020 in./yr) is frequently adopted, with a corrosion allowance of 3 to 6 mm (1/8 to 1/4 in.). This should provide a safe life of ten years or more. The charts summarize the limits of usefulness for the various alloys. Let us now deal with each of the three corrosive environments.

Chlorine

Gaseous chlorine at low temperatures and in the absence of moisture is not severely corrosive and is commonly handled by carbon steel. Usually, a more resistant material such as Alloy 400 is specified for critical parts such as valve trim, instrumentation and orifice plates

Alloys commonly used in Cl_2 and HCl systems — Table I

Materials	Reference in text	Nominal composition, % Ni	Cr	Mo	Cu	Fe	ASTM/ ASME B number	VDM tradenames	Comparable products
Nickel									
Nickel	Alloy 200	99.6	–	–	–	–	161-163	VDM-Nickel 200	Nickel 200
Low-carbon nickel	Alloy 201	99.6	–	–	–	–	161-163	VDM-Nickel 201	Nickel 201
Nickel-copper alloys									
Nickel-copper alloy	Alloy 400	67	–	–	31	1.5	163-165	Nicorros 400	Monel 400
Nickel-chromium-iron alloys									
Nickel-chromium-iron alloy	Alloy 600	76	15	–	–	8	163-168	Nicrofer 600	Inconel 600
Nickel-iron-chromium alloy	Alloy 800	32	21	–	–	46	163-407	Nicrofer 800	Incoloy 800
Nickel-iron-chromium-molybdenum-copper alloy	Alloy 825	42	21	3	2.3	30	163-423	Nicrofer 825	Incoloy 825
High-molybdenum alloys									
Nickel-chromium-molybdenum-iron alloy	Alloy 625	61	21.5	9	–	4	443-446	Nicrofer 6020 HMo	Inconel 625
Nickel-chromium-molybdenum-iron alloy	Alloy C-4	63	16	15	–	2	575-622	–	Hastelloy C-4
Nickel-iron-chromium-molybdenum-copper alloy	Alloy 825 HMo	42	21	4.5	2.3	28	–	Nicrofer 4221 HMo	–
Nickel-iron-chromium-molybdenum-copper alloy	Alloy 904 HMo	25	20.5	6	1.5	46	–	Cronifer 1925 HMo	Avesta 254
Nickel-molybdenum-iron alloy	Alloy B-2	68	–	28	–	1.5	333-622	–	Hastelloy B-2

Note: Nicorros and Nicrofer are registered trademarks of Vereinigte Deutsche Metallwerke AG, West Germany.
Monel, Inconel, Incoloy are registered trademarks of The International Nickel Co.
Hastelloy is a registered trademark of Cabot Corp., Stellite Div.
Avesta 254 SMO is a trademark of Avesta Jernverks AB, Sweden.

Originally published March 10, 1980

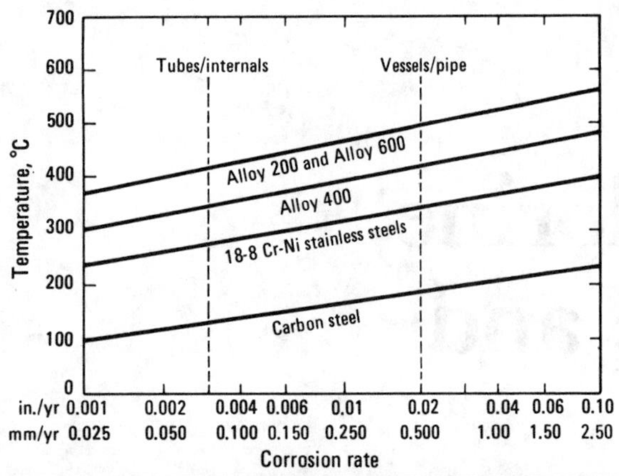

Upper design limits for various alloys in chlorine Fig. 1

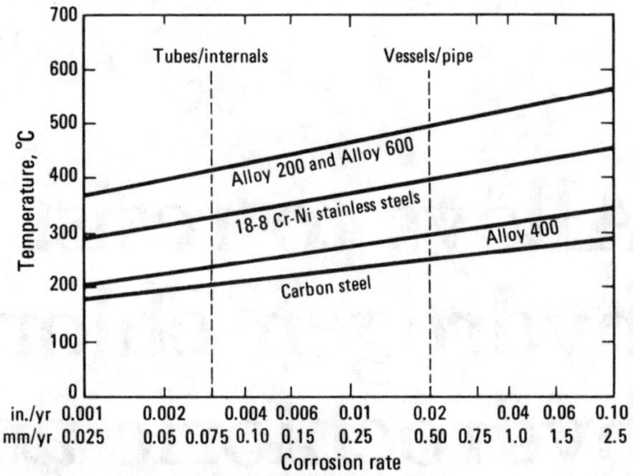

Upper design limits for various alloys in hydrogen chloride Fig. 2

in chlorine pipelines. In contrast, wet chlorine is extremely corrosive on steel and nickel alloys, and requires Alloy C-4 or titanium. The upper limit of usefulness is around 200°C, at which temperature the protective effects of the corrosion products disappear.

Fig. 1 provides a guide to the selection of various alloys for dry chlorine, and indicates design parameters for tubes/internals and vessels/pipe-components.

The surface coating of chlorides on the alloys tends to provide protection up to a temperature level at which melting, vaporization or decomposition removes such films. The corrosion rate appears proportional to the vapor pressure of the metal chlorides.

Alloy 200 and Alloy 600 are those most commonly used for reactors, coils, agitators and piping in the 250-500°C range. Carbon steel can still be used below 150°C.

When the plant is not in operation, proper shutdown procedures should be used to keep the units dry or free from chlorine, so as to prevent attack by wet residual chlorine on the steel or nickel.

Example—Ethylene is to be reacted with chlorine in the presence of a ferric chloride catalyst to produce ethylene dichloride (EDC). The reactor temperature is 60-100°C. The process is exothermic; water cooling removes the heat of reaction.

Fig. 1 indicates that carbon steel can be used for the reactor and auxiliary equipment, provided that the chlorine feedstock is dry, and that proper control of temperature is maintained by thorough mixing of the reactants to prevent hot spots and runaway temperatures. Intimate mixing can be ensured by using EDC as a reaction medium. Alloy 200 should be considered for reactor internals and critical components if experience shows difficulty in controlling temperatures below 150°C, or if it is desirable to operate above this temperature.

Hydrogen chloride

Dry HCl behaves similarly to Cl_2, and carbon steel will suffice up to 250°C, above which Alloy 200 is usually specified. Fig. 2 provides a guide to the selection of various alloys in dry HCl gas. Upper corrosion limits of 0.075 and 0.50 mm/yr are shown as design parameters for certain components. In operations above the dewpoint the presence of moisture does not appreciably increase corrosion rates, unless the temperature drops and the moisture condenses.

It must be pointed out that there are many variables and that even small amounts of addition agents to control catalyst activity, for example, may exert an influence on the tenacity and vapor pressure of the protective corrosion scales. Therefore, absolute corrosion rates for different temperatures in both Cl_2 and HCl systems are difficult to predict. Even so, Fig. 1 and 2 should be sufficiently accurate to serve as a guide.

The performance of Alloy 200 in dry as well as wet HCl gas has been consistently good. In cyclic operating conditions, particularly in the presence of air or oxygen, Alloy 600 and Alloy 825 offer good all-round resistance. It is prudent to assume that Types 304 and 316 stainless steels would be subject to chloride stress-corrosion-cracking conditions during shutdown, despite various precautions that may be taken.

Example—Ethylene is to be reacted with dry hydrogen chloride and oxygen in the presence of copper chloride catalyst in a fixed-bed reactor to produce EDC. The temperature is 275°C and the pressure is 10 atm. The process is exothermic; reaction heat is removed by the generation of steam on the shell side of the reactor.

Fig. 2 indicates that stainless steel, Alloy 200 and Alloy 600 are candidate materials and resist dry and even moist hydrogen chloride. Usually, Alloy 200 is used for the reactor tubes; the tubesheets and heads of the reactor are clad with nickel on steel; and the interconnecting piping between the reactors is made of Alloy 200. Temperatures should be carefully controlled in this exothermic reaction because of byproduct formation and deactivation of the catalyst above 325°C. Alloy 200 has an upper-temperature limit of 550°C, and with localized hot spots of, say 750°C, catastrophic rates of corrosion and tube failure will occur.

Types 304 and 316 stainless steels are subject to chloride stress-corrosion-cracking below the dewpoint and during shutdown, unless extreme precautions are taken to ensure a bone-dry feed to the unit and to maintain shutdown and startup precautions of gas-blanketing and keeping the unit dry. Alloys 800 and 825 resist the

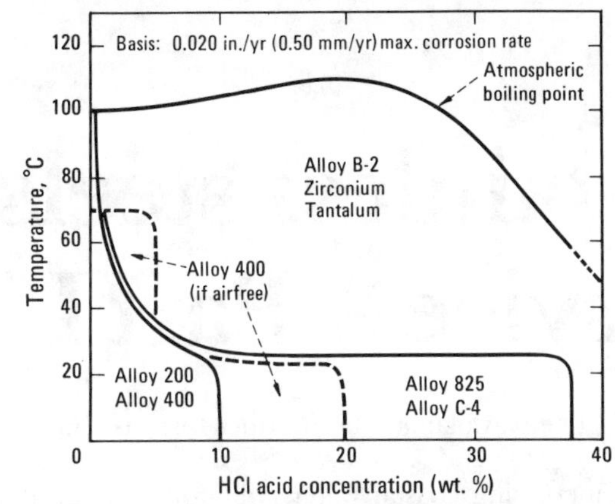

Hydrochloric acid—alloy selection guide **Fig. 3**

chloride stress-cracking phenomenon and have been used, respectively, for EDC-pyrolysis furnace tubing and fluid-bed oxychlorination reactor internals.

Hydrochloric acid

HCl is a typical reducing acid over its entire concentration range. Its strongly acidic character is harmful to steel. Alloy 200 and Alloy 400 behave similarly and find application at ambient temperatures up to 20% concentration, and at higher temperatures below 5% concentration. Fig. 3 shows isocorrosion lines for 0.050 mm/yr (0.020 in./yr), which is generally considered the upper design limit of an alloy selection. The graph immediately delineates the conditions suitable for handling with Alloy 400, and those where Alloy B-2 is required.

At ambient temperature, Alloy 825 and Alloy C-4 can be used over the full range of HCl acid concentration. These alloys are also selected to resist chloride stress-corrosion-cracking—and when oxidizing salts such as ferric and cupric chlorides are present in the system, because these salts strongly accelerate attack on Alloy 200, Alloy 400 and Alloy B-2.

In recent years, titanium [*1*] and zirconium have been increasingly used as their price structure in tubular components has become competitive with the high-nickel alloys. Commercially-pure titanium has provided useful life under mild reducing conditions with a pH above 1.5, and has performed excellently in the presence of oxidizing salts. Performance can be further improved by using palladium-modified titanium. Zirconium tubing is a useful alternative to Alloy B-2 tubing in HCl acid applications. Tantalum [*2*] is used for the sheathing of certain critical components, though its application is limited by high costs.

Example—In the distillation process, entrained water is carried along, and dilute hydrochloric acid is formed by the hydrolysis of organic chlorides, when the stream is cooled below 125°C. Excessive corrosion occurs on carbon steel in the condensers, piping, and in the bottom of the accumulator.

Depending on temperature and HCl acid concentration, corrosion rates on carbon steel frequently run 0.25-4.0 mm/yr. It is extremely difficult to ensure a bone-dry system without inadvertent moisture pickup at flanges and seals, or to prevent entrainment of water. It should be noted that acid concentrations in such cases are mostly less than 0.5% and that, as Fig. 3 shows, Alloy 400 can withstand such conditions satisfactorily.

Alloy 400 has been used at ambient temperature in reducing, airfree systems up to 20% HCl acid concentration. Alloy B-2 and zirconium can resist the full range of concentration and temperature.

Technology trends

No major changes in the use of established alloys of construction are foreseen except for the wider application of titanium and zirconium in tubular components. At higher temperatures in chlorination (and bromination) processes, there is a trend toward specifying special liners beyond the 550°C upper limit of usefulness of Alloy 200 and Alloy 600.

Emission-control standards are becoming more severe, necessitating the scrubbing and neutralization of HCl, the use of wastewater strippers to remove volatile organics, and the enclosing and collecting of emissions for incineration. Corrosion in parts of this equipment can be very severe, and the rate of attack is not always predictable. Some of the previously mentioned alloy-selection guidelines do apply, and new high-molybdenum alloys, such as Alloy 625, Alloy 825 HMo and Alloy 904 HMo very probably will find increasingly wider application (see Table 1).

First-cost considerations will continue to favor the use of plastic coatings and liners for various equipment and piping, in addition to glass-lined equipment, as an alternative to alloys, even though special care is required to maintained such coatings and prevent damage at joints.

Additional information is contained in the references.

References

1. Newman, J., Fighting corrosion with titanium castings, *Chem. Eng.*, June 4, 1979, pp. 149-152.
2. Flanders, R. B., Try tantalum for corrosion resistance, *Chem. Eng.*, Dec. 17, 1979, pp. 109-110.
3. Schillmoller, C. M., Alloy selection for vinyl chloride monomer plants, *Hydrocarbon Process.*, March 1979.
4. "Resistance of nickel and high-nickel alloys to corrosion by hydrochloric acid, hydrogen chloride and chlorine," INCO corrosion engineering Bull. CEB-3, July 1975.
5. Gladis, G. P., Effects of moisture on corrosion in petrochemical environments, *Chem. Eng. Prog.*, October 1960.
6. Schillmoller, C. M., and Mason, J. F., What to do about corroding isomerization units, *Pet. Refiner*, July 1958.

The author

C. M. Schillmoller directs technical marketing activities for VDM (Vereinigte Deutsche Metallwerke AG, West Germany) in the U.S. He is located at Suite 209, 10900 Northwest Freeway, Houston, TX 77092, telephone 713-682-2670. He spent 19 years with International Nickel in the U.S., Australia and Europe, followed by a period of management consulting in Brussels, Belgium. Mr. Schillmoller holds a degree in chemical engineering from the U. of Sydney, Australia. He is a member of AIChE, the National Assn. of Corrosion Engineers and the American Petroleum Institute.

High-performance stainless steels for high-chloride service—Part I

These materials have exceptional pitting and crevice corrosion resistance, and, in most cases, immunity to chloride stress-corrosion cracking.

James D. Redmond, Climax Molybdenum Co., and Kurt H. Miska, Climax Molybdenum Co. of Michigan

☐ Since 1980, the high-performance stainless steels have become the first choice for retubing seawater-cooled utility steam-condensers. Alloys such as AL-6X, Sea-Cure and AL 29–4C account for about 40 million feet of condenser tubing; some of these tubes have been in service for about ten years. A list of these stainless steels, along with their compositions and suppliers, appears in Table I. Their availability by product form is shown in Table II.

Part II will appear in the Aug. 22 issue.

With increasing coastal water-pollution, copper-alloy tubes have failed prematurely due to sulfide attack and erosion-corrosion; but the new stainlesses resist such failures and are economical as well. These new alloys are also suited for brackish water and high-chloride process-stream service. Here is an alternative to copper alloys, nickel alloys and titanium for heat-exchanger tubing, piping and equipment.

Excellent track record

The new alloys have been chosen by 24 utility companies, either to retube condensers or for new construction, at 55 generating stations, including both coal-fired and nuclear plants. Most of these installations are along the U.S. Atlantic and Gulf coasts, but several are in Europe. The seawater stainlesses have been used in steam con-

There is a variety of stainless steels to use for seawater and other high-chloride service — Table I

Tradename	Supplier	UNS* number	Microstructure	Nominal composition, wt% Cr	Mo	Ni	Other
Sea-Cure	Trent Tube Div., Colt Industries Universal-Cyclops Specialty Steel, Div. of Cyclops Corp.	S44660	Ferritic	27.5	3.5	1.2	Very-low C and N, Ti stabilized
AL 29-4-2	Allegheny Ludlum Steel Corp.	S44800	Ferritic	29	4	2	Ultra-low C and N
AL 29-4C	Allegheny Ludlum Steel Corp.	S44735	Ferritic	28.5	3.7	–	Very-low C and N, Ti stabilized
Monit	Uddeholm Steel Corp. Carpenter Technology Corp.	S44635	Ferritic	25	4	4	Very-low C and N, Ti stabilized
AL-6X	Allegheny Ludlum Steel Corp. AL Tech Specialty Steel Corp.	N08366	Austenitic	20	6	24	Very-low C
254 SMO	Avesta Jarnverks AB (Sweden) Ingersoll Johnson Steel Co.	S31254	Austenitic	20	6	18	0.7Cu, 0.20N
Cronifer 1925 HMO	VDM Technologies Corp.	–	Austenitic	20	6	25	1.5Cu, 0.12N
Ferralium alloy 255	Cabot Corp.–wrought products; Wisconsin Centrifugal, Inc., Esco Corp., Janney Cylinder Co.–castings	S32550	Duplex	26	3	5	2.0Cu, 0.17N

*Unified Numbering System (Amer. Soc. for Testing and Materials, and SAE–formerly Soc. of Automitive Engineers).

Note: This is not a comprehensive listing; there are other grades and suppliers.

Originally published July 25, 1983

Selecting an alloy depends, of course, partly on its available product-forms **Table II**

Stainless steel	Product forms
Sea-Cure	Welded tubing, sheet, strip
AL 29-4-2	Welded tubing, sheet, strip, plate up to 0.5 in., welding wire
AL 29-4C	Welded tubing, sheet, strip
Monit	Welded tubing, sheet, strip, welding wire
AL-6X	Welded tubing, sheet, strip, light plate, bar, wire
254 SMO	Welded tubing and pipe, sheet, strip, bar, plate up to 2 in., wire, castings
Cronifer 1925 HMO	Welded tubing and pipe, sheet, strip, bar, wire, plate
Ferralium alloy 255	Welded tubing and pipe, sheet, strip, bar, plate, wire, castings

Evaporator being retubed with AL-6X after pitting failure of Alloy 904L

densers by such utilities as Florida Power and Light, Houston Lighting and Power, and Long Island Lighting Co. Since they already have an excellent track record in the power industry, these alloys can now be used in chemical process industries (CPI) plants with some measure of certainty.

The new high-performance stainless steels are covered by a variety of ASTM specifications and ASME codes (see Table III), which are enabling designers to use these alloys in CPI heat exchangers, tanks, vessels and other equipment.

Improved alloying technology

The highly alloyed stainless steels offer a previously unavailable combination of corrosion resistance, weldability, toughness and economy. This is due to changes in melting and refining technology. Methods such as argon-oxygen decarburization (AOD), vacuum-oxygen decarburization (VOD), and vacuum-induction melting (VIM) have become standard. The use of such technology improves chromium recovery and provides excellent control of alloying conditions. Also, these methods yield low levels of carbon, nitrogen and sulfur, resulting in excellent mechanical properties and processibility.

These new steelmaking practices have made possible chromium-molybdenum-alloy stainlesses that have the necessary pitting and crevice-corrosion resistance for seawater service. This sets them apart from conventional stainless steels such as Types 304 and 316. Type 316 shows a low weight-loss corrosion in seawater; however, it suffers extensive pitting and crevice attack unless kept scrupulously clean—not a practical requirement for heat exchangers and condensers.

Ferritic, austenitic and duplex high-performance stainless steels are used in seawater service. Each type has its advantages and limitations, and will be discussed now.

Ferritic seawater stainlesses

These alloys are highly resistant to chloride stress-corrosion cracking (SCC). Table IV summarizes SCC resistance for the high-performance stainless steels and compares it with that of two conventional austenitics. The Ni-containing seawater ferritics fail the boiling $MgCl_2$ test (except for AL 29-4C), but pass two other common SCC tests. These latter two tests simulate practical service conditions—and these seawater stainlesses do, in fact, perform well under actual conditions.

The seawater stainless steels meet a variety of ASTM specifications and ASME codes **Table III**

Stainless steel	ASTM specification	ASME Boiler and Pressure Vessel Code, Sect. VIII, Div. 1
Sea-Cure	A 240, A 268, A 176, A 763	Code Case No. 1922
AL 29-4-2	A 176, A 240, A 268, A 276, A 314, A 479, A 731	SA 240, SA 268, SA 479, SA 731
AL 29-4C	A 268	Code Case No. 1921
Monit	A 176, A 240, A 268	Code Case No. 1900
AL-6X	B 675, B 676, B 688, B 690, B 691	Div. 1, Sect. III, Class 2 & 3, Code Case N-304, Code Case 1871
254 SMO	A 167, A 182, A 240, A 249, A 269, A 312, A 358, A 409, A 473, A 479	SA 240
Cronifer 1925 HMO	(B 625, B 649, B 673, B 674, B 677) application pending	–
Ferralium Alloy 255	A 240, A 479, A 789, A 790	Code Case No. 1883

Tests indicate how the seawater stainlesses withstand chloride stress-corrosion cracking Table IV

Stainless steel	Stress-Corrosion-Cracking Test: Boiling 42% $MgCl_2$	Wick test	Boiling 25% NaCl, pH=7
Sea-Cure	F	P	P
AL 29-4-2	F	P	P
AL 29-4C	P	P	P
Monit	F	P	P
AL-6X	F	P	P
254 SMO	F	P	P
Ferralium alloy 255	F	N	P
Type 304	F	F	F
Type 316	F	F	F

P = Passed F = Failed N = Not tested

The ferritics show good hot workability, and those having chromium levels of about 30% are easier to process than austenitics with 20% Cr. High chromium levels mean that less molybdenum is needed to achieve corrosion resistance. Some ferritics are alloyed with nickel to improve toughness, processibility and resistance in reducing acid environments.

When alloyed together, chromium and molybdenum have a synergistic effect. As a result, ferritics typically have lower alloy contents than do austenitics of comparable corrosion resistance, and are more efficient users of alloying elements. Since alloying costs are lower for ferritics, they can have an economic advantage over other grades.

Ferritics produced by AOD—e.g., AL 29–4C, Sea-Cure and Monit—contain small amounts of titanium, and preferentially form titanium carbonitrides during welding. This prevents sensitization (chromium carbide precipitation along grain boundaries) and subsequent loss of corrosion resistance. AL 29–4–2, produced by VIM, has such ultra-low carbon and nitrogen levels that no stabilizing element is required.

Ductile-brittle transition

Ferritics are characterized by a ductile-brittle transition temperature (DBTT). A crack initiated below this temperature propagates in a brittle fashion, while one above propagates in a ductile way. The DBTT is a function of many variables, including chromium, molybdenum, nickel, carbon and nitrogen content, grain size and thickness.

For thicknesses above about 0.12 in. (3 mm), the AOD-produced seawater ferritics have a DBTT above room temperature. For this reason, these alloys are typically used in thin sections, such as for tubing, light-wall pipe, sheet and strip. VIM-produced seawater ferritics have extremely low carbon and nitrogen levels and, thus, improved toughness. These materials are usually used in all product forms up to plate about 0.50 in. (12 mm) thick.

Austenitic seawater stainlesses

These materials resist SCC much better than do Types 304 and 316, but are not immune to stress cracking, as are the ferritics. The high total-alloy content of the high-performance austenitics makes them susceptible to precipitation of embrittling phases (e.g., chi and sigma phases), if they are either improperly heat-treated or subjected to high heat input during welding. Precipitation of these phases can significantly lower corrosion resistance and cause embrittlement.

Alloying with nitrogen retards sigma-phase formation, and allows for the production of thicker pieces. For example, AL-6X is available in light plate, while 254 SMO and Cronifer 1925 HMO are made in plates over 2 in. (50 mm). Also, 254 SMO is available as castings.

Generally, the austenitics are produced in a wider variety of forms than are the ferritics. These forms include strip, sheet, plate, tube, pipe, bar, other shapes, wire and wire rope.

Duplex seawater stainlesses

These alloys are a mixture of ferrite and austenite. Resistance to SCC is between that of the nickel-free ferritics and the 300-series austenitics, and decreases with increasing cold working. The duplex stainless steels can be cold-worked to strengths unachievable by either of the other types of stainlesses.

While all of the ferritics and most of the austenitics in Table I are available only as wrought products, Ferralium alloy 255 (a duplex) is made in both wrought and cast forms. This makes it particularly attractive for use in pumps, valves and fittings. (Note that there are several duplex alloys; however, Ferralium alloy 255 is the only one regarded as a seawater stainless steel.)

The authors

James D. Redmond is manager, stainless steel development, for Climax Molybdenum Co., 3072 One Oliver Plaza, Pittsburgh, PA 15222. Tel: (412) 281–6238. He is responsible for stainless-steel market development. He is the liaison between his company and U.S. stainless makers. The author of numerous articles on stainlesses, he holds a B.S. degree in metallurgical engineering from the University of Notre Dame and a Ph.D. degree in materials science from Northwestern University. He belongs to several societies, including the Natl. Assn. of Corrosion Engineers, the American Soc. for Metals, and ASTM.

Kurt H. Miska is supervisor, technical information, for Climax Molybdenum Co. of Michigan, P.O. Box 1568, Ann Arbor, MI 48106. Tel: (313) 761–2300. Before joining the company in 1978, he was an associate editor of *Materials Engineering* for seven years. There he covered developments in metals, composites, adhesives and welding. He has written hundreds of papers on a wide range of technical subjects.

High-performance stainless steels for high-chloride service—Part II

Here is the conclusion of this two-part series. Physical and mechanical properties and corrosion resistance are highlighted.

James D. Redmond, Climax Molybdenum Co., and
*Kurt H. Miska, Climax Molybdenum Co. of Michigan**

☐ Part I summarized the types of seawater stainless steels. Now, we will detail how they perform—including physical and mechanical data, and corrosion test results.

Physical and mechanical properties

Table I lists the physical and mechanical properties of these alloys, and compares them with those of other alloys used in seawater service.

*To meet the authors, see p. 126.

Generally, the ferritics have higher thermal conductivities than do the austenitics; the duplexes fall between the two. The coefficient of thermal expansion of the ferritic and duplex grades is similar to that of carbon steel, and is significantly less than that of the austenitics. Because of this, heat exchangers with ferritics or duplexes often eliminate the need for expansion bellows and floating heads.

The modulus of elasticity (or stiffness) of the austenitics is somewhat higher than that of many copper alloys, and the ferritics and duplexes are even stiffer. All of the seawater stainlesses have a significantly higher modulus of elasticity than do the titanium alloys. This can be important when copper-alloy tubes are replaced. Because of the high modulus of elasticity of the stainless steels, thin-walled tubes normally can replace copper tubes without modification of the tube-support-plate

Nominal mechanical and physical properties of the seawater stainlesses and other tube alloys — **Table I**

Alloy	Yield strength, ksi*	Tensile strength, ksi	Elongation, %	Hardness, HRB†	Thermal conductivity, Btu/(in.)(ft²)(h)(°F), at 68° F	Coefficient of thermal expansion, 10^{-6} in.(in.)(°F), at 68-200° F	Modulus of elasticity in tension, ksi
Sea-Cure	75	90	32	95	114	5.4	31.2
AL 29-4-2	85	95	22	97	105	5.2	—
AL 29-4C	75	90	25	95	119	5.2	30
Monit	95	105	26	98	118	5.9	30
AL-6X	40	90	45	80	95	8.5	29
254 SMO	44	95	35	96	95	9.4	28
Cronifer 1925 HMO	43	86	35	95	98	9.9	27
Ferralium alloy 255	87-99	116-130	25-34	22 HRC†	94	6.1	31
Type 304	35	85	50	80	108	9.6	29
Type 316	35	85	50	80	108	8.8	28
Titanium - Grade 2	50	70	20	92	114	4.8	15
Alloy 706 (90Cu-10Ni)	15	44	40	10	312	9.5	18
Alloy 715 (70Cu-30Ni)	21	57	43	44	204	9.0	22

*1,000 psi

†Rockwell B; Rockwell C

This table is a compilation of data from various sources. The values given may not necessarily be typical of a specific commercial material and should not be relied upon as representative for the purpose of application.

The critical crevice temperature gives an indication of overall corrosion resistance **Table II**

Stainless steel	Highest temperature exhibiting no crevice corrosion, °C
AL 29-4C	52.5
Sea-Cure	50.0
Monit	42.5
AL 29-4-2	40.0
254 SMO	32.5
Ferralium alloy 255	22.5
Cronifer 1925 HMO	15.0
Alloy 904L	0.0
Type 316	-2.5
Type 304	<−2.5

Critical crevice temperature for 1-d exposure in 10% ferric chloride, $FeCl_3 \cdot 6H_2O$, pH = 1

spacing. On the other hand, titanium requires much closer spacing to avoid vibration and the possibility of fatigue failure.

Corrosion resistance

Compared with the conventional 300-series stainless steels, the seawater types have greatly improved pitting and crevice-corrosion resistance. A common measure used to rank the corrosion resistance of materials is their critical crevice-corrosion temperature. Test coupons with a standard artificial crevice are immersed in a 10% ferric chloride solution of pH 1.0 for 24 h at a controlled temperature. The initial temperature selected depends on the composition of the steel and on experience with similar alloys. If no attack occurs, the temperature is raised by 2.5°C (4.5°F), and the test is repeated. The highest temperature at which no attack occurs is the critical crevice temperature. Values are given in Table II for the seawater stainlesses and some conventional stainlesses.

Although laboratory tests indicate seawater corrosion resistance, they are not a substitute for experience under actual conditions, and tests made with seawater. Climax Molybdenum has evaluated corrosion resistance in ambient seawater for up to 2 yr, and in filtered seawater at 25°C (77°F) for 60 d.

For both tests, specimens were 4 × 6 in. (100 × 150 mm), with multiple crevices, which were formed by grooved washers bolted to each side at a torque of 8.5 Nm (75 in.-lb). Each washer had 20 grooves, which gave 40 crevices per specimen. Triplicate specimens were exposed.

In the ambient-seawater tests, specimens were hung vertically in a wooden flume, in which there was a gravity flow of well-aerated seawater at 2 ft/s (0.6 m/s). Temperature was ambient, which ranged from 5 to 31 °C (41 to 88°F). Specimens were removed at 2, 9 and 24 mo.

In the filtered tests, the same-geometry specimens were used, and temperature was 25°C (77°F) at a flowrate of less than 0.32 ft/s (0.1 m/s). The water was filtered to 5 μm to prevent macrofouling, but allow microfouling. The results of both tests appear in Table III. Some conventional stainlesses were tested as well.

60 d in filtered seawater was found to be a more severe test than 2 yr in ambient seawater. Comparing seawater corrosion test results with actual condenser service showed the tests to be significantly more severe than normal operating conditions.

For austenitics with 19–20% Cr, a minimum of about 6% Mo was required for seawater corrosion resistance. For ferritics with about 25–26% Cr, around 3.0–3.5% Mo was needed.

Polluted seawater often contains hydrogen sulfide, which is generated by decaying organic matter. The seawater stainless steels have exceptionally high resistance to sulfide attack, unlike the seawater copper alloys, which can suffer pitting at sulfide levels as low as 20–50 ppb. Table IV shows the effect of sulfides in ASTM D1141 Substitute Ocean Water. Additional corrosion

Two tests show how these stainlesses and other alloys stand up to attack by seawater **Table III**

	Maximum penetration, mm			
		Ambient seawater		Filtered seawater, 25°C
Stainless Steel	2 mo*	9 mo†	24 mo†	2 mo
Sea-Cure	N	0	0	0
AL 29-4-2	0.02	0	0	0
AL 29-4C	N	N	N	0
Monit	0	0	0	0
AL-6X	0	0.01 (0.02)¶	0.02	0.23
254 SMO	0	0‖	0	0
Ferralium alloy 255	N	N	N	0.05
Type 304	1.47‡	N	N	N
Type 316	0.23	N	N	N
Alloy 904L	0	0.06	0.05 (0.08)¶	0.83

*Temperature range 15.6-25°C
†Temperature range 5-31°C
‡Perforated
¶ Attack at fouling sites
‖186 d
N = Not tested

Resistance to sulfides of Sea-Cure, AL-6X and some other condenser-tube alloys — Table IV

Alloy	Corrosion rate (in./yr) in ASTM D1141 Substitute ocean water (pH=8) at ambient temperature: No sulfide	10 ppm Na_2S
Sea-Cure	0.00010	0.00010
AL-6X	0.00010	0.00010
Type 304	0.00010	0.00012
Type 316	0.00010	0.00010
90Cu-10Ni	0.00137	0.00708
70Cu-30Ni	0.00064	0.00490

Data by Trent Tube, Div. Colt Industries

Corrosion rates of AL 29-4-2 compared with those of other alloys — Table V

Boiling solution, wt %	Corrosion rate, mpy*: Type 316	Hastelloy C	Titanium	AL 29-4-2
65% Nitric	11	450	1	3
10% Sulfamic	75	8	285	2
45% Formic	520	5	873	0
20% Acetic	2	0	0	0
10% Oxalic	96	0	950	0
10% Sodium bisulfate	170	8	250	2
10% Sulfuric	371	17	6,290	5†
1% Hydrochloric	210	N	N	0

*Testing time varied with corrosion rate, 10 min for high rates and 10 d for low rates.

†Alloy is self-repassivating.

N = Not tested

Data by Allegheny Ludlum Steel Company

Corrosion rates in 10% sulfuric acid — Table VI

Alloy	Corrosion rate, $mg/dm^2/d$: 50°C	70°C	90°C
AL-6X	5	38	343
254 SMO	1.7	101	642
Alloy 904L	6	6	204
Type 304	3,600	N	N
Type 317L	17	146	1,040

N = Not tested

data on several of the high-performance seawater stainless steels are found in Tables V and VI.

Weldability

All of these steels are produced as autogenously (i.e., no filler wire is added) welded tubing (among other forms). Welding during fabrication and assembly is possible if appropriate precautions are taken.

The ferritic and duplex grades have very low impurity levels, which, in part, give them excellent ductility and toughness. Welding practices must be used that preserve this high level of purity. Carbon, nitrogen, oxygen and hydrogen must be prevented from entering the weld. Complete shielding with a dry, inert gas is necessary.

Grease, oil and solvents must be removed before welding to prevent contamination by carbon. Careful cleaning before and after welding helps to assure corrosion resistance and weld quality. Optimum corrosion resistance is obtained by chemically or mechanically removing the heat tint after welding. (The heat tint is discolored metal in the vicinity of the weld. The discoloration is due to oxide formation.) Chemical means include pickling and acid cleaning; mechanical means include gritblasting, and stainless-steel wire brushing. In welding the low-interstitial ferritics, preheating and postweld heat-treating are usually unnecessary; in fact, they are usually undesirable.

In general, ferritics are less forgiving to weld than austenitics. The ferritics have a greater notch sensitivity, and are more prone to cracking if weld defects develop.

Welding consumables of matching composition are available for some of the seawater ferritics. Nickel-base filler wires of Alloy 625 chemistry (AWS designations ENER NiCrMo—3 or E NiCrMo—3) are recommended for the seawater austenitics, and also work for the ferritic and duplex grades, in most environments.

Tubesheet selection

Ideally, a tubesheet material should match the mechanical properties and corrosion resistance of the tubing. However, the seawater stainlesses often are used to replace other materials, frequently copper-based alloys. Since the stainless steels are stronger than Muntz metal and other frequently encountered tubesheet materials, and because the ferritics have limited ductility, special care is needed in the rolling-in of the tubing.

Rolling-in is a process in which the tubes are secured to the tubesheet. The tubesheet has holes in it and the tubes are placed through these holes. A device with rollers is inserted inside of the tubes, and pressure is used to expand the tubes into the tubesheet, creating a seal. Welding may be used afterward for a tighter seal. Equipment such as five-roller expanders is used here.

For new construction, it is often desirable to match the tube and tubesheet materials. This can be done with austenitics and duplexes. However, the ferritics are unavailable in sufficiently heavy thicknesses. Nonetheless, the plate alloys (i.e., those available in plates) can be used as tubesheets for the other seawater stainlesses, as long as they are compatible with the process streams.

If the tubesheet can have a lower corrosion resistance than that of the tubes, probable tubesheet materials would include Types 316L, 317L and 317LM, and Alloy 904L. For ferritics, if the tubesheet has to match the tubing, a material clad with a ferritic can be used.

The seawater stainless steels are usually used to replace copper-alloy tubes in Muntz metal tubesheets. Often, there is a chance of galvanic corrosion. Depending upon the stainless steel and tubesheet material, it may be necessary to use cathodic protection or coatings to protect the tubesheet and waterbox.

Selecting materials for chlorine-gas neutralization

In designing neutralization systems, proper materials specification is critical, requiring compatibility with chlorine, caustic and sodium hypochlorite.

Neal C. Horowitz, SunOlin Chemical Co.

☐ Modern-day caustic/chlorine plants are usually equipped with systems capable of neutralizing chlorine gas under plant upset conditions to prevent releases. Although designing such a system is not a particularly difficult task for process engineers, proper materials selection is extremely important.

Fig. 1 is a simplified flow diagram of a chlorine neutralization system, [*1*], showing major equipment, and listing materials of construction, the selection of which is discussed later. Vented dry chlorine gas is directed to the scrubbers via carbon steel piping. Wet chlorine gas is piped with properly designed bisphenol-fumarate polyester fiberglass piping. Twenty percent sodium hydroxide, used for neutralization makeup, is handled in carbon steel tanks and piping. The packed scrubbers can be constructed of vinyl ester or polyester fiberglass reinforced plastic (FRP), if careful pH and temperature controls are imposed. In Europe, much success has been reported with polyvinyl chloride (PVC) lined FRP scrubbing equipment, which has proven more resistant to varying process conditions than FRP alone.

The scrubber reservoir is constructed of ethylene propylene rubber-lined steel. Circulation pumps and plate heat exchangers, used to limit process temperatures, are made of titanium. Recirculation and hypo-

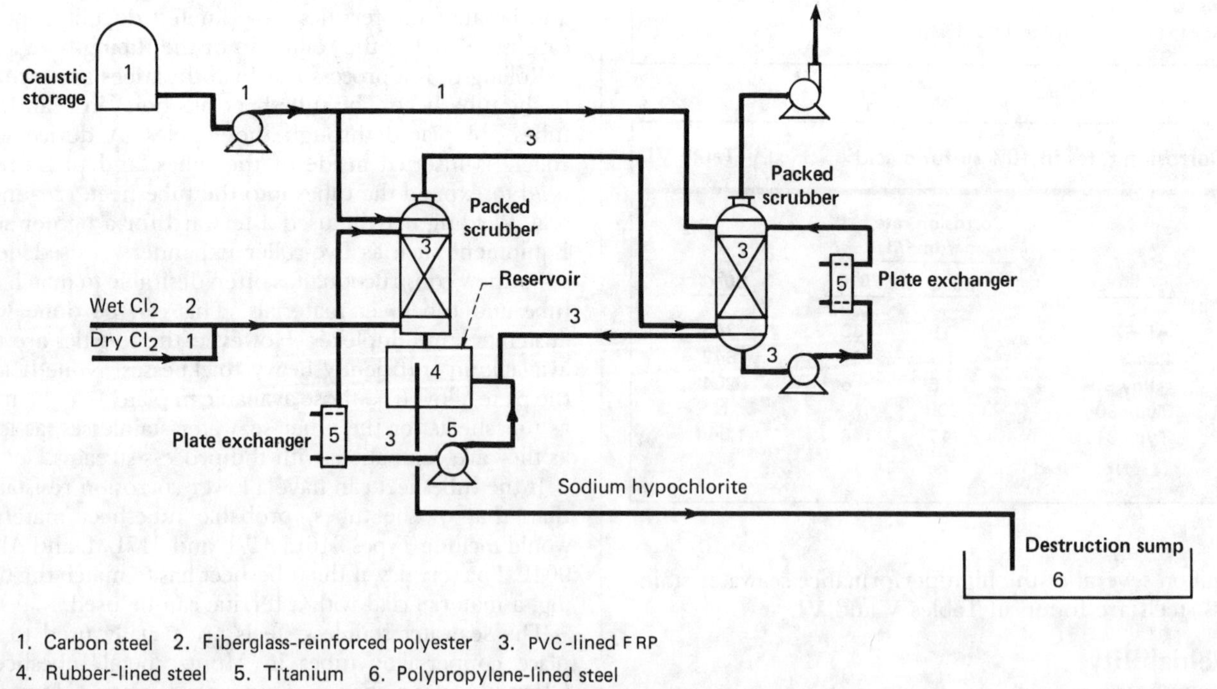

1. Carbon steel 2. Fiberglass-reinforced polyester 3. PVC-lined FRP
4. Rubber-lined steel 5. Titanium 6. Polypropylene-lined steel

Flow diagram of chlorine neutralization system, with typical materials of construction **Fig. 1**

Originally published April 6, 1981

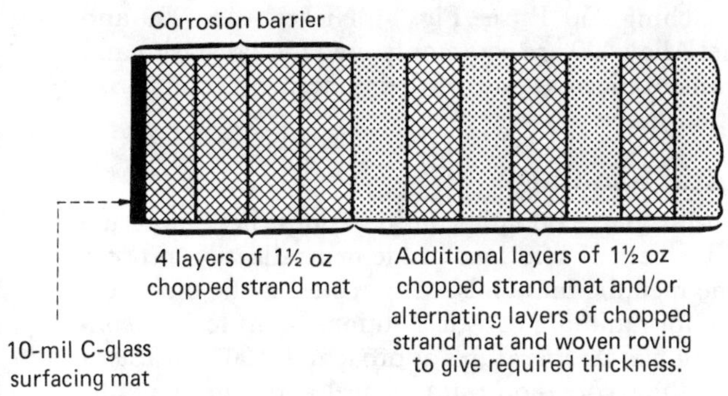

Typical laminate sequence for FRP pipe in chlorine service **Fig. 2**

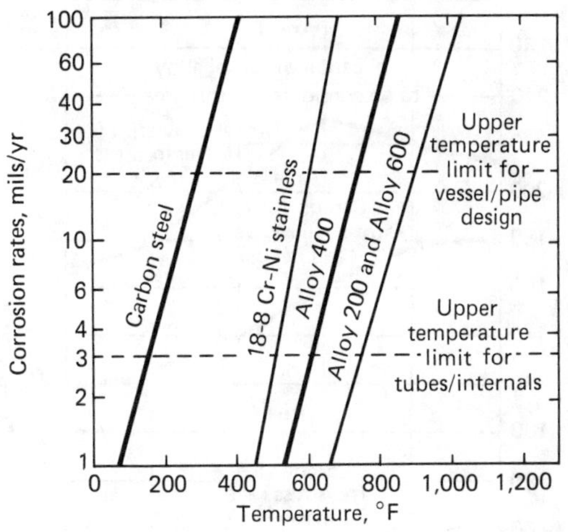

Chlorine alloy selection **Fig. 3**

chlorite piping are usually polypropylene-lined steel or PVC-lined FRP. The destruction tank is polypropylene-lined steel.

Process overview

The neutralization system must be able to absorb high- or low-pressure gas releases throughout the plant. Neutralization is accomplished by contact with caustic solution (sodium hydroxide at approximately 20%) to form sodium hypochlorite via the following very rapid and exothermic reaction:

Sodium hydroxide		Chlorine		Sodium hypochlorite		Sodium chloride	
2NaOH	+	Cl_2	→	NaOCl	+	NaCl	+

$$H_2O + \Delta H = 626\ \text{Btu/lb}\ Cl_2 \quad (1)$$

Many systems use scrubbing towers for neutralization vessels. The towers have recirculating caustic that can absorb the full production rate of chlorine. Some systems use heat exchangers to remove the heat of reaction and prevent temperature surges. After neutralization, the sodium hypochlorite solution is transferred to a decomposition tank for destruction of hypochlorite, and final disposal.

Based on this process description, it is clear that choosing materials of construction for chlorine neutralization requires a basic knowledge of materials selection for handling chlorine, caustic and sodium hypochlorite.

Chlorine handling

Wet chlorine gas (defined as having >50 ppm by volume water) formed at the anode of the diaphragm cell is extremely corrosive to most materials of construction, particularly metals. Titanium is virtually the only commercial metal that resists it at most process temperatures. Hastelloy C-276 has useful resistance from ambient temperature to 140°F.

Of the nonmetallic construction materials, the bisphenol-fumarate polyester fiberglass-reinforced plas-

Corrosion rates for metals in 5 – 14% NaOH **Table I**

Metal	Corrosion rate, mils/yr [9] 5–10%	10%[2]	14%[3]
Titanium	0.04	Nil	–
Zirconium	0.2[4]	0.07	–
Nickel	0.2	0.003	0.02
Monel	0.3	Nil	0.05
Inconel	0.05	Nil	0.03
Mild steel	4[4]	0.6[5]	8.2
Cast iron	–	–	8.2
Ni-Resist Type 1	–	–	2.9

1. Duration of test: 124 d. Temperature: 70°F
2. Test conducted with effluent from electrolytic chlorine cell containing 15% NaCl. Duration of test: 207 d. Temperature: 180°F.
3. Exposure was in first effect of multiple-effect evaporator. Duration of test: 90 d. Temperature: 190°F.
4. Slight pitting attack.
5. Slight attack under spacer (crevice test).

Corrosion rates for metals in 30 – 50% NaOH **Table II**

Metal	Corrosion rate, mils/yr [9]
Nickel	0.1
Monel	0.2
Copper-nickel-zinc (75-20-5)	0.5
Copper	2.3
Mild steel	3.7
Cast iron	7.0
Chromium steel (14% Cr)	33.0

Note: Exposure was in single-effect evaporator concentrating caustic soda from 30 to 50%. Duration of test: 16 d. Temperature: 179°F average.

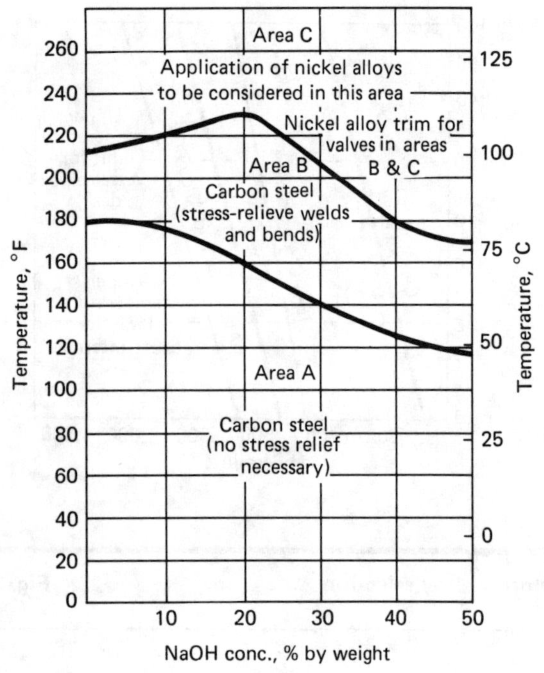

Caustic soda service chart **Fig. 4**

tics (e.g., Atlac 382) offer excellent resistance at an economic price. Considerable experience regarding the use of FRP in chlorine service has been compiled over the last ten years [*2*]. Optimum performance is achieved by making the inner corrosion barrier 1/8 in. thicker than the standard liner, as described in NBS-PS15-69 [*3*]. Fig. 2 shows a typical liner and laminate sequence for FRP construction in chlorine service. [*4*]. Most common thermoplastic and elastomeric materials other than the fluorocarbons are attacked by wet chlorine gas and are not recommended for use.

Dry chlorine gas (<50 ppm water) is relatively noncorrosive and can be handled by most materials of construction. Carbon steel is normally used at temperatures approaching 250°F (see Fig. 3). Nickel Alloy 200 and Monel Alloy 400 are commonly used for valve trim and instrumentation because of their extremely good corrosion resistance.

Caustic handling

Process and economic considerations dictate the use of dilute caustic solutions in the neutralization process. Of the metallic materials, carbon steel is the most economic for handling caustic solutions of up to 50% concentration at temperatures approaching 190°F. Tables I and II show corrosion rates of steel and other metallic alloys in various concentrations of caustic [*5,6*]. Two factors, however, must be considered before using bare steel in caustic service: (1) the possible occurrence of caustic-induced stress corrosion cracking, and (2) the system's tolerance for iron contamination.

Fig. 4 indicates the temperatures and concentrations at which carbon steel can be used safely without the threat of stress corrosion cracking. The curve shows that carbon steel may be used at higher concentrations and temperatures if highly stressed areas, such as welds and bends, are stress-relieved. (The ASME Boiler and Pressure Vessel Code Section VIII, Div. 1, specifies in detail the stress-relief procedure.)

Usually, there is no concern about iron contamination when chlorine is neutralized for environmental reasons, since the sodium hypochlorite formed will ultimately be destroyed. However, some processes may be designed to collect and use sodium hypochlorite as product. In these instances, iron contamination is extremely important, since iron can catalyze hypochlorite decomposition [*7*]. To avoid contamination, spray-applied organic linings are often used inside tanks. Suitable lining materials include the epoxy-phenolics, chlorinated polyether and chlorosulfonated polyethylene [*8*]. For piping systems, Type 304 stainless steel or nickel alloys are specified.

Many nonmetallics offer excellent resistance to caustic within their normal temperature limits. The more common materials include PVC, polypropylene, epoxy FRP, fluorocarbons such as tetrafluoroethylene (TFE),

Corrosion of metals in 16% sodium hypochlorite **Table III**

Material	Corrosion rate, mils/yr	Pitting
Titanium	<0.1	None
Zirconium	<0.1	None
Hastelloy C	0.1	None
Chlorimet 3	0.3	None
Durichlor	0.8	None

Note: Based on repeated exposures to batch manufacture of sodium hypochlorite. Starts with 18 to 20% caustic, ends up with 16% NaOCl. Temperature: 70°F; Test duration: 170 d [*9*].

Corrosion of metals in 1 – 4% sodium hypochlorite **Table IV**

Material	Corrosion rate, mils/yr	Pitting
Titanium	0.1	None
Zirconium	4	Severe
Hastelloy C	46	Severe
Type 316 SS	>100 (consumed)	
Duriron	12	Severe
Durichlor	7	Severe
Mild steel	>200 (consumed)	

Note: Environment contained 1.5 to 4% NaOCl, 12 to 15% NaCl, 1% NaOH. Temperature: 150 to 200°F; Test duration: 72 d [*9*].

Registered trademarks in article

Atlac	ICI Americas, Inc.
Chlorimet 3	Duriron Co.
Durichlor	Duriron Co.
Duriron	Duriron Co.
Hastelloy	Cabot-Stellite Div., Cabot Corp.
Hypalon	Du Pont Co.
Inconel	International Nickel Co.
Kynar	Pennwalt Co.
Monel	International Nickel Co.
Neoprene	Du Pont Co.
Ni-Resist Type 1	International Nickel Co.
Saran	Dow Chemical Co.

fluorinated ethylene propylene (FEP) and Kynar, natural rubber, and Neoprene and Hypalon elastomers.

Sodium hypochlorite handling

Sodium hypochlorite solutions are extremely aggressive, particularly at higher temperatures, when they become unstable. Tables III and IV show corrosion rates of various metals in sodium hypochlorite solutions at 70°F and at 150–200°F [*9*].

At higher temperatures, titanium is virtually the only metallic material of construction with useful resistance. It should be carefully noted, however, that titanium is attacked by dry chlorine gas; therefore, this material's usefulness in a chlorine neutralization system is limited. Fires have been experienced in titanium neutralization towers, resulting from violent reaction between titanium and dry chlorine. (Dry gas may be present at inlet branches and crevices where evaporation can occur.) However, titanium is used successfully in applications where complete wetting is assured.

Many nonmetallics have excellent resistance to stable sodium hypochlorite. Polyester and vinyl ester fiberglass-reinforced plastics [*10*], chlorobutyl and ethylene propylene rubbers, polyvinyl chloride and polypropylene are among the most notable and economic. Unfortunately, stable sodium hypochlorite is not always present in chlorine neutralization towers. The following highly exothermic reaction will occur if excess caustic is not maintained during neutralization [*7*]:

Sodium hydroxide, Chlorine → Sodium chlorate, Sodium chloride

$$6NaOH + 3Cl_2 \rightarrow NaClO_3 + 5NaCl + 3H_2O + \Delta H = 2{,}390 \text{ Btu/lb } Cl_2 \quad (2)$$

This reaction involves the decomposition of sodium hypochlorite via:

Hypochlorite, Hypochlorous acid → Chlorate

$$ClO^- + 2HOCl \rightarrow ClO_3^- + 2H^+ + Cl^- \quad (3)$$

Reaction (3) will result in rapid deterioration of most nonmetallic construction materials, including those previously mentioned as being suitable for handling stable sodium hypochlorite. It appears that the tremendous amount of heat released during this reaction can localize, resulting in hot spots and failures in vessel walls. The thermosetting FRP materials suffer general deterioration, and cracking failures have been reported in many thermoplastics as well, especially when used as linings. Only PTFE, FEP and PFA (perfluoroalkoxy) fluoropolymers have demonstrated complete resistance in hypochlorite-handling systems.

Thermal decomposition of sodium hypochlorite must also be considered in the selection of nonmetallic materials of construction. When temperatures exceed 140°F, free chlorine is liberated, with a resulting drop in pH, and decomposition reactions (1) and (2) may proceed. As previously discussed, only PTFE, FEP and PFA are considered completely resistant in this situation.

Catalytic decomposition

Before disposal, the sodium hypochlorite solution must be destroyed for environmental compliance. A common method is via the following catalytic reaction:

Sodium hypochlorite —Catalyst→ Sodium chloride + Oxygen

$$2NaOCl \rightarrow 2NaCl + O_2 \quad (4)$$

Reaction (4) is carried out in the presence of nickel and iron catalysts. The decomposition rate may double for each 10°F increase in temperature.

Several nonmetallics are suitable for handling catalyzed hypochlorite decomposition, provided the temperature is within their normal working limits. These include PVC, polyvinylidene chloride (Saran), polypropylene, and types of synthetic rubber. Unlined concrete tanks have been used, but deterioration is likely.

References

1. *Speaking Out,* No. 22, Atlas Chemical Industries, Wilmington, Del.
2. Roberts, R. C., Experience in Design and Operation of FRP Vessels and Pipe Systems in a Chlorine Plant Complex, "Managing Corrosion with Plastics," Vol. III, NACE, 1977, pp. 270–275.
3. NBS Voluntary Product Standard PS15-69, Custom Contact-Molded Reinforced Polyester Chemical-Resistant Process Equipment.
4. Schillmoller, C. M., Alloy Selection for VCM Plants, *Hydroc. Proc.,* March 1979.
5. International Nickel Co., Corrosion Engineering Bulletin CEB-2, 1969.
6. Gegner, P. J., and Wilson, W. C., *Corrosion,* Vol. 15, 1959, p. 3,412.
7. Hypochlorite Solutions, "Encyclopedia of Chemical Technology," 2nd ed. Vol. 5, Wiley-Interscience, New York. pp. 12–15.
8. Coatings and Linings for Immersion Service, NACE TPC Publication No. 2, 1972.
9. Gegner, P. J., Corrosion Resistance of Materials in Alkalies and Hypochlorites, "Process Industries Corrosion," NACE, 1975.
10. Miller, J. C., and Longnecker, D. M., Factors Affecting Performance of Reinforced Plastics in Sodium Hypochlorite Environments, 25th Annual Technical Conf., 1970 Reinforced Plastics/Composites Div., The Soc. of The Plastics Industry.

The author

Neal C. Horowitz is Supervisor of Metallurgy and Inspection for SunOlin Co., Claymont, DE 19703. Tel: 302—798-6801. His responsibilities include plant material selection, failure analysis and equipment inspection. When Mr. Horowitz wrote this article he was a Senior Materials Engineer at ICI Americas, Inc., Wilmington, Del. He holds a B.S. in chemical engineering from Ohio University, and an M.S. in materials science from University of Dayton. He is a NACE accredited corrosion specialist.

Choosing materials for desalting by distillation

Some major problems in seawater desalting plants have been caused by improper materials selection, resulting in expensive repairs and shutdowns.

G. Stern, B. J. Bayles, O. H. Chukumerije,
Gibbs & Hill, Inc.

☐ Several processes have attained commercial status for the removal of impurity salts from water. The materials selection process considers environment, cost, availability and previous operating experience.

Desalting processes

Of the commercial processes [*1*], electrodialysis and reverse osmosis are best adapted to the desalting of brackish waters (1,000 to 5,000-ppm dissolved solids), while distillation can handle all ranges of salinities up to seawater having 43,000 ppm (4.3%) of dissolved matter [*2*]. Desalting by distillation has remained the most common large-scale technique in potable-water production from seawater. In addition, distillation has the advantage of being able to utilize the waste heat from conventional power-generating plants, and so becomes preferred when it can be made satellite to a power plant.

There are two methods of distillation—multistage flash distillation and multieffect evaporation.

The multistage flash distillation (MSF) process involves heating seawater within tubes, progressively up to about 250°F, then flashing this superheated seawater in a number of successive flash chambers operating under progessively lower pressures. In each chamber, the vapor is condensed by heat exchange with the incoming feedwater. The distillate stream becomes the product water.

Seawater feed is first pumped through a pretreatment section to remove impurities that would damage the system. Sulfuric acid additions can be made to control scale carbonate. Part of the brine is recycled to improve system performance. Low-temperature, low-pressure steam is used to heat the seawater, and the steam condensate can be pumped back to the steam system.

The thermal efficiency of multieffect evaporators is higher than that of flash distillation stages. But because they use a larger number of stages, they conserve heat and therefore have higher total thermal efficiencies.

Photo: Wide World Photos, Inc.

MSF unit at Point Loma, near San Diego, CA

Originally published September 22, 1980

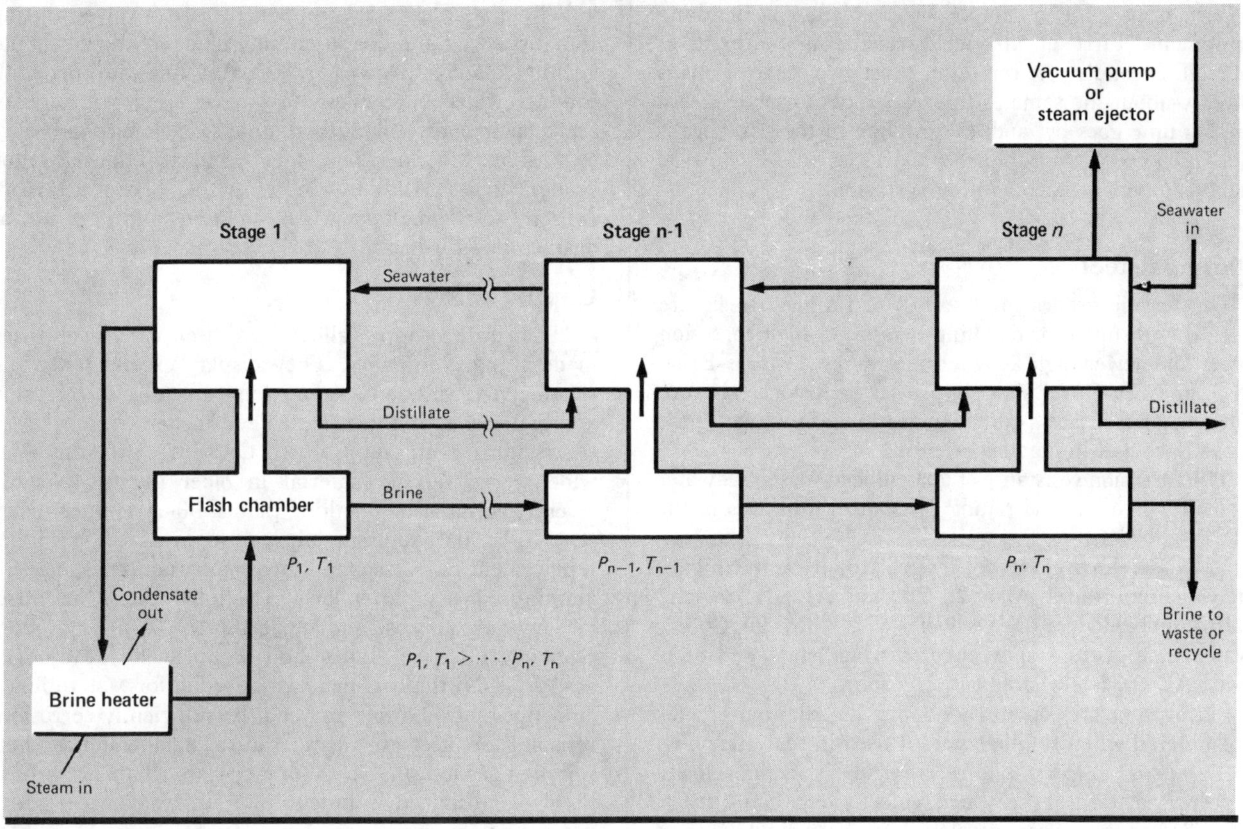

Multistage flash distillation

Multieffect evaporation will not be discussed here. The remainder of this article deals with the selection of materials for MSF equipment (see photo and fig.).

Carbon steel and cast iron

Carbon steels in seawater are subject to complex corrosion reactions that depend on several variables, including the presence of marine growth or slime. Corrosion rates increase with temperature, oxygen, low pH, and water velocity. Also, steel is generally subject to galvanic corrosion when coupled with most other metals.

In fairly heavy thicknesses, mild steel has proven economical when: good water treatment has been assured, including corrosion-inhibition and fouling-control; pH is maintained near or above 7; oxygen content is kept below about 50 ppb; and velocities are kept below 5 ft/s. A proper corrosion allowance must be provided.

There are various other methods for making use of low-cost steels possible. The most economical protection for steel structural components is galvanizing. For those many structural components that do not lend themselves to galvanizing, a good paint system seems to offer the best protection. One example is catalyzed epoxy over an inorganic zinc-rich primer applied to well-sandblasted steel surfaces (Steel Structures Painting Council, SSPC-SP 5 for immersion and No. 6 or 10 for other components and structures). A relatively high-gloss, high-temperature-resistant (250-400°F) paint system such as the silicone-alkyds or silicone-acrylics is sometimes needed in extremely severe cases.

The lining or cladding of carbon steel with corrosion-resistant metals is good practice where the strength of the carbon steel in the required thickness can be furnished on the outside, while the corrosion-resistant metal is exposed to the seawater or brine. This is done in thick-walled steam chests and flash chambers.

Stainless steel Types 304L and 316L, copper-nickel 90/10, and aluminum bronze alloys can be used as the corrosion-resistant lining or cladding material. The choice of the liner or cladding depends on the galvanic relationship to other metals present in the system, the corrosion/erosion conditions, and the ability to join the cladding or lining to the base metal.

Linings may be attached using screws, bolts, spot welds, plug welds, seal welds, or a combination of the above. The identifying feature of a metal lining is that it is not continuously bonded to the backing metal; it is loose. Cladding, on the other hand, results in a continuous, integral bond between the backing and the cladding. Cladding may be accomplished by roll bonding, explosive bonding, or weld overlaying.

Cathodic protection by sacrificial anodes (magnesium, aluminum or zinc) or impressed current can reduce the corrosion of steel components when these parts are to be completely immersed.

Corrosion allowances for carbon steel in seawater at normal temperatures vary from 10 mils per year (mpy), in less than 2 ft/s seawater, to 30 mpy [*6*]. At 250°F in treated seawater, depending on velocity and turbulence, corrosion rates of 30 to 100 mpy should be expected.

Cast irons pose a couple of problems in desalting equipment. First of all, weld repair, if required, is difficult. Secondly, the corrosion resistance of cast irons is initially about the same as that of wrought carbon steels. But, as time goes by, and the graphite in the structure is more exposed, the graphite can act cathodically toward metal components, accelerating corrosion.

Stainless steel

The ferritic grades (400 series) of stainless steels are generally unsuited for desalting because of high corrosion rates. The newer high-alloy ferritic grades (26-1, E-Brite, 26-3, and Sea-Cure) show improved corrosion resistance and resistance to stress-corrosion cracking. However, these alloys have had limited experience.

The corrosion resistance of austenitic grades to seawater (general corrosion and pitting resistance) improves in the following order:

Molybdenum-free grade; Type 316; then the higher molybdenum grades; Alloy 20 Cb3 and AL 6X (3% and 6.25% molybdenum, respectively) [*7*]. Alloy 20 Cb3 is available as castings or wrought product, sheet and tube, while AL 6X is available only as tubing.

Pitting and crevice corrosion are the main problems encountered when stainless steel is used in seawater. The 316 grade is subject to pitting, particularly as the velocity is decreased. In quiet seawater (velocity less than 2 ft/s), Type 316 in three years developed an average pit depth of 72 mils and a deepest pit of 113 mils [*6*]. Crevice corrosion similarly can be a problem in the austenitic grades. Substituting Alloy 20 Cb3 for 316 offers improvement in resistance to pitting, especially where stainless steels are used in stagnant, untreated seawater.

Austenitic stainless steels are subject to stress-corrosion cracking, a complicated phenomenon basically characterized as failure due to the combined action of tensile stress and a corrosive medium [*8-13*].

Where possible, solution heat treatment at 1,950°F, followed by rapid cooling to below 800°F, is specified for stainless steel to minimize tensile stresses. This treatment also places carbides in the stainless steel in solution and helps to prevent another deleterious condition in stainless known as "sensitization."

Sensitization can occur when austenitic stainless steels are held in the temperature range of 800 to 1,650°F (427 to 899°C), which causes carbides to precipitate along grain boundaries. Since these precipitated carbides contain chromium, the grain boundary region is depleted of chromium, making it less corrosion-resistant. Such sensitization is inevitable in the regular (0.03 to 0.08% C) 304 and 316 grades of stainless, in the areas adjacent to the welds. There are at least three ways to prevent stress-corrosion cracking [*6*]:

- Restrict the carbon content of the alloy to 0.03% maximum. Examples are 304L and 316L.
- Stabilize the carbon with titanium or columbium. Examples of this are Type 321 and Type 347 stainless steels and iron-nickel-chromium Alloy 20 Cb3 (which also contains 3% molybdenum).
- Prevent or limit the precipitation of carbides by solution heat treatment and control the heat input during any subsequent welding.

Where stainless steel components are involved during commissioning and operation, any plant shutdown should be immediately followed by draining and flushing with product water or condensate to avoid leaving stagnant seawater in contact with the stainless. Prolonged exposure to stagnant seawater should be avoided. The high-alloy grades, such as Alloy 20 Cb3 or AL-6X, are much more resistant to stagnant seawater conditions and need not be drained and flushed.

Copper alloys

Historically, copper alloys have been extensively used in desalting equipment. They display excellent thermal conductivity, and good-to-excellent resistance to corrosion and biofouling, and are available at moderate cost [*1,5*].

Arsenical aluminum brass (CA 687) was the most widely used tubing material in older plants. In more recently constructed desalting plants, 90-10 copper-nickel (CA 706) has replaced aluminum brass. The higher replacement frequency and lower corrosion resistance for aluminum brass, particularly in polluted waters, has offset its lower initial cost and made the 90-10 alloy a more economical choice. Bates and Popplewell [*14*] judged 90-10 Cu-Ni the best material selection for MSF tubing.

Copper-nickel alloys may not be sufficiently corrosion-resistant for the higher-temperature chambers of the heat rejection section and in the air ejectors. Both areas have highly corrosive environments, and it may be necessary to use a more corrosion/erosion-resistant material for these applications.

Oxygen is one of the most important factors influencing the life of copper alloys in operation. Oxygen accelerates corrosion by reacting with the electrons that flow to cathodic sites, while copper enters the solution as cuprous ions at anodic sites. This electron removal (cathodic depolarization) must continue for corrosion to proceed. Therefore, effective deaeration in pretreatment, accompanied by adequate steps to prevent subsequent air "leak-in," is essential if copper alloys are to reach their potential 30-year design life in desalting plants.

Because most heat-exchanger flows are designed for fluid velocities of less than 8 ft/s, velocity should not present any serious [problems with most copper alloys in desalting plants [No. 706(90/10 Cu/Ni) is designed for 12 ft/s and No. 715(70/30 Cu/Ni) for 15 ft/s] [*4*].

To assure continuous neutralization after the acidification step in pretreatment, pH controls are required. If these controls are inadequate, this can materially reduce the life of copper-base-alloy components. Temperature effects can be slight (corrosion increases with temperature) in the absence of dissolved-oxygen control, but are not significant with most copper alloys.

Aluminum bronze and nickel/aluminum bronze alloys have given excellent service in seawater. They resist fouling and pitting, do not suffer crevice attack and are erosion/cavitation-resistant [*15,16,17*]. In more than 20 yr of industrial use at Ampco Metal Div., there has never been a reported failure of aluminum bronze alloy CA 614 (containing tin) due to stress-corrosion cracking [*16*]. Laboratory tests show that this alloy did not fail after being exposed to a 500-ppm ammoniacal solution at 230°F for 2,400 h. It has been reported that, after two years of field experience, aluminum bronze alloy CA 954 resisted sand erosion better than stainless steel alloy CF8M, Type 316 [18].

The advantage that aluminum bronze material has over Ni-Resist and stainless steel is that aluminum bronze alloys are not subject to pitting attack or crevice corrosion when exposed to stagnant seawater.

Titanium

Titanium is considerably more noble than most other metals. No general corrosion, crevice corrosion, or stress cracking has been noted with titanium in desalting environments up to 250°F, regardless of oxygen content, chlorination, and even in situations where seawater has been deliberately contaminated with H_2S or ammonia. A. D. Little Inc. [1] reported that after six years of operation in one desalting plant, there were no corrosion or erosion failures of titanium tubing.

In 1965, 432,000 ft of titanium seawater/surface condenser-tubing was installed in a plant in St. Croix, Virgin Islands. No failures were reported in five years of operation [19]. Titanium maintains its excellent corrosion resistance in stagnant or slow-moving water, as well as in high-velocity flow. In addition, titanium is at least 20 times more resistant to erosion than the best copper-base alloys [20]. The use of titanium in desalting plants has been limited by high initial costs. In those areas of the plant, such as the vent condensers and heat-rejection section, where no other material will last 30 yr, the use of titanium provides an economical alternative. This is especially true if one considers the inevitable replacement of copper-alloy tubing (at least once, and possibly twice, in a 30-yr lifetime), plus inflation and the resulting increase in copper-alloy costs in the future.

Other materials

The nickel-copper alloys, such as Monel 400 and the age-hardening version of this alloy, Monel K500, have been extensively used in pump shafts, impellers, valve trim, and heat-exchanger materials. These metals have better resistance to cavitation damage than 70-30 Cu-Ni and the austenitic nickel cast-irons [6]. The Monels are subject to pitting in seawater, and although the pits tend to be relatively shallow, cathodic or other forms of protection are required on critical surfaces.

Solid plastic, plastic-lined piping, and fiber-reinforced plastic (FRP) are considered when the temperature, pressure, chemical compatibility, and costs of labor and installation are all acceptable. Polypropylene, polyester and vinyl ester materials are used as linings and piping materials. For higher temperatures and more-aggressive liquids, polytetrafluorethylene (PTFE) or polyvinyl fluoride can be used. FRP piping has been used in raw seawater streams, and blowdown and distillate streams of desalting plants [4].

References

1. Newton, E. H., Birkett, J. D., and Ketteringham, J. M., Survey of Materials Behavior in Large Desalting Plants Around the World, A. D. Little, Inc., Cambridge, Mass., Report No. C-72872, March 1972.
2. Spiegler, K. S., Principles of Desalination, Academic Press, New York, 1966.
3. Proceedings of The First International Symposium on Water Desalination, U. S. Dept. of Interior, Washington, D.C., Oct. 3-9, 1965, Vol. 3.
4. George, P. F., Manny, J. A., Jr., and Schrieber, C. F., Desalination Materials Manual, Dow Chemical Co., May 1975, pp. 4-83; 5-22 to 5-47.
5. Alloys for Desalting Plants, Copper Development Assn., Inc., New York, Technical Report, June 1966.
6. Guidelines for Selection of Marine Materials, International Nickel Co., New York, N.Y., Publication No. 5M-2-76-5303.
7. Deverell, H. E., and Maurer, J. R., *Mater. Perform.*, Mar., 1978, pp. 15-20.
8. Berry, W. E., Some Facts About Stress Corrosion of Austenitic Stainless Steels in Reactor Systems, *React. Mater.*, Spring, 1964.
9. Standard Specification for Detecting Susceptibility to Intergranular Attack in Stainless Steels, Amer. Soc. for Testing and Materials (ASTM), Philadelphia, Pa., A262-77, 1977.
10. Bates, J. F., and Longinow, A. W., Principles of Stress-Corrosion Cracking as Related to Steels, *Corrosion,* Houston, Vol. 20, p. 189, 1964.
11. Hoar, T. P., Stress Corrosion Cracking, *Corrosion,* Houston, Vol. 19, p. 3311, 1963.
12. Barnartt, S., General Concepts of Stress-Corrosion Cracking, *Corrosion,* Houston. Vol. 18, p. 322, 1962.
13. Robertson, W. D., Stress-Corrosion Cracking and Embrittlement, John Wiley and Sons, New York, 1956.
14. Bates, J. F., and Popplewell, J. M., Corrosion of Condenser Tube Alloys in Sulfide Contaminated Brine, Natl. Assn. of Corrosion Engineers (Houston), Corrosion Forum, March, 1974, Chicago.
15. Materials for Seawater and Brine Recycle Pumps, International Nickel Co., New York, NY, Publication No. 38C-7-76-5345.
16. Severson, R., private communication, Ampco Metal Div., Milwaukee, Wis., Jan 23, 1979.
17. Harding, K., and Bridle, D. A., Corrosion of Ampco 8, 483 and 280 in an Operating and Closed Seawater Distiller, United Kingdom Atomic Energy Authority Report, Jan. 1978.
18. Aluminum Bronze Alloy Used in Large Cast Water-Wheels to Resist Pitting by Cavitation, *Western Metals,* Oct. 1953, Ampco Metal Div., Milwaukee, Wis.
19. McCue, D. M., private communication, Timet, Pittsburgh, Pa., Mar. 6, 1979.
20. Cotton, J. B., and Downing, B. P., Corrosion Resistance of Titanium to Seawater, *Trans. Inst. Mar. Eng.,* Aug. 1957, pp. 311-319.
21. Phillips, I. I., Poole, P., and Sheen, I. I., *Corros. Sci.,* Vol. 14, 1974, pp. 533-542.
22. Thomas, N. T. and Nobe, K., *J. Electrochem. Soc.,* Vol. 117, No. 5, 1970, p. 623.
23. Covington, L. C., Parris, W. M., and McCue, D. M., The Resistance of Titanium Tubes to Hydrogen Embrittlement in Surface Condensers, *Corrosion* 76, Houston, National Assn. of Corrosion Engineers, paper No. 79.

The Authors

George Stern

Byron J. Bayles

Ozz H. Chukumerije

George Stern is Director of New Technology at Gibbs & Hill, Inc., 393 Seventh Avenue, New York, NY 10001, telephone 212-760-4045. A graduate of City College of New York and the University of Michigan, with B.Ch.E and M.S. degrees, Mr. Stern is a Professional Engineer in New York State, author of 18 technical publications, co-author of a book on powder metallurgy, and holder of seven U.S. patents. He is a member of the Amer. Soc. of Mechanical Engineers, Amer. Institute of Mining and Metallurgical Engineers, Amer. Welding Soc., Amer. Soc. of Metals, Amer. Nuclear Soc., and the Amer. Powder Metal Institute.

Byron J. Bayles joined Gibbs & Hill, Inc., as Senior Metallurgist in 1978. He was formerly a materials engineer for Stone & Webster Engineering, New York, and a research scientist with United Technology Corp. Laboratories, East Hartford, Conn. Mr. Bayles received his B.Met.Eng. from New York University and his M.Eng. Sci. from Rensselaer Polytechnic Institute, and has authored and co-authored eighteen technical publications and reports. He is a member of the Ameri. Soc. for Metals, the Metallurgical Soc. of the Amer. Institute of Mining and Metallurgical Engineers, and of Sigma Xi, the scientific research society.

Ozz H. Chukumerije is a consulting engineer at Gibbs & Hill, Inc. in the New Technology department. Previously he worked with Columbia University's Arc Research Laboratory and with Exxon Research & Engineering Co. in Florham Park, N.J. A graduate of State University of New York at Stony Brook, where he received a B.E. degree, and Columbia University, where he received the degrees of M.S., Ch.E. (Professional Engineer degree) and Eng.Sc.D., Dr. Chukumerije has written a number of papers and reports for UN-based conferences. He is a member of AIChE and the Amer. Institute of Mining and Metallurgical Engineers.

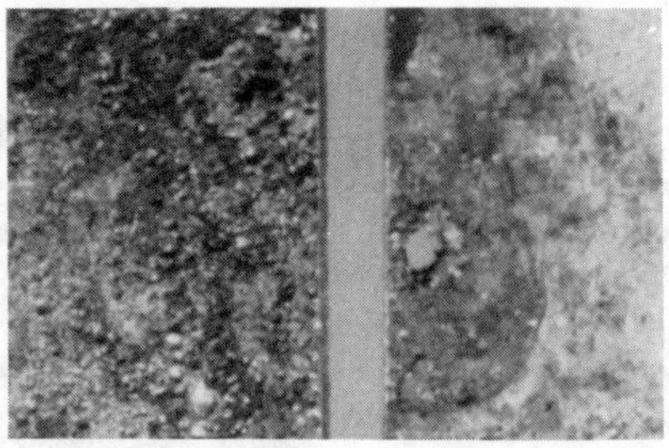

Selecting alloys for chloride service — PART 1

With a large variety of chloride types and process conditions, finding the best metal or alloy to fight corrosion is not always easy. Here are guidelines toward making optimum choices.

Gary N. Kirby, Ciba-Geigy

Aqueous chloride solutions are severely corrosive. If the pH is alkaline or neutral, carbon steel may be suitable. However, even then, the steel may discolor the chemical product. Furthermore, steel requires exterior maintenance painting.

In many plants, the alloys next considered would be Types 304 and 316 stainless steels, the most common austenitics. Austenitics have a face-centered cubic structure [*1*]. (The structure affects corrosion resistance, as will be explained.) Under certain conditions, austenitic stainlesses suffer localized chloride attack (e.g., pitting). The above photographs show ductwork made of Type 304 that pitted due to HCl fumes. More-highly-alloyed stainlesses can resist some or all of these types of attack, but they are expensive [*2*] (see Table I). Even more expensive are the high-nickel alloys, which resist even hydrochloric acid.

There are suitable alloys with lower nickel contents. Containing zero to 5.5% max Ni, these include the ferritics and the new duplexes. Titanium-base alloys compete with the iron- and nickel-base ones. So do cupro-nickel alloys. Each alloy offers its own advantages and disadvantages. Here, we will offer guidelines on when to select a particular alloy, depending on chemistry, pH and temperature. Tables II and III compare alloys as to their resistance in various tests.

Carbon steel and chloride solutions

Carbon steel is suitable for chlorides if the solution is alkaline or chemically inhibited (see Table IV). For the alkali-metal chlorides, the corrosive effects increase in the order: Li, Na and K [*3*]. The alkaline-earth chlorides, such as $CaCl_2$ and $SrCl_2$, are slightly less corrosive than are the alkali-metal chlorides. The acidic, oxidizing chlorides, such as ferric and cupric chloride, are especially corrosive, attacking steel at rates above 50 mils/yr at ambient temperature. The alkaline, oxidizing hypochlorites also are very corrosive.

NaCl deserves special attention. Its corrosion rate increases with increasing salt content up to about 3% NaCl, then decreases linearly until saturation (26% NaCl at room temperature) is reached. Saturated brine is less corrosive than distilled water. The low rate is due to the reduced solubility of O_2.

Seawater contains about 3.4% NaCl at a pH of 8. In full, quiet immersion, it corrodes steel at about 4–16 mils/yr. Above 50°C (122°F), this rate jumps to over 50 mils/yr. Impingement and turbulence also increase the rate. For example, flowrates over 5 ft/s raise the corrosion rate of steel in seawater to over 20 mils/yr. In the splash zone, seawater corrodes steel at about 15 mils/yr, with pitting depth at 45 mils/yr.

Calcium chloride is used widely as a brine coolant and can be handled in steel. In one steel pipe used for such service at –15°C (+5°F), the author measured the inside and outside diameters of the pipe after 19 years of service. Internal brine corrosion averaged only 1 mil/yr; external, 4 mils/yr. The external corrosion was due to moisture condensation under insulation. For 14 years, the brine was chemically inhibited with chromates. This was followed by nitrite inhibition for the remaining five years.

In any aqueous solution, the corrosion rate of steel will increase with increased acidity, temperature, dissolved oxygen, and solution velocity [*4*].

Overall corrosion of stainlesses

In neutral or alkaline chloride solutions, the overall corrosion of stainless steels will be low. Even in acidic solutions,

Originally published February 4, 1985

Relative cost is a major factor in selecting metals and alloys for aqueous chloride service — **Table I**

Relative plate costs, comparing equal areas

Alloy	Base	Approximate composition, %	Relative cost
C-276	Ni	59 Ni-16 Cr-16 Mo	7.3
625	Ni	61 Ni-21 Cr-9 Mo	5.5
Zirconium R60702	Zr	99.2% min (Zr + Hf)-4.5% max Hf	5.0
Titanium-Grade 7	Ti	99 Ti-0.2 Pd	4.5
Nickel 200	Ni	99 Ni min	3.5
600	Ni	72 Ni-15-Cr-8 Fe	3.3
Titanium-Grade 12	Ti	98 Ti-0.3 Mo-0.8 Ni	3.0
Ferritic 29-4-2 SS*	Fe	64 Fe-29 Cr-4 Mo-2 Ni	2.8
400	Ni	65 Ni-32 Cu	2.5
20Cb-3	Fe	34 Fe-20 Cr-33Ni-2.2 Mo	2.5
Titanium-Grade 2	Ti	99 Ti	2.5
Ferritic 26-1 SS	Fe	71 Fe-26 Cr-1 Mo	1.9
904L SS	Fe	47 Fe-20 Cr-25 Ni-4.5 Mo	1.8
Duplex 255 SS	Fe	62 Fe-25 Cr-5.5 Ni-3 Mo-0.2 N	1.7
317L SS	Fe	63 Fe-19 Cr-12 Ni-3.2 Mo	1.3
316L SS	Fe	67 Fe-17 Cr-11 Ni-2.2 Mo	1.0

*SS is stainless steel.

chloride attack will most likely be limited to localized phenomena such as pitting, crevice corrosion or stress-cracking corrosion. However, in percent-level (vs. ppm-level) HCl solutions, overall corrosion can be severe.

In one 30-h test of stressed U-bend samples of stainless steels, corrosion rates were 8.1 *inches*/yr for Type 304 and 0.165 in./yr for Type 316L, for 36% HCl at room temperature. Examination at 30× magnification showed no pitting or cracking.

However, greatly reduced corrosion rates were achieved by chemical inhibition. A complex amine dyestuff, which acted as an inhibitor, was added to a mixture of this acid at a pH less than 0. The rate for Type 304 was 13–96 mils/yr; for Type 316L, it was only 5.6–6.1 mils/yr. Again, there were no signs of pitting or cracking. However, such attack can occur months or years after the start of service, and caution is warranted. Tests were run for up to 30 h in anticipation of a 1-h exposure; these two stainless steels were deemed suitable for use. Still, hydrochloric acid should be avoided with the 300 Series of stainless steels for any extended service.

How various alloys stack up in tests in different types of chloride solutions — **Table II**

Successful alloy tests in chloride solutions

Chloride solution	Carbon steel	Duplex 255 SS*	Ferritic 26-1 SS	Alloy 20 (% Ni >30) 2-3% Mo	Alloy 20 (% Ni >30) 6-9% Mo	Alloy C-276	Titanium Grade 2
Alkaline	Seawater pH = 8 50° C	Seawater pH = 8 50° C	26% NaCl + Na_2CO_3 pH = 11 boiling	Seawater pH = 8 NR† at 25° C‡	Sat'd. NaCl pH = 8.5-10 120° C	Sat'd. NaCl pH = 8.5-10 120° C	Seawater pH = 8 boiling
Neutral pH	22% NaCl 25° C	26% NaCl 200° C	2/% NaCl 25° C	No data	26% NaCl 200° C	Sat'd. NaCl 125° C	26% NaCl 113° C
Acidic	NR†	4% NaCl + 280 ppm Fe^{+3} pH = 2 35° C	6% $FeCl_3$ pH = 1 25° C	4% NaCl + 280 ppm Fe^{+3} pH = 2 15° C	4% NaCl + 280 ppm Fe^{+3} pH = 2 25° C	40% $MgCl_2$ pH = 0.5-1.0 boiling	20% $MgCl_2$ 100° C
Oxidizing	NR†	6% $FeCl_3$ 25° C	10% $FeCl_3$ 38° C	6% $FeCl_3$ Below –5° C	4% NaCl + 280 ppm Fe^{+3} pH = 2 25° C	15% $FeCl_3$ 25° C	40% $FeCl_3$ boiling
Hypochlorite	NR†	No data	5.25% NaOCl 70° C	NR†	10% NaOCl +2% NaOH 30° C	10% NaOCl 25° C	16% NaOCl 25° C

*SS is stainless steel.
†NR is not recommended.
‡Alloy 20Cb-3 pits in seawater at 25°C.

Adapted from Refs. [8-12]

Titanium and nickel alloys' resistance in hypochlorite, sodium chloride and hydrochloric acid solutions — **Table III**

Alloy resistance

Alloy	Hypochlorite solutions	NaCl solutions: Min pH	NaCl solutions: Max pH	Hydrochloric acid: Diluted	Hydrochloric acid: Concentrated
Titanium-Grade 2	Good	6.5 at 200° F*	14 at 180° F†	Good under 5% HCl‡	Poor
Titanium-Grade 12	Good	2 at 200° F*	14 at 180° F†	Good under 8% HCl‡	Poor
Titanium-Grade 7	Good	0.5 at 200° F*	14 at 180° F†	Good under 23% HCl‡	Poor
Nickel 200	Good	4-5ξ	No limit	Questionable	Poor
400	Poor	0-3ξ	No limit at moderate temperatures	Good if solution is nonoxidizing	Poor
B-2	Poor	No limit	No limit	Good if solution is nonoxidizing	Good if solution is nonoxidizing
C-276	Good	0.2ξ	No limit	Good up to 2.5% HCl	Good under 150° F

*Crevice corrosion limit.
†Possible hydrogen embrittlement beyond limit.
‡Corrosion rate of 5 mils/yr (125 μm/yr).
ξEstimate.
Adapted from Refs. [11, 13].

Carbon steel is able to resist some aqueous chloride solutions under certain conditions — Table IV

Chloride	Corrosion resistance of steel
Aluminum	Poor
Ammonium	Adequate, if solution is ammoniacal or buffered
Barium	Good, if pH exceeds 7
Beryllium	Poor
Calcium	Good, especially if brine is chemically inhibited
Cupric	Poor
Ferric	Poor
Lithium	Poor
Magnesium	Adequate, especially if solution is chemically inhibited or pH exceeds 7
Mercuric	Poor
Sodium	Adequate, especially if solution is alkaline or concentrated
Stannic	Poor
Zinc	Poor

Adapted from Ref. [*14, 15*].

Critical threshold temperatures are lower for crevice corrosion than they are for pitting — Table V

Alloy	10-d test in 10% ferric chloride Critical threshold temperature Pitting, °C	Crevice corrision, °C
316 SS	20	Below –5
20Cb-3	20	Below –5
825	25	Below –5
317LM SS*	25	10
904L SS	45	20
Duplex 255 SS	50	35

*SS is stainless steel. Adapted from Ref. [*10*].

Pitting of stainlesses

On the surface of stainless steels, the chloride ion may penetrate and disrupt the oxide coating that provides corrosion resistance. A locally exposed site on the surface becomes a rapidly attacked anode, which is surrounded by a large oxide-covered cathodic area that drives the corrosion into the pit. In the pit, the solution concentration may become increasingly acidic, even if the bulk-solution pH is neutral or even alkaline. The pit progresses quickly through the alloy. Since pitting may not initiate for months or years, it is difficult to evaluate. Electrochemical potentiodynamic pitting-scans are helpful in making short tests [*5*].

Pitting is favored by stagnant solutions. It often progresses downward through vessel bottoms. In chloride solutions, Types 304 and 316 pit most severely in pHs of 4–8.

For each alloy, there is a critical pitting temperature, below which this form of attack will not occur [*6*]. This temperature is mainly a linear function of the alloy's molybdenum content. In one electrochemical scan of a sodium-chloride/methanol solution at 60°C (140°F), Alloy 904L (4.5% Mo) resisted pitting, while Alloy 825 (3% Mo) did not.

Type 304 has no Mo, and should not be used in chloride solutions. In one case, it was used for cooling coils in $CaCl_2$ brine at –15°C (+5°F), and pitted after only a few months. Pitting could have occurred at ambient temperatures during weekends or shutdowns. For stainlesses, pitting will be worse for oxidizing chlorides than for the milder NaCl solutions. Ti alloys are more resistant in oxidizing conditions.

For the ferritic and duplex alloys, pitting resistance results from a combination of high chromium and lowered molybdenum (see Table I). A pH of 12 should stop many NaCl pitting problems. In one test at 90°C (195°F), even Type 304 did not pit in 4% NaCl at a pH of 12 [*7*].

Crevice corrosion of stainlesses

Since crevices may concentrate chlorides and exclude oxygen, crevice corrosion occurs more readily than pitting in stainless steels. Thus, crevice corrosion has lower critical threshold temperatures (see Table V).

In Table II, the successful tests were mostly short ones that showed no crevice corrosion. The temperatures listed are thus considered critical crevice-corrosion temperatures. Since some of the tests ran for only 1 or 10 days, longer tests should be run for more-reliable guidelines.

Short tests are widely used for alloy comparisons. They yield useful results because the crevices employed in them are tight and deep, much more so than those found in industrial equipment. Note that in chloride service, crevices accelerate corrosion dramatically. Good design minimizes the number of crevices. Similarly, this reinforces the need for high-quality welds with full penetration and without excessive porosity. As with pitting, temperature accelerates crevice corrosion, and Mo in stainless steels inhibits it.

References

1. Redmond, J. D., and Miska, K. H., High-Performance Stainless Steels for High-Chloride Service, *Chem. Eng.*, Part I, July 25, 1983, p. 93, Part II, Aug. 22, 1983, p. 91.
2. Schillmoller, C. M., and Althoff, H. J., How to Avoid Failures of Stainless Steels, *Chem. Eng.*, May 28, 1984, p. 119.
3. Uhlig, H. H., "Corrosion and Corrosion Control," 2nd ed., Chap. 5, John Wiley & Sons, New York, 1971.
4. Kirby, G. N., Corrosion Performance of Carbon Steel, *Chem. Eng.*, Mar. 12, 1979, p. 72.
5. Pitting Corrosion Overcome by Rapid-Scan Technique, *Chem. Proc.*, Mar. 1977, Reprint 500-5-78 R-64.
6. Brigham, R. J., and Tozer, E. W., Effect of Alloying Additions to the Pitting Resistance of 18%-Cr Austenitic Stainless Steels, *Corrosion*, Vol. 30, No. 5, May 1974, p. 161.
7. Corrosion Resistance of the Austenitic Chromium-Nickel Stainless Steels in Chemical Environments, pamphlet pub. by International Nickel Co., New York, 1963.
8. Ferralium Alloy 255, brochure pub. by Cabot Corp., Kokomo, Ind., 1983.
9. Hastelloy Alloy G-3, brochure pub. by Cabot Corp., Kokomo, Ind., 1983.
10. Information presented at INCO Chemical and Petrochemical Industries Materials Conference, held by International Nickel Co. at Wrightsville Beach, N.C., Sept. 19-21, 1983.
11. Covington, L. C., and Schutz, R. W., Corrosion Resistance of Titanium, brochure pub. by Timet Div., Titanium Metals Corp. of America, Pittsburgh, Pa.
12. Bond, A. P., and Dundas, H. J., Resistance of Stainless Steels to Crevice Corrosion in Seawater, paper presented at Corrosion/84, New Orleans, La., Apr. 2-6, 1984, NACE, Houston, Tex.
13. Manning, Paul, Cabot Corp., Kokomo, Ind., private communication, 1984.
14. Rabald, E., "Corrosion Guide," 2nd rev. ed., Elsevier, New York, 1968.
15. NACE, "Corrosion Data Survey, Metals Section," 5th ed., NACE, Houston, Tex., 1974.

The author

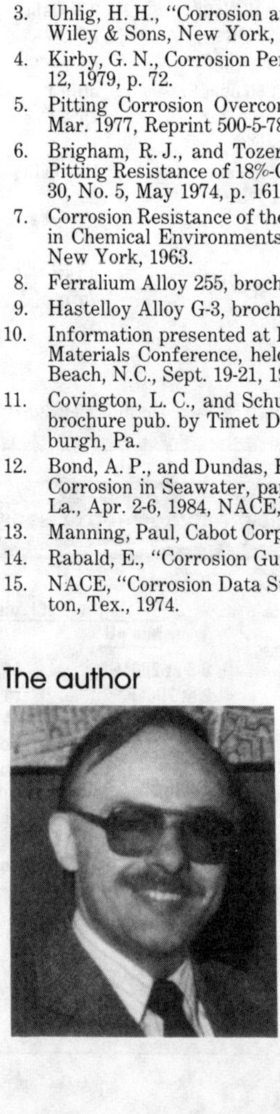

Gary N. Kirby is materials and metallurgical engineer for Ciba-Geigy Corp., at its Toms River plant, P.O. Box 71, Toms River, N.J. 08753. Tel: (201) 349-5200. He consults regularly with other Ciba-Geigy plants. Previously, he was with Crawford & Russell, Inc., American Metals Climax Co., International Nickel Co. and Union Carbide Metals Co. A member of the Amer. Soc. for Metals, he holds a Ph.D. degree from the University of Michigan, an M.S. from Lehigh University and a B.S. from Cornell University, all in metallurgical engineering.

Selecting alloys for chloride service — PART 2

Stainless steels, and alloys of titanium, zirconium and copper are used to resist chloride solutions, depending upon conditions. Also, a less-noble metal or alloy can be selected, if galvanic protection is used.

Gary N. Kirby, Ciba-Geigy

In Part 1 of this two-part series (*CE*, Feb. 4, p. 81), we mentioned the types of metals and alloys suitable for chloride service, and discussed how carbon steel stands up under chlorides. Also, a review of the corrosion of stainless steels in chlorides was begun. Here, we shall conclude this review, and look at other metals and alloys (e.g., titanium, see Fig. 1), and galvanic protection, as well.

Stress-corrosion cracking of stainlesses

The austenitic (face-centered cubic) phase is susceptible to chloride stress-corrosion cracking (SCC), while the ferritic (body-centered cubic) phase is not. The alloy's nickel content determines the phase. Ferrritic alloys such as 26-1 and 29-4 contain little nickel and are highly resistant to SCC. However, ferritic structures have limited low-temperature toughness, especially in heavy sections. Also, they require special high-purity welding methods to keep oxygen and nitrogen impurity levels low. An experienced fabricator is needed to perform such welding.

The duplex (two-phase) alloys contain about 4.5% Ni, which results in a 50-50 ferrite-austenite mixture. The theory is that cracks started in the austenite will be halted when they reach a ferrite region. However, in one case, after an improper heat treatment, one of these alloys cracked in boiling 4% and 25% NaCl at pH values of 1.5–7.0. So, some caution is needed. Obviously, experienced suppliers and fabricators are a must.

Types 304 and 316, widely used in the chemical process industries, contain 9–11% Ni, which makes them predominantly austenitic. Due to their nickel level, these stainlesses are highly susceptible to chloride SCC, at least in speed of cracking in the boiling magnesium chloride test [*1*]. This effect of nickel level was essentially confirmed in 1981 for 22% NaCl solutions at 105°C [*2*]. The variation of resistance to SCC with nickel level is found in Fig. 2. Such chloride SCC occurs most readily at places where these ions accumulate or concentrate — crevices or surfaces under sediment, plus heated and alternately-wet-and-dry surfaces. Now let us cover the critical factors affecting chloride SCC:

1. Effect of chlorides—In water at higher temperatures, chlorides present in parts-per-million levels can cause SCC. For example, 40 ppm can cause SCC of Types 304 and 316 at 80°C (176°F) after 1½–2 yr.

However, in such cases, chloride accumulation or concentration is the culprit. An extensive study of SCC showed that cracking occurred with stagnant solutions, low flow velocities, rust or scale and sedimentation [*3*]. In a test of stressed 304 samples, cracking did not take place until the chloride concentration reached nearly 10,000 ppm (1%).

2. Effect of pH—Acid chloride solutions are more often encountered when SCC occurs (see Table I). However, alka-

Not all industrial cases of chloride SCC occur in the acid range — Table I

pH	Number of cases	Percent	
Described as "low" or "acid"	12	20.3	72.8% in acid pH range
pH 2 to 4	7	11.9	
pH 4 to 6	14	23.7	
pH 6 to 7	10	16.9	
pH 7 to 8	6	10.1	27.1% in alkaline pH range
pH 8 to 10	5	8.5	
pH 10 to 11	2	3.4	
pH 14 and over	1	1.7	
Described as "alkaline"	2	3.4	
Total	59	99.9	

Note: Total is 99.9% due to roundoff error.

Adapted from Ref. [*3*].

Stainless steels can show chloride SCC even at room temperature — Table II

Temperature, °F	Number of cases	Percent
0 - 100	8	8.7
100 - 200	31	33.7
200 - 300	38	41.3
300 - 400	9	9.8
400 - 500	1	1.1
500 - 600	2	2.2
600 - 700	1	1.1
over 700	2	2.2
Total	92	100.1

Note: Total is 100.1%, due to roundoff error.

Adapted from Ref. [*3*].

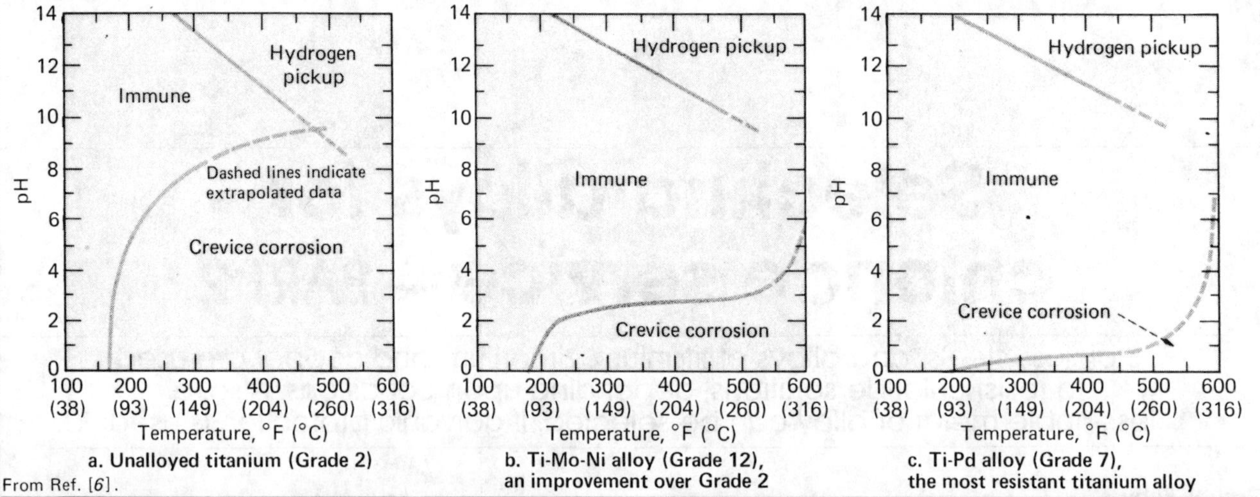

From Ref. [6].

The effects of pH and temperature on Ti alloys with saturated NaCl brine **Fig. 1**

line solutions, too, can cause SCC. Caustic solutions alone can result in SCC of most alloys except Nickel 200 (commercially pure Ni alloy).

Caustic SCC usually takes place above 200°C (392°F). Thus, alkaline conditions in chloride solutions raise the threshold temperatures for SCC. As an example, Type 316L can thus be used in a diaphragm cell for chlor-alkali manufacture where there is an alkaline chloride solution (13% NaOH and 13% NaCl) at 99°C (210°F).

Remember that pH varies with temperature. At elevated temperatures, the pH will be lower than at ambient temperatures.

3. Effect of temperature—Most industrial cases of SCC occur above 80°C (176°F), but, as Table II shows, cracking can take place at or near room temperature. The metal temperature is more important than the process-side, water-in or water-out temperatures.

4. Effect of solution velocity—Stagnant conditions or low solution velocities are conducive to SCC. Thus, in shell-and-tube heat exchangers, water solutions containing parts-per-million levels of chlorides are less likely to cause SCC if the water is put on the tubeside of the unit. The worst situation is the water on the shellside of a vertical exchanger. Any air pockets in the upper shell provide places for chlorides to accumulate. Two guidelines address these problems:

- Water on the shellside. For cooling-water containing 20–60 ppm chlorides, SCC of Types 304 and 316 is rare if the maximum water temperature is below 60°C (140°F).
- Water on the tubeside. For water below 60°C (140°F), Type 316 can tolerate about 1,000 ppm chlorides; Type 304, about 200 ppm. This assumes no scale, sedimentation or stagnation.

These guidelines are certainly not absolute. They assume a neutral pH.

Other stainless steels

The higher nickel levels (e.g., alloys 20Mo-6 and 20Cb-3) resist chloride SCC. At 33% Ni and higher there are few, if any, industrial cases of chloride SCC. Alloy 20Cb-3 (33% Ni) has survived 864 h without cracking in a glass-wool wick test, which concentrates chlorides drawn from a 1,500-ppm solution. The alloy sample was stressed and heated to 100°C (212°F). In this test, Type 304 stainless steel cracked after a 72-h exposure. Solutions such as magnesium chloride may require at least 40% Ni content.

The ferritic alloys are also very resistant. Alloy 26-1 has survived 200 h without cracking in boiling 42% $MgCl_2$. In comparison, Type 304 failed after 8 h and Type 316 lasted only 24 h. Alloy 29-4C has not cracked in U-bend tests after 500 h in boiling 26% NaCl.

Thus far in tests, the duplex alloys are resistant. In U-bend tests, Alloy 255 survived 30 d in a modified wick test at 100°C (212°F), using 1,650 ppm NaCl plus 500 ppm $FeCl_3$.

Titanium in chlorides

Commercially pure titanium (Grades 1 and 2) and its most common alloys (Grades 7 through 12) have good chloride resistance. This resistance does, however, drop off at high temperatures. Fig. 1 shows this problem for NaCl solutions of various pHs. For acidic conditions, crevice corrosion can occur when the pH is low enough. A rule of thumb for Grade 1 and 2 Ti is to keep the pH above 1 at higher temperatures.

For alkaline solutions, use is limited to temperatures below 80°C (176°F) to avoid excessive hydrogen pickup. Above alloy hydrogen levels of 800–900 ppm, titanium can suffer gross embrittlement. Some hot, concentrated chloride solutions corrode Grades 1 and 2. Aluminum or calcium chlorides can result in excessive overall corrosion. Magnesium or zinc chlorides can produce severe pitting. Grades 12 or 7 are more resistant in such cases (see Table III).

A major advantage of titanium alloys is their excellent resistance to oxidizing chlorides, such as $FeCl_3$ and $CuCl_2$. These chlorides present a problem for the austenitic stainless steels. However, oxidizing conditions, even if resulting from dissolved chlorine, help the corrosion resistance of titanium alloys. Grades 1 and 2 titanium can also withstand 5–20% $MgCl_2$ at 100 C° (212°F). Magnesium chloride is extremely aggressive to stainless steels.

Unlike stainless steels, titanium alloys withstand hydrochloric acid under some conditions. At ambient temperatures, Grades 1 and 2 titanium are resistant to about 10% HCl. Fig. 3 shows two views of a pump impeller that was

How unalloyed zirconium and grades of titanium resist various aqueous chloride solutions — Table III

Solution	Concentration, %	Temperature, °C	Alloy*	Corrosion rate, mils/yr
$AlCl_3$	5-25	35-100	Zr	Under 1
	40	Boiling	Titanium Grade 2	4,300
NH_4Cl	1-sat'd	35-100	Zr	Under 1
	Sat'd	100	Titanium Grade 2	Nil
$CuCl_2$		35 to boiling	Zr	Over 50
	55	Boiling	Titanium Grade 2	0.1
$FeCl_3$	10	25	Zr	9.0
	10	25	Zr welded	Over 50
	50	Boiling	Titanium Grade 2	0.7
$MgCl_2$	47	Boiling	Zr	Under 1
	42	Boiling	Titanium Grade 2	Nil
$ZnCl_2$ slurry	–	130-150	Zr welded	Nil
	–	130	Titanium Grade 7	Nil
	–	150	Titanium Grade 7	Pitting

*For Zr, unalloyed zirconium R60702 was tested.

Adapted from Ref. [*5*].

In hydrochloric acid, oxidizing agents inhibit corrosion of titanium — Table IV

HCl, wt %	Oxidizing inhibitor	Temperature, °C	Corrosion rate, mils/yr
4	–	60	43.0
5	–	190	1,120
5	Chlorine gas	190	Less than 1.0
10	–	65	350 - 800
10	0.1% $CuSO_4$	65	1.0
10	1.0% CrO_3	65	0.4
10	0.7% HNO_3	65	1.3

Adapted from Ref. [*4*].

severely etched after exposure to HCl that was too concentrated. This resistance can be extended to higher temperatures or even concentrations if oxidizing chemical species are present. These include dissolved chlorine gas, $CuSO_4$, CrO_3 and HNO_3 (see Table IV) [*4*]. Grade 7 alloy (0.15% Pd) also provides greater corrosion resistance.

SCC is generally not a problem with the titanium alloys that are used in the chemical process industries. Some high-strength aerospace alloys suffer SCC in hot, dry sodium chloride, methanolic chloride solutions, chlorinated solvents, seawater and other environments.

Zirconium alloys in chlorides

Zirconium has good corrosion resistance to chloride solutions, even at high temperatures [*5*]. For example, one test showed that zirconium and Grade 2 titanium have about equal resistance to overall corrosion in seawater at 101°C (214°F). However, zirconium is less susceptible to chloride crevice corrosion in hot chloride solutions.

Furthermore, unlike titanium alloys, zirconium alloys resist chloride crevice attack regardless of the pH, temperature and chloride concentration.

Zirconium alloys are highly resistant to chloride SCC. U-bends in boiling seawater have shown no cracking after 1 yr. However, oxidizing media can cause SCC.

As a generalization, zirconium alloys have excellent resistance to reducing chloride solutions, but may be attacked by oxidizing ones. This behavior is opposite to that of titanium alloys. Unalloyed zirconium is attacked by oxidizing media, such as ferric or cupric chloride (see Table III). However, zirconium is more resistant to zinc chloride slurries than even Grade 7 titanium.

In highly oxidizing chloride solutions, zirconium welds are less corrosion-resistant than is the base metal. In such cases, tests should be made with welded samples.

Both tungsten-inert-gas (TIG) and metal-inert-gas (MIG) arc welding are used for joining zirconium [*7*]. An inert-gas chamber containing argon or an argon-helium mixture is commonly used.

Copper alloys

Copper and its nickel-containing alloys are widely used in seawater and brine service. Unalloyed copper is less expensive, but has some vulnerability. It starts to deteriorate in seawater at velocities above 3–4 ft/s [*8*]. Also, ammonia contamination can readily cause SCC of copper.

Cupro-nickel (90 Cu-10 Ni) is more resistant to the effects of velocity and ammonia. It is used as a standard seawater alloy for piping, at design velocities of 5–12 ft/s. Corrosion rates are less than 5 mils/yr. However, if small amounts of hydrogen sulfide (or other sources of reactive sulfur) are present in the solution, a copper-nickel alloy may not offer an advantage over copper.

A 70-30 cupro-nickel alloy and Alloy 400 (65 Ni-32 Cu) also are used in seawater. However, they are more expensive than the 90/10 alloy, and at solution velocities under 3 ft/s, may pit more deeply than does the cheaper alloy.

Copper-base alloys containing over 60% copper resist fouling by biological organisms. Steel, stainless steel and titanium offer little such resistance. Biological fouling is less likely above solution velocities that are around 3 ft/s.

Galvanic protection

In the marine industry, metals have been protected galvanically, following these rules:

1. Whenever possible, construct equipment from just one alloy.

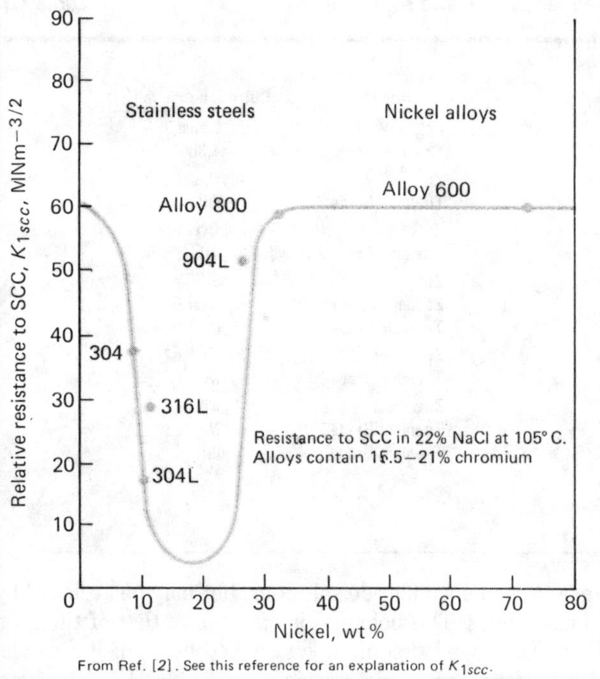

Variations in Ni content affect the resistance to stress chloride cracking of Fe-Ni-Cr alloys Fig. 2

2. When this cannot be done, make sure that the critical components are more noble.

3. Allow for increased corrosion of the less-noble alloy by providing a large area or a heavy thickness.

4. If painting is to be done to a structure, paint the parts made of the more-noble alloy and leave those of the corroding alloy bare. If parts made of the corroding alloy are painted, pinholes in the coating will result in rapid corrosion of the exposed less-noble alloy.

Examples of galvanic protection

Some examples:

- In seawater, heavy bronze valve bodies sacrificially protect Alloy 400 trim.
- In heat exchangers used for desalting, nickel-aluminum bronze tubesheets are employed with titanium tubing. For other seawater uses, Alloy 400 tubesheets are combined with titanium tubing.
- Types 304 and 316 stainless steel heat-exchanger tubes can be protected from attack by carbon steel tubesheets and shells. With a 1:10 ratio of steel to stainless steel area, Types 304 or 316 can be completely protected from what would otherwise be severe chloride crevice corrosion in ambient-temperature seawater [9]. Such protection does not extend over large distances.

Acknowledgements

The author wishes to thank the following people for their help and contributions: Paul Bond of Climax Molybdenum Co. of Michigan, Ivan Francen of Allegheny Ludlum Steel Corp., Paul Manning of Stellite Div., Cabot Corp., James Martin of Carpenter Technology Corp., Ron Schutz of Timet, a Div. of Titanium Metals Corp. of America, and Te-Lin Yau of Teledyne Wah Chang Albany. None of the above reviewed this article, which represents the experience and opinions of the author alone.

The author wishes to express his appreciation to Kathy Davenport for her fine secretarial help.

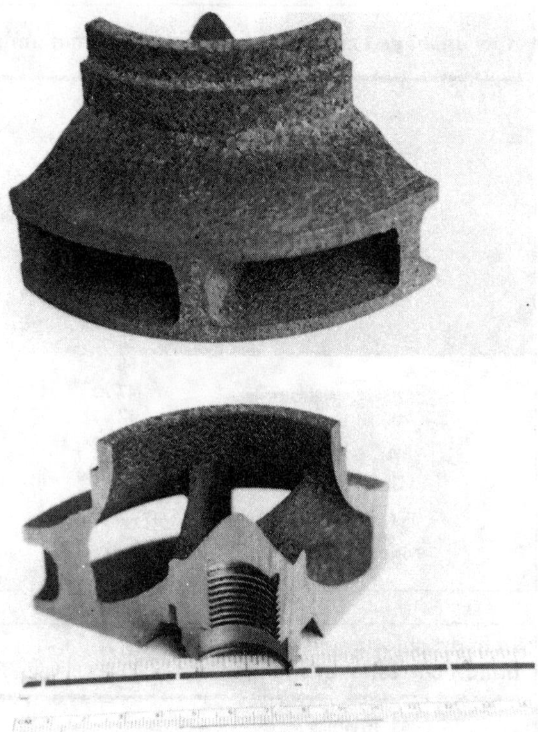

Section of pump impeller severely etched by upset of strong hydrochloric acid in process stream Fig. 3

References

1. Kirby, G. N., How to Select Materials, *Chem. Eng.*, Nov. 3, 1980, p. 86.
2. Speidel, M. O., Stress Corrosion Cracking of Stainless Steels in NaCl Solutions, *Met. Trans.*, Vol. 12A, p. 779, May 1981.
3. Report on Stress-Corrosion Cracking of Austenitic Chromium-Nickel Stainless Steels, ASTM Special Technical Publication No. 264, ASTM (formerly Amer. Soc. for Testing and Materials), Philadelphia, 1960. Available from University Microfilms International, Ann Arbor, Mich., 1976.
4. Bomberger, H. B., The General Corrosion Resistance of Titanium, "Titanium and Titanium Alloys," Amer. Soc. for Metals, Metals Park, Ohio, 1982, p. 161.
5. Yan, T., The Corrosion Properties of Zirconium Alloys in Chloride Solutions, paper presented at Corrosion/83, Anaheim, Calif., Apr. 18-22, 1983, Natl. Assn. of Corrosion Engineers (NACE), Houston, Tex.
6. Covington, L. C., and Schutz, R. W., Corrosion Resistance of Titanium, brochure pub. by Timet Div., Titanium Metals Corp. of America, Pittsburgh, Pa.
7. Cary, H. B., "Modern Welding Technology," Prentice-Hall, Inc., Englewood Cliffs, N.J., 1979.
8. Guidelines for Selection of Marine Materials, 2nd ed., pamphlet pub. by International Nickel Co., New York, 1971.
9. Lee, T. S., and Tuthill, A. H., Use of Carbon Steel to Mitigate Crevice Corrosion of Stainless Steels in Sea Water, *Materials Perf.*, Jan. 1983, p. 48.

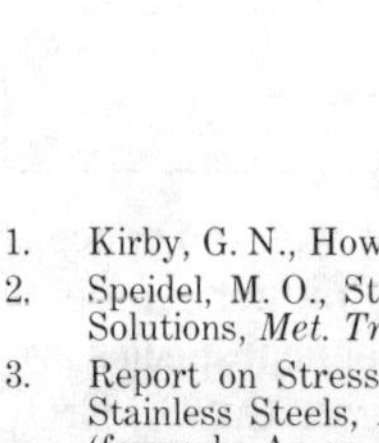

The author

Gary N. Kirby is materials and metallurgical engineer for Ciba-Geigy Corp., P.O. Box 71, Toms River, N.J. 08753. Tel: (201) 349-5200. He consults regularly with other Ciba-Geigy plants. Previously, he was with Crawford & Russell, Inc., American Metals Climax Co., International Nickel Co. and Union Carbide Metals Co. A member of the Amer. Soc. for Metals, he holds a Ph.D. degree from the University of Michigan, an M.S. from Lehigh University and a B.S. from Cornell University, all in metallurgical engineering.

Section V
METALS AND ALLOYS

How to select material properties for wear resistance

A guide is provided for determining material selection criteria to combat wear in operating equipment, based on observed wear characteristics.

K. C. Ludema, The University of Michigan

☐ Chemical engineers design and select machinery and components for complex services often involving combinations of corrosion and wear. For example:

- Centrifugal pumps for SO_2 stack-gas scrubbers.
- Bends, elbows and valves in synfuel systems.
- Seals and shafts in slurry pumps.
- Pumps and valves for hydraulic systems.
- Stirring paddles in mixing tanks
- Scrapers in sedimentation tanks, etc.
- Pistons, cylinders and orifices in plastic-molding machines.

The following is a short explanation of some mechanisms of wear. A table is given that aids in defining terms, and connects these terms with the modes of material loss. A figure is given that shows the influence of such parameters as sliding speed, applied load, temperature and time on wear rate for most of the modes of material loss.

Designing for wear

Loss of material from a surface is called wear, and material can be removed by only a few well recognized mechanisms (discussed later). Most materials wear by several mechanisms, either in turn or simultaneously. A prominent mode of wear begins with surface corrosion, which forms soft products that are in turn rubbed or eroded away. Also, many pumped solutions are poor lubricants, or contain abrasive species, both contributing factors to adhesive and abrasive wear. In addition, materials may suffer a shortening of use because of the influence of pH on fatigue life, as frequently happens with bearings.

Numerical values for the wear resistance of materials are not yet available. The major reason is that wear is very dependent on the system in which surfaces function. The paucity of data on wear resistance is probably hidden by successful design of long-lasting mechanical components, and publication of numerous articles containing wear data. A cursory review of the literature reveals, however, that available wear data are derived from tests that usually involve a very narrow range of conditions.

Designers can often select catalog items to design products for a given wear life, because such components as bearings, gear boxes, belts, etc., are made to consistent quality and have been observed in use in almost every possible situation. It is in the components that are less standard or completely new that the designer must select a material with little outside aid.

Sometime it is possible to benefit from experience with devices somewhat similar to the proposed design. More often, however, the proposed design must function under conditions beyond those of previous experience. Not only must materials be selected from among several generic classes, but specifications must be written so that the chosen material may be prepared to a particular surface finish, dimensional tolerance, and final mechanical properties.

A helpful step in designing for wear resistance is to locate equipment that most nearly approximates the proposed design. If this cannot be found, it may be possible to modify existing equipment to operate under the new conditions, or to construct prototype components. From these, it is very useful to identify the types of wear occurring, by careful observation of worn surfaces, wear debris, and general behavior of operating equipment.

The hazard in doing so, however, is that it then becomes necessary to communicate the findings to others, and, in so doing, specialized terms must be used. Inevitably, certain subjective terms are used by designers and engineers in describing worn surfaces. Worn surfaces are often described as manifesting either "abrasive wear" or "adhesive wear." Some will further describe wear in terms of "cutting wear," "metal-to-metal wear" or "fatigue wear." Few of these terms are well understood or agreed upon.

Material loss mechanisms

The modes of material loss in wear are here suggested to be limited to: corrosion, cutting, brittle fracture, ductile fracture, low-cycle fatigue, high-cycle fatigue, and melting.

The table connects descriptions of observed debris and wear surfaces with the seven possible material-loss mechanisms. A procedure is given in the table whereby

Guide for determining material

How to use the table:

I. Observe the nature of wear in existing equipment, or of similar materials from appropriate wear-testing machines.

II. Check the lists in **Section A** below for an applicable description of worn surfaces or type of service, and note the code that follows the selected term.

III. Proceed to **Section B** and verify that the code listing is an adequate description of the worn surface. (It is possible to use Section B without reference to Section A.) From **Section B**, find the major term (capitalized) in Columns **a, b, c,** and **d**. Columns **e** and **f** are added to complete the description of the surface.

IV. In **Section C,** find the detailed description of the capitalized term from **Section B**, and note which *material-loss mechanism* is applicable, and confirm from the nature or description of wear debris.

V. Find the *material-loss mechanism* in **Section D** and note the material characteristics and microstructure that should influence wear resistance of material, and note the precautions in material selection to prevent failure.

VI. Select materials in conjunction with materials specialists.

Section A—Description of worn surface and type of service

Surface appearance

- Stained — **f**
- Polished, or smooth wear — **a1+c+e**, or **a2+c+e**
- Scratched (short grooves) — **b3+c+e**
- Gouged — **b3+d1**
- Scuffed — **a1**+initiated and periodically perpetuated by **d3**, +**e**
- Galled — **b1+d3+e** (usually very rough)
- Grooved (smooth or rough) — **a1**+periodically advanced by **d1**, +**e**
- Hazy — **b2**
- Exfoliated, or delaminated — **d4+e**
- Pitted — **b2** and/or **d5**
- Spalled — **d4**
- Melted — **a3**
- Fretted — **a1+d5+f**

Type of service

- Surface corrosion or Erosion/corrosion: in solid machinery — **a1+c**; in fluids — **a2+d2**
- Abrasive wear (multiple scratches) — **b3+c**
- Gouging — **b1+d1+e**
- Dry wear or unlubricated sliding — **b1+d3+e**, or **a1+c+e**
- Metal-to-metal wear, or adhesive wear — **b1+d3+e**
- Erosion at high angle — **b2+d4**
- Erosion at low angle — **b3+d1** or **d2**
- Fretting — **a1+d5+f**

Note: Rigorous connection cannot always be made between the terms in the two columns because of the wide diversity of use and meaning of terms.

Section B—Code listing

a—Micro-smooth

1. Progressive loss and reformation of surface films by fine **ABRASION** and/or tractive stresses, mutually imposed by **ADHESIVE** or viscous interaction
2. Very fine **ABRASION**, with loss of substrate in addition to loss of surface film, if any
3. From **MELTING**

b—Micro-rough

1. Due to tractive stresses resulting from **ADHESION**
2. Micro-pitting by **FATIGUE**
3. **ABRASION** by medium-coarse particles

c—Macro-smooth

By abrasive held on or between solid backing

d—Macro-rough

1. **ABRASION** by coarse particles, including carbide and other hard inclusions in the sliding materials that are removed by sliding action as wear of matrix progresses
2. **ABRASION** by fine particles in turbulent fluid, producing scallops, waves, etc.
3. Severe **ADHESION**, at least as an initiator of damage
4. Local **FATIGUE** failure resulting in pits or depressions, by repeated rolling-contact stress, gradients, high-friction sliding, or impact by hard particles as in erosion
5. Advanced stages of micro-roughening, where little unaffected surface remains between pits

e—Shiny

Very thin (or no) surface film of oxide, hydroxide, sulfide, chloride, or other species

f—Dull or matte

Thick films of perhaps greater than 25-nm thickness (resulting from aggressive environments including high temperatures) due to **CORROSION**

properties for wear resistance

Section C—Material-loss mechanisms and nature of debris

Material-loss mechanisms (underlined)	Nature of debris*
CORROSION (of surfaces)—chemical combination of material surface atoms with passing or deposited active species to form a new compound, i.e., oxide, etc.	Newly formed chemical compound, usually agglomerated, and sometimes mixed with fragments of the original surface material
Abrasion—involves particles (of acute angular shapes but mostly obtuse) that cause wear debris, some of which forms ahead of the abrasive particle, which mechanism is called CUTTING, but most of which is material that has been plowed aside repeatedly by passing particles, and breaks off by LOW-CYCLE FATIGUE	Long, often curly chips or strings
Adhesion—a strong bond that develops between two surfaces (either between coatings and/or substrate materials) that, with relative motion, produces tractive stress that may be sufficient to deform materials to fracture. The mode of fracture will depend on the property of the material, involving various amounts of energy loss, or ductility to fracture—i.e.:	
low energy and ductility → BRITTLE FRACTURE	Solid particles, often with cleavage surfaces
or	
high energy and ductility → DUCTILE FRACTURE	Severely deformed solids, sometimes with oxide clumps mixed in
Fatigue—due to cyclic strains, usually at stress levels below the yield strength of the material, also called HIGH-CYCLE FATIGUE	Solid particles, often with cleavage surfaces and ripple pattern
MELTING from very high-speed sliding	Spheres, solid or hollow, and "splat" particles

*Useful in indicating trends, new events, progressions; sometimes reveals unexpected causes of wear

Section D—Material selection characteristics

Material-loss mechanisms	Appropriate material characteristics to resist wear	Precautions to be observed when selecting a material*
Corrosion	Reduce corrosiveness of surrounding region; increase corrosion resistance of material by alloy addition, or by selection of soft, homogeneous material	Total avoidance of new surface species can result in high adhesion of contacting surfaces; soft materials tend to promote galling and seizure
Cutting	Use material of high hardness, with very hard particles or inclusions such as carbides, nitrides, etc. and/or overlaid or coated with materials that are hard or contain very hard particles	All methods of increasing cutting resistance cause brittleness or lower fatigue resistance
Ductile fracture	High strength achieved by any method other than by cold working or by heat treatments that produce internal cracks or large and poorly bonded intermetallic compounds	
Brittle fracture	Minimize tensile residual stress, for cold temperature; ensure low-temperature brittle transition; temper all martensites; use deoxidized metal; avoid carbides such as in pearlite, etc.; effect good bond between fillers and matrix to deflect cracks	In essence, soft materials will not fail through brittleness and will not resist cutting
Low-cycle fatigue	Use homogeneous and high-strength materials that do not strain-soften; avoid overaged materials or two-phase systems with poor adhesion between filler and matrix	(see above)
High-cycle fatigue	For steel and titanium, use stresses less than half the tensile strength (however achieved); for other materials to be load-cycled $< 10^8$ times, allow stresses less than 1/4 the tensile strength (however achieved); avoid retained austenite; use spherical pearlite rather than plate structure; avoid poorly bonded second phases; avoid decarburization of surfaces; avoid platings with cracks; avoid tensile residual stresses, or form-compressive residual stresses by carburizing or nitriding	Calculation of stress should include the influence of tractive stress
Melting	Use material of high melting point and/or high thermal conductivity	

*Materials of high hardness or strength usually have decreased corrosion resistance, and all materials with multiple and carefully specified properties and structures are expensive.

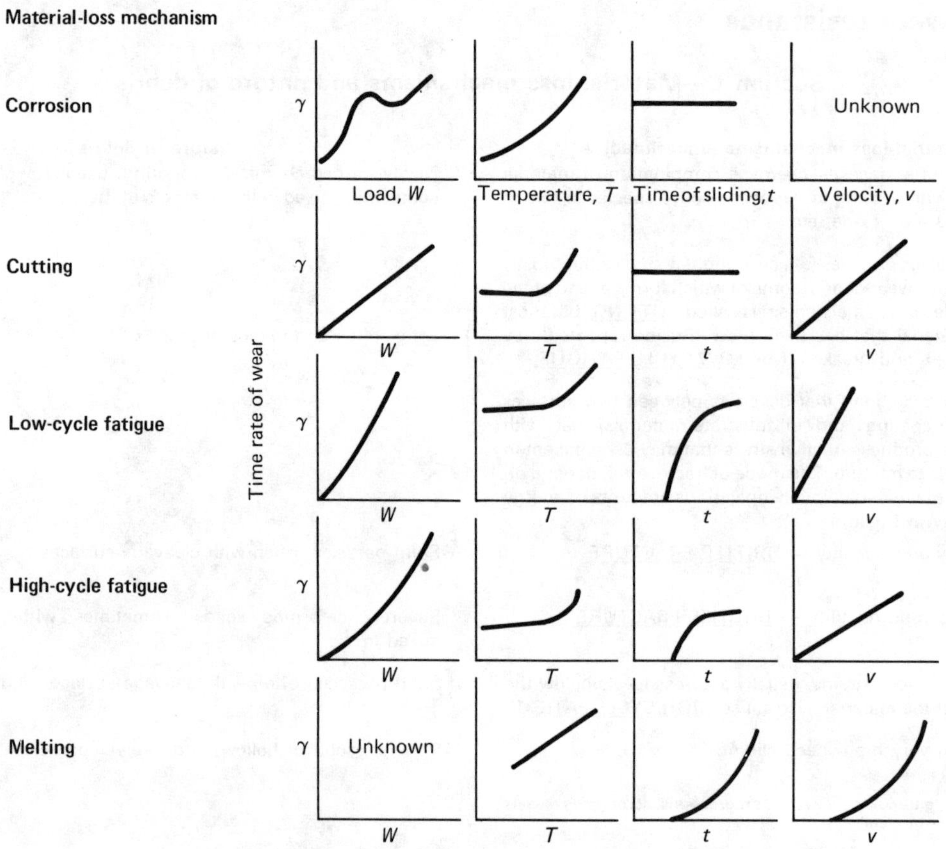

Note: The relationships for ductile fracture and brittle fracture are, as of yet, unknown.

Wear rate curves for material-loss mechanisms

wear-resisting material characteristics may be determined, from which specific materials of construction can be selected.

Wear models

The approach taken in the table avoids a direct statement on the exact combination of material removal modes for each wear mechanism; conversely, no single or combined property of materials is claimed to control wear. For example, in some applications, hardness may be the single most useful indicator of wear resistance, but in other instances, fatigue strength or toughness or oxidation resistance may be more important.

The expected form of wear models for each mode of material removal if operating alone is shown in the above figure.

These curves are somewhat speculative because definitive wear experiments that involve single material-loss mechanisms have not been done. The figure serves, however, to show why simple or linear wear models are not always found for specific materials. The figure also suggests the reason why wear models are unavailable for many sliding systems: We do not yet know how much material properties enter into the wear process.

References

1. Ludema, K. C., A Perspective on Wear Models, ASTM Standardization News, Vol. 2, No. 9, 1974, p. 13.
2. McClintock, F. A., and Argon, A. S., "Mechanical Behavior of Materials," Addison-Wesley, 1964.
3. Suh, N. P., Wear Mechanisms: An Assessment of the State of Knowledge, Proc. of Conf. on Fundamentals of Tribology, M.I.T., Cambridge, Mass., June 1978, p. 443.
4. Jahanmir, S., On the Wear Mechanisms and the Wear Equations, *ibid.,* p. 455.
5. Rice, S. L., A Review of Wear Mechanisms and Related Topics, *ibid.,* p. 469.
6. Quim, T. F. J., The Classifications, Laws, Mechanisms and Theories of Wear, *ibid.,* p. 477.

This paper was presented at the Third International Conference on Wear of Materials, ASME, San Francisco, 3/30-4/1, 1981.

The author

Kenneth C. Ludema is Professor of Mechanical Engineering at the University of Michigan, Ann Arbor, MI 48109. He holds a B.S. in production engineering and an M.S. in mechanical engineering from the University of Michigan, and a Ph.D. in Physics from the University of Cambridge, England. He is a member of the Amer. Soc. of Lubrication Engineers, ASME, SAE, ASTM and ASM. He was the Editorial Chairman for the International Conference on the Wear of Materials, ASME, 1981 in San Francisco.

A stainless steel alternative to cobalt wear alloys

Though cobalt alloys are the leaders in wear and galling resistance, there is a stainless steel that can compete in some applications.

William J. Schumacher, Armco, Inc.

☐ Cobalt alloys are essential in valves, pumps and fittings used in the chemical process industries to combat corrosion, wear and galling. These alloys have the unique capability of minimizing all three ravages simultaneously in many sliding systems. However, in the last two years, supply uncertainty and high prices have caused users of cobalt alloys to look for alternative materials.

Nickel-base alloys are receiving close attention because they offer as good or better corrosion and wear resistance as cobalt alloys; however, their galling resistance is far inferior.

Austenitic stainless steels offer good corrosion resistance, but the wear and galling characteristics of these alloys have been inadequate, with the exception of Nitronic 60. For many wear and galling applications in poorly lubricated or unlubricated systems, this alloy may be an alternative to cobalt alloys. A direct comparison of Nitronic 60 and Cobalt Alloy No. 6 (Stellite 6)—the most widely used cobalt alloy in the hardfacing of valve seats and disks—will follow.

Galling

An insidious form of adhesive wear, galling is most damaging at high stresses. Seizure of mating surfaces is frequently the result of severe galling, and usually occurs early in the life of equipment. When two surfaces are loaded together, the contacting asperities may form strong bonds due to the extremely high localized pressure and heat generated by subsequent motion. If these bonds sever at the interface, little damage occurs and the parts run smoothly; however, if fracture takes place in either material, gross damage—galling—results.

Poor galling-resistant alloys usually display a distinct mound of metal displaced from grooves; this is the result of weld-junction formation separating the two surfaces and producing large local stresses that can ultimately cause seizure. Alloys with better galling resistance generally do not exhibit large, distinct weld junctions. Types 416 and 440C stainless steel, for example, exhibit gradual surface deterioration by scoring, and much flatter or flake-like weld junctions, until the surfaces are considered as "galled."

Modified Taber Met-Abrader wear machine with crossed-cylinder geometry Fig. 1

Galling may be minimized by:

(1) decreasing contact stress to below the threshold galling stress of the couple;

(2) using protective surface layers such as lubricants, electrodeposited coatings, weld overlays, etc;

(3) using material couples with inherently better galling resistance (higher threshold galling stresses).

Most valves require good anti-galling materials in the stem area as well as the seat and disk areas. Although there is no standard laboratory test method, Armco developed the "Button and Block Galling Test" about 30 years ago to simulate these types of sliding [*2*]. The test has since been modified to better simulate actual service conditions [*3*].

Since common components such as valves and fasteners undergo reversing motion, this type of sliding was selected for comparing Nitronic 60 and Cobalt Alloy 6. Specimens were prepared with a 3/0 emery grit finish and hand-rotated under load 360° clockwise; this procedure was repeated two additional times under the dead load weight

Originally published September 21, 1981

of the test machine, a Brinell Hardness floor-model type. Despite the very slow speed, the specimens became warm to the touch and fine wear debris was observed.

A summary of all the galling tests is recorded in Table I. Both Nitronic 60 and Alloy 6 performed well against fully hardened 17-4PH (HRC 43) with neither couple galling at 42.4 ksi. Another exceptional couple was Alloy 6 mated with Nitronic 60, with no galling up to 51.2 ksi. In applications where Alloy 6 is used on both mating surfaces, Nitronic 60 may be substituted on one face—if superior corrosion and galling resistance are required.

While Alloy 6 exhibited better galling resistance than Nitronic 60 against Types 410 and 304, the opposite was true against Type 316. The higher nickel content of Type 316 decreases its galling resistance (based on unpublished work).

Comparative galling resistance of Cobalt Alloy 6 and Nitronic 60 — **Table I**

Button-and-block galling test (3 cycles-reverse motion)

Mating alloy (Rockwell hardness)	Threshold galling stress (ksi) Cobalt alloy 6 (R_C 48)	Nitronic 60 (R_B 93)
17-4PH (R_C 43)	42.4 +	42.4 +
Nitronic 60 (R_B 93)	51.2 +	–
Alloy 6 (R_C 48)	–	51.2 +
Type 410 (R_C 23)	35.3 +	<23.6
Type 410 (R_C 37)	35.3 +	34.0
Type 304 (R_B 78)	35.3 +	21.0
Type 304 (R_C 37)	34.1 +	24.0
Type 316 (R_B 93)	< 3.6	9.0

+ Did not gall

The threshold galling stress of standard stainless steel couples would most likely be less than 1 ksi based on unidirectional, single-rotation tests causing galling at 2 or 3 ksi

Metal-to-metal wear resistance of Cobalt Alloy 6 and Nitronic 60 — **Table II**

Mating alloy (Rockwell hardness)	Total couple wear volume (mm^3) Cobalt Alloy 6 (R_C 48)	Nitronic 60 (R_B 95)
Alloy 6 (R_C 48)	1.3	2.4
Nitronic 60 (R_B 95)	2.4	3.7
Type 304 (R_B 99)	3.9	7.6
Type 316 (R_B 91)	7.0	5.4
Type 416 (R_C 24)	54.4	7.0
17-4PH (R_C 43)	4.8	6.8
Nitronic 32 (R_B 95)	2.5	4.1
Nitronic 50 (R_B 99)	3.7	4.4
K-500 Monel (R_C 34)	23.8	29.0
Haynes 25 (R_C 28)	1.8	2.5
Waspaloy (R_C 36)	3.0	3.9
IN 718 (R_C 38)	3.4	3.9
IN X750 (R_C 36)	10.1	7.0

Conditions: 0.5-in. crossed cylinders, 16-lb load, 105 rpm (7 cm/s), 10,000 cycles, room temperature, air, dry, density-corrected

Metal-to-metal wear tests for coal-gasification valve service — **Table III**

Alloy	Atmosphere	Wear volume (mm^3)
Nitronic 60	air + steam	10.6
Cobalt Alloy 6	air + steam	28.0
Nitronic 60	helium	6.9
Nitronic 60	air + steam*	8.7
Mild steel	air + steam	266.0
Type 304	air + steam	106.0

*Preoxidized – 1,000°F, 3 in air

Conditions: Self-mated thrust washers, 500°F, 500 rpm, 100-lb load, 4,000 cycles

Alloys and trademarks mentioned in this article

Alloy	Company
17-4PH, Nitronics 32, 50, 60	Armco Inc. Middletown, Ohio
Stellite 6, Haynes 25, Tribaloy 700	Cabot Corp. Kokomo, Indiana
IN 718, IN X750, K-500 Monel	International Nickel Co. New York, N.Y.
Waspaloy	Pratt & Whitney Aircraft Co. Hartford, Conn.
Colmonoy 6	Wall Colmonoy Corp. Detroit, Mich.

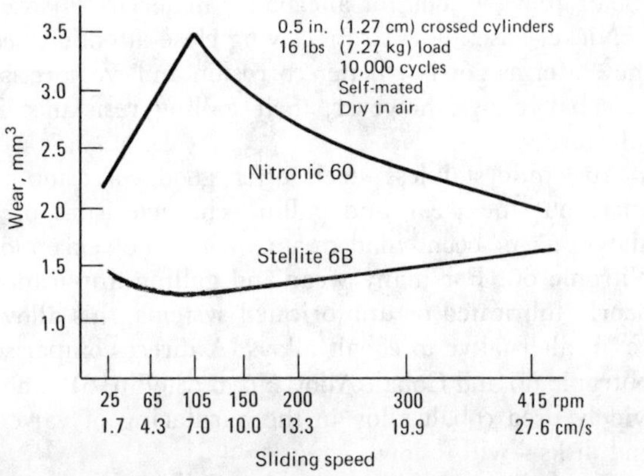

Effect of speed on the wear of Nitronic 60 and Stellite 6 B — **Fig. 2**

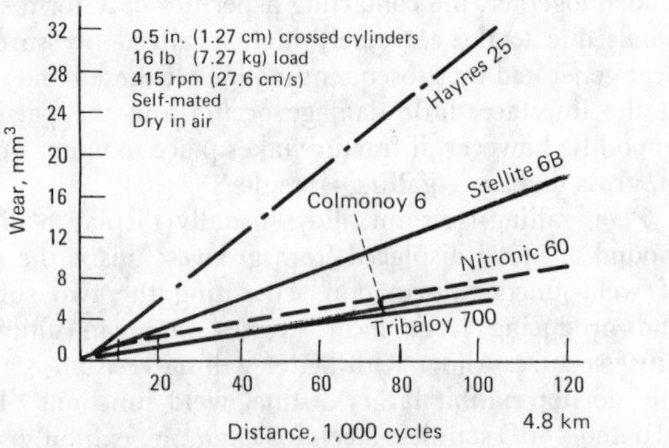

Wear resistance of Nitronic 60 compared to nickel and cobalt alloys — **Fig. 3**

Wear Resistance

When stresses do not cause early catastrophic damage to sliding components, dimensional loss by wear occurs until the component is termed *worn out* and must be replaced. This process may be due to: (1) metal-to-metal, (2) metal-to-abrasive, or (3) metal-to-liquid wear (cavitation). Cobalt Alloy 6 and Nitronic 60 were compared under the first two wear modes by a modified Taber Met-Abrader. This machine is shown in Fig. 1 with its crossed cylinder (0.5-in. dia.) geometry. Results are listed in Table II for tests at 105 rpm. Alloy 6 was less than 100% more wear-resistant than Nitronic 60 in 10 couples and actually was inferior in three couples.

Use of a variable-speed motor with the modified Taber Met-Abrader permitted study of the effects of different interfacial temperatures. Most of the speed studies were conducted on self-mated couples because previous work showed this to be a good method to rank alloys [*4*]. Indeed, the ASTM G-2 Committee on Wear and Erosion is developing a crossed-cylinder test method using self-mated specimens to rank metallic materials.

The importance of speed as a variable is shown in Fig. 2. Alloy 6 exhibited only a slight, but measurable, increase in wear going from 25 rpm to 415 rpm. In contrast, Nitronic 60 displayed lower wear above and below 105 rpm. This fact makes the data at 105 rpm in Table II appear conservative for Nitronic 60.

Since the 415-rpm tests took a relatively short time to run 10,000 cycles, longer sliding distances were studied. Once steady-state conditions were reached, Nitronic 60 wore in the oxidative wear regime with a sharp decline in wear rate. As shown in Fig. 3, Nitronic 60 was much better than Alloy 6 and Haynes 25, and almost as good as two nickel-base wear alloys.

The improved wear resistance of Nitronic 60 over Alloy 6 also has been verified by the U.S. Bureau of Mines. Metal-to-metal wear tests were conducted at 500°F to screen alloys for coal-gasification valve service with the results in Table III. Nitronic 60 was 10 times better than Type 304 and almost 3 times more wear-resistant than Alloy 6. One actual service history has also demonstrated the use of Nitronic 60. In butadiene-processing equipment, Nitronic 60 outlasted Alloy 6 by 55% in gate-valve trim operating at 750°F.

Abrasion tests were also conducted on several alloys. High-density alumina and solid tungsten carbide were held fixed in severe abrasion tests. The results in Table IV show the speed effect noted in the metal-to-metal tests. The cobalt- and nickel-base wear alloys were 2 to 4 times more abrasion-resistant than Nitronic 60 at the lower speed (105 rpm), but at the higher speed (415 rpm), Nitronic 60 outperformed Alloy 6 and was only slightly inferior to Colmonoy 6.

The tungsten carbide abrasion results in Table V show Nitronic 60 to be superior to all stainless grades except the poorly corrosion-resistant Type 440C. The cobalt- and nickel-base wear alloys were 2 to 3 times better than Nitronic 60 at both 105 and 415 rpm.

Alumina abrasion tests with variable speed — Table IV

Alloy	Rockwell hardness	Wear volume (mm^3) 105 rpm	Wear volume (mm^3) 415 rpm
Tribaloy 700	R_C 45	0.92	–
Colmonoy 6	R_C 56	1.10	0.84
Cobolt Alloy 6	R_C 48	1.63	2.10
Type 440 C	R_C 56	2.10	0.73
Nitronic 60	R_B 95	3.54	0.98
Type 301	R_B 90	4.66	–
Nitronic 32	R_B 94	5.76	–
Type 304	R_B 79	6.76	5.06
Type 310	R_B 72	8.84	–
17-4PH	R_C 43	24.13	1.80

Test condtions: Taber Met-Abrader machine, 0.5-in.-dia. specimen mated to 0.25-in.-wide flat alumina in fixed position, 16-lb load, room temperature, 10,000 cycles, air, density-corrected

Tungsten carbide abrasion tests with variable speed — Table V

Alloy	Rockwell hardness	Wear volume (mm^3) 105 rpm 10,000 cycles	Wear volume (mm^3) 415 rpm 40,000 cycles
Colmonoy 6	R_C 56	1.08	3.12
Cobalt Alloy 6	R_C 48	1.35	4.94
Tribaloy 700	R_C 45	1.43	3.90
Type 440 C	R_C 56	1.50	1.51
Nitronic 60	R_B 95	2.82	9.04
Nitronic 32	R_B 94	4.20	17.39
Type 304	R_B 79	6.18	52.80
Type 316	R_B 74	7.70	34.06
Nitronic 50	R_B 99	8.72	30.18
17-4PH	R_C 43	9.92	22.37
Type 310	R_B 72	15.26	39.09

Test conditions: Taber Met Abrader machine, 0.5-in.-dia. crossed cylinders, 16-lb load, room temperature, tungsten carbide in fixed position, dry, air, density-corrected

References

1. Bhansali, K. J., and Miller, A. E., Role of Stacking Fault Energy on the Galling and Wear Behavior of a Cobalt Base Alloy, *Proc.*, Third International Conference on Wear of Materials, 1981, pp. 179-185.
2. Tanczyn, H., Stainless Steel Galling Characteristics Checked, *Steel*, Vol. 132, April 1953, pp. 150-152.
3. Schumacher, W., New Galling Data Aids in Selecting Stainless Steels, *Materials Eng.*, April 1973, pp. 60-61.
4. Schumacher, W., Wear and Galling Can Knock Out Equipment, *Chem. Eng.*, May 9, 1977, pp. 155-160.

The author

William J. Schumacher is a Senior Staff Engineer at Armco, Inc., Middletown, OH 45043. Tel: 513-425-2768. For the past eleven years, he has worked primarily on evaluating the galling and wear characteristics of stainless steels and has published several articles on the subject. He received his B.S. in metallurgical engineering from Drexel University, and belongs to the American Soc. for Metals and ASTM.

Alloys to protect against corrosion and wear

Certain nickel or cobalt-chromium alloys stand up to corrosion and the various types of wear found in processing plants.

Paul Crook and Aziz Asphahani, Cabot Corp.

☐ Wear of equipment is a major problem in industry. Not only does it mean replacement or repair of components but it often involves substantial and costly downtime. In the chemical process industries (CPI), wear is hastened by corrosion and by limitations imposed upon lubrication. To minimize wear, one must look at the causes of wear and at what types of materials work best under various conditions.

Before these topics are discussed, let us very briefly cover how various types of alloys perform in preventing wear:

The commodity alloys—for example, Type 316 stainless steel—have limited resistance to wear. The alternatives to using commodity alloys are either to coat critical components with wear-resistant alloys or to manufacture the components entirely out of these special alloys.

Wear-resistant alloys best suited for corrosive environments are based on cobalt and nickel. These alloys are, of course, relatively hard and can be applied by welding as coatings. They are therefore known as hardfacing materials.

Then there are alloys that are chiefly known for corrosion resistance, but have moderate resistance to wear. These high-performance alloys include formulations of Ni-Mo, Ni-Cr-Mo, and Ni-Si [*1*].

Now, we will discuss the types of wear: abrasion, metal-to-metal wear, and erosion.

Abrasion

Abrasion is caused by hard particles or protuberances moving and pushing against a solid surface. Such wear is common in oil-extraction machinery used in the food-processing industry. The abrasion process can be described as plowing (for ductile surfaces) or chipping (for brittle surfaces).

Since plowing involves mainly plastic deformation, the resistance to abrasion of ductile, single-phase materials is closely related to their hardness. For brittle materials, the fracture characteristics determine the extent of material removal.

Hardfacing materials are typically made up of a hard phase that is dispersed throughout a softer metallic matrix—e.g., chromium boride in a Ni-rich matrix. For such alloys, abrasion is more complex, and depends upon the size, shape and hardness of the abrading species, as well as the hard phase.

Contrary to popular belief, abrasion resistance of these alloys is not necessarily related to their overall hardness. A high volume fraction of the hard phase and a coarse structure, in fact, generally promote abrasion resistance [*2*].

Metal-to-metal wear

Traditional theory states that adhesion is the cause of metal-to-metal wear. Such wear, which is caused by the sliding of unlubricated metal surfaces over each other, is often therefore called adhesion. Strong interfacial bonds may occur at deformed surface asperities (surface high spots), and mechanical degradation results from shear failure of the weaker of the two metal mating surfaces [*3*].

Newer theory postulates another mechanism—that subsurface crack nucleation and growth follow asperity shearing and flattening [*4,5*].

Oxide films may protect most metals and alloys from corrosion and metal-to-metal wear. In sliding systems where oxide films do not break down, true metal-to-metal contact never happens.

Mild wear occurs at low loads, and is controlled by oxide-film breakdown and repair; its debris is an oxide.

Nozzle that failed due to high-velocity erosive attack in chlorine dioxide service **Fig. 1**

Originally published January 10, 1983

Automated hardfacing machine at work on a seat valve Fig. 2

Severe wear is characterized by metallic debris at higher loads. Gross damage is termed galling.

For wear alloys, the softer matrix seems to be most important in controlling metal-to-metal wear [6,7].

Erosion

Erosion may be caused by solid particles or by a liquid, which may cause cavitation erosion or impingement erosion (see Fig. 1). Cavitation can take place wherever there are high-pressure turbulent fluids. Liquid-impingement erosion and cavitation erosion are closely related. Cavitation is the formation and collapse of bubbles within a liquid. When the bubbles collapse, liquid jets arise from implosion to damage surfaces.

The response of a surface to solid particles—whether they are in a gas or a liquid—depends on the nature and momentum of the particles, as well as their angle of attack. Particles result in craters on ductile surfaces, and craters that are surrounded by cleavage cracks on brittle surfaces for normal impact.

For materials that contain a hard phase—i.e., wear-resistant alloys—it is this phase that controls impingement by solids. In liquid impingement/cavitation erosion, the softer matrix is most important [8].

Methods of protection

Wear-resistant alloys are applied by welding them to surfaces. Welding technique affects cost and performance. In welding, high deposition rates cause high dilution of the molten weld pool by the substrate. Thus, large components, which require high deposition rates for reasons of cost, require multilayer deposits.

Plasma-transferred arc welding offers the best balance of deposition rate and dilution. This method allows for the use of prealloyed powder consumables with mechanized welding systems specifically designed for hardfacing (see Fig. 2).

Wear-resistant alloys have limited ductility and are rarely made into structural components. Usually, cast, powder-metallurgy or wrought limited-size parts are inserted at critical points in a process.

For example, cast or powder-metallurgy valve-seat inserts are brazed into valve bodies, and wrought cobalt-based alloys are used as cutting edges in the plastics processing industry.

Cobalt-base alloys

The first Co-rich wear alloys were developed at the turn of the century by Elwood Haynes [9]. Experimenting with cobalt and chromium, he discovered that binary alloys containing more than about 10% Cr had excellent resistance to oxidation and corrosion plus outstanding hardness at elevated temperatures to about 1,000°C (1,800°F). Tungsten and molybdenum were added to increase strength.

The Co-Cr alloys used today are similar to those designed by Haynes (see Table I). The alloys listed in the table contain carbon, which promotes the formation of Cr-rich carbides during solidification.

These alloys differ by carbon level, hence, by carbide volume-fraction. The best protection from abrasion is provided by alloys having a high percentage of carbides and by those hardfacing techniques (or parameters) that promote a coarse carbide grain-structure [2].

The properties that distinguish Co-Cr alloys from all other hardfacing materials are their outstanding resistance to liquid impingement and cavitation erosion, and

Nominal composition of prime cobalt-chromium alloys, and typical processing applications Table I

Alloy	American Welding Soc. rod (R) designation	% by weight Co	Cr	W	Mo	C	Typical hardfacing applications
Alloy No. 1	R Co Cr-C	Bal.	31	13	–	2.45	Screw components, pump sleeves
Alloy No. 6	R Co Cr-A	Bal.	28	4.5	–	1.0	Valve, pump and screw components
Alloy No. 12	R Co Cr-B	Bal.	30	8.5	–	1.5	Cutting edges
Alloy No. 21*	–	Bal.	27	–	5.5	0.25	Valve seats

*Aerospace Material Specification AMS 5385D

Nominal composition of nickel-base hardfacing alloys. They are for mildly aggressive applications **Table II**

Alloy	American Welding Soc. rod (R) designation	% by weight Ni	Cr	Si	B	C	Fe	Typical hardfacing applications
Alloy No. 50	R Ni Cr-B	Bal.	12	4	2.5	0.45	4	Cutting edges (e.g., pulp knives)
Alloy No. 60	R Ni Cr-C	Bal.	15	4	3.5	0.75	4	Pump parts and pipe elbows

CO-Cr wear alloys generally offer superior corrosion resistance, compared to the Ni-based formulations **Table III**

Alloy	Concentration and temperature	Gas-tungsten-arc deposits: Acetic acid 30% boiling	Formic acid 80% boiling	Nitric acid 65%, 150°F (66°C)	Phosphoric acid 50%, 150°F (66°C)	Sulfuric acid 5%, 150°F (66°C)
Alloy No. 1		G	–	S	E	E
Alloy No. 6		E	E	U	E	E
Alloy No. 12		G	E	E	E	E
Alloy No. 21		E	E	E	E	E
Alloy No. 50		U	S	U	–	U
Alloy No. 60		U	G	U	–	U
Ni-17Cr-17Mo-6Fe-5W		E	E	S	–	E

Note:
Five 24-h test periods. Determined in laboratory tests. It is recommended that samples be tested under actual plant conditions.

Code:
E Less than 5 mils/yr (mpy) (<0.13 mm/yr)
G 5 mpy (0.13 mm/yr) to 20 mpy (0.51 mm/yr)
S Over 20 mpy (>0.51 mm/yr) to 50 mpy (1.27 mm/yr)
U More than 50 mpy (>1.27 mm/yr)

their excellent self-mated antigalling behavior. Both features are due to the same matrix characteristics—an extremely low stacking-fault energy (a property of the atomic structure that influences deformation behavior) and a tendency to transform or "twin" under stress [*7,8,10,11*]. (Time transformation is from a face-centered cubic to a hexagonal close-packed structure.)

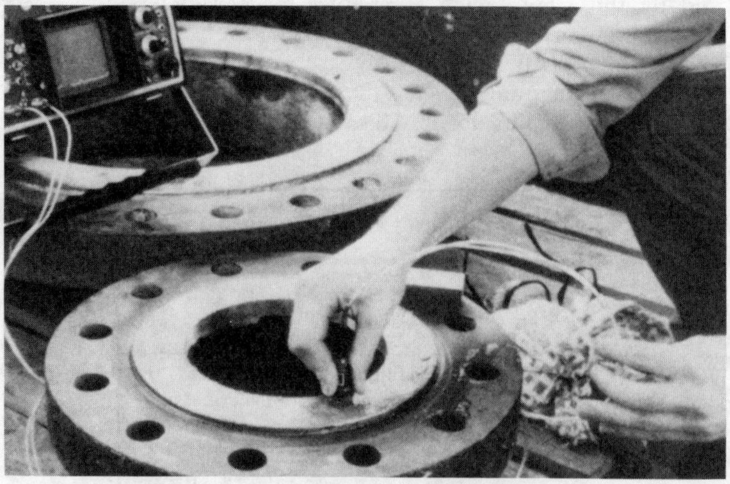

Flanges that have been hardfaced with alloy No. 60 in bore and alloy 6 in sealing face **Fig. 3**

It is not well understood how the above properties affect the behavior of the alloys to mechanical degradation. Planar slip of the structure during deformation and twinning tendencies, however, may restrict crack initiation and propagation [*5,10,12*].

Of the hardfacing alloys, the Co-Cr ones respond best to welding. They are readily deposited by arc processes, and are typically applied to components, by means of the oxyacetylene process (with a 3x flame).

Many alloys have been derived from the Co-Cr mixtures for specific uses and to conserve cobalt, which has periodically been scarce and expensive. Among the alloys designed to conserve cobalt are those containing cobalt-iron-chromium, which have many of the same properties as do the Co-Cr alloys [*13*].

The iron-containing alloys have relatively high amounts of molybdenum to compensate for the increase in matrix stacking-fault energy brought about by the partial replacement of Co by Fe.

Also, there are alloys that, during solidification, form hard metallic compounds rather than hard carbides. Two such are Tribaloy* T-400 (8.5Cr-28.5Mo-2.6Si-balance Co) and Tribaloy T-800 (17.5Cr-28.5Mo-3.4Si-bal. Co). The high Mo and Si levels help to form the hard metallic phase. T-800 has a higher chromium content and resists corrosion better.

*Tribaloy is a registered trademark of Cabot Corp.

Teeth of this mixer are protected with Alloy No. 12; the flight has Alloy No. 6 Fig. 4

Nickel-base hardfacing alloys

Of the Ni-rich alloys designed to resist wear, the most widely used are those containing Ni-Cr-Si-B (Table II). Compared with the Co-Cr alloys, these only moderately resist corrosion, and are therefore used in mildly aggressive environments (Table III). Carbon steel flanges that are hardfaced with Alloy No. 60 (and also Alloy No. 6) are shown in Fig. 3.

These alloys have a complex microstructure, which is described in Ref. [*14*]. However, they essentially contain a high volume of boride—which imparts excellent resistance to abrasion—in a Ni-rich matrix. Si and B affect welding characteristics: The molten weld pool is highly fluid and these alloys are self-fluxing during gas welding—hallmarks of this class of alloys.

To conserve cobalt, attempts have been made to develop Ni-Cr equivalents to the Co-Cr alloys. However, such alloys lack sufficient resistance to galling and cavitation erosion. Their abrasion resistance is equivalent to the Co-Cr alloys. The Ni-Cr alloys have better corrosion resistance than do the Ni-Cr-Si-B alloys.

Rather than relying on hard-phase formation during solidification of the alloys, it is possible to protect surfaces by using composite welding consumables. One such material is 60% (wt.) tungsten carbide and 40% Ni-Mo alloy.

Ni-base high-performance alloys

These alloys, as mentioned earlier, are designed primarily for corrosion resistance, not for wear—although they do have moderately good wear resistance. The Ni-Mo alloys are typically used in reducing environments; the Ni-Cr-Mo alloys are used in oxidizing and reducing media. The Ni-Si alloys have excellent resistance to sulfuric acid at nearly all concentrations and temperatures.

Applications

In the CPI, wear and its mechanisms are varied, and it is difficult to generalize about when specific materials should be used. Therefore, it is hard to precisely measure the economic advantages of using wear-resistant alloys. However, some typical applications are:

Co-Cr alloys Nos. 6 and 12 have been used to protect the teeth and flights of screw-type mixers (see Fig. 4). These alloys are also used as valve-trim materials, particularly for handling petroleum products. Alloy No. 3 is used in proportioning pumps. Co-Cr alloys are used in the pulp and paper industry for spray nozzles, defiberizing-unit blades and agitator-shaft sleeves.

The high-performance Ni-base corrosion-resistant alloys are used for wear and corrosion, when corrosion is the main problem. Applications include valves and pumps—here the alloy 16Cr-16Mo-4W-bal. Ni is often used.

References

1. Asphahani, A. I., and Hodge, F. G., "Proceedings, Alloys for the 80s," Climax Molybendum Co. symposium, p. 329, 1980.
2. Silence, W. L., *J. of Lubrication Technology,* Vol. 100, No. 3, 1978, p. 428.
3. Rabinowicz, E., "Friction and Wear of Materials," John Wiley & Sons, New York, 1965.
4. Suh, N. P., *Wear,* Vol. 25, 1973, p. 111.
5. Rigney, D. A., and Glaeser, W. A., "Wear of Materials—1977," Amer. Soc. of Mechanical Engineers, p. 41.
6. Bhansali, K. J., "Wear of Materials—1979," ASME, p. 146.
7. Bhansali, K. J., and Miller, A. E., "Wear of Materials—1981," ASME, p. 179.
8. Antony, K. C., and Silence, W. L., "Proc. Fifth Intl. Conf. on Erosion by Liquid and Solid Impact," Cambridge University Press, 1979, p. 671.
9. Gray, R. D., A History of the Haynes Stellite Co., Cabot Corp. pub., p. 18.
10. Woodford, D. A., *Metallurgical Transactions,* Vol. 3, p. 1137, 1972.
11. Heathcock, C. J., et al., "Wear of Materials—1981," ASME, p. 597.
12. Remy, L., and Pineau, A., *Materials Science and Engineering,* Vol. 26, 1976, p. 123.
13. Crook, P., "Wear of Materials—1981," ASME, p. 202.
14. Knotek, O., and Lugscheider, E., *Welding Research Supplement,* Oct. 1976, p. 314.

The authors

Paul Crook is senior engineering associate in the Technology Dept. of Cabot Corp., 1020 West Park Ave., Kokomo, IN 46901. Tel: (317) 456-6241. He is a member of the firm's Wear Group, and his responsibilities are design and testing of wear-resistant materials. Crook holds metallurgy and physics degrees (including a Ph.D.) from the University of Manchester (England), and is a member of the Amer. Soc. for Metals and the Amer. Soc. for Testing and Materials.

Aziz Asphahani is director of the Technology Dept. at Cabot Corp., 1020 West Park Ave., Kokomo, IN 46901. Tel: (317) 456-6230. He joined Cabot in 1975, upon completion of his doctoral work at M.I.T. He serves as chairman of the Natl. Assn. of Corrosion Engineers (NACE) Committee T-3E (Stress Corrosion Cracking), and is a member of the NACE Strategic Planning Committee. He is also active in the Materials Technology Inst. as manager of the Corrosion Engineering Section and as a member of the Technical Advisory Council.

High-temperature effects on metallic materials

This article provides some insight into the metallurgical changes that occur when various of the more common metals and alloys are exposed to high temperatures.

Dwight C. Deck, *Celanese Chemical Co.*

☐ Exposure to temperatures above that of design could occur during a process upset, fire, inadvertent flame impingement, heat of welding or any other condition that could cause overheating during normal operations (for example, accumulation of deposits on the internal surface of an externally fired tube).

Should materials be replaced?

Many articles have been written about metal failures resulting from exposure to such elevated temperatures. These articles usually detail the mechanism of failure and show microstructural conditions that were present at the time. But what effect does elevated temperature exposure have on the metallurgical characteristics of parts that do *not* fail? What can be done to assure that these metal parts are suitable for continued use?

Prior to examination of the *metallurgical* condition of the material, the *physical* condition of the part should be determined. Has deformation resulting from the occurrence limited its usefulness? Has oxidation or corrosion reduced the sound metal thickness to less than allowable? Is the resulting surface-finish acceptable? Close visual inspection, accurate measurements, and nondestructive testing can assist in resolving these questions.

Nondestructive testing and some understanding of metallurgy can also be helpful, often eliminating replacement costs and many problems for the maintenance engineer. But prior to final judgment for continued use, it is imperative that the requirements of strength and corrosion resistance of the metal be known. First, one should understand that exposure to elevated temperature is only part of what determines the final properties. Equally important is the *rate* at which the metal cools from the maximum temperature of exposure.

We will now look at some examples of changes that occur to different metals when they are exposed to elevated temperatures, and subsequently cool to ambient temperatures.

Carbon steels

Carbon steels are the most common metals used in construction because of low cost and ease of fabrication. They are basically solid solutions containing approximately 98% iron, with carbon and manganese, plus other trace constituents such as silicon, phosphorus, and sulfur. As in all other metals, the microstructure consists of crystals or individual grains (see Fig. 1). Carbon steel is most often used in the annealed or normalized condition with a hardness of 70-90 Rockwell B. This material can be used almost indefinitely at temperatures below 750°F. Its microstructure in the annealed or normalized condition is of ferrite (white area in Fig. 2), which is almost pure iron, and pearlite (the gray area), which is the carbide-containing phase, usually present in lamellar form.

When heated above 750°F, the microstructure may alter. This depends on the maximum temperature that is reached and, in some cases, the method by which the steel is cooled. Such microstructural changes may considerably

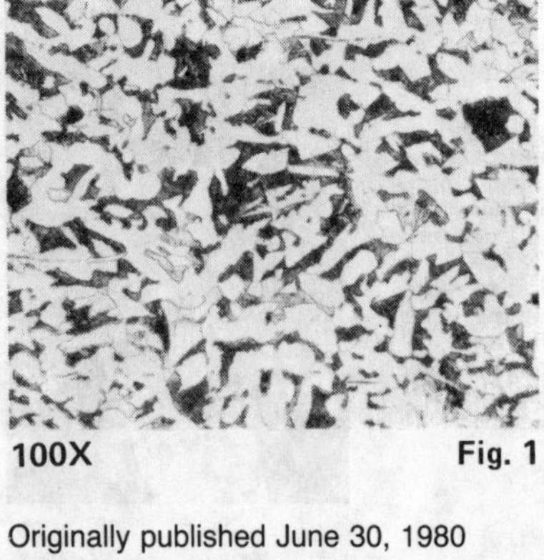

100X Fig. 1

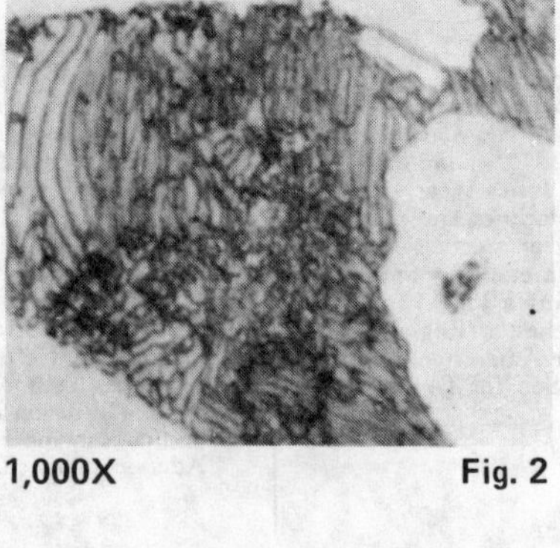

1,000X Fig. 2

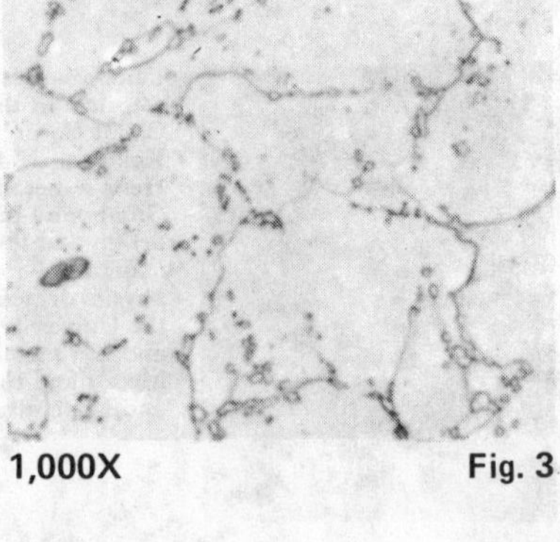

1,000X Fig. 3

Originally published June 30, 1980

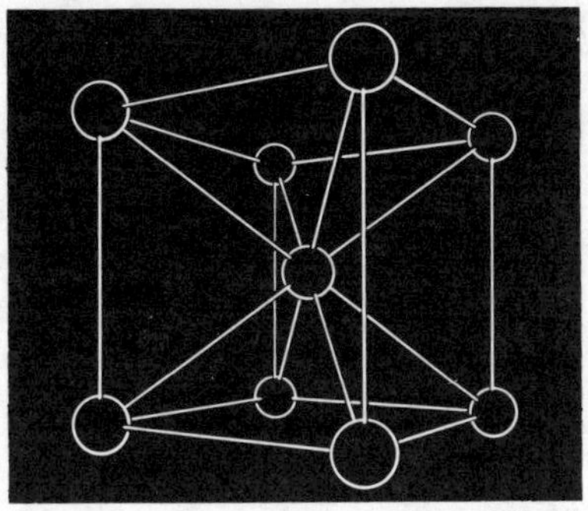

Fig. 4

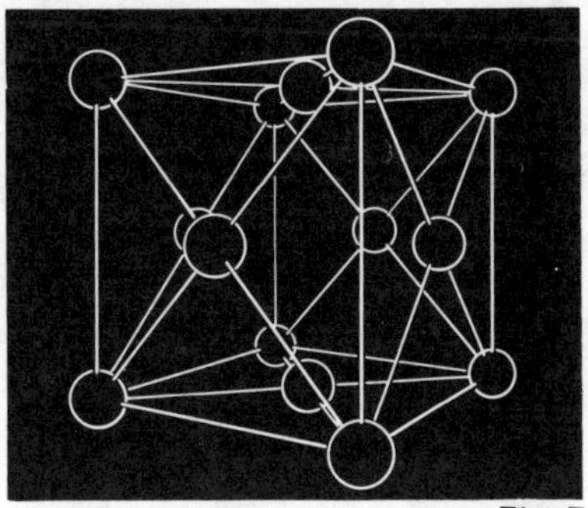

Fig. 5

1,000X Fig. 6

affect the hardness and other mechanical properties. Because of this, a hardness test can often be used to gain insight as to whether or not the steel was adversely affected during the temperature exposure.

When heated between 750°F and approximately 1,330°F, the carbides of the pearlite phase will coalesce and form spheroidal iron carbides, which can be seen in Fig. 3. Thousands of hours are required for this change to occur near 750°F, but when the metal is heated near 1,330°F, only a few minutes are required. This microstructural change results in a reduction of hardness and tensile properties. In most applications, however, this decrease is tolerable, and the steel part is considered satisfactory for continued use.

The rate at which carbon steel is cooled from temperatures below 1,330°F has little effect on its properties. However, when it is exposed above this temperature, the cooling rate is important. At approximately 1,330°F, carbon steel undergoes a crystal structure change or phase transformation. The crystal lattice changes from a body-centered cubic (BCC) structure (Fig. 4) to a face-centered cubic (FCC) one (Fig. 5).

Carbon is more soluble in the FCC structure, thus the carbon present in the pearlite phase dissolves and goes into solution. The new FCC grains continue to grow in size as the temperature increases. If the maximum temperature of exposure is below 1,600°F, and the steel is cooled slowly, the phase transformation will reverse, and the microstructure will exist as it was before the exposure; but more than likely, the grain size will be smaller. There will be a slight increase in hardness because of this. If the temperature of the steel exceeds 1,600°F and the cooling rate is slow, a slight decrease in hardness is noted, due to an increase in grain size.

The cooling rate is very important when carbon steel is heated above 1,330°F. When cooled rapidly, as during water quenching, the carbon does not have time to diffuse from the FCC structure before transformation to BCC structure. As a result, the carbon is trapped, and the resulting crystal lattice is distorted and highly strained due to formation of martensite (see Fig. 6). Martensite is hard and has little resistance to impact loading. Parts containing untempered martensite are not recommended for use. Hardness values to Rockwell C-60 are possible in high-carbon steels.

However, in most low-to-medium carbon steels, hardness tests give values of less than Rockwell C-30. Values higher than Rockwell B-100 or Rockwell C-20 signify that more-extensive testing should be performed before continued use can be safely recommended.

Low-alloy and tool steels

Metals of this type are used because of their strength and/or resistance to wear. They depend primarily on the previously described phase transformation for their exceptional properties. (Fig. 7 shows the microstructure of a high-carbon, low-alloy steel in the annealed condition.) During controlled heat treatment, these alloys are hardened by quenching from above their transformation temperatures to form martensite (Fig. 8). The alloy is then exposed to a temperature below the transformation temperature, in order to relieve some of the "locked in" stresses. This intermediate temperature exposure, or tempering treatment, allows some of the carbon to diffuse out of the distorted crystal structure, increasing the toughness of the material, though with some sacrifice in tensile strength (Fig. 9). The final properties of the alloy depend on the tempering temperature and the time spent at that temperature.

Onsite hardness tests can be used to determine what

500X Fig. 7

500X Fig. 8

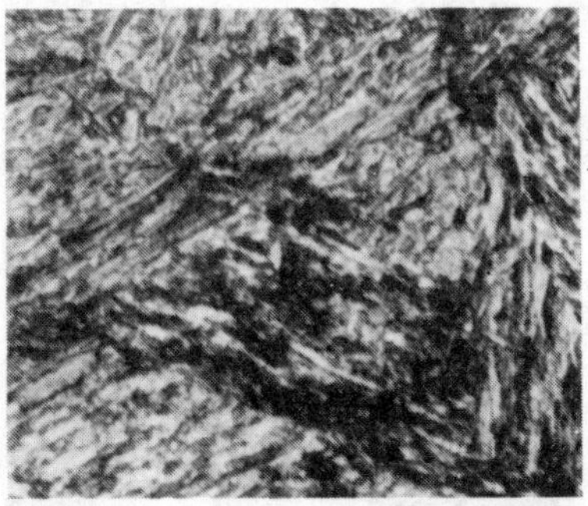

500X Fig. 9

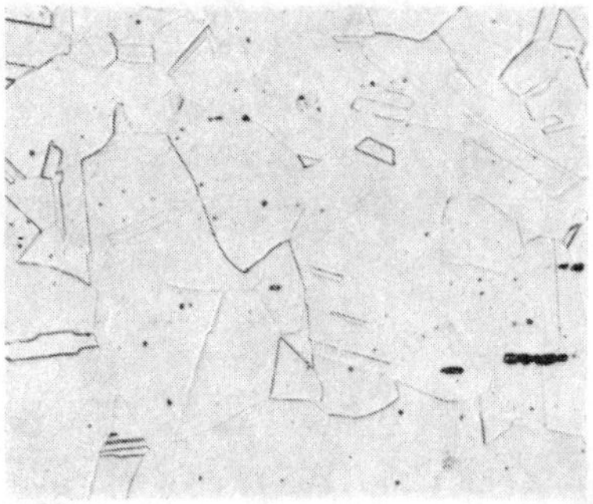

500X Fig. 10

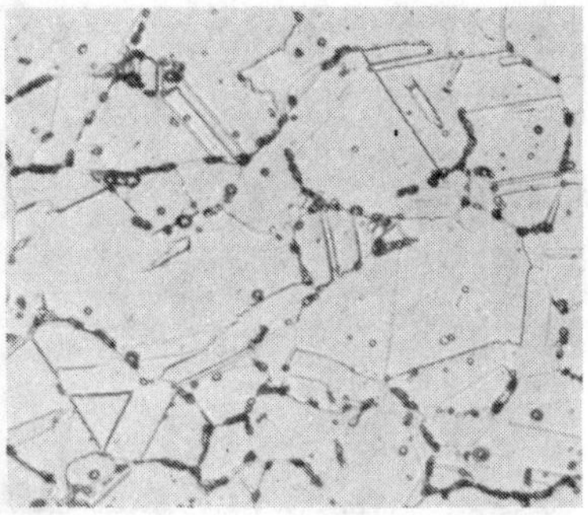

500X Fig. 11

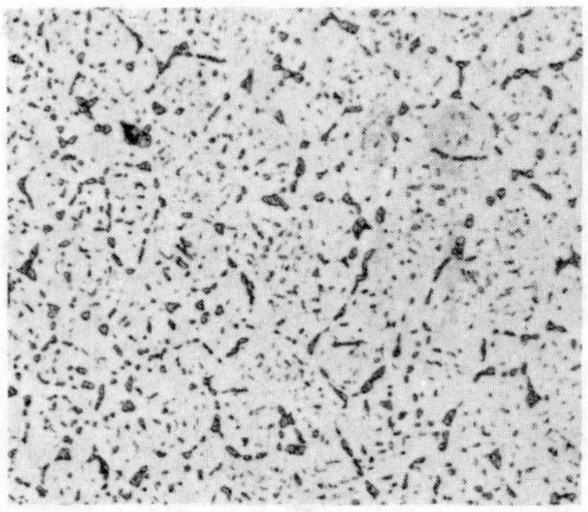

500X Fig. 12

effect the elevated temperature has on the mechanical properties of heat-treated low-alloy and tool steels. In almost all cases, an increase in hardness would suggest that the material was exposed above the transformation temperature and cooled at a rate fast enough to form free martensite in the microstructure. Because of the loss in toughness with the presence of untempered martensite, an increase in hardness generally signifies that reuse without subsequent tempering is not advisable.

Loss of hardness indicates that the exposure temperature was below the transformation temperature. The decision for continued use must then be based on the strength requirement of the material after the elevated temperature exposure.

Regardless of the effect on mechanical properties, the general corrosion resistance of these materials does not change appreciably with change of the microstructural condition.

Stainless steels

Superior resistance to corrosion is the primary characteristic of stainless steels. There are three basic types: martensitic, ferritic and austenitic.

The martensitic stainless steels are hardenable by heat treatment due to the crystal structure change discussed previously. General guidelines regarding continued use are very similar to those discussed for low-alloy and tool steels.

The ferritic and austenitic stainless steels (Types 405, 430, 442, 446, and all Series 300 stainless steels) are not hardenable by heat treatment because their microstructures do not undergo phase transformations when exposed to elevated temperatures (Fig. 10). As a result, exposure to elevated temperature will not cause appreciable change in their hardness or tensile strength. However, some microstructural changes that occur do have a pronounced effect on the corrosion resistance and toughness of these alloys. When exposed in the 800-1,600°F temperature range, the carbon that is normally present in solution will diffuse to the grain boundaries and unite with chromium to form carbides. As a result, the zone adjacent to the grain boundaries becomes depleted of chromium and will be preferentially attacked when exposed to corrosive environments. Stainlesses with carbides in the grain boundaries are referred to as being "sensitized" (Fig. 11). A different phase, known as the sigma phase, can form due to prolonged exposure between 950°F and 1,600°F (Fig. 12). This phase reduces the ductility of the alloy and, when present in large quantities, makes it brittle.

Hardness tests are of little use in determining the presence of intergranular carbides or sigma phase—it is necessary to examine the microstructure. Portable equipment is available to perform such examination at the site where the elevated temperature exposure occurred.

Aluminum, copper, nickel and their alloys

Heat-transfer characteristics and resistance to specific environments are the primary reasons these materials are selected. For the most part, they are considered "nonhardenable" by heat treatment. There are exceptions and these will be discussed later. In many applications, these materials are used in the annealed condition (Fig. 13, aluminum; Fig. 14, copper), and exposure to elevated temperature has only limited effect on room-temperature properties. In most applications where these metals are used in the annealed condition, the loss of strength is insufficient to cause concern. Normally, the physical damage—distortion, oxidation—overshadows the effect of the change in metallurgical condition.

Aluminum, copper, nickel and their alloys are often

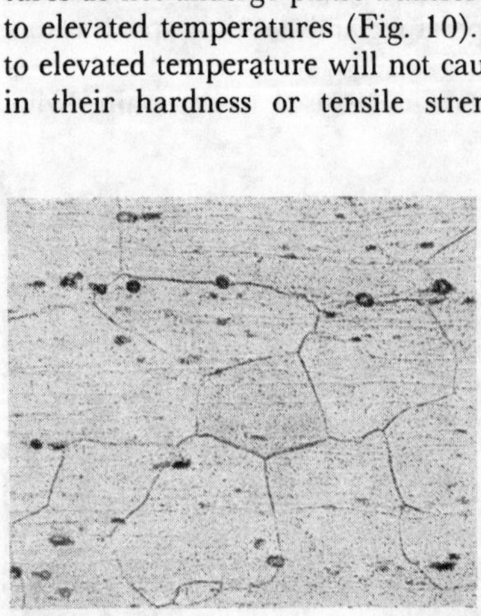

100X Fig. 13

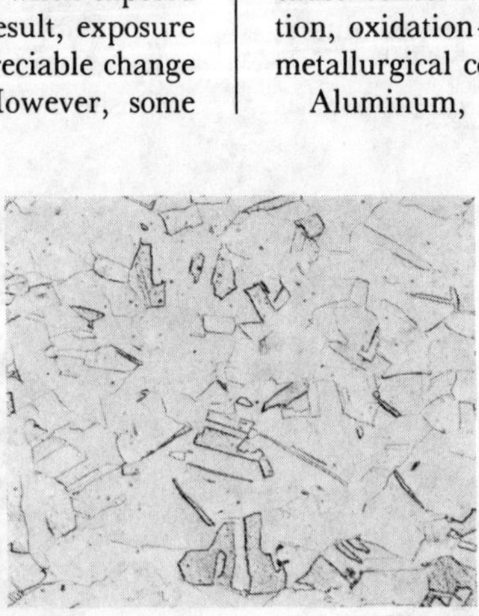

100X Fig. 14

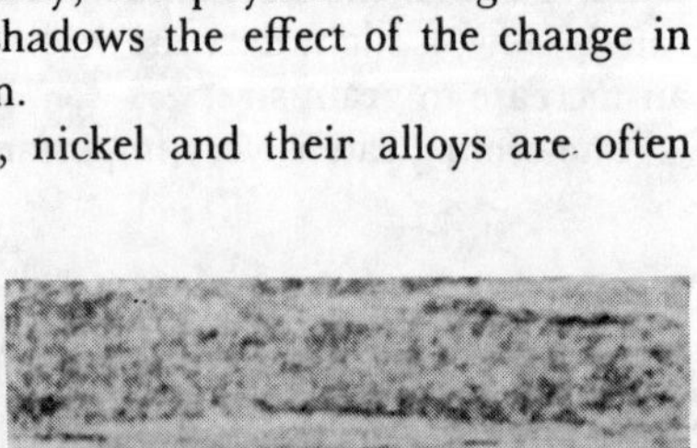

100X Fig. 15

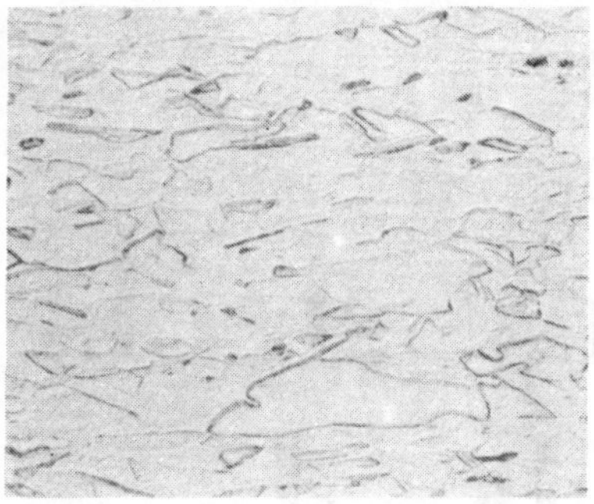

100X Fig. 16

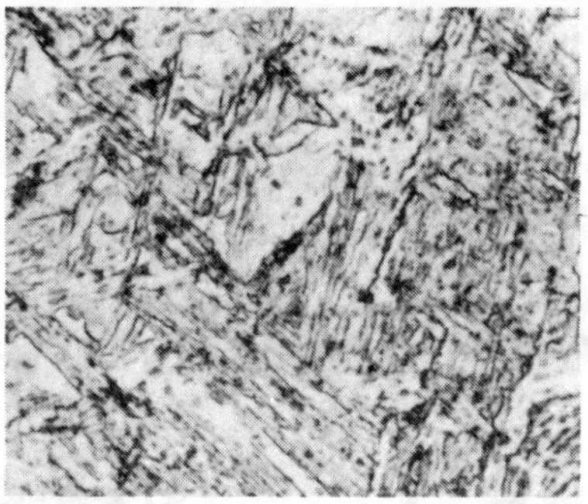

1,000X Fig. 17

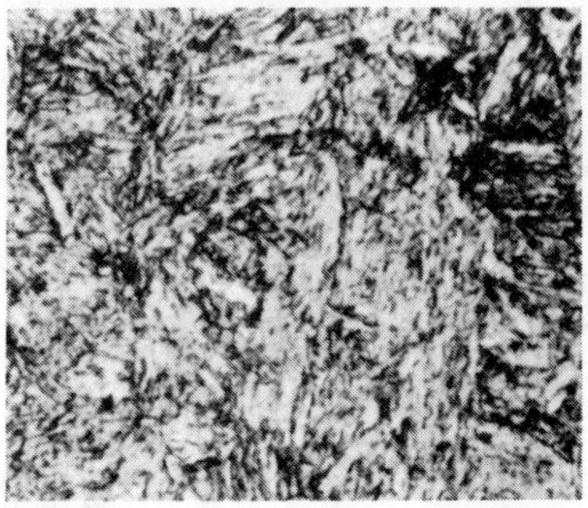

1,000X Fig. 18

used in the "cold worked" condition (Fig. 15, aluminum; Fig. 16, copper). Strength is determined by the degree of cold work imparted in the metal by forging, drawing or rolling. An improvement in properties is attributed to internal barriers to flow, such as displaced impurities and interaction between the deformed crystal structures. Exposure of a cold-worked metal to elevated temperature will cause the deformed microstructure to recrystallize into new grains. In most cases, this greatly reduces the hardness and strength of the material to that of the annealed material. Hardness tests readily reveal the loss of mechanical properties resulting from recrystallization. If, after the elevated temperature exposure, the material is not strong enough to withstand service requirements, it must be replaced.

Precipitation-hardened alloys

High mechanical strength coupled with corrosion resistance makes these alloys very valuable to the engineer. They gain improved hardness and strength due to precipitation of one or more of their alloying constituents from solid solution. The precipitates act as keys to prevent slip and strain within the structure. Heat treatment of these alloys involves three steps:

- Solution annealing to put all alloying constituents in solid solution.
- Quenching to form a supersaturated solid solution.
- Hardening treatment to allow controlled precipitation of one or more of the alloying constituents.

A phase change often accompanies the heat treatment of precipitation-hardened (PH) stainless steels, and the resulting properties are due both to this phase change (as in carbon and low-alloy steels) and to the precipitation-hardening treatment. A magnification of 17-4 PH stainless steel is shown in Fig. 17 (solution-annealed) and Fig. 18 (precipitation-hardened).

This treatment must be precisely controlled to obtain the desired properties. Proper balance between strength, hardness and toughness must be obtained to fulfill the service requirement.

During the treatment, submicroscopic precipitates form. With time or exposure to higher temperature, more precipitates form and some begin to coalesce and form larger "particles." Spacing between these becomes greater and, as a result, slips or cracks can bypass them and the overall strength is reduced to near that of the matrix material. Such loss of strength in these alloys is referred to as "overaging."

Onsite hardness checks can provide valuable information that may be used in judging the suitability for continued use of parts made from precipitation-hardened alloys that include those mentioned, and some aluminum and copper ones.

Titanium, zirconium, and specialty alloys

Use of these metals is usually limited, due to their cost or difficulty of fabrication. Resistance to corrosion in many environments is better than most of the previously discussed metals. These alloys are used in the annealed condition.

Discussion of the interaction between metals and their environment at elevated temperatures has been purposely avoided, other than to state that corrosion or oxidation may limit continued use of the metal.

It is recognized that air must be excluded during welding of titanium, zirconium, and some highly-pure specialty alloys. This is not only for better control and to produce a better-looking weld, but also to prevent adverse conditions in the microstructure. These metals will absorb oxygen, nitrogen and hydrogen at elevated temperatures. The absorbed gases form oxides, nitrides and hydrides in the microstructure. These undesirable constituents make the metal harder and stronger, but will greatly reduce toughness.

In most instances, the loss in toughness is much more pronounced than that resulting from the martensitic formation in low-alloy steels and, as a result, the material must be replaced. Onsite hardness tests can detect evidence in cases where absorption of these gases is moderate to severe, but microstructural examinations are required when results obtained from hardness testing are questionable. The difficulty of preparation of these metals for microstructural studies often necessitates the removal of samples for laboratory study.

The author

Dwight C. Deck received a B.S. in metallurgical engineering from Washington University, St. Louis, Mo., and did graduate work at the University of Missouri. His job experience has included work with Shell Oil Co. and Monsanto Co. His present duties include managing the metallurgical section of the Corrosion Control Laboratory of Celanese Chemical Co., Box 9077, Corpus Christi, TX 78408, telephone 512-241-2343. He is a member of the American Soc. for Metals, the Natl. Assn. of Corrosion Engineers and the American Welding Soc.

How to select alloy steels for pressure vessels—I

Credit: Riley-Beaird

This review examines the range of alloy steels now available for such vessels, the fabrication techniques, and the specification requirements.

W. C. Hagel and K. H. Miska, Climax Molybdenum Co.

☐ Hydrocrackers, urea autoclaves, methanol converters and high-pressure, large-capacity storage tanks all have to resist internal-pressure expansion or external-pressure buckling so that contained gases and liquids will not leak. The decision about the right alloy steel to be used for a pressure vessel often has to be made by specialists from a number of different disciplines. What follows is the metallurgist's view of the critical considerations.

Steel specifications

Pressure-vessel steels are usually produced to meet precise standard specifications prepared by one of several specification-writing bodies. For example, ASTM A 20 covers most general requirements, and Table I gives a list of individual American Soc. for Testing and Materials (ASTM) specifications for various types of alloys. All but the last three of these specifications are also published by the American Soc. of Mechanical Engineers (ASME), which adds an "S" in front of the "A," e.g., SA 202. Any steel is considered to be an alloy steel when a definite range or minimum quantity of the following elements is included in its composition: aluminum, boron, chromium, cobalt, molybdenum, niobium, nickel, titanium, tungsten, vanadium and zirconium. Each specification may contain numerous grades with similar or different composition.

Static tensile properties of the various grades, types and classes of pressure-vessel plate covered by ASTM specifications are listed in Table II. It should be noted that some of these values vary with plate thickness and/or width. Mechanical properties are influenced by several variables. Chemical composition is the most influential, and others include deoxidation practice, finishing temperature, plate thickness, and overall processing history. Greater thicknesses require minor alloy modifications and/or improvements in quenching rates. In Europe, the 0.2% yield strength or the elastic limit is used as a stress calculation basis; in the U.S., tensile strength is used. Both tensile and yield strengths decrease as temperature increases.

In addition, allowances must be made for the creep

Originally published July 28, 1980

ASTM specifications for pressure-vessel alloy steels — Table I

A202	Cr-Mn-Si alloy-steel plates
A203	Nickel alloy-steel plates
A204	Molybdenum alloy-steel plates
A225	Mn-V alloy-steel plates
A302	Mn-Mo and Mn-Mo-V alloy-steel plates
A353	Double normalized and tempered (N&T) 9% nickel for cryogenic use
A358	Cr-Mo alloy-steel plates for elevated-temperature service
A517	Quenched and tempered (Q&T) plates of high tensile strength
A533	Q&T Mn-Mo and Mn-Mo-Ni plates
A542	Q&T Cr-Mo alloy-steel plates
A543	Q&T Ni-Cr-Mo alloy-steel plates
A553	Q&T 8% and 9% nickel plates
A645	Specially heat-treated 5% nickel for cryogenic use
A658	36% nickel plate of low thermal expansion
A734	Q&T alloy and high-strength low-alloy for cryogenic use
A735	Low-carbon Mn-Mo-Nb for low and moderate temperatures
A736	Age-hardening low-carbon Ni-Cu-Cr-Mo-Nb plates

factor in strength calculations if long-term high-temperature operation of a pressure vessel is anticipated. The usual factors to be considered in steel selection are: availability, high-temperature strength, notch toughness, weldability and heat treatment. The material should be available in a full range of sizes to allow the fabricator to build vessels without excessive weld seams. The classical boiler equation stipulates that the wall thickness of a vessel must be directly proportional to its internal pressure. Thus, a high pressure usually does not affect choice of material but simply calls for greater wall thickness.

Strength and toughness

Regardless of the method used for calculating allowable stresses, higher-strength steels do allow higher design stresses and lighter sections. Since yield strength will increase more rapidly than tensile strength, the design stress advantage of higher-strength steels is more marked with codes based on yield strength. While it might seem logical to calculate the economy of various grades of steel merely on the basis of the ratio of design stress divided by price per unit weight, this approach is too simple and does not reflect the complex economic significance of strength. On that basis, carbon steels would be the only ones selected for light- and moderate-wall vessels operating at room temperature.

Each pressure vessel must be evaluated individually. The fact that thinner sections can be used without loss of load capacity not only reduces the weight of steel but also foundation and transportation costs. The lighter gages can also decrease costs of forming, welding, heat treatment and stress relief, erection, nondestructive testing, and control. Finally, thinner sections generally have improved resistance to brittle fracture. The weight and difficulty of management of some of the large heavy-section pressure vessels have already reached a point where only the use of high-strength alloy steel makes it feasible to fabricate and transport them.

There is an increasing trend in Japan and the U.S. toward quenched-and-tempered (Q&T) bainitic, martensitic and maraged materials to obtain higher strength and more toughness, even in sections exceeding 300 mm (12 in.). Without alloying elements, it is impossible to quench to martensite in any appreciable thickness.

It has long been appreciated that increasing attention must be paid to ferritic materials and their susceptibility to brittle fracture as service temperature drops below room temperature. It is common practice to specify that steel for pressure vessels designed to operate in this range must meet certain toughness requirements, usually in terms of a Charpy V-notch impact transition-temperature.

Such impact requirements generally result in killed carbon steels and normalized and tempered (N&T) low-alloy steels being limited to temperatures above −45°C (−50°F). Q&T steels covered by ASTM A 517 are notch-ductile down to at least −60°C (−75°F) and thus are used for applications such as liquefied-gas containers. For temperatures ranging down to −100°C (−150°F), a 3.5%-nickel steel can be utilized, and either a 9%-nickel or an austenitic steel is used at still lower temperatures. Normally, austenitic steels and nonferrous alloys retain high toughness down to cryogenic temperatures.

Generally speaking, impact transition temperature increases with carbon content, although this effect can be masked by a number of factors. However, it is important to keep carbon content low to obtain optimum toughness, especially at high strengths. A tempered martensite microstructure will display a lower transition temperature than one that is bainitic or contains a mixture of pearlite and ferrite.

Low-temperature impact tests are particularly sensitive in revealing one of the main contributions of molybdenum to pressure-vessel steels, namely, minimizing susceptibility to temper embrittlement. This form of embrittlement is likely to occur in heavier plates and forgings during their inherently slower cooling from tempering or during the course of stress relieving.

Fracture mechanics has been applied with considerable success to vessels built with materials of relatively low ductility. Its usefulness results from the coordination of concepts involving flaw size, stress and transition temperature in terms meaningful to the designer.

One of the major problems relates to the thickness effect. The inherently poorer properties in heavy sections have long been known and have been one reason for the trend to higher-strength steels. Increased thickness promotes greater internal restraint at a defect and thus increases the likelihood of defect growth by brittle cracking. The limiting flaw size must be large enough that it can be detected by nondestructive testing. Design and fabrication procedures should be selected to minimize stress concentrations, particularly in critical areas. Consideration should also be given to the proper choice of welding and stress-relieving conditions in order to avoid considerable effects on toughness.

Many special environments influence the choice of pressure-vessel materials, but three of particular importance are corrosion, high temperature, and hydrogen.

Corrosion

When corrosion is encountered, the complex interaction of stress, temperature, fluid velocity, and time does not allow development of specific design procedures for load.

Mechanical properties of pressure-vessel alloy steels — **Table II**

Specification	Grade or type	Tensile strength* ksi	Yield strength* ksi	Elongation, %, in 50 mm or 2 in.
A202	A	75-95	45	19
	B	85-110	47	18
A204	A	65-85	37	23
	B	70-90	40	21
	C	75-95	43	20
A225	A	70-90	40	21
	B	75-95	43	20
	C	105-135	70	20
A302	A	75-95	45	19
	B	80-100	50	18
	C, D	80-100	50	20
A353	–	100-120	75	20
A387	2, 12 (class 1)	55-80	33	22
	11 (class 1)	60-85	35	22
	22, 21, 5, 7 and 9 (class 1)	60-85	30	18
	2 (class 2)	70-90	45	22
	11 (class 2)	75-100	45	22
	12 (class 2)	–	–	22
	22, 21, 5, 7 9 (class 2)	75-100	45	18
A517	All	115-135	100	16
A533	1	80-100	50	18
	2	90-115	70	16
	3	100-125	83	16
A542	1	105-125	85	14
	2	115-135	100	13
	3	95-115	75	20
	4	85-105	60	20
A543	1	105-125	85	14
	2	115-135	100	14
	3	90-115	70	16
A553	I, II	100-120	85	20
A645	–	95-115	65	20
A658	–	65-80	35	30
A734	A	77-97	65	20
A735	1	80-100	65	18
	2	85-105	70	18
	3	90-110	75	18
	4	95-115	80	18
A736	1	72-92	65	20
	2	85-105	75	20

*In SI units, 1 ksi (1,000 psi) = 6.89 megapascals (MPa)

Frequently, neither the problem nor the remedy can be stated in quantitative terms that would guide the designer on how to alter vessel construction for longer life. The practical approach must, therefore, rely extensively on experience. Laboratory and pilot-plant data must always be confirmed by service testing. Economics is a major factor in determining whether corrosion is to be combatted by a corrosion allowance, a change in the vessel material, or use of a different material on the inner surface in contact with the corrosive medium.

If the corrosion rate is not too high, provision of a corrosion allowance in the form of extra wall thickness may be an adequate design precaution where contamination of the product and ease of cleaning are not involved. This, however, may not be an economic solution for vessels intended for long service. Solid vessels of corrosion-resistant steels or superalloys are less common, but they may be necessary where severe corrosion takes place or where the high-temperature strength of these materials can be utilized.

On the other hand, many vessels use corrosion-resistant linings, claddings or weld overlays. With non-nuclear applications, the strength of the cladding in integrally clad plate can be taken into account in designing according to the ASME Boiler and Pressure Vessel Code, but linings are not considered to contribute to wall strength. With nuclear vessels, no structural strength is attributed to cladding except where bearing stress is involved. Corrosion damage prior to service, resulting from improper pickling, acid-wash procedures, testing fluids, and storage can be eliminated by proper control. In the chemical industry, the general attitude is that steels with up to 5% alloying elements can be grouped together with respect to wet corrosion. Low-alloy steels, however, may be much better than unalloyed steels in resistance to oxidation and some types of sulfur and sulfides.

The authors

William C. Hagel is Manager of High Temperature Materials Development at Climax Molybdenum Co., a division of Amax, Inc, 2929 Plymouth Rd., Ann Arbor, MI 48106, tel: (313) 761-2300. He has a B.Met.E. from Cornell University and M.S. and Ph.D. degrees in physical metallurgy from Carnegie-Mellon University. He is a member of the American Soc. for Metals and the American Institute of Mining, Metallurgical and Petroleum Engineers. He has authored over 70 technical publications and has a dozen patents to his credit.

Kurt H. Miska is Supervisor of Technical Information at Climax Molybdenum Co., One Greenwich Plaza, Greenwich, CT 06830, tel: (203) 622-3598. Before joining the company in 1978, he was an Associate Editor of *Materials Engineering* for seven years. There he covered developments in metals, composites, adhesives and welding in news and feature articles. He has written hundreds of papers on a wide range of technical subjects.

How to select alloy steels for pressure vessels—II

Credit: Devine Mfg. Co.

In fabricating pressure vessels of alloy steels, the operating temperature range and the effect of hydrogen attack are two prime design considerations.

W. C. Hagel** and **K. H. Miska**, Climax Molybdenum Co.*

We have already discussed steel specifications, strength and toughness; and the effect of corrosion on the selection of alloy steels for these vessels (see Part I). This concluding article looks at two other important criteria and then examines some fabrication techniques and specific applications.

Temperature

For the petrochemical industry, service conditions in terms of temperature and pressure vary as follows:

- Hydrodesulfurizers—425°C (800°F) and 10.5 MPa (megapascal), or 1,500 psi.
- Hydrocrackers—425 to 480°C (800 to 900°F) and 7 to 28 MPa (1,000 to 4,000 psi).
- Catalytic reformers—425 to 540°C (800 to 1,000°F) and 7 MPa (1,000 psi).
- Hydrodealkylation—595 to 760°C (1,100 to 1,400°F) and 5.6 to 8.5 MPa (800 to 1,200 psi).

In other words, most vessels operating above room temperature are at temperatures below 540°C (1,000°F). While some large chemical vessels operate at 600 to 1,200°C (1,110 to 2,190°F), these are normally cold-wall vessels, lined with a refractory, so the shell temperature is below 200°C (400°F). A few fairly small vessels operate up to 900°C (1,650°F) but with quite low internal pressures. Thus, the most significant range is from about 95 to 540°C (200 to 1,000°F).

For reasons associated with material economics and scaling behavior, it is unlikely that shell operating temperatures for most of these vessels will increase beyond 540°C (1,000°F). There is an increasing trend toward larger vessels (whose finished weight can be 800 metric tons) for greater efficiency and more economical construction. The life expectancy of such vessels varies but is generally around 20 yr. For the above, elevated-temperature strength is a most important selection criterion.

Stress-rupture and creep-strength considerations govern when the vessel is to operate in the creep range, which

*To meet the authors, see p. 165.

may be set between 315 and 540°C (650–1,000°F), depending largely on grade and heat treatment. Rupture and creep characteristics may also have to be factored in for vessels operating at lower temperatures, since large vessels may be exposed in the creep range for 50 to 100 h during tempering and stress relieving. Design stress is usually based on the average stress for rupture in 100,000 h, or on the average stress for 1% creep in 100,000 h, with suitable factors of safety. The most popular of the steels for use in the creep range are the chromium-molybdenum grades, particularly 1Cr-½Mo and 2¼Cr-1Mo.

As indicated by ASME Boiler and Pressure Vessel Code values for N&T plate, these steels show marked strength advantages over carbon steel. The increased strength may be sufficient so that the alloy steels will be less expensive for a given unit of strength than carbon steel, in addition to all the other cost benefits inherent in the thinner alloy-steel sections. The higher-chromium grade has higher strength in some ranges, as well as better resistance to corrosion and hydrogen attack. Improved hardenability is also a factor for heavy sections.

There was a time when designers used room-temperature properties for all temperatures below the creep range. This was feasible when economics was somewhat less important, and excessively high factors of safety were applied. With better knowledge of creep properties, increasing emphasis is being given to strength at service temperatures. Section III (Nuclear) of the ASME Boiler and Pressure Vessel Code is based on using a safety factor of ⅓ times minimum tensile strength at operating temperature, and of ⅔ times minimum yield strength at operating temperature, whichever is lower; over about 200°C (400°F), yield strength is generally limiting.

All major pressure-vessel codes in Western Europe now permit stresses up to ⅔ of yield strength for designs below the creep range. This, in turn, means a greater number of alloy steels to choose from, since they are generally more economical than carbon steels for high-pressure elevated-temperature vessels designed on a yield-strength basis. Within the temperature range of interest in the near future, up to 540°C (1,000°F), Q&T alloy steels show a definite strength advantage over N&T steel.

Hydrogen

Hydrogen absorbed at room temperature can cause serious mechanical embrittlement, or at elevated temperatures can attack the carbon contained in the steel. Before a steel is placed in service, contained hydrogen can be minimized by using vacuum melting or degassing, employing low-hydrogen weld rods, and exercising suitable control of hydrostatic testing fluids. It has been observed that increased hardness, cold working and the presence of notches or other stress concentration regions accentuate the effect of hydrogen at room temperature. Martensite is usually more susceptible to embrittlement than pearlite, but high-alloy maraging steels are less likely to be embrittled than low-alloy martensitic steels.

Hydrogen stress cracking occurs when hydrogen is produced either by a corrosion reaction or during cathodic protection. This phenomenon is loosely related to stress-corrosion cracking, and field failures resulting from hydrogen stress cracking usually take place only in steels having a yield strength above 1,030 MPa (150 ksi).

As soon as engineers started working with hydrogen under high pressures and temperatures, they became aware that carbon steel had severe limitations above about 200°C (400°F). The hydrogen diffusing into the steel combines with the carbon present to form methane and other products. This results first in decarburization, followed by fissuring due to high gas pressure at localized sites.

The most suitable alloying elements conferring resistance to high-temperature hydrogen attack are carbide-forming ones, particularly chromium and molybdenum. A classic Nelson diagram can be used to plot necessary alloy content (as a function of temperature) versus hydrogen partial pressure.

With suitable selection of chromium and molybdenum contents, the steel will resist hydrogen attack with respect to decarburization, fissuring, and loss of strength. Molybdenum is four times as effective as chromium. Small amounts of vanadium, titanium and niobium (up to 0.1%) are as effective as molybdenum in resistance to hydrogen (although not in respect to other properties), while silicon, nickel and copper are neutral.

Steels with 0.5% Mo find considerable use in the petroleum industry, where they have replaced carbon steel because their greater resistance to hydrogen permits use of higher temperatures. The possibility of graphitization limits their maximum temperature to 470°C (875°F). Chromium-molybdenum steels are selected for more-severe conditions. In the petroleum industry, many hydrocracking units have an inner cladding of austenitic stainless steel for resistance to hydrogen sulfide, and this cladding is backed up by chromium-molybdenum steel for resistance to hydrogen.

Fabrication

All pressure vessels require care in fabrication. This is particularly true with those high-strength materials where optimum fabrication and inspection are necessary to take full advantage of their properties. While such a precaution is expensive, fabrication costs do not increase in proportion to strength, as there are compensating benefits from the decreased section size. For the highest-quality vessels, degassed or even vacuum-melted steel, and close control of deoxidation to minimize inclusions, is generally required. In some cases, special processes may be applied. There are differences in formability among the various materials used for pressure vessels. Generally, however, weldability requirements are more demanding than those for formability. Any material having adequate weldability usually has adequate formability. Increasing strength can be advantageous, as it enables use of thinner material, and lighter components.

Formed parts are increasingly replacing forgings. The limits between hot and cold forming are generally set by available equipment. Hot forming roughly doubles the plate thickness that can be bent in a specific machine. There is, however, a distinct difference in heat treatment. For hot forming, the steel is purchased hot-rolled. After forming at around 1,065°C (1,950°F), parts must be fully heat-treated, sometimes annealed or normalized prior to final heat treatment. Cold-formed parts are formed in fully-heat-treated condition. Plate is usually formed to shape in one continuous operation without intermediate reheating. Spinning and its various modifications are often used to form heads.

In most applications, increasing the pressure-vessel size will mean more-efficient operation. Even with the use of high-strength steels, the pressure-vessel weight can reach the point where there are problems in transporting the finished vessel to its final site. Therefore, field erection must come into the picture, much as fabricators would prefer to remain with shop fabrication. Each case has to be considered individually. With suitable precautions, the final product is as good and reliable as shop-fabricated vessels, but careful planning is necessary. Basically, the pressure vessel is temporarily assembled in the shop to establish dimensional tolerances as well as to insert nozzles and provide cladding. Then it is broken down into large subassemblies for shipment to the destination. The assembly site must provide for welding, machining, stress-relieving and inspection facilities.

Most pressure vessels are welded rather than brazed or riveted. Weldability, therefore, must be considered a criterion in steel selection. Moreover, in the development of new grades of steel, weldability is one of the factors that must be weighed in arriving at the final composition. Weldment characteristics are just as important as those of the parent metal in determining whether the vessel will be satisfactory. One of the difficulties in trying to nail down weldability is its dependence on so many other factors: steel, parent metal, filler metal, welding conditions, and method of evaluation.

Possibly the most serious hazard in welding high-strength steels is cold cracking of the heat-affected zone. Since this type of cracking occurs only when at least some martensite is formed, transformation stress appears to be a primary cause. This stress can be lowered by reducing the martensite lattice strain, by decreasing carbon content, or by raising the temperature range of martensite formation so that self-tempering takes place. Consequently, there is a high degree of correlation between the martensite-formation temperature and cold cracking. The pressure-vessel standards established in most countries generally specify permissible design features, including position and type of welds.

Filler-metal composition depends not only on strength requirements but also on whether the filler is expected to respond to an austenitizing treatment and whether it is to be stress-relieved. If the weldment is to be reheat-treated (as is necessary in most cases for electroslag welds, and sometimes for submerged-arc welds), the composition must be selected to respond to heat treatment in the same way as the parent metal.

Generally, a closely matching composition is chosen. Usually, however, filler metal is selected so that when it is deposited by a multipass technique, a complete reheat treatment is unnecessary. There is seldom any trouble in obtaining the necessary strength, but notch-toughness properties will be poor unless a suitable composition is used with good welding practices. Mechanical properties of deposits generally improve with an increasing number of passes and weld refinement. One effective means of minimizing strain is preheating, which serves to eliminate cold cracking, but there are many views as to whether preheating is necessary, and as to the amount required.

Except for the necessity of observing code restrictions, heat treatment is basically no different for pressure vessels than for other parts. It is not feasible to include detailed instructions or to repeat information contained in codes, but the heat treatment given pressure-vessel steels in general consists of: some form of cooling from the austenitizing temperature, tempering, and stress-relieving. Because of the large mass of many of the vessels concerned, the above treatments are of prime importance in arriving at desired combinations of strength and toughness.

Applications

For some time, burning of residual products in closed containers to produce petroleum coke has shown increasing growth within the petroleum market. This delayed coking process uses drums up to 9.8 m (30 ft) in diameter, fabricated from roll-bonded clad plates. Although early coke drums were AISI 405 stainless-clad on carbon steel, the trend today is to clad AISI 410S on N&T 1Cr-½Mo and 1¼Cr-½Mo steels such as A 387-12 or A 387-11; C-½Mo (A 204-C) backing steels have also been used. A 1¼Cr-½Mo steel in accordance with A 387-11 will give the highest allowable design stress at 480°C (900°F) and above, close to the possible oil inlet-temperature. At this temperature, A 387-11 backing steel clad with AISI 410S is a more economical material.

With the advent of low- and non-lead gasolines, processes providing high-octane aromatics are of extreme interest for gasoline upgrading. One of the predominant processes is catalytic reforming. As is usual in processes where varying degrees of corrosion, erosion and various hydrogen effects are present, there are several choices of materials for the major process vessels. Usually, reformer vessels are constructed of alloy steels ranging from C-½Mo (A 204-B, A 204-C) to 1Cr-½Mo (A 387-11, A 387-12) to 2¼Cr-1Mo (A 387-21, A 387-22). These grades are also available as roll-bonded clad.

Vaporizing and condensing characterize the operations in crude-oil-unit atmospheric and vacuum towers. Since corrosive conditions are pervasive, clad steels are often used for crude units. While the materials for the cladding are usually selected because of the corrosive environment, and vary by refinery, typical crude units include Monel cladding in the top of the atmospheric unit, AISI 316L or 410S cladding in the bottom section, and possibly carbon steel in the middle portion. The vacuum tower could include unclad carbon steel in combination with AISI 316L or 410S clad material for the bottom portion. Usually, the carbon-steel unclad areas have considerable corrosion allowance added to their thicknesses. Backing steels can be carbon or low-alloy steels.

Acknowledgements

The authors appreciate assistance provided by Alloy Crafts Co., Charles Ross & Son Co., De Dietrich (USA) Inc., Devine Mfg. Co., Nooter Corp., Patterson Industries Inc., and Riley-Beaird, Inc.

Parts of this article were condensed from Dunn, R. G., Whitley, G. F., and Fairhurst, W., Molybdenum's Place in the Pressure-Vessel Field, 262 pp. (1969), Code M-188 (available free of charge from Climax Molybdenum Co., Greenwich, Conn.).

The basics of stainless steels

To pick the right stainless steel, you must know the grades and types available, their corrosion resistance, their physical and mechanical properties, and how they perform during fabrication and welding.

***James D. Redmond**, Climax Molybdenum Co., and **Kurt H. Miska**, Climax Molybdenum Co. of Michigan*

☐ Over 70 years ago, it was discovered that a minimum of about 11% chromium would impart corrosion and oxidation resistance to steel; and by about 1914, stainless steels had become commercial. By definition, stainless steels are those ferrous alloys that contain a minimum of about 11% Cr.

Since that time, much development and commercialization of stainless steels has taken place and today there are hundreds of stainlesses for countless applications. These range from general-purpose grades to alloys created for specific severe environments.

Activity and passivity

Stainless steels resist corrosion chiefly because they are passive. As an example of passivity, it has been known for a long time that iron is not attacked by dilute nitric acid if it has first been immersed in concentrated nitric acid. This procedure shifts the surface of the iron from the active state to the passive one. The phenomenon is general, and metals are, depending on circumstances, in either the active or passive state. In the case of iron in concentrated nitric acid, the passive state lasts only briefly and is easily destroyed.

Compared to this, the passive state of some other metals, especially chromium, is very stable, and this is true in many media. Also, elements that have a stable passivity can transmit this characteristic to other metals with which they are alloyed in sufficient amounts.

Passivity is due to the formation of a protective layer on the surface of the metal. This layer is generally an oxide, but its exact nature is debatable. It may be a layer of oxygen, or another element, bound by chemisorption; that is, a surface layer of ions or fractions of molecules that are bound by electronic forces to the metal.

Electrochemistry of corrosion

The passive film can cause considerable change in the electrochemical behavior of the metal.

All wet corrosion phenomena consist of an electrochemical exchange between the electrolyte and the surface of the metal. The basis for this is the mixed-potential theory. In essence, this theory states that the oxidation and reduction reactions of corrosion occur at separate sites, and that the total rate of all the oxidation reactions equals the total rate of all the reduction reactions on a corroding surface.

Oxidation reactions—called anodic, because they occur at the anodic sites on a corroding metal (or at the anode in an electrochemical cell)—can be represented by:

$$M \rightarrow M^{+n} + ne$$

This is the generalized corrosion reaction, in which metal atoms are removed by oxidizing them to ions. In this reaction, the number, n, of electrons produced, e, equals the valence, +n, of the metal ion produced, M^{+n}. The mixed-potential theory assumes that all electrons generated by anodic reactions are consumed by corresponding reduction ones. Reduction reactions are known as cathodic, because they occur at the cathodic sites of a corroding metal (or at the cathode of an electrochemical cell). The more common cathodic (electron-consuming) reactions encountered in aqueous corrosion are:

1. Reduction of hydrogen ions:

$$2H^+ + 2e \rightarrow H_2$$

2. Oxygen reduction (acid solutions):

$$O_2 + 4H^+ + 4e \rightarrow 2H_2O$$

3. Oxygen reduction (basic or neutral solutions):

$$O_2 + 2H_2O + 4e \rightarrow 4OH^-$$

4. Metal-ion reduction:

$$M^{+n} + e \rightarrow M^{+(n-1)}$$

5. Metal deposition (plating):

$$M^{+n} + ne \rightarrow M$$

During corrosion, more than one anodic and more than one cathodic reaction may be taking place. For example, consider the corrosion of a stainless steel in a hydrochloric acid solution that is contaminated with ferric ions. By means of the anodic reaction, all the component elements (i.e., iron, chromium, etc.) of the alloy go into the solution as their respective ions.

The electrons produced by these anodic reactions will be consumed by the cathodic reactions 1 and 4 above, with Reaction 4 in this instance being represented by:

$$Fe^{+3} + e \rightarrow Fe^{+2}$$

Removing one of the available cathodic reactions

Originally published October 18, 1982

Registered trademarks and tradenames that appear in the tables of alloy compositions

Registered Trademark or Tradename	Producer
Carpenter 20Cb-3	Carpenter Technology
Carpenter 20Mo-6	Carpenter Technology
Carpenter Project '70 182-FM	Carpenter Technology
Carpenter Stainless No. 7-Mo	Carpenter Technology
Carpenter Custom 450	Carpenter Technology
Carpenter Custom 455	Carpenter Technology
904L	Nyby Uddeholm (Sweden)
JS904L[a]	Jessop
Monit	Nyby Uddeholm (Sweden)
Sea-Cure/SC-1	Trent Tube Div. of Colt Industries
Ferralium Alloy 255[b]	Langley Alloys (UK)
DP-3	Sumitomo Metal Ind. (Japan)
254 SMO	Avesta (Sweden)
Nitronic steels	Armco
18SR	Armco
AL-4X	Allegheny Ludlum
AL-6X	Allegheny Ludlum
E-Brite 26-1	Allegheny Ludlum
AL 29-4-2	Allegheny Ludlum
AL 29-4C	Allegheny Ludlum
Alloy 6X[c]	Al Tech Specialty Steel Corp.
2RK65	Sandvik (Sweden)
3RE60	Sandvik (Sweden)
SAF 2205	Sandvik (Sweden)
Sanicro 28	Sandvik (Sweden)
JS-700	Jessop
JS-777	Jessop
AF-22	Mannesmann (W. Germany)
Tenelon	U.S. Steel
Cryogenic Tenelon	U.S. Steel
Cronifer 2328	VDM (W. Germany)

[a]Nyby Uddeholm 904L produced under license
[b]Produced in U.S. by Cabot Corp.
[c]AL-6X produced under license

Courtesy: Fluor Corp., designer and builder

Nuclear-waste calcining facility uses stainless steel

(e.g., Reaction 4, by the removal of the ferric ions) will reduce the corrosion rate. Thus, the observation that hydrochloric acid contaminated with ferric ions is more corrosive than the pure acid can be easily explained.

Role of alloying elements

In addition to the fundamental role played by chromium in making steel stainless, nickel and molybdenum are probably the next most vital elements here. Other elements are sometimes found in stainless steels, and will be briefly mentioned later. These may be present in residual levels from processing or may be deliberately added.

Chromium

Chromium concentrations lower than about 11% are sufficient in some cases to retard attack, but it is only with this amount that a stable, transparent, passive film spontaneously forms, and the alloy can be considered as stainless steel. Higher amounts cause a marked increase in corrosion resistance by strengthening the film and causing it to repair itself rapidly if broken. Additions to 30% Cr are used in some commercial stainless steels.

Chromium is called a ferrite former because it tends to suppress the ferrite-austenite transformation observed during heating of iron or carbon steel. (The various types of steel are briefly described in Table I. Tradenames used in the tables here appear in the box.) In the constitution diagram for very-low-carbon alloys (Fig. 1), steels with over 13% Cr remain in the ferrite region, regardless of the temperature to which they are heated.

These alloys do not undergo any transformation and on heating do not pass into the austenitic region. On the other hand, steels with lower chromium levels behave like ordinary steels; that is, above a certain temperature (depending on chromium content), they become austenitic, and during cooling this austenite is transformed into an aggregate of ferrite and carbides. Slow cooling gives a dispersion of more-or-less spheroidal carbides in a ferritic matrix, but rapid cooling leads to a hard martensitic structure. In other words, the steel hardens on rapid cooling.

When the carbon content of iron-chromium alloys is increased, the chromium content above which the steel no longer becomes austenitic on heating also increases. With a carbon content of 0.6%, for example, this would mean 18% Cr. Chromium stainless steels that have no other alloying elements fall into two groups—those that do, or do not, become austenitic on heating and, consequently, can be hardened, or not, on cooling.

Another characteristic of the iron-chromium alloys is the existence of an intermetallic iron-chromium compound named the sigma phase. An intermetallic phase is a hard, brittle precipitate that forms in the metal. This phase is rich in Cr, Mo and Ni. It can cause problems when the metal is stressed or subjected to corrosive conditions.

The sigma phase has low toughness, as is true for almost all intermetallic compounds. It may be formed in chromium steels when they are maintained at temperatures between 500 and 900°C (930 and 1,650°F). The lower the chromium content of the steel the longer it takes for this phase to form. However, even with 27%

Here is a short guide to identifying the various types of stainless steels **Table I**

Type	Composition or alloy content	Microstructure	Mechanical properties	Physical properties	Applications
Austenitic	15-27% Cr, 8-35% Ni, 0-6% Mo, Cu, N	Austenite	Tensile strength: 490-860 MPa Yield strength: 205-575 MPa Elongation in 50 mm: 30-60%	Non heat-treatable; nonmagnetic	Most widely used in general applications
Ferritic	11-30% Cr, 0-4% Ni, 0-4% Mo	Ferrite	Tensile strength: 415-650 MPa Yield strength: 275-550 MPa Elongation in 50 mm: 10-25%	Non heat-treatable; magnetic; good resistance to chloride stress-corrosion-cracking	Parts requiring combination of good general corrosion resistance with good stress corrosion resistance, seawater applications
Martensitic	11-18% Cr, 0-6% Ni, 0-2% Mo	Martensite	Tensile strength: 480-1,000 MPa Yield strength: 275-860 MPa Elongation in 50 mm: 14-30%	Hardenable by heat treatment; high strength	High-strength parts, pumps, valves and paper machinery
Duplex	18-27% Cr, 4-7% Ni, 2-4% Mo, Cu, N	Austenite and ferrite	Tensile strength: 680-900 MPa Yield strength: 410-900 MPa Elongation in 50 mm: 10-48%	Non heat-treatable	Shell-and-tube heat exchangers, wastewater treatment and cooling coils.
Precipitation hardening	12-28% Cr, 4-25% Ni, 1-5% Mo, Al, Ti, Co	Austenite and martensite	Tensile strength: 895-1,100 MPa Yield strength: 276-1,000 MPa Elongation in 50 mm: 10-35%	Hardenable by heat treatment; very high strength	Parts requiring high strength, corrosion- and/or high-temperature resistance

Cr, several thousand hours at 550°C (1,020°F) are necessary before the sigma phase becomes metallographically discernible. Its rate of formation may be greatly accelerated when certain other alloying elements are present or when the temperature is increased.

Nickel

Nickel is present in many stainless steels. In contrast to chromium, it promotes the stability of austenite and, therefore, is called an austenite former. Low-carbon steels containing 24% Ni remain austenitic at ambient temperatures. The amount of nickel needed to retain the austenite structure decreases as the carbon content increases.

When nickel is added to straight-chromium stainless steels, the structure changes from ferritic to austenitic. Toughness, ductility and ease of weldability improve. The addition of nickel aids the formation of the passive film and increases resistance to strong acids, especially reducing acids. This element is used in concentrations to about 40% in stainless steels.

Molybdenum

Molybdenum is added to stainless steels to improve pitting and crevice-corrosion resistance (these terms will be discussed later on). Molybdenum increases the strength of the passive film, especially with respect to attack by chlorides that may poison the surface and disrupt the protective oxide film. Molybdenum also improves the elevated-temperature mechanical properties of austenitic stainless steels and the strength and resistance to tempering of martensitic stainless steels. Levels used are as high as 6–7%.

Minor elements

The structure, mechanical properties and corrosion resistance of stainless steels are determined principally by the alloy elements—chromium, nickel or molybdenum—used. However, some properties can also be improved and modified by the intentional addition of small amounts of minor elements, which include carbon, nitrogen, columbium (niobium), tantalum, titanium, sulfur, selenium and lead (all of these typically at levels less than 1%), and copper, manganese, silicon and aluminum, which may be present in the steel in concentrations to a few percent.

Though inclusion of carbon and nitrogen cannot be avoided during melting, these elements are also deliberately added to some grades, primarily to increase

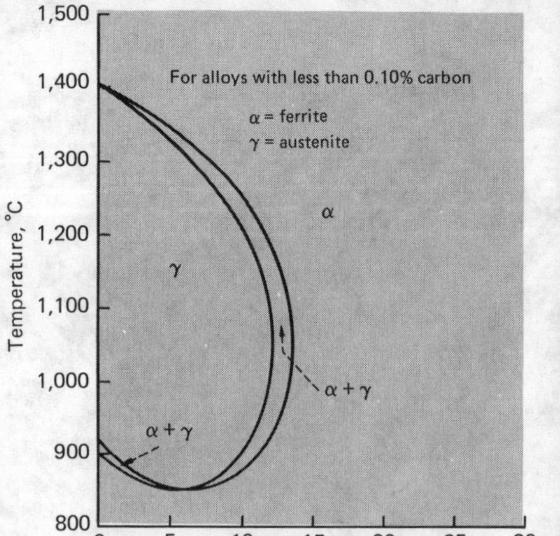

Above a certain chromium content, the ferrite-to-austenite transformation is suppressed **Fig. 1**

Austenitic stainless steels have excellent corrosion resistance and good formability **Table II**

Designation or type	UNS no.	Composition, %*							
		C	Mn	Si	Cr	Ni†	P	S	Others
201	S20100	0.15	5.5-7.5	1.00	16.0-18.0	3.5-5.5	0.06	0.03	0.25 N
202	S20200	0.15	7.5-10.0	1.00	17.0-19.0	4.0-6.0	0.06	0.03	0.25 N
205	S20500	0.12-0.25	14.0-15.5	1.00	16.5-18.0	1.0-1.75	0.06	0.03	0.32-0.40 N
216 (XM-17)	S21600	0.08	7.5-9.0	1.00	17.5-22.0	5.0-7.0	0.045	0.03	2.0-3.0 Mo; 0.25-0.50 N
301	S30100	0.15	2.00	1.00	16.0-18.0	6.0-8.0	0.045	0.03	–
302	S30200	0.15	2.00	1.00	17.0-19.0	8.0-10.0	0.045	0.03	–
302B	S30215	0.15	2.00	2.0-3.0	17.0-19.0	8.0-10.0	0.045	0.03	–
303	S30300	0.15	2.00	1.00	17.0-19.0	8.0-10.0	0.20	0.15 min	0.06 Mo (‡)
303 Plus X (XM-5)	–	0.15	2.5-4.5	1.00	17.0-19.0	7.0-10.0	0.20	0.25 min	0.6 Mo
303 Se	S30323	0.15	2.00	1.00	17.0-19.0	8.0-10.0	0.20	0.06	0.15 min Se
304	S30400	0.08	2.00	1.00	18.0-20.0	8.0-10.5	0.045	0.03	–
304H	S30409	0.04-0.10	2.00	1.00	18.0-20.0	8.0-10.5	0.045	0.03	–
304HN	S30452	0.04-10.0	2.00	1.00	18.0-20.0	8.0-10.5	0.045	0.03	0.10-0.16 N
304L	S30403	0.03	2.00	1.00	18.0-20.0	8.0-12.0	0.045	0.03	–
304LN	S30453	0.03	2.00	1.00	18.0-20.0	8.0-10.5	0.045	0.03	0.10-0.15 N
304N	S30451	0.08	2.00	1.00	18.0-20.0	8.0-10.5	0.045	0.03	0.10-0.16 N
S30430	S30430	0.08	2.00	1.00	17.0-19.0	8.0-10.0	0.045	0.03	3.0-4.0 Cu
305	S30500	0.12	2.00	1.00	17.0-19.0	10.5-13.0	0.045	0.03	–
308	S30800	0.08	2.00	1.00	19.0-21.0	10.0-12.0	0.045	0.03	–
308L	–	0.03	2.00	1.00	19.0-21.0	10.0-12.0	0.045	0.03	–
309	S30900	0.20	2.00	1.00	22.0-24.0	12.0-15.0	0.045	0.03	–
309 Cb	S30940	0.08	2.00	1.00	22.0-24.0	12.0-15.0	0.045	0.03	8 X %C min Cb
309 Cb + Ta	–	0.08	2.00	1.00	22.0-24.0	12.0-15.0	0.045	0.03	8 X %C min Cb+Ta
309S	S30908	0.08	2.00	1.00	22.0-24.0	12.0-15.0	0.045	0.03	–
310	S31000	0.25	2.00	1.50	24.0-26.0	19.0-22.0	0.045	0.03	–
310 Cb	S31040	0.08	2.00	1.50	24.0-26.0	19.0-22.0	0.045	0.030	Cb + Ta; 10 X C min 1.10 max
310S	S31008	0.08	2.00	1.50	24.0-26.0	19.0-22.0	0.045	0.03	–
312	–	0.15	2.00	1.00	30.0 nom	9.0 nom	0.045	0.03	–
316	S31600	0.08	2.00	1.00	16.0-18.0	10.0-14.0	0.045	0.03	2.0-2.0 Mo
316 Cb	S31640	0.08	2.00	1.00	16.0-18.0	10.0-14.0	0.045	0.030	Cb + Ta; 10 X C min 1.10 max; N 0.10 max; Mo 2.0-3.0
316F	S31620	0.08	2.00	1.00	16.0-18.0	10.0-14.0	0.20	0.10 min	1.75-2.5 Mo
316H	S31609	0.04-0.10	2.00	1.00	16.0-18.0	10.0-14.0	0.045	0.03	2.0-3.0 Mo
316L	S31603	0.03	2.00	1.00	16.0-18.0	10.0-14.0	0.045	0.03	2.0-3.0 Mo; 0.10-0.30N
316LN	S31653	0.03	2.00	1.00	16.0-18.0	10.0-14.0	0.045	0.03	2.0-3.0 Mo; 0.10-0.30N
316 Ti	S31635	0.08	2.00	1.00	16.0-18.0	10.0-14.0	0.045	0.030	Ti 5 X (C+N) min, 0.70 max; N 0.10 max; Mo 2.0-3.0
317	S31700	0.08	2.00	1.00	18.0-20.0	11.0-15.0	0.045	0.03	3.0-4.0 Mo
317L	S31703	0.03	2.00	1.00	18.0-20.0	11.0-15.0	0.045	0.03	3.0-4.0 Mo
317LM, 317LX, 317L Plus	–	0.03	2.00	1.00	18.0-20.0	12.0-16.0	0.045	0.03	4.0-5.0 Mo
321	S32100	0.08	2.00	1.00	17.0-19.0	9.0-12.0	0.045	0.03	5 X %C min Ti
321H	S32109	0.04-0.10	2.00	1.00	17.0-19.0	9.0-12.0	0.045	0.03	5 X %C min Ti
330	N08330	0.08	2.00	0.75-1.5	17.0-20.0	34.0-37.0	0.04	0.03	–

(continued) Table II

Designation or type	UNS no.	C	Mn	Si	Cr	Ni†	P	S	Others
		Composition, %*							
330HC	–	0.40	1.50	1.25	19.0 nom	35.0 nom	–	–	–
332	–	0.04	1.00	0.50	21.5 nom	32.0 nom	0.045	0.03	–
347	S34700	0.08	2.00	1.00	17.0-19.0	9.0-13.0	0.045	0.03	10 X %C min Cb + Ta‡
347 H	S34709	0.04-0.10	2.00	1.00	17.0-19.0	9.0-13.0	0.045	0.03	10 X %C min Cb + Ta
348	S34800	0.08	2.00	1.00	17.0-19.0	9.0-13.0	0.045	0.03	0.2 Cu; 10 X %C min Cb + Ta‡
348H	S34809	0.04-0.10	2.00	1.00	17.0-19.0	9.0-13.0	0.045	0.03	0.2 Cu; 10 X %C min Cb + Ta‡
384	S38400	0.08	2.00	1.00	15.0-17.0	17.0-19.0	0.045	0.03	–
385	–	0.08	2.00	1.00	11.5-13.5	14.0-16.0	0.045	0.03	–
18-18-2 (XM-15)	S38100	0.08	2.00	1.5-2.5	17.0-19.0	17.5-18.5	0.03	0.03	0.08-0.18 N
18-18 Plus	–	0.15	17.0-19.0	1.00	17.5-19.5	–	0.045	0.03	0.5-1.5 Mo; 0.5-1.5 Cu; 0.4-0.6 N
20Cb-3	N08020	0.07	2.00	1.00	19.0-21.0	32.0-38.0	0.045	0.035	2.0-3.0 Mo; 3.0-4.0 Cu; 8 X %C min Cb‡
20 Mo-6	–	0.03	1.00	0.50	22.0-26.0	33.0-37.0	0.03	0.03	5.0-6.7 Mo; 2.00-4.00 Cu
254 SMO	S31254	0.02	1.00	0.80	19.5-20.5	17.5-18.5	0.030	0.010	6.00-6.50 Mo; 0.50-1.00 Cu; 0.18-0.22 N
904L, JS904L, AL-4X, 2RK65	N08904	0.02	2.00	1.00	19.0-23.0	23.0-28.0	0.045	0.035	4.0-5.0 Mo; 1.0-2.0 Cu
AL-6X	N08366	0.03	2.00	0.75	20.0-22.0	23.5-25.5	0.030	0.003	6.0-7.0 Mo
Cronifer 2328	–	0.04	0.75	0.75	22.0-24.0	26.0-28.0	0.030	0.015	Cu 2.5-3.5; Ti 0.40-0.70; Mo 2.5-3.0
Cryogenic Tenelon (XM-14)	S21460	0.12	14.0-16.0	1.00	17.0-19.0	5.0-6.0	0.06	0.03	0.35-0.50 N
JS-700	N08700	0.04	2.00	1.00	19.0-23.0	24.0-26.0	0.04	0.03	4.3-5.0 Mo; 0.5 Cu; 8 X %C min Cb ¶; 0.005 Pb; 0.035 Sn
JS-777	–	0.04	2.00	1.00	19.0-23.0	24.0-26.0	0.045	0.035	4.0-5.0 Mo; 1.9-2.5 Cu
Nitronic 32‖	S24100	0.10	12.0	0.5	18.0	1.6	–	–	0.35 N
Nitronic 33	S24000	0.08	11.50-14.50	1.00	17.0-19.0	2.25-3.75	0.060	0.030	0.20-0.40 N
Nitronic 40	S21900	0.08	8.0-10.0	1.00	18.0-20.0	5.0-7.0	0.06	0.03	0.15-0.40 N
Nitronic 50	S21910	0.06	4.0-6.0	1.00	20.5-23.5	11.5-13.5	0.04	0.03	1.5-3.0 Mo; 0.2-0.4 N; 0.1-0.3 Cb; 0.1-0.3 V
Nitronic 60	S21800	0.10	7.0-9.0	3.5-4.5	16.0-18.0	8.9-9.0	0.04	0.03	–
Sanicro 28	N08028	0.020	2.0	1.0	26.0-28.0	29.5-32.5	0.020	0.015	Mo 3.0-4.0; Cu 0.6-1.4
Tenelon	S21400	0.12	14.5-16.0	0.3-1.0	17.0-18.5	0.75	0.045	0.03	0.35 N

*Single values are maximum values unless otherwise noted.
†For some tubemaking processes, the Ni content of certain grades must be slightly higher than shown.
‡Optional.
¶0.50% maximum.
‖Nominal composition; limits not available.

strength and hardness. Nitrogen is sometimes put in austenitic and duplex stainless steels to improve pitting resistance and lessen the tendency to form the sigma phase.

Sulfur, selenium and lead are used to improve machinability. These elements prolong tool life and reduce machining power-requirements.

Titanium, columbium and tantalum are included to preferentially combine with carbon and nitrogen so as to prevent sensitization and thereby eliminate susceptibility to intergranular corrosion. Sensitization is the combining of chromium with carbon along grain boundaries to form chromium carbides. This can give rise to corrosion in chromium-depleted areas (intergranular corrosion).

Columbium additions can improve high-temperature creep strength. Copper is added for improved resistance to sulfuric acid. Manganese is used as a partial substitution for nickel. Silicon and aluminum improve oxidation resistance.

Types of stainless steels

Although several hundred stainlesses are produced, each can be classified as either: austenitic, ferritic, duplex, martensitic or precipitation-hardening.

Stainless steels are sometimes also classified by the American Iron and Steel Institute (Washington, D.C.) as AISI "standard types," and proprietary or "nonstandard types." The criteria used by the AISI to establish the standard types are loosely defined, but include tonnage produced during a specific period, number of producers, and alloy composition-limits.

Increasingly, alloys are being designated by the Unified Numbering System (UNS) of the American Soc. for Testing and Materials (Philadelphia) and SAE, formerly the Soc. of Automotive Engineers (Warrendale, Pa.). Stainless steels with a UNS number beginning with the letter S are alloys with more than 50% iron. A UNS number beginning with N denotes a nickel-bearing alloy with less than 50% iron. In the accompanying tables, UNS numbers are given, as well as commonly used designations or types.

Austenitic

The austenitic stainless steels available to the chemical engineer are listed in Table II. This table may not be complete, but it does contain the major austenitics available, including the newer formulations. This is true for all such tables in the report.

Courtesy: Badger America, Inc. Built by Canadian Badger for Polysar Ltd.

Ethylbenzene/styrene plant employs stainless

The stainless steels in this group have many common characteristics. They can be hardened by cold working, but not by heat treatment. Annealed, all are nonmagnetic, although some may become slightly magnetic by cold working. They have excellent corrosion resistance, very good formability, and show an increase in strength as a result of cold work; most are readily weldable.

Type 304 (sometimes referred to as 18-8 stainless) is the most frequently used austenitic alloy. Its nominal composition is 18Cr–8Ni. This stainless is used extensively for chemical processing equipment and in the food and beverage industries. Type 304 has good atmospheric corrosion resistance, and thus is often specified for architectural applications.

Type 316 is alloyed with 2% molybdenum for improved pitting and crevice corrosion resistance compared with 304. It is used extensively for chemical-processing, pulp-and-paper and food-and-beverage equipment.

Ferritic

Generally, the ferritic stainless steels (Table III) are straight-chromium types containing from 11 to 30% chromium, and no other major alloying elements. However, some of the grades contain up to 4% molybdenum for improved pitting and crevice corrosion resistance. As a class, the ferritics are highly resistant to chloride stress-corrosion-cracking. They cannot be hardened by heat treating and only moderately so by cold working. They are magnetic, have good ductility, and resist corrosion and oxidation.

Type 430, with about 17% Cr, is a general-purpose grade. It is used frequently for automotive trim and for other such applications in which weldability is not a requirement.

In recent years, a family of low-interstitial ferritics has been developed. These materials contain carbon and nitrogen at 0.030% or less each, and offer better ductility, toughness and weldability than do the conventional ferritic stainless steels. The new materials range from the general-purpose grade—Type 444 (18Cr–2Mo) with pitting and crevice corrosion resistance equivalent to Type 316 in many environments—to highly alloyed grades, such as AL 29-4-2. The latter competes with titanium- and nickel-base superalloys in some environments. High-performance ferritics such as

Ferritics are highly resistant to chloride stress-corrosion-cracking **Table III**

Designation or type	UNS no.	Composition, %*							
		C	Mn	Si	Cr	Ni	P	S	Others
405	S40500	0.08	1.00	1.00	11.5-14.5	–	0.04	0.03	0.10-0.30 Al
409	S40900	0.08	1.00	1.00	10.5-11.75	–	0.045	0.045	6 X %C min Ti†
430	S43000	0.12	1.00	1.00	16.0-18.0	–	0.04	0.03	–
430F	S43020	0.12	1.25	1.00	16.0-18.0	–	0.06	0.15 min	0.6 Mo‡
430FSe	S43023	0.12	1.25	1.00	16.0-18.0	–	0.06	0.06	0.15 min Se
434	S43400	0.12	1.00	1.00	16.0-18.0	–	0.04	0.03	0.75-1.25 Mo
436	S43600	0.12	1.00	1.00	16.0-18.0	–	0.04	0.03	0.75-1.25 Mo; 5 X %C min Cb + Ta¶
442	S44200	0.20	1.00	1.00	18.0-23.0	–	0.04	0.03	–
446	S44600	0.20	1.50	1.00	23.0-27.0	–	0.04	0.03	0.25 N
430Ti	S43036	0.10	1.00	1.00	16.0-19.5	0.75	0.04	0.03	5 X %C min Ti†
444 (18-2)	S44400	0.025	1.00	1.00	17.5-19.5	1.00	0.04	0.03	1.75-2.5 Mo; 0.035 max N; 0.2 + 4 (%C + %N) min (Ti + Cb)
18SR‖	–	0.04	0.30	1.00	18.0	–	–	–	2.0 Al; 0.4 Ti
182-FM	S18200	0.08	2.50	1.00	17.5-19.5	–	0.04	0.15 min	–
E-Brite 26-1 (XM-27)	S44627	0.01	0.40	0.40	25.0-27.5	0.50	0.02	0.02	0.75-1.5 Mo; 0.015 N; 0.2 Cu; 0.5 Ni + Cu; Cb 0.05-0.20
AL 29-4-2	S44800	0.010	0.30	0.20	28.0-30.0	2.0-2.5	0.025	0.02	3.5-4.2 Mo
Monit	S44635	0.025	1.00	0.75	24.5-26.0	3.5-4.5	0.04	0.03	3.5-4.5 Mo; 0.03-0.06 (Ti + Cb)
Sea-Cure/SC-1	S44660	0.025	1.00	0.75	25.0-27.0	1.5-3.5	0.04	0.03	2.5-3.5 Mo; 0.2 + 4 (%C + %N) min (Ti + Cb)
439 (XM-8)	S43035	0.07	1.00	1.00	17.00-19.00	0.50	0.040	0.03	Ti 0.20 + 4 (%C + %N) min 1.10 max; Al 0.15 max; N 0.04 max
AL 29-4C	S44735	0.025	1.00	0.75	28.00-30.00	0.50	0.040	0.03	3.50-4.50 Mo; Ti 0.2 + 4 (%C + %N) min

*Single values are maximum values unless otherwise noted.
†0.75% max.
‡Optional.
¶0.70% max.
‖Nominal composition; limits not available.

Duplex stainless steels are fairly tough and can be made in plate-thicknesses **Table IV**

Designation or type	UNS no.	Composition, %*							
		C	Mn	Si	Cr	Ni	Mo	P	S	Other
329, No. 7Mo	S32900	0.10	2.00	1.00	25.00-30.00	3.00-6.00	1.0-2.0	0.045	0.03	–
3RE60	–	0.030	1.50	1.7	18.5	4.9	2.7	0.030	0.030	–
SAF2205/AF22	–	0.030	2.0	0.8	22	5.5	3.0	0.030	0.020	0.14N
Ferralium 255	S32550	0.030	2.0	1.0	26	5.0	3.0	0.030	0.020	0.17N, 2.0Cu
DP-3	–	0.030	2.0	1.0	25	6.5	3.0	0.030	0.020	0.5Cu, 0.15N

*Single values are maximum unless otherwise indicated.

Sea-Cure, AL 29-4C and Monit have been used increasingly in seawater-cooled condensers and heat exchangers since 1980.

Duplex

The duplex stainless steels (Table IV) are a mixture of austenite and ferrite. Their resistance to chloride stress-corrosion-cracking is in between that of the ferritics and austenitics, and decreases with increasing cold working. Because the duplexes have better toughness than do the ferritics, they are available in plate thicknesses and may therefore be used for tubesheets, for example. The crevice corrosion resistance of these steels is somewhat less than that of ferritic or austenitic grades of equivalent chromium and molybdenum compositions.

Martensitic stainlesses resist corrosion in mild environments and have fairly good ductility **Table V**

Designation or type	UNS no.	Composition, %* C	Mn	Si	Cr	Ni†	P	S	Others
403	S40300	0.15	1.00	0.50	11.5-13.0	–	0.04	0.03	–
410	S41000	0.15	1.00	1.00	11.5-13.0	–	0.04	0.03	–
414	S41400	0.15	1.00	1.00	11.5-13.5	1.25-2.50	0.04	0.03	–
416	S41600	0.15	1.25	1.00	12.0-14.0	–	0.04	0.03	0.6 Mo‡
416Se	S41623	0.15	1.25	1.00	12.0-14.0	–	0.06	0.06	0.15 min Se
420	S42000	0.15 min	1.00	1.00	12.0-14.0	–	0.04	0.03	–
420F	S42020	0.15 min	1.25	1.00	12.0-14.0	–	0.06	0.15 min	0.6 Mo‡
422	S42200	0.20-0.25	1.00	0.75	11.0-13.0	0.5-1.0	0.025	0.025	0.75-1.25 Mo; 0.75-1.25W; 0.15-0.3 V
431	S43100	0.20	1.00	1.00	15.0-17.0	1.25-2.50	0.04	0.03	–
440A	S44002	0.60-0.75	1.00	1.00	16.0-18.0	–	0.04	0.03	0.75 Mo
440B	S44003	0.75-0.95	1.00	1.00	16.0-18.0	–	0.04	0.03	0.75 Mo
440C	S44004	0.95-1.20	1.00	1.00	16.0-18.0	–	0.04	0.03	0.75 Mo
Type 410Cb (XM-30)	S41040	0.18	1.00	1.00	11.5-13.5	–	0.04	0.03	0.05-0.30 Cb
Type 410S	S41008	0.08	1.00	1.00	11.5-13.5	0.60	0.04	0.03	–
Type 414L	–	0.06	0.15	0.15	12.5-13.0	2.5-3.0	0.04	0.03	0.5 Mo; 0.03 Al
416 Plus X (XM-6)	S41610	0.15	1.00	1.00	12.0-14.0	–	0.06	0.15 min	0.6 Mo

*Single values are maximum unless otherwise indicated.
†For some tubemaking processes, the Ni content of certain grades must be slightly higher than shown.
‡Optional.

Precipitation-hardening stainless steels can be hardened to high strengths by treatment and aging **Table VI**

Designation or type	UNS no.	Composition, %* C	Mn	Si	Cr	Ni†	P	S	Others
PH 13-8 Mo	S13800	0.05	0.10	0.10	12.25-13.25	7.5-8.5	0.01	0.008	2.0-2.5 Mo; 0.90-1.35 Al; 0.01 N
15-5 PH	S15500	0.07	1.00	1.00	14.0-15.5	3.5-5.5	0.04	0.03	2.4-4.5 Cu; 0.15-0.45 Cb+Ta
17-4 PH	S17400	0.07	1.00	1.00	15.5-17.5	3.0-5.0	0.04	0.03	3.0-5.0 Cu; 0.15-0.45 Cb+Ta
17-7 PH	S17700	0.09	1.00	1.00	16.0-18.0	6.5-7.75	0.04	0.03	0.75-1.5 Al
AM-350 (Type 633)	S35000	0.07-0.11	0.5-1.25	0.50	16.0-17.0	4.0-5.0	0.04	0.03	2.5-3.25 Mo; 0.07-0.13 N
AM-355 (Type 634)	S35500	0.10-0.15	0.5-1.25	0.50	15.0-16.0	4.0-5.0	0.04	0.03	2.5-3.25 Mo
AM-363‡		0.04	0.15	0.05	11.0	4.0			0.25 Ti
Custom 450 (XM-25)	S45000	0.05	1.00	1.00	14.0-16.0	5.0-7.0	0.03	0.03	1.25-1.75 Cu; 0.5-1.0 Mo; 8 X %C min Cb
Custom 455 (XM-16)	S45500	0.05	0.50	0.50	11.0-12.5	7.5-9.5	0.04	0.03	0.5 Mo; 1.5-2.5 Cu; 0.8-1.4 Ti; 0.1-0.5 Cb
PH 15-7 Mo (Type 632)	S15700	0.09	1.00	1.00	14.0-16.0	6.5-7.75	0.04	0.03	2.0-3.0 Mo; 0.75-1.5 Al
Stainless W (Type 635)	S17600	0.08	1.00	1.00	16.0-17.5	6.0-7.5	0.04	0.03	0.4 Al; 0.4-1.2 Ti
17-10 P‡	–	0.07	0.75	0.5	17.0	10.5	0.28	–	–

*Single values are maximum unless otherwise noted.
†For some tubemaking processes, the Ni content of certain grades must be slightly higher than shown.
‡Nominal composition; limits not available.

A recent development in the duplex alloys is the addition of nitrogen. These stainless steels can be cold worked to strengths unobtainable by ferritic and most austenitic grades. These stainlesses are characterized by reduced alloy-element segregation between the ferrite and austenite phases; consequently, they exhibit better as-welded corrosion resistance than do the conventional duplex grades.

Martensitic

Martensitic stainless steels (Table V) have an austenitic structure at elevated temperatures that can be transformed into martensite (i.e., hardened) by suitable cooling to room temperature. They generally contain 11–18% chromium. The lower limit is governed by corrosion resistance, and the upper limit by the requirement for the alloy to convert fully to austenite on heating. Martensitic stainlesses are magnetic, resist corrosion in mild environments and have fairly good ductility. Some can be heat-treated to tensile strengths exceeding 200,000 psi (1,400 MPa).

Type 410 is a general-purpose martensitic, and the most widely produced grade. Martensitics are used for parts such as bolts, pump shafts, cutlery, valves and bearings.

Precipitation-hardening

Precipitation-hardening stainless steels (Table VI) are chromium-nickel types containing other alloying elements such as copper or aluminum. These other elements result in formation of precipitates during processing. They can be hardened to high strengths by solution treatment and aging. They are used for gears, fasteners, cutlery and aircraft parts.

Corrosion resistance

Stainless steels resist corrosion over a broad range of conditions, but they are not immune to every environment. For example, they perform poorly in reducing environments, such as with 50% sulfuric and hydrochloric acids at elevated temperatures. Another example is a highly corrosive environment that is abrasive, i.e., a slurry. Here, the particles would continually remove the protective film. Such forms of attack result in a breakdown of the protective film over the entire metal surface, and a corresponding general weight loss.

The above misapplications of stainless steels are relatively rare. The types of attack that are more likely to happen are localized—pitting, crevice corrosion, stress-corrosion-cracking and intergranular corrosion.

Pitting

Pitting occurs when the protective film breaks down in small isolated spots, such as when halide salts are in contact with the surface. Once started, attack may accelerate because of differences in electric potential between the large area of passive surface and the active pit.

In general, such attack is avoided by matching the corrosion resistance of the stainless steel to the severity of the environment. Better corrosion resistance results from greater amounts of chromium and molybdenum. Because of limited hot workability, few austenitic stainless steels contain more than about 20% chromium; thus, further improvement in pitting resistance is achieved by increased molybdenum levels. These can be as high as 6%, as in AL-6X and 254 SMO.

The ferritic stainless steels, on the other hand, can be worked easily, even at chromium levels as high as 30%. Because of the synergistic effect of chromium and molybdenum on corrosion resistance, the high-chromium ferritics often demonstrate better pitting resistance than do the austenitics, even though they contain less molybdenum. Since they do not need nickel to maintain an austenitic structure, the ferritics have a lower total alloy content than do comparable austenitics. In a sense, the ferritics are more efficient in their use of alloying elements.

Crevice corrosion

Crevice corrosion results from local differences in concentration of oxygen or other oxidizing agents associated with deposits on metal surfaces, gaskets, crevices, or under bolt or rivet heads where small amounts of liquid can collect and become stagnant.

The material responsible for the formation of a crevice need not be metallic. Wood, plastics, rubber, glass, concrete, asbestos, wax and microorganisms can cause crevice corrosion. Once attack begins within the crevice, it may progress rapidly.

The corrosion environment is made more severe by tighter crevices, higher temperatures and greater chloride levels. The best solution to this problem is to choose a design that eliminates the crevice. Because this is often impractical, stainless steels containing molybdenum are used to minimize the problem.

Stress-corrosion-cracking

This is caused by the combined effects of tensile stress and corrosion. Numerous alloys experience stress-corrosion-cracking (SCC)—brass in ammonia, carbon steel in nitrate solutions, titanium in methanol, and some aluminum alloys in seawater. Many stainless steels are susceptible to SCC in chloride environments.

Tensile stress, chlorides, and elevated temperature all must be present for chloride stress-corrosion-cracking of stainless steels to occur. A temperature of 54–60°C (130–140°F) is sometimes enough to produce SCC. Wet-dry (i.e., a splash zone) or heat-transfer conditions, which promote the concentration of chlorides, are particularly strong initiators of SCC.

While the mechanism of stress-corrosion-cracking is not fully understood, laboratory tests and service experience have given rise to methods to minimize the problem. In one severe test to evaluate resistance to chloride SCC, using boiling 42% $MgCl_2$, it has been demonstrated that maximum susceptibility to cracking occurs when the nickel content is 8%. Either increasing or reducing this content results in significant improvement in SCC resistance. The findings of this test can be generalized for stainlesses with chlorides.

Austenitic stainless steels such as 20Cb-3 (35Ni) have shown excellent SCC resistance in a variety of process environments. The duplex stainless steels, which typically contain 5 to 6% nickel, are significantly more resistant to SCC than is Type 304 (18Cr-8Ni).

The ferritics, particularly those that are nickel-free, are practically immune to SCC. Even the nickel-containing ferritics such as Sea-Cure and AL 29-4-2, which will crack in the severe boiling-$MgCl_2$ test, do not crack in boiling 25% NaCl or in the wick test.

The wick test was designed to simulate contaminated insulation on steam pipes (see Fig. 2). In the test, a boiling chloride solution is brought to a piece of stressed metal by the wicking action of fiberglass wool.

Intergranular corrosion

For austenitic stainless steels, it is heating or cooling through about 430–900°C (800–1,650°F), that causes the chromium along grain boundaries to combine with

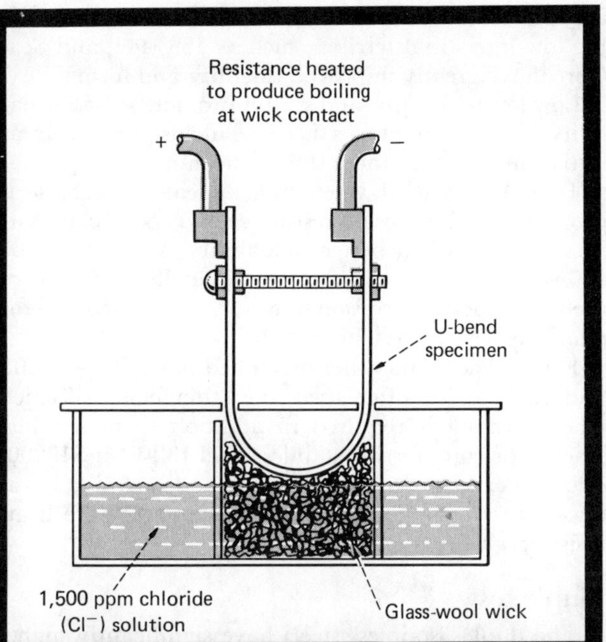

The wick test is used to simulate contaminated insulation on steam pipes **Fig. 2**

carbon to form chromium carbides. This, of course, lowers corrosion resistance in areas adjacent to the boundaries. This is time-temperature dependent.

Sensitization may result from slow cooling due to annealing, stress-relieving within the sensitization range, or welding. Due to the longer times at annealing or stress-relieving temperatures, it is possible that the entire piece of material will become sensitized, whereas the shorter times at welding temperatures can result in sensitization of a narrow band adjacent to, but slightly removed from, the weld.

Intergranular corrosion depends upon the amount of sensitization and the aggressiveness of the environment. Many environments do not cause intergranular corrosion in sensitized austenitic stainless steel. For example, glacial acetic acid at room temperature does not, nor does fresh clean water; strong nitric acids do.

Carbide precipitation and subsequent corrosion in austenitics have been thoroughly investigated; the causes are understood and methods of prevention have been devised. These include:

1. Use of stainless steel in the annealed condition. (Although stainless steels can become sensitized by annealing, proper control of the cooling rate and temperature will eliminate this problem.)

2. Selection of the low-carbon (0.030% maximum) stainless steels for weld fabrication. Low-carbon grades commonly used are Types 304L, 316L and 317L (L indicates low carbon). The less carbon available to combine with the chromium, the less likely will carbide precipitation occur. However, even the low-carbon grades may become sensitized at extremely long exposures to temperatures in the sensitization range.

3. Choice of a stabilized grade, such as Type 321 (Ti-stabilized), Type 347 (Cb-stabilized), Type 316Ti or Type 316Cb, for service at 430–900°C (800–1,650°F). Protection obtained with these grades is due to the greater affinity of titanium and columbium for carbon as compared with chromium.

Columbium stabilization is sometimes preferred because the carbides are more readily retained in the welds, and columbium is easier to control in the steelmaking process. However, the use of columbium-stabilized steel requires additional care in welding.

4. Redissolving of carbides by annealing, after fabrication; however, this is not always practical.

The above steps are necessary only if the service environment is likely to cause intergranular corrosion.

Although sensitization can also occur in the ferritics, the problem is eliminated by keeping carbon and nitrogen at low levels, as in the low-interstitial ferritics that have been developed in the last several years. For example, 444, Sea-Cure and E-Brite are really L-grades, even though L does not appear in their designation. In addition, these stainless steels typically contain stabilizing elements such as titanium and columbium.

Other forms of corrosion

Other corrosion phenomena that should be considered when using stainless steels are:

- *Corrosion fatigue,* which is encountered in cyclic loading in a corrosive environment.
- *Hydrogen embrittlement,* which is caused by hydrogen impregnation of an alloy during processing, and can lead to brittle failure when the alloy is subsequently loaded.
- *Hydrogen-assisted stress-cracking,* which can result when components are stressed in a hydrogen or hydrogen sulfide environment.

The austenitic stainless steels resist the effects of hydrogen, but the ferritic, martensitic and precipitation-hardening types may be susceptible.

Sulfide ions, and selenium, phosphorus and arsenic compounds increase the propensity for hydrogen to enter hardenable stainless steels and cause hydrogen stress-cracking. Their presence should warn of a possible failure. Cathodic protection can also cause hydrogen stress-cracking of high-strength alloys, if the material is over-protected. Therefore, in corrosive environments, use cathodic protection or coupling of stainless steels to less-noble materials with caution.

Physical and mechanical properties

There are relatively few applications for stainless steels in which physical properties are the determining factors in selection. However, there are many uses where consideration of the physical properties is important in product design.

For example, stainless steels are selected for numerous elevated-temperature applications, often along with steels of lower alloy content. Because many stainlesses have higher coefficients of thermal expansion and lower thermal conductivities than do carbon steels, these properties must be considered in designing such products as heat exchangers.

Some of the physical properties of the austenitic stainless steels are similar to those of the martensitic and ferritic steels. For example, for Type 304 the modulus of elasticity is 193 GPa (28×10^6 psi) and density is 8,060 kg/in.^{3} (0.29 lb/in.^{3}). The modulus for the martensitics is 200 GPa (29×10^6 psi), which is slightly lower than that of carbon steel, 207 GPa (30×10^6 psi).

The densities of the martensitics are a little lower (7,780 kg/m^3; 0.28 lb/in.^{3}) than those of carbon and alloy steels. On the other hand, the thermal conductivity of the ferritics is higher than that of the austenitics. The coefficient of thermal expansion of the ferritics is very similar to that of carbon steel, and typically is 30 to 40% less than that of the austenitic stainless steels.

In general, physical properties may vary slightly with product form and size, but these variations are usually not critical.

Typical physical properties of some common stainless steels are given in Table VII.

Mechanical properties

Mechanical properties of most stainless steels, especially ductility and toughness, are superior to those of carbon steels. Strength and hardness can be raised by cold working of ferritic and austenitic types, and by heat treating of precipitation-hardening and martensitic types.

Austenitics

The austenitic stainless steels cannot be hardened by heat treatment but can be strengthened by cold working. At room temperature, austenitics exhibit yield strengths between 207 and 1,380 MPa (30 and 200×10^3 psi), depending on composition and amount of cold working. They also show good ductility and toughness even at high strengths, and these properties are retained down to cryogenic temperatures.

Because austenitic stainless steels can be cold-worked to high tensile and yield strengths, yet retain good ductility and toughness, they meet a wide range of design criteria.

In structural applications, the toughness and fatigue strength of these steels are important. At room temperature, annealed austenitics can have impact strengths exceeding 135 J (100 ft-lb). Type 301 stainless has good toughness, even after cold rolling to high tensile strengths.

Ferritics

Ferritic stainless steels have a relatively large grain size after welding, which can have a detrimental effect on mechanical properties. This and a relatively low toughness at low temperatures have limited their use in structural applications. Generally, toughness in the annealed condition decreases as the chromium content increases. As a result, ferritics are used for nonstructural applications that require good resistance to corrosion, and bright, highly polished finishes.

In comparison to low-carbon steels such as SAE 1010, the conventional ferritic stainless steels, such as Types 430, 434 and 446, exhibit somewhat higher yield and tensile strengths and low elongations; thus they are not as formable as the low-carbon steels. On the other hand, the low-interstitial ferritics, such as 439, 444, and Sea-Cure, have greatly improved ductility and formability.

Low levels of sulfur and silicon promote good formability of the ferritic types, because inclusions can act as initiation sites for cracks during forming.

Type 405 is used where the annealed mechanical properties and corrosion resistance of Type 410 are satisfactory but when better weldability is desired. Type 430 is used for formed products. Types 434 and 436 are used when better corrosion resistance is required and for relatively severe stretching.

For fasteners and other machined parts, Types 430F and 430F Se are often used, the latter being specified when forming is required in addition to machining. The proprietary ferritic stainless steel 182-FM (S18200) offers an excellent combination of good corrosion resistance and machinability. Types 443 and 446 are heat-resisting grades.

Duplexes

The duplex stainless steels have significantly higher yield strengths than do the austenitic grades.

Martensitics

In the hardened condition, the martensitic stainless steels have very high strength and hardness, but to obtain optimum ductility and impact strength, these are given a stress-relieving or tempering treatment, usually in the range of 150 to 370°C (300 to 700°F). As-quenched martensitics usually show better corrosion resistance than do tempered steels.

Most martensitic stainless steels fall into two groups in regard to mechanical properties: low-carbon compositions with a maximum hardness of about HRC 45 (Rockwell), and the higher-carbon compositions, which can be hardened to about HRC 60. The maximum hardness of both groups in the annealed condition is about HRC 24. The dividing line between the two groups is a carbon content of approximately 0.15%.

For a low carbon content, the steel must also have a low chromium content or it will not harden. At higher carbon levels, the chromium content can be raised to about 18%. However, because the higher-carbon martensitic stainless steels are often used for cutlery and other applications that require high hardness, they are not usually tempered to the same degree as the low-carbon ones.

In the low-carbon class are Types 403, 410 and 416. The properties, performance, heat treatment, and fabrication of these stainless steels are similar—except for the better machinability of Type 416. On the high-carbon side are Types 440A, B and C.

Types 420, 414 and 431, however, do not fit into either category. Type 420 has a minimum carbon content of 0.15% and is usually produced to a carbon specification of 0.2–0.4%. While it will not harden to such values as the 440 types, it can be tempered without substantial loss in corrosion resistance so that a combination of hardness and adequate ductility results.

Types 414 and 431 contain 1.25–2.50% nickel, which is enough to increase hardenability, but not enough to make them austenitic at ambient temperature. The

Typical physical properties of common wrought stainless steels in the annealed condition **Table VII**

Designation or type	UNS no.	Density		Elastic modulus		Mean coefficient of thermal expansion*		Thermal conductivity*		Specific heat*	
		Mg/m^3	lb/in.^{3}	GPa	10^6 psi	μm/(m)(°C)	μin./(in.)(°F)	W/(m)(K)	Btu/(h)(ft)(°F)	J/(kg)(K)	Btu/(lb)(°F)
201	S20100	7.8	0.28	197	28.6	15.7	8.7	16.2	9.4	500	0.12
202	S20200	7.8	0.28	—	—	17.5	9.7	16.2	9.4	500	0.12
205	S20500	7.8	0.28	197	28.6	—	—	—	—	500	0.12
301	S30100	8.0	0.29	193	28.0	17.0	9.4	16.2	9.4	500	0.12
302	S30200	8.0	0.29	193	28.0	17.2	9.6	16.2	9.4	500	0.12
302B	S30215	8.0	0.29	193	28.0	16.2	9.0	15.9	9.2	500	0.12
303	S30300	8.0	0.29	193	28.0	17.2	9.6	16.2	9.4	500	0.12
304	S30400	8.0	0.29	193	28.0	17.2	9.6	16.2	9.4	500	0.12
304L	S30403	8.0	0.29	—	—	—	—	—	—	—	—
S30430	S30430	8.0	0.29	193	28.0	17.2	9.6	11.2	6.5	500	0.12
304N	S30451	8.0	0.29	196	28.5	—	—	—	—	500	0.12
305	S30500	8.0	0.29	193	28.0	17.2	9.6	16.2	9.4	500	0.12
308	S30800	8.0	0.29	193	28.0	17.2	9.6	15.2	8.8	500	0.12
309	S30900	8.0	0.29	200	29.0	15.0	8.3	15.6	9.0	500	0.12
310	S31000	8.0	0.29	200	29.0	15.9	8.8	14.2	8.2	500	0.12
314	S31400	7.8	0.28	200	29.0	—	—	17.5	10.1	500	0.12
316	S31600	8.0	0.29	193	28.0	15.9	8.8	16.2	9.4	500	0.12
316L	S31603	8.0	0.29	—	—	—	—	—	—	—	—
316N	S31651	8.0	0.29	196	28.5	—	—	—	—	500	0.12
317	S31700	8.0	0.29	193	28.0	15.9	8.8	16.2	9.4	500	0.12
317L	S31703	8.0	0.29	200	29.0	16.5	9.2	14.4	8.3	500	0.12
321	S32100	8.0	0.29	193	28.0	16.6	9.2	16.1	9.3	500	0.12
329	S32900	7.8	0.28	—	—	—	—	—	—	460	0.11
330	N08330	8.0	0.29	196	28.5	14.4	8.0	—	—	460	0.11
347	S34700	8.0	0.29	193	28.0	16.6	9.2	16.1	9.3	460	0.11
384	S38400	8.0	0.29	193	28.0	17.2	9.6	16.2	9.4	460	0.11
405	S40500	7.8	0.28	200	29.0	10.8	6.0	27.0	15.6	460	0.11
409	S40900	7.8	0.28	—	—	11.7	6.5	—	—	—	—
410	S41000	7.8	0.28	200	29.0	9.9	5.5	24.9	14.4	460	0.11
414	S41400	7.8	0.28	200	29.0	10.4	5.8	24.9	14.4	460	0.11
416	S41600	7.8	0.28	200	29.0	9.9	5.5	24.9	14.4	460	0.11
420	S42000	7.8	0.28	200	29.0	10.3	5.7	24.9	14.4	460	0.11
422	S42200	7.8	0.28	—	—	11.2	6.2	23.9	13.8	460	0.11
429	S42900	7.8	0.28	200	29.0	10.3	5.7	25.6	14.8	460	0.11
430	S43000	7.8	0.28	200	29.0	10.4	5.8	26.1	15.1	460	0.11
430F	S43020	7.8	0.28	200	29.0	10.4	5.8	26.1	15.1	460	0.11
431	S43100	7.8	0.28	200	29.0	10.2	5.7	20.2	11.7	460	0.11
434	S43400	7.8	0.28	200	29.0	10.4	5.8	—	—	460	0.11
436	S43600	7.8	0.28	200	29.0	9.3	5.2	23.9	13.8	460	0.11
440A	S44002	7.8	0.28	200	29.0	10.2	5.7	24.2	14.0	460	0.11
440C	S44004	7.8	0.28	200	29.0	10.2	5.7	24.2	14.0	460	0.11
444	S44400	7.8	0.28	200	29.0	10.0	5.6	26.8	15.5	420	0.10
446	S44600	7.5	0.27	200	29.0	10.4	5.8	20.9	12.1	500	0.12
PH13-8Mo	S13800	7.8	0.28	203	29.4	10.6	5.9	14.0	8.1	460	0.11
15-5 PH	S15500	7.8	0.28	196	28.5	10.8	6.0	17.8	10.3	420	0.10
17-4 PH	S17400	7.8	0.28	196	28.5	10.8	6.0	18.3	10.6	460	0.11
17-7 PH	S17700	7.8	0.28	204	29.5	11.0	6.0	16.4	9.5	460	0.11

*At 0 to 100°C (32 to 212°F).

addition of nickel serves two purposes: It improves corrosion resistance because it permits a higher chromium content, and it enhances notch toughness.

Martensitic stainless steels are subject to temper brittleness and should not be heat-treated or used in the range of 430 to 565°C (800 to 1,050°F) if toughness is important.

Martensitic grades have a ductile-brittle transition temperature at which notch ductility drops very suddenly. The transition is near room temperature; and at low temperature, about −185°C (−300°F), these steels become very brittle. The effect depends on composition, heat treatment and other variables.

Another important property is abrasion or wear resistance. Generally, the harder the material, the more its abrasion resistance. In applications where corrosion occurs, however, such as in coal-handling operations, this general rule may not hold, because the oxide film is continuously removed, resulting in a high corrosion rate.

Other mechanical properties of martensitic stainless steels, such as compressive yield-shear strength, are generally similar to those of carbon and alloy steels of the same strength level.

The martensitic stainless steels are generally selected for moderate resistance to corrosion, relatively high strength, and good fatigue properties after suitable heat treatment.

High-carbon martensitics are generally not recommended for welded applications, although Type 410 can be welded with relative ease. Hardening heat treatments should follow forming operations because of the poor forming qualities of the hardened steels.

Precipitation-hardening steels

The principle of precipitation-hardening is that a supercooled solid solution changes its metallurgical structure on aging. The main advantage to this is that products can be fabricated in the annealed condition and then strengthened by a relatively low-temperature, 480 to 620°C (900 to 1,150°F), treatment, minimizing the problems associated with high-temperature treatments, such as warping. Tensile-strength levels to 1,790 MPa (260×10^3 psi) can be achieved—exceeding even those of the martensitics, while corrosion resistance is usually superior. Ductility is similar to corresponding martensitic grades at the same strength level.

Precipitation-hardening stainless steels have high strength, relatively good ductility, and good corrosion resistance at high temperatures.

Fabrication

Generally, stainless steels are selected first for corrosion resistance and second for strength or other mechanical properties. A third consideration is fabrication. While the general-purpose stainless steels are most frequently used, variations of these are better suited to certain manufacturing operations.

Hot forming

Stainless steels are readily formed by hot operations such as rolling, extrusion and forging.

Hot rolling is used to produce standard mill forms and special shapes. Relatively small quantities of extruded shapes are both feasible and economical. Virtually any shape whose cross-section will fit into a 165-mm (6-½ in.) dia. circle can be extruded. Hollow shapes as well as solids are readily produced.

Forging is used extensively for stainless steels of all types, and in sizes from a few ounces to thousands of pounds. Special operations, such as drawing, piercing and coining further enhance forging capabilities.

Cold forming

The mechanical properties of stainless steels indicate how readily they can be formed at ambient temperature. Because of excellent mechanical properties, stainless steels have excellent cold-forming characteristics.

Sheet and strip

The bending characteristics of annealed austenitic stainless steels are excellent. Many types will withstand a free bend of 180 deg. with a radius equal to one-half the material's thickness, or less. As the hardness of the stainless increases, bending becomes more restrictive.

The 400 series also bends easily; however, it shows less elongation than austenitic grades, so the minimum bend radius is less. The new low-interstitial ferritic stainless steels are superior to the 400 series alloys in this regard.

The ferritics are not as ductile as the austenitic grades, and they have a lower work-hardening coefficient. Their formability is thus more like that of carbon steel in that they cannot be stretched without thinning and fracturing. Formability usually decreases with increasing chromium content. In addition, these grades can show brittle tendencies that become more pronounced with more chromium. To offset this factor, moderate warming of the higher-chromium types is often recommended prior to drawing.

In simple bending operations, there is little need to consider variations of the general-purpose alloys, since all stainless steels within a metallurgical group tend to behave in a similar manner.

Bar and wire

Many stainless-steel components are made from bar or wire. Cold heading is the most widely used forming method here. Many types of stainless steels are available as cold-heading wire. However, for multiple-blow operations in which a number of forming steps are performed in rapid sequence, a material with good ductility but with a low work-hardening rate is required.

Three stainless steels work well: Types 305, 384 and UNS S30430. These three are similar to Type 304 in terms of corrosion resistance and mechanical properties.

With a chromium-nickel ratio less than that of Type 304, Type 305 has less tendency to work-harden; accordingly, a greater amount of deformation is possible before annealing is necessary. In terms of corrosion resistance, Type 305 is interchangeable with Type 304. Type 305 resists attack by nitric acid, and is used with a wide range of organic and inorganic chemicals, foodstuffs and sterilizing solutions. It also has a good high-temperature scaling resistance, and is specified for continuous service to 870°C (1,600°F).

The Cr-Ni ratio in Type 384 is even lower than in Type 305. As a result, Type 384 work-hardens the least of the austenitic stainless steels, and like Type 305 remains nonmagnetic during cold working.

Type 384 is widely used for fasteners, cold-headed bolts, screws, upset nuts, and instrument parts, also for severe coining (stamping), extrusion, and swaging (rotary-forming). Because Type 384 is similar in alloy content to Type 304, it generally can be used anywhere that the latter is used.

Both Types 305 and 384 are subject to carbide precipitation if heated or cooled slowly in the range of 427–

900°C (800–1,650°F). This can be corrected, however, by annealing and water quenching from at least 1,040°C (1,900°F).

Type UNS S30430 is one of the most widely used cold-heading stainless steels and is often identified by a popular tradename, 304 HQ. Its composition is similar to that of Type 304 except that it contains 3–4% copper. This eliminates cracking, especially in recessed heads, and results in improved tool performance. It becomes mildly magnetic after severe cold working.

Machining

The machining characteristics of stainless steels are substantially different from those of carbon or alloy steels and other metals. In varying degree, most stainless steels without composition modification are tough, rather gummy, and tend to seize and gall.

While the 400 series stainless is the easiest to machine, a stringy chip produced during operations can slow productivity. The 200 and 300 series, on the other hand, are the most difficult to machine, primarily because they work-harden very rapidly.

Free-machining types

Some stainless steel compositions contain sulfur, selenium, lead, copper, aluminum or phosphorus—either separately or in combination—in sufficient quantity to improve machining characteristics. These elements reduce the friction between the workpiece and the tool, thereby minimizing the tendency of the chip to weld to the tool. Also, sulfur and selenium form inclusions that reduce the friction forces and transverse ductility of the chips, causing them to break off more readily.

Grades produced specifically for machining operations include 303, 303 Se, 430F, 416 Se and UNS S18200 (18Cr–2Mo). These grades are used for items such as shafts and screws.

The alloying elements used to improve the machining characteristics of stainless steels can adversely affect corrosion resistance, traverse ductility and weldability. Consequently, the free-machining grades should be used only after careful consideration; when used, however, they offer significant productivity advantages.

Joining

Stainless steels are commonly joined by welding, brazing and soldering. Arc welding is the overwhelming choice for joining these steels to each other because it produces a good, integral, relatively crevice-free joint. Resistance welding is also used often because it quickly and inexpensively produces a mechanically strong joint.

Procedures and precautions appropriate for the various types of stainless steels are important if optimum corrosion resistance and mechanical properties are to be attained in the completed assembly. Brazing usually is preferred for joining stainless steels to other metals.

Welding

Nearly all stainless steels can be welded by most industrial methods. Because of differences between stainless and carbon or low-alloy steels, precautions in welding practices must be observed.

First, procedures should be followed to preserve corrosion resistance in the weld and in the area immediately adjacent to it, known as the heat-affected zone. Second, optimum mechanical properties in the joint must be maintained, and, third, certain steps are needed to minimize problems of heat distortion.

In welding, it is necessary to select a weld rod or wire of a weld-filler metal that has corrosion-resistant properties nearly identical to—or better than—the base metal. This is not always as obvious to accomplish as some might expect.

For instance, Type 308 weld-rod is specified for welding Type 304, and a 300 series rod is often used for joining 400 series steels. The best suggestion is to follow AWS (American Welding Soc., Miami) practices for filler-metal selection and weld procedures, or to consult manufacturers of welding consumables.

Proper weld selection not only ensures preservation of corrosion-resistant properties, it also helps to achieve optimum mechanical properties.

Another principal difference between stainless and carbon or low-alloy steels is thermal conductivity, with stainless being about half as conductive as other steels. Hence, heat is not dissipated as rapidly.

There are four methods to minimize this limitation:

- Lower weld-current settings.
- Skip-weld techniques that will minimize heat concentration.
- Backup chill bars or other cooling techniques to dissipate heat.
- Proper joint design.

The first three methods are welding-shop procedures that are often adequately covered by AWS recommended practices and welding-shop standard practices. It is always good policy, however, for designers to double-check that the shop follows proper procedures. Also, it is desirable to provide specimen welds to establish quality standards.

These precautions are important because corrosion problems often begin in weld areas. For example, one problem, discussed earlier, is carbide precipitation. Also, improper heat dissipation can distort the finished product and require costly straightening operations.

From the designer's viewpoint, the configuration of the joint can encourage heat dissipation. For this reason, the use of beveled joints is common in thinner gages, which in carbon steel might be welded as a square-edge butt. Beveling permits the use of several light passes, thus avoiding the high temperature that would be reached in a single, heavy pass.

Cleaning the edges to be welded is needed to remove contamination from grease and oil. Such contamination can lead to carburization in the weld area, and subsequent reduction of corrosion resistance. Post-weld cleanup should not be done with carbon-steel files and brushes. Carbon-steel cleaning tools, as well as grinding wheels that are used on carbon steel, can leave fine particles imbedded in the stainless steel surface that will later rust, and may even lead to active corrosion if not removed by chemical cleaning.

Welding martensitics

Because of the limited ductility of most martensitic grades, preheating and post-weld heat-treating are used frequently to reduce the tendency to crack during welding. The filler metal for welds can be identical to the base metal or it can be an austenitic stainless steel.

Welding ferritics

Three major difficulties associated with welding the conventional ferritic stainless steels are: (1) excessive grain growth; (2) sensitization; and (3) lack of ductility. Heat treating after welding can minimize some of these problems. Alternatively, the low-interstitial grades can be used because they were developed to provide major improvements in as-welded ductility and toughness.

The filler metal can be of either a similar or austenitic composition (Types 308, 309, 316L or 310), which is helpful in improving ductility and toughness.

Welding austenitics

The 200 and 300 series are the most weldable of the stainless steels. If corrosion results from sensitization in the heat-affected zone, the problem can usually be eliminated by using the low-carbon or stabilized grades of these stainlesses.

Preheating is not required; post-heating is necessary only to redissolve precipitated carbides and to stress-relieve components that are to be used in SCC-causing environments.

The coefficient of expansion of austenitic types is higher than that of carbon steels; hence, thermal contraction is greater. Precautions are thus necessary to avoid weld-bead cracking and minimize distortion. These include using tack welding, skip welding, copper chill-bars, minimum heat input, and small weld passes.

Welding precipitation-hardening steels

The precipitation-hardening grades are suited to welding, with little need for pre- or post-heat-treatment, except to restore or improve mechanical properties.

Welding free-machining steels

Problems of porosity and segregation arise when free-machining types are welded. However, special welding consumables are available that, with careful exclusion of hydrogen from the weld, will assist the welding operation.

Soldering

Stainless steels are readily soldered with relatively few temperature-related problems. Aggressive fluxes, however, are necessary to first prepare the surface. Phosphoric-acid-type fluxes are preferred because they are not corrosive at room temperature.

Brazing

All stainless steels can be brazed, but because all brazing alloys are usually composed of copper, silver and zinc, rather high temperatures are required. This can lead to such high-temperature problems as carbide precipitation. For the nickel-free stainless steels, special braze compositions are required to prevent interfacial corrosion.

Fasteners

Although fasteners are available in many materials, stainless fasteners are a good first choice. They are easy to make, in both standard and special designs, and are readily available. Since corrosion resistance is an important aspect of product reliability, equipment is usually supplied with proper fastener materials. Generally, fasteners are made of metals or alloys that have equal or better corrosion resistance than do the materials they join.

This practice is justified because the fasteners may have to withstand higher loads with greater stress than do the parts being held together, and they are usually considerably smaller, too. Also, corrosion-weakened fasteners may lead to a more immediate failure with more serious consequences, compared with attack elsewhere on the assembly.

Corrosion protection for a fastened joint encompasses much more than a consideration of the corrosion resistance of the fastener itself. It includes an analysis of the entire assembled joint taken as a system. This means evaluating structural design, material stresses, product life-expectancy and environmental conditions.

The authors

James D. Redmond is manager, stainless steel development, for Climax Molybdenum Co., 3072 One Oliver Plaza, Pittsburgh, PA 15222. Tel: (412) 281-6238. He is responsible for stainless-steel market development. He is the liaison between his company and U.S. stainless makers. The author of numerous articles on stainlesses, he holds a B.S. degree in metallurgical engineering from the University of Notre Dame and a Ph.D. degree in materials science from Northwestern University. He belongs to several societies, including the Natl. Assn. of Corrosion Engineers, the American Soc. for Metals, and ASTM.

Kurt H. Miska is supervisor, technical information, for Climax Molybdenum Co. of Michigan, P.O. Box 1568, Ann Arbor, MI 48106. Tel: (313) 761-2300. Before joining the company in 1978, he was an associate editor of *Materials Engineering* for seven years. There he covered developments in metals, composites, adhesives and welding. He has written hundred of papers on a wide range of technical subjects.

Selecting stainless steel for pumps, valves and fittings

This guide is designed to assist the engineer in specifying the proper stainless steel for a particular service environment.

Robert S. Brown, *Carpenter Technology Corp.*

☐ Many pumps, valves and fittings find service in severely corrosive environments. Frequently, this equipment or portions of it is, or should be, made from stainless steel. In determining which alloy to use, the primary consideration is the anticipated service environment and its effect on materials. Once this is established, a specific stainless can be chosen that will provide the best combination of corrosion resistance and mechanical properties.

For an alloy to be considered a stainless steel, it should contain at least 11% chromium. Many of the stainless steels contain more than that, as well as other alloying elements such as nickel, molybdenum, copper and nitrogen. It is the chromium content that gives the alloy its resistance to corrosion, by combining with oxygen in the environment to form a chromium-rich oxide at the surface of the part. The chemical compositions of alloys identified in this article are given in the table.

Corrosive attack of stainless steels can be of a general or localized nature.

General corrosion is the uniform removal of metal that occurs as a result of the electrochemical/chemical reactions between the environment and the metal. Usually, resistance to general corrosion increases as the alloy content of the stainless steel increases.

Fig. 1 shows the relative resistance to general corrosion of many of the most widely used stainless steels. This type of corrosion is the least troublesome because it is predictable. It is easily combated through selection of a stainless steel with adequate resistance, and through proper corrosion allowance in design.

Localized corrosion is the corrosive attack that occurs in a specific area. It may cover a large portion of the part or an isolated spot. This type of corrosion, including pit/crevice corrosion, erosion corrosion, cavitation corrosion, intergranular corrosion and stress-corrosion cracking, cannot be compensated for through the use of thicker material.

Pit/crevice corrosion

Pitting and crevice corrosion occurs when stainless steels are exposed to halogen ions, particularly chloride. Resistance depends on the amount of molybdenum and chromium present in the material. Increasing these generally improves resistance, as will increased nitrogen content, which also improves the strength of the stainless. Fig. 2 illustrates the resistance of a number of stainless steels to this form of corrosion.

Pit/crevice corrosion usually occurs in areas where there is relatively little fluid flow; thus, designs that reduce the occurrence of crevices are very helpful.

When a unit is used for handling fluids that contain solids, component design should eliminate (as much as possible) stagnant or dead areas. This reduces the chance of a deposit or buildup of solids under which crevice corrosion could occur. A flowrate of 10 ft/s is considered the minimum for preventing buildup in most systems. Likewise, it is important to design for full fill; that is, to prevent vapor pockets where there is a liquid/vapor interface. If such an interface does exist, concentration of halogen salts can occur, increasing the likelihood of pitting.

Erosion corrosion

Erosion corrosion is a combination of general corrosion and mechanical abrasion in a localized area. Mechanical

Originally published March 9, 1981

Chemical composition of stainless steel alloys identified in article

UNS[2] No.	AISI[3] type	C	Cr	Mn	Mo	Ni	P	S	Si	Misc.
		Chemical composition, %[4]								
		High nickel								
N08020[1]		0.07 max.	19-21.0	2.0 max.	2-3.0	32-38.0	0.045 max.	0.035 max.	1.0 max.	**Cb**: 8x**C**-1.0 **Cu**: 3-4.0
		Nitrogen-strengthened								
S20910[1]		0.06 max.	20.5-23.5	4-6.0	1.5-3.0	0.2-0.4	0.04 max.	0.03 max.	1.0 max.	**Cb**: 0.1-0.3 **N**: 0.2-0.4 **V**: 0.1-0.3
S21904[1]		0.04 max.	18-21.0	8-10.0	–	5-7.0	0.06 max.	0.03 max.	1.0 max.	**N**: 0.15-0.4
S24100[1]		0.15 max.	16.5-19.5	11-14.0	–	0.5-2.5	0.06 max.	0.03 max.	1.0 max.	**N**:0.2-0.45
S28200[1]		0.15 max.	17-19.0	17-19.0	0.75-1.25	–	0.04 max.	0.04 max.	1.0 max.	**Cu**: 0.75-1.25 **N**: 0.4-0.6
		Austenitic								
S30200	302	0.15 max.	17-19.0	2.0 max.	–	8-10.0	0.045 max.	0.03 max.	1.0 max.	–
S30400	304	0.08 max.	18-20.0	2.0 max.	–	8-10.5	0.045 max.	0.03 max.	1.0 max.	–
S30403	304L	0.03 max.	18-20.0	2.0 max.	–	8-12.0	0.045 max.	0.03 max.	1.0 max.	–
S31600	316	0.08 max.	16-18.0	2.0 max.	2-3.0	10-14.0	0.045 max.	0.03 max.	1.0 max.	–
S31700	317	0.08 max.	18-20.0	2.0 max.	3-4.0	11-15.0	0.045 max.	0.03 max.	1.0 max.	–
S32100	321	0.08 max.	17-19.0	–	–	9-12.0	0.045 max.	0.03 max.	1.0 max.	**Ti**: 5x**C** min.
S34700	347	0.08 max.	17-19.0	2.0 max.	–	9-13.0	0.045 max.	0.03 max.	1.0 max.	**Co**: 10x**C** min.
		Austenitic-ferritic								
S32900	329	0.20 max.	23-28.0	1.0 max.	1-2.0	2.5-5.0	0.04 max.	0.03 max.	0.75 max.	–
		Ferritic								
S40500	405	0.08 max.	11.5-14.5	1.0 max.	–	–	0.04 max.	0.03 max.	1.0 max.	**Al**: 0.1-0.3
$43000	430	0.12 max.	16-18.0	1.0 max.	–	–	0.04 max.	0.03 max.	1.0 max.	–
		Martensitic								
S41000	410	0.15 max.	11.5-13.5	1.0 max.	–	–	0.04 max.	0.03 max.	1.0 max.	–
S42000	420	over 0.15	12-14.0	1.0 max.	–	–	0.04 max.	0.03 max.	1.0 max.	–
S43100	431	0.20 max.	15-17.0	1.0 max.	–	1.25-2.5	0.04 max.	0.03 max.	1.0 max.	–
S44004	440C	0.95-1.20	16-18.0	1.0 max.	0.75 max.	–	0.04 max.	0.03 max.	1.0 max.	–
		Precipitation-hardening								
S45000[1]		0.05 max.	14-16.0	1.0 max.	0.5-1.0	5-7.0	0.03 max.	0.03 max.	1.0 max.	**Cb**:8x**C** min. **Cu**: 1.25-1.75
S45500[1]		0.05 max.	11-12.5	0.5 max.	0.5 max.	7.5-9.5	0.04 max.	0.03 max.	0.50 max.	**Cb**: 0.1-0.5 **Cu**: 1.5-2.5 **Ti**: 0.8-1.40

[1]Proprietary alloy
[2]Unified Numbering System, adopted by Soc. of Automotive Engineers and Amer. Soc. for Testing and Materials
[3]Amer. Iron and Steel Institute
[4]Balance is Fe.

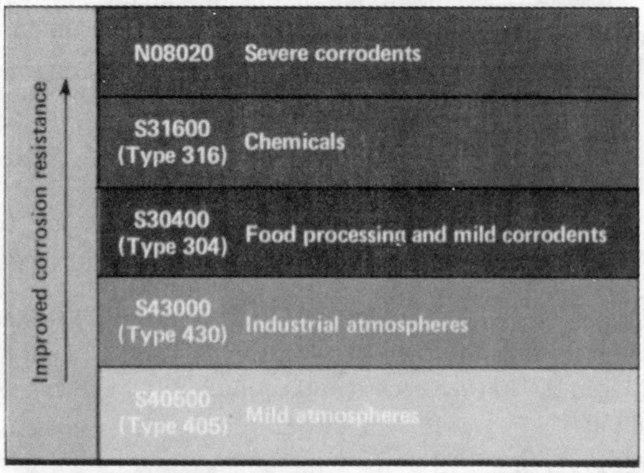

General corrosion resistance of widely used stainless steels **Fig. 1**

abrasion results from solid particles impinging on a part, or from turbulence in the process stream.

Three factors should be considered in anticipating or facing an erosion problem. These are: material of construction, physical design, and flowrate or velocity.

In materials that rely on a protective film, such as the chromium-rich oxide one on stainless steels, the general corrosion resistance of the alloy and the flowrate of the process stream are very important. Increased solid-solution (annealed) hardness is beneficial at a given level of corrosion resistance in combating erosion corrosion. But increasing hardness through heat treatment will not always prove helpful. As a rule of thumb, the greater the general corrosion resistance to the specific environment, and the harder the alloy, the better the alloy's resistance to erosion corrosion.

Fig. 3 illustrates the relative resistance to erosion

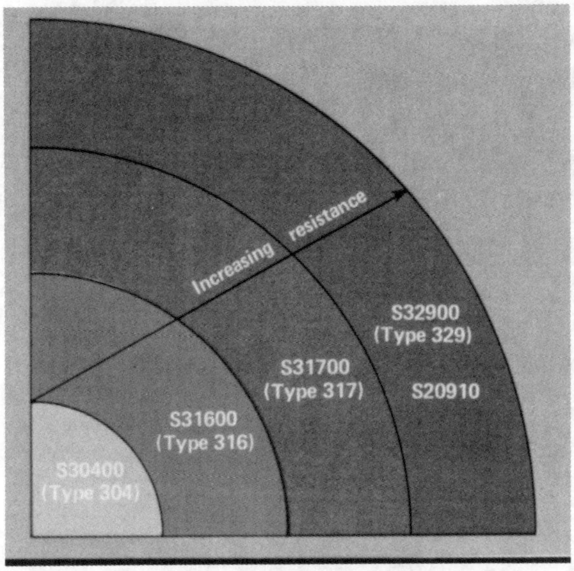

Increasing Mo and Cr content generally improves resistance to pit/crevice corrosion **Fig. 2**

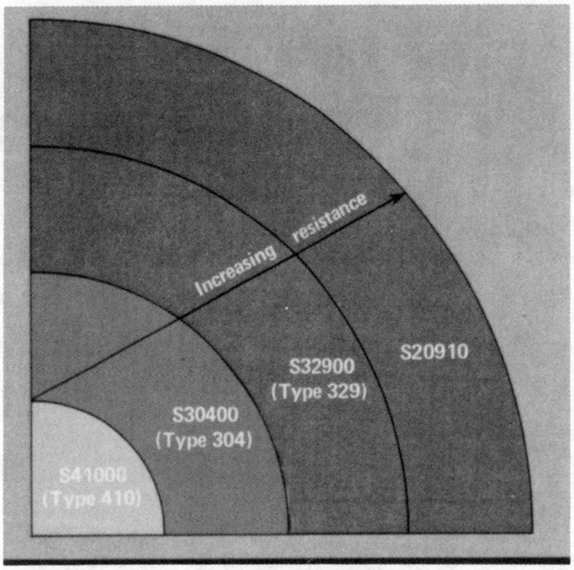

The harder the alloy, the better the resistance to cavitation/erosion corrosion **Fig. 3**

corrosion of a number of stainless alloys in the annealed condition under similar service conditions. Streamlining the flow pattern of the unit or system to prevent impingement or turbulence is as important in preventing erosion corrosion as is material selection.

Cavitation corrosion

A combination of mechanical and electromechanical/chemical material deterioration, cavitation corrosion occurs when equipment design or flow conditions allow formation of bubbles on the surface of a part, usually a rotating part. Once the bubble forms it will break with enough force to rupture the protective film of the stainless steel. Thus the unprotected material is instantaneously exposed to the corrosive environment and general corrosion can occur. Even though the film will be very rapidly regenerated, a finite amount of general corrosion will occur before that takes place. Some believe that very small amounts of metal can actually be removed by the explosive force of the bubbles rupturing.

Control of cavitation corrosion is achieved through design that prevents bubble formation, by selecting alloys with greater corrosion resistance and strength, and through polishing the surfaces of rotating parts to remove formation sites. Stainless steel selection is essentially the same as for erosion corrosion (see Fig. 3).

Intergranular corrosion

Intergranular corrosion is a localized type that occurs adjacent to the material's grain boundaries when a contin-

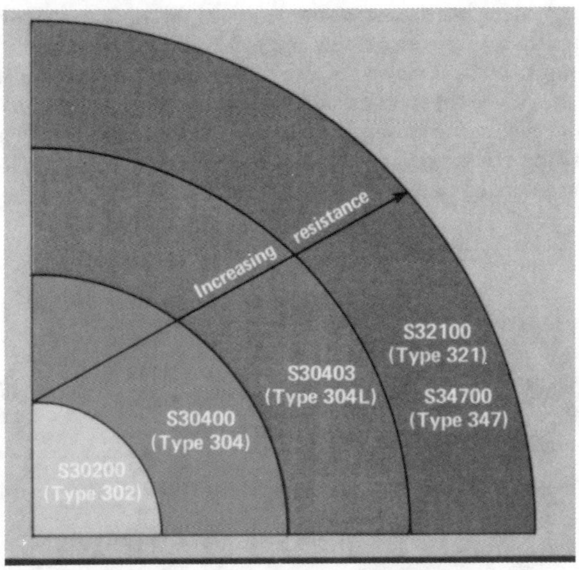

Type 304 stainless steel family is designed to combat intergranular corrosion **Fig. 4**

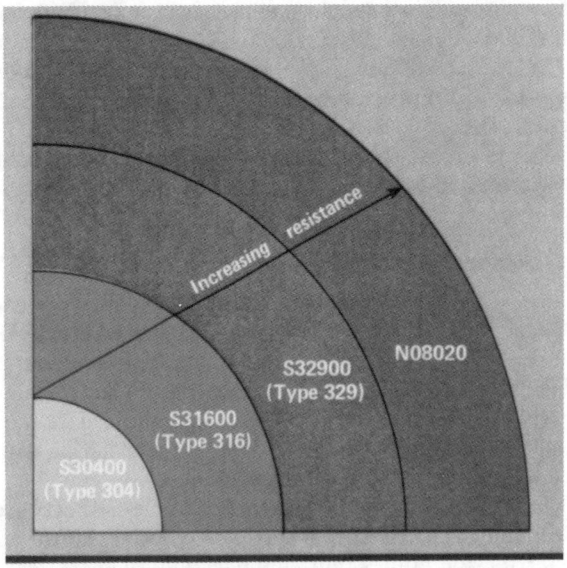

Increasing Ni content improves resistance to chloride stress-corrosion cracking **Fig. 5**

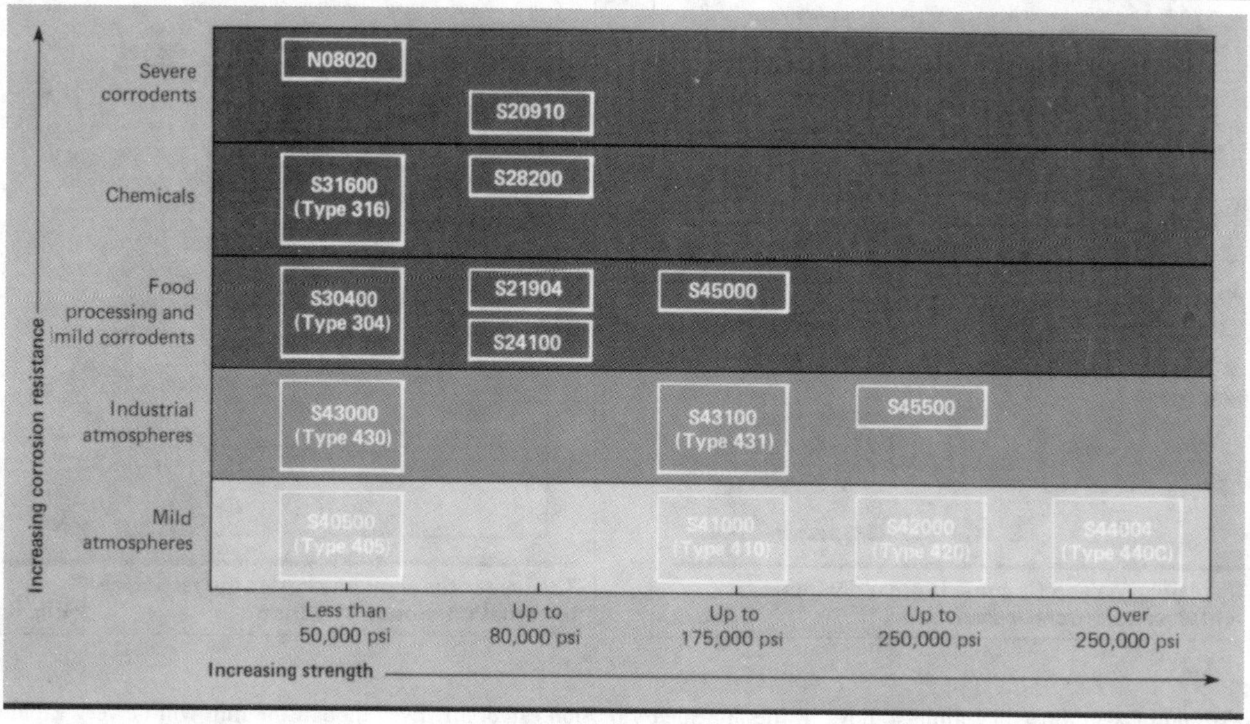

Though corrosion resistance is the prime factor, adequate mechanical properties are also essential **Fig. 6**

uous network of intergranular chromium carbide is present. The chromium carbide can be precipitated intergranularly when a stainless steel is subjected to a temperature range of 800-1,500°F. Such temperatures are encountered during welding, brazing and stress relieving. The formation of the intergranular chromium carbide results in a chromium-depleted zone adjacent to the grain boundary.

Because stainless steels depend to a large degree on chromium for corrosion resistance, the chromium-depleted zone has reduced corrosion resistance. Reducing carbon content or chemically combining the carbon with a stabilizing element, such as titanium or columbium, will reduce susceptibility to intergranular carbide precipitation and thus intergranular corrosion.

Fig. 4 shows the Type 304 stainless steel family designed to combat intergranular corrosion. Through proper selection of a stainless steel within a family, it is possible to expose the part to elevated temperature during fabrication without having it lose resistance to intergranular corrosion.

Stress-corrosion cracking

Stress-corrosion cracking is a particularly troublesome form of localized corrosion, usually unpredictable in nature. It occurs when a piece under tensile stress is subjected to a corrosive environment, most often containing chloride ions, at an elevated temperature. The part usually fails through brittle cracking of the material, with very little evidence of general corrosion.

Most of the common austenitic stainless alloys are susceptible to chloride stress-corrosion cracking. However, the ferritic and duplex (austenitic/ferritic) stainless steels, as well as a few of the more highly alloyed austenitic steels, do offer good resistance, as shown in Fig. 5.

When immunity to chloride stress-corrosion cracking is required, along with good overall protection against general corrosion, alloys with more than 45-47% nickel should be considered. However, commercially available alloys with nickel content over 30% are often deemed virtually immune.

Mechanical properties

Fig. 6 offers a good overview of the relationship between yield strength and corrosion resistance for many of the stainless steels.

While corrosion resistance is the most important factor to consider in selecting a stainless steel for a pump or valve application, it is extremely important that the mechanical properties be reviewed before a final material selection is made. Even the most corrosion-resistant material available will not perform satisfactorily if its strength, hardness, galling resistance or abrasive resistance is inadequate for the intended part.

The author

Robert S. Brown is a senior metallurgist, stainless alloy metallurgy, Carpenter Technology Corp., Reading, PA 19603. Tel: 315-371-2000. His responsibilities include overseeing studies on corrosion and welding of stainless steel, and alloy review and fabrication. Mr. Brown holds a B.S. in metallurgical engineering from Lafayette College, Pa. He is a member of Amer. Welding Soc., Amer. Soc. for Metals, Amer. Soc. for Testing and Materials, and Natl. Assn. of Corrosion Engineers. An NACE-accredited corrosion specialist, he is the author of several articles in the field.

How to avoid failures of stainless steels

Preventing stress-corrosion cracking, pitting and crevice corrosion is the key to success. Knowing when to use a stainless, and when not to, is crucial.

*C. M. **Schillmoller**, Schillmoller Associates,*
*and **Heinz J. Althoff**, Vereinigte Deutsche Metallwerke AG*

☐ For stainless steels, almost 60% of equipment failures in the chemical-process industries (CPI) are due to stress-corrosion cracking (SCC), or localized attack in the form of pitting or crevice corrosion.

Operating parameters and environmental factors greatly affect alloy performance, especially pH, temperature, and chloride and oxygen levels. Failures of stainlesses can be avoided by understanding their limits in light of the above variables.

This discussion will focus on stainless steels, but will include some other alloys used for chloride service—high-molybdenum nickel alloys. It is not possible to review all of the specialty stainlesses, chrome and Ni-Cr-Mo alloys. Coverage will be limited to several of the austenitics. Included are Types 304L, 316L and 317L. These alloys are easy to fabricate and offer proven performance. They are the standard alloy selections.

The new ferritic and duplex stainless steels will not be discussed. These were dealt with recently in this department by Redmond and Miska [*1*].

The table lists some of the high-technology stainless steels and other alloys used in CPI plants. The alloys listed, as well as some others, will be looked at in this article in terms of SCC, and pitting and crevice corrosion.

SCC: Environmental factors

Generally, for stainless steels to suffer SCC, chloride concentrations must be 30 ppm or higher. However, when chlorides are present in trace amounts, high enough concentrations may develop locally that will cause cracking. This happens, for example, via chloride enrichment on hot surfaces, such as at vapor-liquid interfaces under deposits and in crevices. Proper design can often eliminate places where chlorides can concentrate. Oxygen is necessary for SCC to occur, and is usually present in small amounts.

Most failures occur above 170°F (75°C), and 120°F (50°C) is generally considered the threshold below which SCC is not a problem.

Tensile stresses are necessary for the propagation of

Stainless steels and high-nickel alloys for use in the chemical process industries

Generic name	316L	317LM	904	904hMo	28	20	825	825 hMo	G	C-276	625
	Stainless steels							High-nickel alloys			
Tradenames and comparable products	Cronifer 1812 LC	Cronifer 1713 LCN	Cronifer 1925 LC, 904L, 2RK65, JS700	Cronifer 1925 hMo, AL-6X, 254SMo	Nicrofer 3127, Sanicro 28	Nicrofer 3620 Cb, Carpenter 20Cb-3	Nicrofer 4221, Incoloy 825	Nicrofer 4221 hMo	Nicrofer 4520 hMo, Hastelloy G	Nicrofer 5716 hMoW, Hastelloy C-276	Nicrofer 6020 hMo Inconel 625
Chemical composition (nominal analysis, %)											
C max.	0.03	0.03	0.02	0.02	0.02	0.03	0.025	0.025	0.015	0.015	0.10
Cr	17.5	18.0	20.5	20.5	27.0	20.0	21.0	21.0	22.0	16.0	21.5
Ni	13.5	13.0	25.0	25.0	31.0	37.5	42.0	42.0	45.0	57.0	61.0
Mo	2.75	4.5	4.7	6.0	3.5	2.5	3.0	6.0	6.4	16.0	9.0
Cu	–	–	1.5	1.5	1.0	3.5	2.5	2.5	2.0	–	–
N	–	0.14	–	0.12	–	–	–	–	–	–	–
Cb/Ta	–	–	–	–	–	0.3	–	–	2.0	–	3.65
Others	–	–	–	–	–	Cb	Al, Ti	Al, Ti	–	Ti	Al, Ti, Co
Wirksumme	26	32	36	40	38	28	31	41	43	N/A	51

Tradenames: Nicrofer and Cronifer—VDM Germany; Incoloy and Inconel—International Nickel Co.; Cartech—Carpenter Technology Corp.; Hastelloy—Cabot Corp.
904L is produced by Uddeholm; 254SMo by Avesta; 2RK65 and Sanrico 28 by Sandvik; AL-6X by Allegheny Ludlum; and JS700 by Jessop Steel.

SCC; residual stresses from welding and fabrication are often near the material's yield point. Stress-relieving of equipment can often do more harm than good because of resultant deformations. Also, there is the chance of sensitizing the stainless, making it prone to intergranular corrosion, unless it is low-carbon or stabilized.

It is difficult to control SCC by modifying the process environment; McIntyre [2] details the options available to prevent SCC. The safest method, actually, is to use materials that are not susceptible to this phenomenon.

Raising the nickel content of austenite is an effective way to prevent chloride SCC. Studies by Copson [3] of various Ni-containing alloys relate their susceptibility to SCC by testing SCC resistance in a boiling magnesium chloride solution. Copson found that alloys containing 42% Ni and over are fully resistant to chloride SCC. This includes Alloy 825, Alloy G and Alloy 625. The zone with 8 – 12% Ni is the most susceptible; this zone includes Types 304 and 316.

The boiling $MgCl_2$ test is very severe and seldom relates to actual operating conditions. Thus, one can say that as a practical guide, alloys with a Ni content of 22% and over seldom experience SCC. This includes Alloy 904, Alloy 28 and Alloy 20. These materials have performed well in many industrial applications containing chlorides.

Pitting and crevice corrosion

Because of their unpredictability, pitting and crevice corrosion are major problems affecting the use of stainless steels in chloride environments. The severity of attack is exponentially related to the chloride content and the acidity, as well as to the pH, and to the presence of oxygen or oxidizers. High chloride levels locally break down the passive layer of a stainless steel, as well as provide a good conductor that promotes corrosion.

Crevice corrosion is more severe than pitting because it involves an oxygen-concentration cell. Good equipment design—without rivets, lapjoints and sharp corners—will reduce the accumulation of solids, and thus lower the risk of crevice corrosion. For heat-exchanger tubing, a minimum velocity of 5 ft/s is recommended to stop solids deposition. Also, regular cleaning will reduce susceptibility to crevice corrosion, as well as to pitting.

Resistance of alloys to pitting and crevice corrosion is enhanced by addition of chromium and molybdenum. Alloys are generally ranked as to pit resistance by the critical pit potential or critical pit temperature.

Similarly, a critical crevice-corrosion temperature can be measured; tests are carried out in a ferrric chloride solution. Schillmoller and Klein [4] reported that an empirical formula is used in Germany to rank alloys as to their pitting and crevice-corrosion resistance. This is the Wirksumme or pitting index, which equals %Cr + 3.3 %Mo. The higher this index, the more resistant the alloy (see Fig. 1).

This relationship is thought to be meaningful for alloys with 16 – 24% Cr and 1 – 7% Mo. In Fig. 1, the alloys in the table are plotted as to pitting resistance. For seawater with chlorides of 20,000 ppm and a pH of 7.8, alloys with an index of 32 or more are resistant to pitting. Those of 36 and above resist crevice corrosion at ambient temperatures.

Performance ranking of alloys

As mentioned, the effects of pH, chloride level, oxygen/oxidizer content and temperature must be considered in selecting alloys for a particular service. Fig. 2, from Ref. [5], is such an attempt for acid-brine systems. The figure was developed to predict the usefulness of various alloys in flue-gas scrubbers. However, it may be used more generally, whenever actual conditions match those of the figure. In Fig. 2, oxidizing conditions prevail, and temperatures are normally 120 – 150°F.

The alloys are listed on a cost/performance basis. The diagonal lines indicate zones in which an alloy is effective

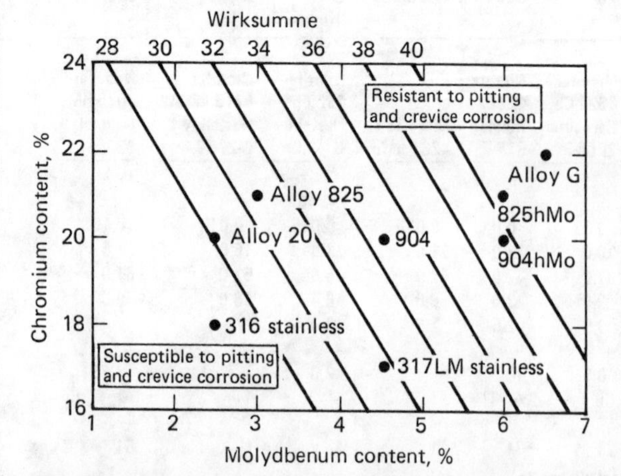

Wirksumme number measures resistance to pitting and crevice corrosion **Fig. 1**

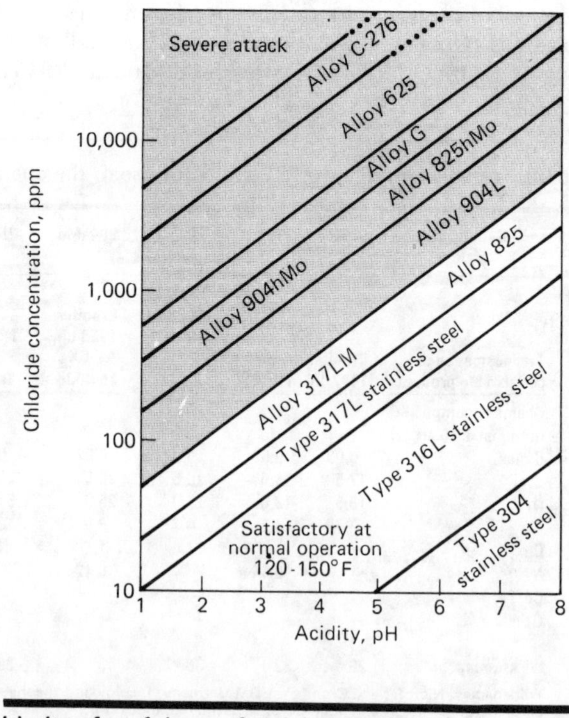

Limits of usefulness of stainless steels and other alloys for acid-brine systems **Fig. 2**

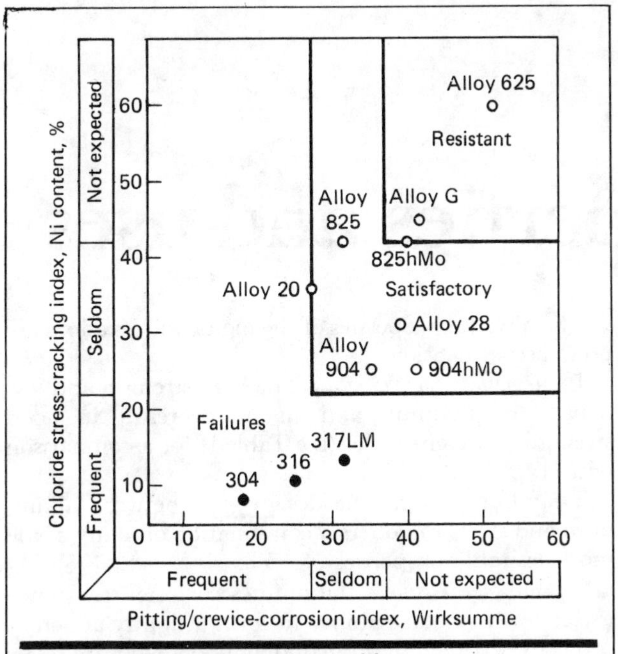

Relative resistance to stress cracking and crevice corrosion in hot brines **Fig. 3**

and economical. For example, for a pH of 5.5 and at 10,000 ppm chlorides, any alloy above the zone for Alloy G may also be specified (i.e., Alloy 625 and Alloy C276). Though these will be more expensive, they will provide more than adequate protection. Any alloy below the zone of Alloy G will be unsatisfactory. The dotted lines indicate ranges where there are no good data.

The unshaded portion of Fig. 2 is for the more common stainlesses. Thus, it provides a quick guide to determine whether these alloys can be used, or whether a higher-performing one is needed.

Because scrubbers recirculate the liquor, chloride levels can become very high. By controlling the pH between 5.5 and 6.0, good performance can be obtained from Alloy 904L, Alloy G and Alloy 625. Ranking of alloys is in regard to Wirksumme number.

However, much research needs to be done before one can accurately define the useful limits of stainless steel in chloride media. Fig. 2 is a conservative guide for the CPI—it shows frequently-encountered alloys.

Note that Types 316 and 317 often have molybdenum contents that are at the low end of the permissible range, say, 2.1% Mo for Type 316. In many applications, however, 2.5 – 2.75% would be needed for pitting resistance. This is especially so when effluent streams containing chlorides are recycled.

Care must be taken in applying general charts such as Fig. 2. Corrosion is a complex process, and determination of environmental factors is difficult. Seemingly unimportant variables, such as small amounts of moisture, metal chlorides or impurities, can change the corrosion picture completely.

Although a generalized chart for alloy selection has been shown, the authors are basically opposed to such simplification if this alone is used for selecting materials. Nonetheless, condensed information is valuable. It presents a bird's-eye view of the situation and can be used to screen out certain alloys. Further, because the chart ranks alloys, the number of materials to be tested can be minimized.

Overall, the Ni-Cr-Mo alloys have consistently offered reliable performance, since they work well in oxidizing and reducing media, and cope with high chloride levels, as well as with oxidizing salts. Such salts break down the protective films on stainless steels. Also, the Ni-Cr-Mo alloys tolerate deviations in design, and resist SCC, and pitting and crevice corrosion.

Fig. 3 ranks these alloys, along with some common stainlesses, relative to both SCC and pitting/crevice corrosion. The ordinate reflects Ni content, which controls SCC resistance; the abscissa is a measure of Wirksumme number—resistance to pitting and crevice corrosion.

As previously mentioned, for each alloy and set of conditions, it often is possible to designate an approximate pitting index, above which no attack is anticipated. "Marginal" alloy selections may have a long initiation time before passivity breakdown occurs. Periodic cleaning and good design may allow such alloys to perform satisfactorily. Alternatively, use of cathodic protection, or adjustment of the reduction-oxidizing (REDOX) potential will enhance performance of marginal alloys.

References

1. Redmond, J. D., and Miska, K. H., High-Performance Stainless Steels for High-Chloride Service, *Chem. Eng.*, July 25, 1983, p. 93.
2. McIntyre, D. R., How to Prevent Stress-Corrosion Cracking in Stainless Steels, *Chem. Eng.*, Apr. 7, 1980, p. 107, and May 5, 1980, p. 131.
3. Copson, H. R., Effect of Nickel Content on the Resistance to Stress-Corrosion Cracking of Iron-Nickel-Chromium Alloys in Chloride Environments, paper presented at First Intl. Congress on Metallic Corrosion, London, Apr. 10 – 15, 1961.
4. Schillmoller, C. M., and Klein, H., Selecting and Using Some High-Technology Stainless Steels, *Metal Progr.*, Feb. 1981.
5. Schillmoller, C. M., and Koehlert, G., Aspects of Alloy Selection for Flue-Gas Desulfurizer Scrubber Systems, paper presented at ACHEMA Conf., Frankfurt, June 9, 1982. Later isssued as VDM Technologies Corp. Report No. 6, March 1983.

The authors

C. M. "Dick" Schillmoller is consulting materials engineer (marketing and technical) for Schillmoller Associates, Chancellors Atrium II, Suite 300, 10103 Fondren, Houston, TX 77076. Tel: (713) 270-6438. He spent nineteen years with International Nickel Co., in the U.S., Australia and Europe, and six with VDM Technologies Corp. as manager, marketing & technical for the U.S. The author of over 200 papers on corrosion and high-temperature problems, he holds a B.S. in chemical engineering from the University of Sydney. He belongs to AIChE, the Natl. Assn. of Corrosion Engineers, and other societies.

Heinz J. Althoff is communications manager of Vereinigte Deutsche Metallwerke AG, Nickel Technology Div., Worthstrasse 171, P.O. Box 100167, D-4100 Duisburg, West Germany. He previously held senior management positions in R&D, quality assurance, application engineering and technical marketing, all in the German nonferrous-metals-manufacturing industry. He received his diploma and doctorate in metallurgical engineering from the Technical University in Clausthal, West Germany. He holds several international patents, and is the author of numerous technical papers.

Titanium: Its properties and uses

When to use titanium and its alloys? How expensive is it? Answers to these and other questions appear here.

Stan R. Seagle and Bruce P. Bannon, RMI Co.

☐ Titanium is gaining acceptance in a wide variety of chemical-process-industries (CPI) applications because of its high strength, light weight, good heat-transfer and corrosion resistance.

Such properties often make the metal an economical solution to engineering problems. Titanium processing equipment often lowers operating costs by minimizing maintenance, limiting downtime and maximizing lifespan.

Properties of titanium

The metal is allotropic and has two crystalline forms—alpha (a hexagonal, close-packed structure) and beta (a body-centered cubic structure). This is analogous to steel's austenite-to-ferrite transformation, and allows for forming alloys with a wide range of strengths, as shown in Table I.

Titanium's properties often make it a logical, cost-effective choice for CPI applications.

The physical properties of the metal and its alloys are now discussed briefly.

1. *Strength-to-weight-ratios.* The high strength and low density of titanium and its alloys result in good strength-to-weight ratios (see Table II for a comparison with Type 403).

Titanium "bridges the design gap" between aluminum and steel, and offers a combination of many of the most desirable properties of each.

2. *Corrosion resistance.* Titanium's corrosion resistance is due to the metal's ability to form a tightly adherent protective-oxide film immediately upon exposure to air or most other oxidizing media. The metal is not readily attacked, even by such aggressive corrosive agents as ferric or cupric chloride.

Welded fabrications of commercially pure titanium offer an additional advantage in corrosive service. Because commercially pure titanium is a single-phase material and not responsive to thermal cycles, the heat-affected zone of the weld, the weld deposit and the parent metal all have similar corrosion resistance.

Titanium demonstrates excellent resistance to attack from chlorine and its derivatives in oxidizing or neutral aqueous solutions even at elevated temperatures. The metal also resists oxidizing acids, nonoxidizing acids in low concentrations or at higher concentrations when oxidizing agents or heavy-metal ions are present, or-

Titanium alloys are available in a wide range of strengths and ASTM Grades — Table I

Nominal composition	Alloy type	Typical minimum yield strength (ksi)	Typical use	ASTM Grade
Ti (high-purity)	Alpha	25	Corrosion resistance	1
Ti (plus oxygen)	"	40	Corrosion resistance	2
Ti (plus oxygen)	"	55	Corrosion resistance	3
Ti (plus oxygen and iron)	"	70	Corrosion resistance	4
Ti-6Al-4V	Alpha-beta	120	Shafts, blades, disks, etc.	5
Ti-0.15Pd	Alpha	40	Corrosion resistance in reducing media	7
Ti-0.3Mo-0.8Ni	Alpha	50	Corrosion resistance in reducing media	12
Ti-5Al-2.5Sn	Alpha	115	For uses requiring weldability and strength at elevated temperatures	–

Grades 1, 2, 3 and 4 apply to commercially pure titanium and differ in the amounts of oxygen available. This has an effect on the metal's strength. The higher the grade, the greater the amount of oxygen and the greater the metal's strength. Grades 5, 7 and 12 are Ti alloys.
ksi = 1,000 lb/in.^{2}

Originally published March 8, 1982

Titanium and its alloys are strong and light, with outstanding strength-to-weight ratios **Table II**

Material	Yield strength, ksi	Density, lb/in.^{3}	Strength-to-weight, 10^{-5}in.
Ti-6A1-4V	125*	0.160	7.81
Ti-8A1-1Mo-1V	130†	0.158	8.23
AMS 5616	100	0.284	3.52
AMS 6304	125	0.280	4.46
403 SS annealed	35	0.280	1.25

*Annealed
†Heat-treated
ksi = 1,000 lb/in.^{2}
AMS = aircraft material specification (for commercially pure Ti)

ganic compounds, and alkaline solutions at moderate temperatures and concentrations.

Addition of small amounts (up to 0.2%) of palladium to titanium (ASTM Grade 7) materially improves corrosion resistance to reducing media and high-temperature chloride salts. This extends the range of titanium's application to hydrochloric, phosphoric and sulfuric acid solutions, and areas of service where the operating conditions vary between oxidizing and mildly reducing. The presence of palladium has no other significant effect on the physical or mechanical properties of titanium.

ASTM Grade 12, containing 0.8% nickel and 0.3% molybdenum, is a lower-cost alternative to Grade 7. Grade 12 offers better performance than commercially pure titanium in reducing media, and provides greater strength at room and elevated temperatures.

3. *Erosion resistance.* Titanium offers superior resistance to erosion, cavitation and impingement attack, making it ideal for pumps, piping and high-velocity heat exchangers. The wear properties of titanium—particularly in electrolytic or galvanic conditions—are uniform, minimizing localized corrosion or pitting, even in high-velocity situations.

4. *Elevated-temperature performance.* Most titanium alloys are satisfactory for continuous service to 800°F, and some for continuous service to near 1,000°F, extending the range for light metals to well above the familiar 250°F limit.

5. *Other properties.* Titanium also has low elastic and shear moduli, low thermal expansion, high resistivity, short radioactive half-life, and good heat transfer; and is nonmagnetic, nontoxic and biologically compatible. Heat-transfer coefficients are shown in the figure.

Titanium is often economical for equipment that should be highly corrosion-resistant, strong and heat-conducting to 600°F.

The metal is effective in corrosive environments involving chemicals such as chlorides, oxidizing acids and organic compounds. However, it is not recommended for hydrofluoric acids or dry chlorine (less than 0.5% water).

Uses of titanium

Briefly, some CPI uses of the metal include:

1. *Equipment.* The material is used in the construction of pulp and paper diffusion bleach-washers and piping systems, electrodes for electrowinning of metals, chlorine and sodium chlorate dimensionally stable anodes, and condensers for liquefaction of natural gas.

2. *Heat exchangers.* Because of its favorable heat transfer characteristics, its corrosion and saltwater resistance and strong resistance to high temperature, titanium is used in the construction of heat exchangers and condensers for power generation.

Although the thermal conductivity of titanium is low compared to other metals, it has excellent heat transfer, as shown in the figure.

The major resistance to heat transfer is associated with surface films, and not the thermal conductivity of the metal. The favorable film characteristics and lack of corrosion result in good heat transfer.

3. *Sour-gas wells.* High-strength titanium alloys possess the strength and corrosion resistance required at elevated temperatures needed for this service. Several titanium alloys have performed well in laboratory and field test programs.

4. *Powder metallurgy.* Titanium powder offers several cost-saving advantages—greater utilization of metal, less waste and lower fabrication costs. Elemental titanium powder is particularly desirable, since the powders are less expensive than prealloyed powder.

5. *Nuclear-waste containment.* Titanium alloy ASTM Grade 12 is proving to be the most viable material for the storage of high-level solidified nuclear wastes. Five years of testing have narrowed a field of nearly 20 alloys to titanium Grade 12 for use on the outer container, generally referred to as the overpack. The overpack will be used for storage of solidified defense-generated wastes scheduled to be stored in salt mines in the western U.S., beginning in the mid-1980s.

6. *Other uses.* Titanium is used for scrubbers and other pollution-control equipment in coal-fired applications. The metal also is being considered for downhole production-drilling applications because of its corrosion-resistant properties.

Manufacture and availability

Titanium is the fourth most plentiful structural metallic element in the earth's crust, exceeded only by alu-

Comparing costs of titanium and its alloys with other metals, including stainless steel **Table III**

	Relative price for plate*
Hastelloy C276	2.9
Inconel 625	2.2
Zirconium	2.0
Ti, ASTM Grade 7	1.8
Nickel 200	1.4
Inconel 600	1.3
Ti, ASTM Grade 12	1.2
Monel	1
Incoloy 800	1
Ti, ASTM Grade 2	1
317L SS	0.7
316L SS	0.4

*Based on equal volume (density adjusted).

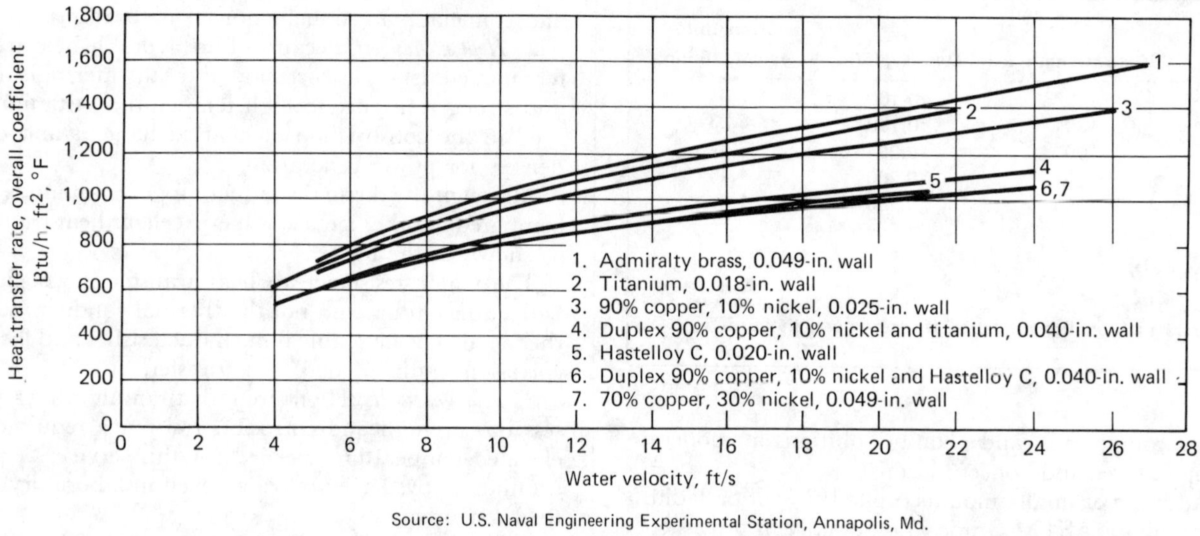

Heat-transfer rate of titanium versus other materials. Titanium offers high heat transfer

minum, iron and magnesium. It has two principal ores—rutile and ilmenite—and 90% of the ore is used in the production of the white, opaque, nontoxic titanium dioxide pigment used in paints and ceramics and in the coating of some grades of paper. Only 10% of the ore is used in the production of titanium sponge, which is then converted into mill products.

Ores of titanium are found throughout the world. As a result, long-range availability should not be subject to the geopolitical pressures common to many strategic nonferrous metals. Because of the large demand of the titanium oxide pigment industry, sufficient supplies of titanium tetrachloride, the intermediate product required to produce titanium sponge, have been and are expected to be available.

Although the price of commerically pure titanium ranges from \$10 to \$12 a pound, its density is about half that of steel and nickel-based alloys. Therefore, its cost, based on volume and adjusted accordingly, would be \$5 to \$6 a pound—the same price or lower than nickel-based alloys (\$6 to \$20 per pound) and competitive with highly alloyed stainless steels (see Table III for relative prices of plate).

Machining and welding

Titanium can be economically machined if shop procedures are set up to allow for the physical characteristics of the metal. Machinability of commercially pure grades is similar to that of 18–8 stainless steel. The alloy grades are somewhat harder to machine than is the pure metal.

Different grades of titanium will not all have identical machining characteristics, any more than all steels or all aluminum grades will. Also, as with stainless steel, the low thermal conductivity of titanium inhibits dissipation of heat within the workpiece itself, thus requiring coolants.

In recent years, welding has come to be the most widely used method of joining metals. Titanium welded assemblies withstand service conditions with high tensile and compressive loads and fatigue-causing stresses.

Commercially pure titanium and many of its alloys can be welded with good results. Satisfactory methods for welding include electron-beam, inert-gas-shielded arc-welding, and spot, seam and flash welding. These methods include provision of shielding the molten weld metal and adjacent heated zones from active gases such as oxygen and nitrogen.

The authors

Stan R. Seagle is vice-president, research and technical development, for RMI Co., Niles, OH 44446. He has had over 25 years experience in researching and developing titanium. He holds bachelor's and master's degrees in metallurgical engineering from Purdue University. The author of numerous papers and the holder of six U.S. patents, he is a member of the American Inst. of Mining Engineers, ASM, the Techl. Assn. of the Pulp and Paper Industry, and the Natl. Assn. of Corrosion Engineers.

Bruce P. Bannon is manager, industrial sales development, for RMI Co. He is responsible for coordinating developmental activities for titanium in non-aerospace markets. He is a *summa cum laude* graduate in geological sciences from Pennsylvania State University. Before joining RMI in 1980, he held various sales positions in the titanium industry.

Zirconium:

A corrosion-resistant material for industrial applications

Thiokol Chemical Corp.

Zirconium has been used in the chemical process industries as a structural material for over twenty-five years. It has excellent corrosion resistance, and is cost-competitive with other specialty metals.

Donald R. Knittel, *Teledyne Wah Chang Albany*

☐ Zirconium is immune to corrosion in all natural waters, even ones flowing at high velocity. It has excellent oxidation resistance up to 400°C against air, steam, carbon dioxide, sulfur dioxide, nitrogen, and oxygen, and is also very resistant to strong basic and acid solutions [*1*,*2*]. This excellent corrosion resistance comes from a tenacious, inert oxide film that forms on the metal's surface when it is exposed to air. Crevice corrosion in zirconium has not been reported.

Let us now look at the corrosion resistance of zirconium to some common corrosive media.

Sulfuric acid

Zirconium is considered one of the most economical construction materials for handling sulfuric acid in the difficult concentration range of 40-60 weight %. Fig. 1 reveals the excellent corrosion resistance—less than 5 mils per year (mpy)—of zirconium in boiling sulfuric acid at concentrations up to 70%. Welds are slightly less resistant than the parent metal.

In the viscose rayon process used at Avtex Fibers, Inc., zirconium replaces impervious-graphite heat exchangers and lead heating coils in a process stream containing sulfuric acid, zinc sulfate, sodium sulfate, and small amounts of hydrogen sulfide and carbon disulfide. Conversion to zirconium units has reduced energy and maintenance costs, and has improved equipment availability. Zirconium has been used in this process for about ten years; the only change in appearance has been a slight darkening of the welds [*3*].

Zirconium is employed in a Rohm and Haas Co. process that contains 50-60 weight % sulfuric acid at 160°C maximum [*4*]. And it has been successfully used by FMC Corporation for many years in hot sulfuric acid at concentrations up to 80% in the manufacture of hydrogen peroxide [*5*]. Even though zirconium normally dissolves in 80% sulfuric acid, in this case it has survived by forming an impervious black film on the metal surface.

Other applications of zirconium in sulfuric acid include acid pickling of steels, alcohol stripping towers, and process equipment for recovery and recycling of spent acid. In another process, zirconium blades wipe an abrasive sulfuric acid slurry off a plastic film. Nooter Corp. produced a zirconium stripper column 7 ft dia. x 100 ft for sulfuric acid service. See also Refs. 5, 6.

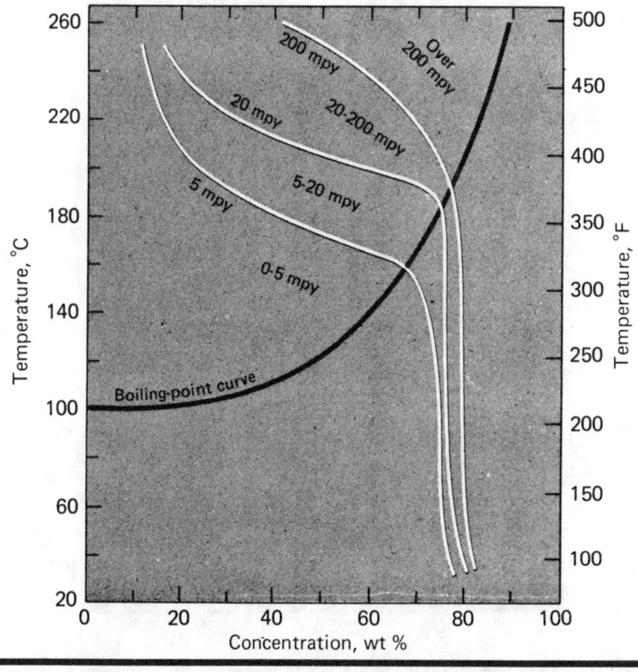

Corrosion resistance of zirconium in sulfuric acid **Fig. 1**

Hydrochloric acid and other inorganic acids

Zirconium is one of the few metals that is corrosion-resistant in hydrochloric acid at all concentrations, as shown in Fig. 2. Corrosion rates, at concentrations to 12N and below the boiling curve, are less than 0.1 mpy.

Zirconium will pit in the presence of oxidizing impurities in chloride solutions, such as ferric and cupric ions. Zirconium heating units, agitators, and various other pieces of hardware have been used for 15 years in the azo-dye coupling reactions that cycle between hot hydrochloric or sulfuric acid, and alkaline solutions. Thiokol Chemical Corp. is using zirconium pump seals and heat exchangers in a hydrochloric acid environment for the production of polymers [7].

Corrosion rates of less than 1 mpy are observed for zirconium exposed in *nitric acid* at all concentrations and temperatures up to 250°C. The metal has been used as a construction material handling nitric acid in nuclear fuel reprocessing at Oak Ridge National Laboratories.

It also shows excellent resistance in *phosphoric acid* at temperatures to boiling, and concentrations below about 50%. Even at higher concentrations and lower temperatures, zirconium demonstrates minimal corrosion. In phosphoric acid a minor amount of fluoride ions in the acid environment can cause severe corrosion.

Zirconium's corrosion resistance to *hydrobromic acid* compares with its resistance to hydrochloric acid. Tests have yielded corrosion rates of less than 5 mpy in 48% hydrobromic acid at the boiling point.

Alkaline solutions

Zirconium completely resists attack from basic solutions and fused alkalis [1]. Its corrosion properties in sodium hydroxide environments are equivalent to those of nickel.

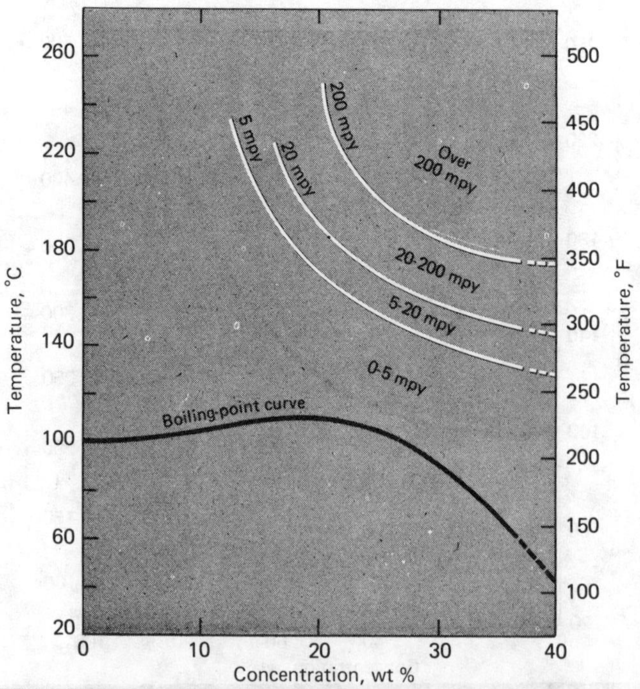

Corrosion resistance of zirconium in hydrochloric acid Fig. 2

Corrosion rates for zirconium in salt solutions [8] Table I

Environment	Concentrations, weight %	Temperature, °C	Corrosion rate, mpy
$AlCl_3$	5, 10, 25	35-100	<0.2
NH_4Cl	1, 10, Sat.	35-100	<0.2
$BaCl_2$	5, 20	35-100	<0.07
$CaCl_2$	5, 10, 25	35-100	<0.03
$CuCl_2$	5, 10, 20	35-100	>50
$FeCl_3$	5, 10, 20	35-100	>50
$MgCl_2$	5, 20, 42	35-Boiling	<0.1
$MnCl_2$	5, 20	35-100	<0.09
$HgCl_2$	1, 5, 10, Sat.	35-100	<0.07
$NiCl_2$	5, 20	35-100	<0.1
KCl	Sat.	60	0.009
KF	0.5 Molar	Boiling	<0.2
NaCl	3, Sat.	35-boiling	<0.03
$SnCl_4$	5, 24	35-100	<0.1
$ZnCl_2$	5, 20	35-boiling	<0.08

Corrosion rates for zirconium in organic solutions [10] Table II

Environment	Concentration, weight %	Temperature, °C	Corrosion rate, mpy
Acetic acid	5-99.5	35-boiling	<0.07
Acetic anhydride	99.5	boiling	0.03
Aniline hydrochloride	5, 20	35-100	<0.01
Chloroacetic acid	100	boiling	<0.01
Citric acid	10-50	35-100	<0.2
Dichloroacetic acid	100	boiling	<20
Formic acid	10-90	35-boiling	<0.2
Lactic acid	10-85	35-boiling	<0.1
Oxalic acid	0.5-25	35-100	<0.5
Tartaric acid	10-50	35-100	<0.05
Tannic acid	25	35-100	<0.1
Trichloroacetic acid	100	boiling	>50
Urea reactor	58% Urea 17% NH_3 15% CO_2 10% H_2O	193	<0.1

Cost of some alloys compared to zirconium Table III

Alloy	Relative price per unit weight	Relative price per unit volume
Zirconium, Grade 702	1.0	1.0
Titanium, Grade 2	0.7	0.5
Hastelloy Alloy B-2	1.1	1.4
Hastelloy Alloy C-276	1.0	1.2
Tantalum	20	50

Saffran Engineering Co.

With additions of impurities—chlorine gas or ammonia, for example—zirconium is more resistant than nickel. In other alkaline environments, namely potassium hydroxide, calcium hydroxide, and ammonium hydroxide, zirconium is also very resistant to attack. Failure in fused caustics is by oxidation rather than chemical corrosion at higher temperatures. One of zirconium's most useful applications is in environments that alternate between strongly acidic and strongly caustic conditions.

Salt solutions

Zirconium has excellent resistance in most salt solutions, as shown in Table I. Of the corrosive chloride salts, only ferric chloride and cupric chloride attack it. It has better corrosion-resistant properties than titanium in zinc chloride solutions and is being considered in that environment by Du Pont for the production of zinc chloride and ammonium chloride [*9*].

Organic solutions

Zirconium resists chemical attack in most organic environments. Corrosion data for some of these are listed in Table II. Corrosive attack is not normally caused by the organic material but by the other compounds present, such as impurities or catalysts; however, zirconium will resist many of these.

In the urea-synthesis reaction, an entire process, designed by T. O. Wentworth of Chemical Process of Ohio Inc., centers around the use of zirconium as the reactor lining. Most commercial urea processes are based on recycling a corrosive carbamate solution. With zirconium-lined vessels, the reactor can operate at higher temperatures, thus permitting high conversion rates of CO_2 to urea—80-85%, compared with 65%-70% for the carbamate recycle method [*11*].

General remarks

As a construction material, zirconium has been considered expensive. However, it is now very competitively priced. Table III gives an approximate comparison in cost, for both weight and volume, between zirconium and some other alloys.

Fabrication properties are very similar to those of commercially pure titanium [*12*]. Zirconium can be formed, machined and welded, using standard machinery and well-established techniques. It is available in plate, foil, bar and tube. Castings and clad plate are also available by special order.

The picture on p. 194 shows a zirconium sieve-tray distillation column, and the one on this page is a photo of a zirconium heating coil.

References

1. Knittel, D. R., and Webster, R. T., Corrosion Resistance of Zirconium and Zirconium Alloys in Inorganic Acids and Alkalies, presented at the Symp. on Industrial Applications of Zirconium and Titanium, ASTM B-10 Meeting, Oct. 15-18, 1979, New Orleans.
2. Schemel, J. H., ASTM Manual on Zirconium and Hafnium, ASTM STP 639, 1977.
3. Bowen, L. B., Use of Zirconium Heat Exchangers in the Viscose Rayon Process, presented at the Symp. on Industrial Applications of Zirconium and Titanium, ASTM B-10 Meeting, Oct. 15-18, 1979, New Orleans.
4. Frechem, B., Morrison, B., and Webster, R. T., Improving the Corrosion Resistance of Zirconium Weldments, presented at the Symp. on Industrial Applications of Zirconium and Titanium, ASTM B-10 Meeting, Oct. 15-18, 1979, New Orleans.
5. Webster, R. T., The Use of Zirconium in Plastic Sheet, Film and Fiber Production Processes, presented at the Natl. Assn. of Corrosion Engineers, Corrosion/79, Mar. 12-16, 1979, Atlanta.
6. Ashbaugh, W. G., Zr Distillation Trays Boost Production, Save $125,000 In Expansion Cost, *Chem. Process.*, Mar. 1966, p. 68.
7. Armstrong, G. W., Zirconium Reduces Polymer Plate Down Time, *Chem. Process.*, Dec. 1967, p. 52.
8. Golden, L. B., Lane, I. R., Jr., and Acherman, W. L., Corrosion Resistance of Titanium, Zirconium and Stainless Steel, *Ind. Eng. Chem.*, Aug. 1952, p. 1930.
9. Smallwood, R. E., The Corrosion of Reactive Metals in Zinc Chloride Solutions, presented at the Symp. on Industrial Applications of Zirconium and Titanium, ASTM B-10 meeting, Oct. 15-18, 1979, New Orleans.
10. Lane, I. R., Jr., Golden, L. B., and Acherman, W. L., Corrosion Resistance of Titanium, Zirconium and Stainless Steel in Organic Compounds, *Ind. Eng. Chem.*, May 1953, p. 1967.
11. McDowell, D. W., Jr., Corrosion in Urea-Synthesis Reactors, *Chem. Eng.*, May 13, 1974, p. 118.
12. Newman, J., Fighting Corrosion With Titanium Castings, *Chem. Eng.*, June 4, 1979, p. 149.

The author

Donald R. Knittel is a corrosion scientist at Teledyne Wah Chang Albany, 1600 N.E. Old Salem Road, Albany, OR 97321, telephone 503-926-4211. He received a B.A. and an M.S. in mathematics, and a Ph.D. in Chemistry, from Arizona State University. Dr. Knittel did postdoctoral research at Arizona State University and SRI International. He is a member of the Electrochemical Soc., the Natl. Assn. of Corrosion Engineers, and the American Soc. for Testing and Materials.

When to use refractory metals and alloys in the plant

For hot and/or highly corrosive service, these materials are often the best choice to resist attack. The author presents guidelines for selecting them.

Fred S. Shuker, KBI Div., Cabot Corp.

☐ In choosing materials of construction, many equipment designers have overlooked the refractory metals and alloys. This may stem from unfamiliarity with these materials or from their reputation for high initial cost.

However, for the most corrosive environments, as well as for high temperatures, nothing surpasses the refractory metals (see Table I). When the total cost is considered over the equipment's lifetime, including process downtime as well as product contamination, a refractory metal or alloy may well prove optimum.

What are refractory metals?

Refractory metals are usually considered those that have extremely high boiling points—above 4,000°C. They are the fifth- and sixth-period elements in groups IVb through VIb: zirconium, columbium (niobium), molybdenum, hafnium, tantalum and tungsten.

From the standpoint of corrosion resistance, the corresponding fourth-period elements—titanium, vanadium and chromium—often are considered refractory metals. Although their chemical properties are similar to those of the fifth- and sixth-period elements, their boiling points and other temperature-related properties are lower.

Corrosion resistance of the refractory metals is second only to that of the noble metals. Unlike the noble metals, however, the refractory metals are inherently reactive.

Such reactivity is a decided plus for corrosion resistance. On contact with air or any other oxidant, refractory metals immediately form an extremely dense, adherent oxide film. This passivating layer prevents access of the oxidant to the underlying metal and renders it resistant to further attack.

Unfortunately, these oxides spall or volatilize at elevated temperatures, leaving the metals susceptible to oxidation at around 300 to 500°C. For high-temperature applications under nonreducing conditions, the refractory metals must be protected by an applied coating, such as a metal silicide.

When considering use of refractory metals, one must look at their fabricability by welding, drawing, bending, etc. All these metals must be protected from the air during welding. However, this is not a major limitation, since the required techniques are well established.

More significant is the low room-temperature ductility of tungsten and chromium. These metals can be almost like glass in their tendency to shatter on impact. Also, they are difficult to draw or bend, except at elevated temperatures.

Tantalum and columbium, on the other hand, approach gold in ductility, as well as corrosion resistance. This makes fabrication extremely easy, though it can be a drawback when strength is required.

Molybdenum can be either brittle or ductile, depending on purity, temperature and the amount of cold work employed.

Tantalum

Tantalum is clearly the top of the line for corrosion resistance. The only media that can affect it are fluorine, hydrofluoric acid, sulfur trioxide (including fuming sulfuric acid), concentrated strong alkalis, and certain molten salts.

The corrosion resistance of tantalum can be compared to that of glass, although tantalum withstands higher temperatures and offers the intrinsic fabrication advantages of a metal. Tantalum equipment is frequently used in conjunction with glass, glass-lined steel and other nonmetallic materials of construction. Tantalum is also used extensively to repair damage and flaws in glass-lined steel equipment.

Because of its high cost and lack of strength on one hand, and its easy fabricability on the other, tantalum is usually used as a lining over a stronger, less-expensive base material. For protecting the inside or outside surfaces of pipes or tubes, tantalum in sheet form is usually applied without metallurgical bond.

Tantalum can be clad in thin sections to substrate metals by using various cladding techniques that produce a metallurgical bond. This, together with specialized fabrication methods, has made it possible to produce very large chemical-processing vessels.

The widespread use of tantalum in heat exchangers for corrosive media is based on the relatively high heat-transfer coefficients it provides because of its:

- Relatively high thermal conductivity, which is several times that of stainless steel or common corrosion-resistant nickel-based alloys. While copper and alumi-

Originally published May 2, 1983

How refractory metals stack up against some widely used high-corrosion alloys **Table I**

	Tantalum	Columbium	Zirconium	Titanium	Hastelloy Alloy C-276 (nickel-chromium-molybdenum)	Hastelloy Alloy B-2 (nickel-molybdenum)	Monel Alloy No. 400 (nickel-copper)	Carpenter Alloy 20 (high-nickel stainless steel)	Type 316 stainless steel
Acetic acid, 50%, boiling	S	S	S	S	S	S	SV	S	S
Aluminum chloride, 5%	S	S	—	S	S	S	SV	S	X
Ammonium chloride, 50%, boiling	S	S	S	—	S	—	SV	S	S*
Ammonium sulfate, saturated, boiling	S	—	—	—	V	S	SV	S	S*
Bromine, dry	S	—	—	—	S	X	S	S	X
Bromine water	S	S	—	—	S	X	X	SV	X
Caustic soda	X	X	—	—	S	S	S	S	S
Chlorine gas, dry, 25°C	S	S	V	X	S	S	S	S	X
Chlorine gas, moist, 25°C	S	S	X	S	S	X	—	X	X
Chlorine gas, moist, 100°C	S	S	X	S	X	X	X	X	X
Chlorosulfonic acid, 10%	S	—	—	—	S	S	SV	X	X
Chromium-plating bath	S	—	—	—	—	—	X	SV	SV
Ferric chloride, 5%, agitated	S	S	X	S	S	X	—	X	X
Flue gases	S	—	—	—	S	V	—	SV	SV
Fluorine	X	X	—	—	—	—	S	VX	X
Hydrochloric acid, 38°C (all concentrations)	S	V	S	X	S	S	—	X	X
Hydrogen peroxide, 25°C	S	V	S	S	S	V	SV	S	S†
Lead, molten	S	—	—	—	S	S	X	—	X
Mine water, acid	S	—	—	—	S	S	X	S	S*
Nitric acid, 5%, 25°C	S	S	S	S	S	X	X	S	S
Nitric acid, conc., boiling	S	S	S	S	X	X	—	S	V
Potassium cyanide	S	—	—	—	S	S	SV	S	S
Sodium chloride, sat., boiling	S	—	—	S	S	S	S	S	S*
Sodium hypochlorite, 5%, 25°C	—	X	V	V	S	X	X	V	S*
Sulfur dioxide, moist, 25°C	S	S	—	S	S	X	X	S	S
Sulfuric acid, 5%, 25°C	S	S	S	V	S	S	SV	S	V
Sulfuric acid, 50%, boiling	S	X	S	X	X	S	X	VX	X

*Subject to pitting at air line or when allowed to dry
†Attack may occur when sulfuric acid is also present

S: Fully resistant
V: Some attack
X: Unsatisfactory
—: No data available

num provide still better conductivity, their corrosion resistance is limited.

■ Essentially zero corrosion rate in most media. Thus, thinner sections can be used, since the thickness need not include a corrosion allowance.

■ Almost total resistance to biofouling, which in other materials creates a thermally insulating layer.

Due to the higher transfer coefficients of tantalum, smaller heat exchangers can be designed. The weight of metal required is correspondingly less. This may well offset tantalum's high per-pound cost and render it cost-effective even on an initial-cost basis.

Other refractory metals

Columbium can be a less expensive alternative to tantalum. However, its corrosion resistance is more limited, since it is sensitive to most alkalis and certain strong oxidants.

Nevertheless, columbium remains totally resistant to such highly corrosive media as wet or dry chlorine, bromine, saturated brines, ferric chloride, hydrogen sulfide, sulfur dioxide, nitric and chromic acids, and, within specific temperature and concentration limits, sulfuric and hydrochloric acids.

Even though the mechanical strength of columbium is less than that of tantalum, it can be used economically where the extreme inertness of tantalum is not required. As with tantalum, cladding processes are often used to apply columbium linings over less-expensive base materials.

Use of **tungsten** in chemical process equipment is limited due to its lack of ductility and the consequent diffi-

KBI-40 fills the corrosion-resistance gap between tantalum and columbium — **Table II**

Solution	Salt	Temperature, °C	Ta	KBI-40	Cb
5% HCl	10% NaCl	101*	nil†	nil†	1†
10% HCl	-	102*	nil	nil	< 1
10% HCl	0.005% $FeCl_3$	190	nil	-	-
10% HCl	10% NaCl	102*	nil	nil	7
15% HCl	10% NaCl	109*	nil	nil	16
15% HCl	0.005-0.040% $FeCl_3$	109*	nil	nil	-
20% HCl	-	110*	nil	nil	-
20% HCl	0.005-0.040% $FeCl_3$	110*	nil	nil	30
20% HCl	0.005-0.040% $FeCl_3$	190	nil	<1	-
20% HCl	-	150	nil	nil	64
30% HCl	-	92*	nil	nil	-
30% HCl	0.005% $FeCl_3$	190	1	31	-
36% HCl	-	81*	nil	<1	-
36% HCl	-	190	8	141	-
36% HCl	0.005% $FeCl_3$	190	8	259	-
10% H_2SO_4	0.1% $FeCl_3$	101*	nil	nil	<1

*boiling †mils/yr

Reagent-grade chemicals used. For temperatures above boiling point, solution is under pressure.

culty of equipment fabrication. Nevertheless, the metal's high strength, especially at elevated temperatures, and its generally good corrosion resistance—superior to titanium or any nickel-based alloy—may make it attractive in some applications.

Molybdenum is relatively brittle in the as-cast condition. However, rolling or drawing at temperatures of a few hundred degrees centigrade makes it reasonably ductile. Fabricability is thus not usually a problem.

Molybdenum provides corrosion resistance that is slightly better than that of tungsten; it particularly resists nonoxidizing mineral acids. Molybdenum is relatively inert to carbon dioxide, hydrogen, ammonia and nitrogen to 1,100°C, and in reducing atmospheres containing hydrogen sulfide.

Molybdenum has excellent resistance to corrosion by iodine vapor, bromine and chlorine up to clearly defined temperature limits, and good resistance to several liquid metals including bismuth, lithium, potassium and sodium.

Zirconium is used primarily in the nuclear industry because it has a very low thermal-neutron cross-section and resists corrosion by high-temperature water and steam. This metal exhibits good resistance to corrosion by alkalis and acids, except for hydrofluoric acid and hot concentrated hydrochloric and sulfuric acids.

Refractory-metal alloys

Alloying of refractory metals is usually intended to improve their intrinsic mechanical properties. One exception is KBI-40 alloy (60% tantalum, 40% columbium—i.e., Ta-40Cb). Here, tantalum is alloyed to fill the wide gap in corrosion resistance between tantalum and columbium (see Table II).

KBI-40 can be substituted for tantalum in many applications, including heat exchangers, reactors, condensers and piping. Although its corrosion resistance is surprisingly close to that of tantalum, it costs considerably less.

The density of KBI-40 is about 25% less than that of tantalum. Thus, less weight of alloy is required to provide the same surface area, saving money.

This alloy is particularly attractive in applications where the choice between tantalum and columbium is not clear. For example, columbium shows slight corrosion in boiling 10–30% sulfuric acid. Whether this slight (a few mils/yr) corrosion rate requires the use of the more expensive tantalum can be a fine exercise in engineering judgment. Many processing equipment designers would choose to err on the side of safety. KBI-40 offers a safe substitute for costly tantalum at an economical price.

Other tantalum alloys are designed primarily for strength. This is accomplished by addition of up to 10% tungsten and other refractory metals, sometimes with oxygen, nitrogen, carbon and hydrogen, which occupy interstitial positions in the crystal lattice. Typical examples are Ta-10W, Ta-7.5W (having a high interstitial content of oxygen, nitrogen, carbon and hydrogen), Ta-2.5W (which retains nearly the full corrosion resistance of the parent metal) and T-111 (Ta-8W-2Hf). This last alloy offers particularly outstanding resistance to liquid-metal corrosion.

Columbium is usually alloyed either with 1% zirconium or with 10% hafnium, 1% titanium and 0.7% zirconium. Cb-1Zr offers the corrosion resistance of pure columbium together with greater high-temperature strength and somewhat easier fabricability, because of an inherently finer grain structure. The mechanical properties of Cb-10Hf-1Ti-0.7Zr are even better, but at the expense of a slight loss in corrosion resistance.

Tungsten alloys of interest to process equipment designers are those in which the metal's brittleness is moderated by addition of rhenium. Rhenium levels ranging from 1.5 to 3% are generally enough to reduce cold fracture tendency to acceptable amounts. While tungsten is softer with rhenium additions up to 5%, further additions increase the yield strength.

Tungsten-molybdenum alloys, such as W-30Mo, are also common. These provide improved machinability in applications where a strength somewhat lower than that of tungsten can be tolerated.

The two principal molybdenum alloys are TZM (Mo-0.5Ti-0.1Z-0.02W) and Mo-30W. TZM offers better high-temperature strength than does pure molybdenum, while Mo-30W has a higher melting point and offers outstanding resistance to corrosive attack by liquid metals, especially liquid zinc.

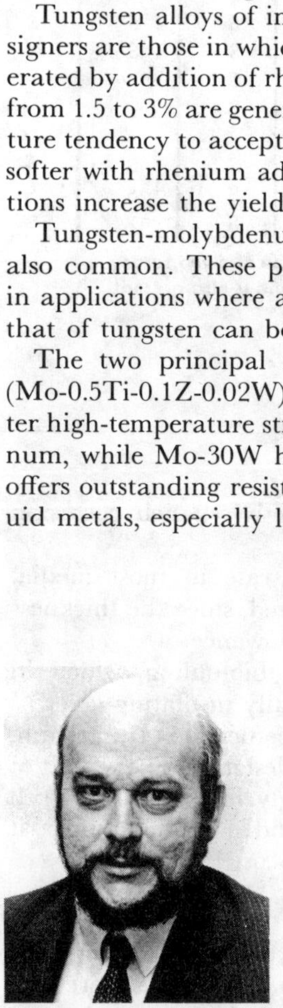

The author

Fred S. Shuker is product manager for refractory metallurgical products for the KBI Div. of Cabot Corp., P.O. Box 1462, Reading, PA 19603. Tel: (215) 371-3612. KBI produces tantalum and columbium. Shuker has more than 20 years' experience in the refractory metals business, in product management, sales, and chemical research and development. He holds a B.S. degree in industrial technology and an A.S. degree in chemistry from Northeastern University.

Refractory-metals roundtable

On Apr. 5, 1984, a roundtable, put together by KBI, a divison of Cabot Corp., was held in New Orleans. The session was moderated by Edward Kubel, associate editor at *Materials Engineering* magazine. [*CE*'s Richard Greene was scheduled to serve as co-moderator, but took ill.]

The roundtable was to serve as a learning experience on refractory metals and their alloys. Panelists included corrosion engineers, corrosion consultants, and fabricators.

The text that follows was condensed from a lengthy transcript. During the editing process, it was not possible to include the views of all of the panelists, or to cover every topic brought up during the meeting.

The version presented here tries to retain as much as possible the original meaning of the speakers. Incomplete statements and missing rebuttals are due more to the editing process than to what was actually said at the roundtable.

The panelists were:

Edward Kubel, associate editor, *Materials Engineering*, 1 Penton Plaza, Cleveland, OH 44114.

Thomas F. Degnan, a consultant in materials engineering, 205 Welwyn Rd., Wilmington, DE 19803. He was principal consultant in Du Pont's Materials Group, part of its Central Engineering Dept.

George V. Dormer, principal engineer, The Pfaudler Co., 1000 West Ave., P.O. Box 1600, Rochester, NY 14692.

Joseph J. Esseff, at the time of the roundtable, was vice-president, Cosmos Minerals Corp., 4725 Calle Alta, Camarillo, CA 93010. Cosmos, which has since been bought by Pfaudler Co., is a fabricator of refractory metals, among other materials, and manufactures vessels.

Dale R. McIntyre, senior engineer, Cortest Laboratories, Inc., 11115 Mills Rd., Suite 102, Cypress, TX 77429. Cortest is a corrosion-testing firm.

Lyman A. Scribner, staff engineer (materials of construction), Union Carbide Corp., Woodbine, GA 31569.

James B. Shriver, director of purchasing, Astro Metallurgical, 3225 Lincoln Way, West, P.O. Drawer 520, Wooster, OH 44691. Astro is a fabricator of chemical-processing equipment and specialty wire, and uses mainly reactive and refractory metals.

Oliver Siebert, senior fellow for materials technology, Corporate Engineering Dept., Monsanto Co., 800 N. Lindbergh Blvd., St. Louis, MO 63167.

From KBI, Div. of Cabot Corp., P.O. Box 1462, Reading, PA 19603, were:

Don Schwartz, product specialist, refractory metals.

Lawrence S. O'Rourke, vice-president, Cabot Corp., and general manager, KBI Div.

Robert Burns, research engineer, Cabot Engineered Products Technology.

☐ [The session opened with the questions: What are refractory metals? Where are they used?]

Schwartz: Refractory metals are generally classified as metals with melting points above 2,000°C. They include tantalum, columbium, molybdenum and, for all practical purposes, zirconium. Of particular interest to KBI are tantalum, columbium and their alloys.

Shriver: In the case of tantalum, for instance, it's used where there really isn't anything else that will work.

Alloys and pure metals

Kubel: Are the metals used in their pure form or are you talking more about alloys? And if so, why?

Degnan: As for tantalum, it's frequently used as a tantalum/2½%-tungsten alloy because that is stronger. Alloying doesn't affect corrosion resistance much. The alloy resists scratches, its greater strength is useful, and it doesn't cost more.

Shriver: The 2½%-tungsten alloy has gained a lot of popularity over the last couple of years. It allows you to use less material for a given application. You get a higher tensile strength; this allows you to use a lighter gauge of material; and there are fewer dollars involved.

Degnan: Using the alloy, flanges are less easily damaged [i.e., scratched], so we prefer to use it.

Esseff: Is there any difference in corrosion resistance between the alloy and the metallurgical grade?

Degnan: The patent on the alloy shows some advantages, but I don't know that they are significant. I think you can assume that the alloy performs about the same as unalloyed tantalum, except for mechanical properties.

Siebert: In my experience, I have not seen it better or worse from a corrosion standpoint. I haven't seen a problem even with the 10%-tungsten alloy, but I haven't tested every combination of alloys and environments.

Scribner: We have tested the 10%-tungsten alloy, and

in the most severe environments, it has shown slight corrosion. The Ta-2½W alloy generally is nonreactive.

Esseff: We don't know all metals and all applications, either from the mill's standpoint or the user's. What we're saying is that you tell us what you want and we know how to fabricate it. But for applications, I depend upon people like Ollie Siebert to give me input.

Siebert: To get back to the original question: Are pure metals and alloys different? So far, we have only addressed tantalum. For zirconium, we used to say that 702 [R60702—zirconium, unalloyed, using the Unified Numbering System of SAE, the Soc. of Automotive Engineers] and 705 [R60705—zirconium alloy] differed only in structure, but now we know that they have different corrosion resistances.

Available data

Kubel: Is the corrosion resistance of these materials still relatively new, or have they been around so long that there is a large data base? Can you thus make a selection?

Siebert: I think it's a mature industry.

Esseff: For tantalum, especially, you have a good data base. A lot of refractories start out in other areas and move into the chemical-process industries (CPI). Titanium, for example, was big in the aerospace industry, and it became economical in the CPI. Zirconium is going that route—its primary use is in the nuclear industry.

Siebert: [He related a story in which his firm needed to build zirconium-clad reactors to manufacture a certain product. But there was not the knowhow to clad the zirconium on a steel substrate that later was welded and formed *without* cracking; the venture was dropped.] So that the zirconium industry does not jump on me for the example I just gave, it is now felt that by proper control of the interstitials in the zirconium, a selected product can be made of 702. This alloy can be explosion-bonded on steel and, in turn, formed and welded. A kiln was built this way and has been in service for two years. I think that the jury is still out as to whether or not this is a success, because the application is not a liquid exposure.

Fabrication

Shriver: What about fabrication of these materials?

Degnan: They're very reactive and, when heated, absorb gases such as N_2, O_2 and H_2, which can cause embrittlement. They have to be fabricated with proper care and the surface has to be very clean—free of iron. That could give problems with hydrogen pickup in service, or embrittlement during welding.

Esseff: Few fabricators work with refractory metals as a specialty. Shop fabrication of these metals is more sophisticated than it was years ago. How do you perform repairs? You can take the same set of standards that you used in-house and transfer them to steel applications [refractory metals clad on steel] in many areas. We've done it on some tantalum-clad-steel equipment.

Siebert: The key to fabricating the reactive metals is shielding. All other things are important, cleanliness and the rest, but the bottom line is shielding.

Esseff: I agree with Ollie about shielding. You can take all the precautions up to doing the welding, but if you don't have proper shielding, you'll wind up with a brittle or contaminated weld. The metal is unforgiving. You can shield perfectly, but if you haven't cleaned properly, contaminant problems can occur. The fatty oils from skin are picked up by tantalum. Handling this metal is difficult because of its affinity for oxygen, even at low temperatures. And, with tantalum, if you have moisture, it can create problems. I might add, as a fabricator, that you can write all the specifications in the world, but you have to depend upon people to carry them out. We train our own operators, and we always wind up better off.

Field repair

Kubel: If you have components that fail for some reason, can they be repaired?

Degnan: As we were saying, you can field-repair and get a pretty good weld. But it is difficult to repair loose-lined equipment in the field because if you get contamination (such as iron salts) on the backside of the lining, it is difficult to remove.

Esseff: However, if you're dealing with a solid tantalum/columbium fabrication, the techniques and preparation for doing repair work are rather straightforward. One set of techniques can be used.

Shriver: Straightforward, but not trivial.

Esseff: Not at all.

Shriver: Cleaning after a piece of equipment has been in service is not a simple thing.

Kubel: Are the more exotic welding processes, such as electron-beam or laser, being used? Or is conventional fusion more satisfactory?

Siebert: With some of these, you are limited by equipment size. We've done laser and electron-beam welding on very small parts, not on big equipment. We have done some things as big as 6 or 8 ft long, but those were straight seams. Mostly when you've got complex welding techniques, you're talking about a very limited size.

Burns: When you have tantalum, what about hydrogen embrittlement and galvanic corrosion? What steps do you take to prevent them?

Siebert: You do things like on a reactor you put a high-alloy flange on the steel nozzles so that you can cut down the potential difference between the tantalum and flange material. Another example: You avoid making any type of bimetallic couple in a liquid medium with tantalum.

Applications

Kubel: Is it possible to look at some specific applications and see what materials are used? Why are they used? What are the advantages?

Scribner: We can talk about it generically. Consider concentrated nitric and sulfuric acid. Historically, silicon cast-iron equipment was used. It worked well, but it was brittle. Tantalum, titanium and zirconium have replaced this metal.

Siebert: In chlorine making, heat exchangers used to be made out of stoneware for the shell, reinforced plastic for the tubesheet top and bottom, and borosilicate glass for the tubes. Titanium came along and revolutionized the business. Titanium and Teflon.

Kubel: How about some of the other alloys—zirconium, columbium and tungsten? What are their uses?

Siebert: Take hydrochloric acid. You can handle it in glass and wood, but with limitations. Or you can use silicon iron, or Hastelloy B and, now, zirconium. These are about all the materials available to handle it.

Degnan: If you're handling concentrated HCl, and a little bit of chlorine is present, you have to use tantalum.

Siebert: Right. Hastelloy B and zirconium will fail in hydrochloric acid with strong oxidizers present.

McIntyre: One area these materials are used in is in oil- and gas-production equipment. Looking at tests we've done, zirconium, tantalum and titanium are among the most resistant to H_2S corrosion, sulfide stress cracking and combined H_2S/CO_2 sour-gas corrosion.

Esseff: Economics plays a big role in what materials are selected. But even if someone would provide unlimited funds and say, "Go ahead and build it the way you want to," tantalum wouldn't be the material that's used everyplace. Tantalum is not required in many areas. Titanium would not do it all, nor would zirconium or glass. Every facility has people, and when you have people, they're going to make mistakes. They're going to "torque down" too much on a flange. And if it's glass, they're going to break it. They're going to not tighten down enough or not align things up. And if it's metal, it's not going to seal. Gaskets can be a whole subject themselves. For example, trying to get sealings and the right type of gaskets. Sometimes, there are corrosion problems. Titanium has real trouble with corrosion in some chlorine applications with gaskets. But you've got a variety of materials that are favorable. In a lot of cases, it's the combined choice of the user, the manufacturer and the fabricator as to what materials are selected.

Availability

Siebert: Many times it's a function of what's available. A materials engineer can be so blind as to evaluate the material of construction and specify it to the process people only to find out that it only comes in pipe but not in plate. So, you'll often find a chemical plant that has five different materials of construction.

Kubel: What sort of problems are there in hot-working or whatever?

Schwartz: Tantalum and almost all of the metals we've been talking about are available in just about any form. The biggest problem, primarily with tantalum and columbium, is size—unusual sizes or large ones. Sheets are available in very large sizes and are reasonable in cost, but piping becomes extremely expensive because of the small quantities needed in unusual sizes. That's why a combination of various materials is used in fabricating a unit. We've done a great deal over the past ten years to eliminate some of the cost problems. One certainly does not build solid tanks out of tantalum. With the development of cladding, more material has gone into construction of larger tanks. Generally speaking, with the exception of tantalum/10%-tungsten, all of the metals are cold- or warm-workable, and most are available.

Welding

Esseff: Generally, the fewer weld joints you have to put into a fabrication the better. Primarily, in tantalum, where tungsten/inert-gas (TIG) welding is more detailed and where the concern for contamination, embrittlement and so forth is stronger, you would like to avoid such situations as much as possible. And up until recently, the width of a sheet was such that you might put three pieces together to make a 6-ft-long tank. Now, we're getting to the point where two pieces will do it. This is where the fabricator can begin to reduce costs, because of savings in labor. We use a resistance-welding method of cladding. The preparation of TIG welding joints and the welds is labor-intensive.

Dormer: [The question was raised: What is a satisfactory weld?] A satisfactory weld, obviously, is one that, in the simplest terms, is going to do the job. This may be an oversimplification but it's what's required. The weld has to have high integrity, some measure of strength, and adequate corrosion resistance for the service intended.

Shriver: In evaluating welds, we can easily determine that a good weld is one that has good reduction in area, which is the most important property to be looked for. It's something that you can't measure after it's made. So, all we have are a few clues.

Siebert: [The point was raised that there are no satisfactory nondestructive methods for testing welds of some clad refractory metals.] As mentioned, we do not have a nondestructive method of testing these welds after fabrication. As an example, I have four tantalum-clad reactors, explosion-bonded. We have had some problems by the mechanism that was just outlined. In practice, it's almost a function of the welder's gut feeling as to whether or not he's making a satisfactory weld. But that's not sensible when you're spending between a half-million and a million dollars for a 1,000-gallon reactor.

Degnan: These refractory metals all have a low coefficient of thermal expansion, compared with steel. So, thermal cycling puts a considerable stress on the weld. But we have not had failures from thermal cycling on a tantalum-clad vessel.

Siebert: The difficulty I had had nothing to do with clad-sheet construction or with the basic batten-strip design. It was because there was no satisfactory nondestructive method to tell what the quality and quantity of the fusion bond was. Because of the nature of tantalum on copper on steel [N.B., this is a three-way clad], there's no eddy-current, ultrasonic, X-ray or other nondestructive method we can use to determine weld total-integrity.

Availability and price

Esseff: [The discussion turned to the availability and cost of certain metals. Talk was not limited to the refractory metals, but included such elements as chromium. The effect of world politics on the supply of certain metals was discussed.] There are users who are, for example, looking at prices of tantalum going back three or four years, when tantalum got up to over $300 a pound. So, materials-specifications people are eliminating tantalum out-of-hand, even before they consider it. But prices have dropped. And today, better processes are available to fabricate these materials. For example, we have a cladding process that fits well into the marketplace and is economical.

O'Rourke: I've been working with tantalum for 20-odd years, and I can assure you that supplies were so enlarged by that high price, that the scarcity period isn't going to come back to haunt us. There's at least a 20% additional supply today vs. the 1979-1980 period. Also, the weight of tantalum required per device has dropped.

Section VI
FIBERGLASS-REINFORCED PLASTICS AND OTHER POLYMERS

Fiberglass composite materials and fabrication processes

Pipes and tanks, ducts and scrubbers are just a few of the equipment items being made out of fiberglass-reinforced plastics. Here is a comprehensive survey of the technology, from raw materials to finished products.

J. Albert Rolston, *Consulting Engineer*

☐ After several false starts and some dramatic failures, fiberglass-reinforced plastic (FRP) technology has reached a level of sophistication and reliability such that it not only stands alone in certain applications, but is competing successfully with steels in others. A whole family of fabrication processes has grown up and standards are being set to protect the quality of the finished product.

This article was adapted from material used in the 1979 short course, "Construction with fiberglass composites for corrosive environments," at the University of Virginia, Div. of Continuing Education, Charlottesville, Va.

Commercial composite materials for corrosion-resistant equipment mainly use fiberglass* reinforcements with polyester or epoxy resin matrices (binders). The principal fabrication process is contact molding, although more and more machine-made parts are being produced.

The purpose of this review is to compare the properties, performance, and economics of the principal mate-

*"Fiberglas" is the registered trademark of Owens-Corning Fiberglas Corp. In this article, glass-fiber material is generically referred to, and is spelled fiberglass.

rials systems, and to compare some of the fabrication processes used for the construction of corrosion-resistant equipment.

Economics

The manufacture of composite materials for chemical process equipment has grown into a billion-dollar annual-sales industry over the last 25 to 30 years. The annual growth rate still approaches 20%. While a major part of the business is in pipe and tanks, a significant portion is in ducts, scrubbers, and wastewater treatment facilities. Markets are developing for pumps, valves and support structures.

The success of any new material (or materials system, in the case of composites) depends upon the cost/performance ratio of the material compared with competing materials. The current position for composites is very favorable, and the outlook bright. In Table I a cost comparison [*1*] for a 7,000-gal storage tank shows the attractive economics of composites in relation to carbon-, stainless- and rubber-lined-steel vessels. Further, a cost comparison for chemical plant piping showed that a system of fiberglass pipe and fittings ranked very close to carbon steel in cost and was far cheaper than a stainless steel system [*2*].

And for many environments (e.g., halogen acids or salts), the corrosion resistance of reinforced plastics is superior to that of the best stainless steels.

Polyester resins

Let us initially note some fundamentals about polyesters.

An ester is a reaction product (a salt) of an alcohol and an acid. One simple ester is ethyl acetate, the product of combining ethyl alcohol and acetic acid.

For thermosetting polyesters, propylene glycol is the usual alcohol, while phthalic and maleic anhydrides (in various ratios) provide the acids. The reaction product is a low-melting-point solid that must be dissolved in a low-molecular-weight monomer to achieve further polymerization. Styrene is the most common monomer; the normal level is 45–55% styrene in the polyester.

The most-used resin in reinforced-plastic corrosion-resistant structures is isophthalic polyester; primarily because of low cost, easy handling, good acid resistance

Price of a 7,000-gal vertical tank* — Table I

Material	Sales price ($)
Bisphenol A polyester/fiberglass	5,850
Carbon steel	6,324
Furan resin/fiberglass	7,753
Rubber-lined steel	9,455
304 Stainless	14,219
316 Stainless	16,985

*These May 1978 prices are for a tank with two 4-in. nozzles and one 12-in. manway.

and fast cure time. More expensive resins provide special properties. The approximate costs of polyester resins for corrosion-resistant equipment ($ per lb, May 1979) are as follows:

Orthophthalic polyester	0.55
Isophthalic polyester	0.61
Bisphenol A fumarate	0.95
Vinyl ester	0.95 and up

Curing of the polyester requires a catalyst—normally, an oxygen supplier such as an organic peroxide. When the organic peroxide decomposes (either from heat or the chemical action of a promoter), the double bond of the styrene molecule opens up and reacts with the polyester molecules, achieving a cross-linked thermoset polymer (in contrast to the linear linkage of thermoplastic polymers). The resulting solid plastic may be thought of as a single giant molecule, because all of the atoms are chemically linked to one another.

Monomers

Styrene is a benzene derivative and is classed as an aromatic solvent. It may also be called vinyl benzene or phenyl ethylene. The chemical formula is $C_6H_5CH{:}CH_2$. The present price is about $0.27 per lb (June 1979).

The effects of an increase in styrene content on polyester resins are: viscosity decreases; gel time decreases (except for the BPO/DMA system); heat-distortion temperature approaches that of polystyrene (may increase or decrease, depending on polyester type); shrinkage increases linearly from 3–4% at 0% styrene (from extrapolation) to 14% at 100% styrene; corrosion resistance increases (especially acid resistance); refractive index increases; brittleness increases. Other mechanical properties and ultraviolet yellowing are not affected by styrene content.

Since styrene accounts for nearly all of the monomer usage in polyester, other monomers are generally used only for special purposes. Some of these other monomers and their characteristics are: *vinyl toluene*—lower shrinkage, next cheapest to styrene; *diallyl phthalate*—less shrinkage, lower exotherm (exothermic polymerization reaction); *methyl methacrylate*—ultraviolet stability (best mixed with styrene); *divinyl benzene*—higher heat-distortion temperature, better chemical resistance.

Curing of polyesters

Inhibitors—Polyesters (with styrene) are inherently unstable and can auto-cure in storage from the action of heat or sunlight. A minute amount of an inhibitor (a hydroquinone compound) is therefore added to the resin by the manufacturer to prevent this. Further inhibitor addition is sometimes desirable for certain heat-cured applications.

Catalysts—There are a large number of peroxide compounds on the market for catalyzing. Three commonly used catalysts in the fabrication of corrosion-resistant equipment are:

- *Methyl ethyl ketone peroxide*—normally sold as a 60% solution in dimethyl phthalate. This is the most commonly used catalyst.
- *Benzoyl peroxide*—sold as powder (pure) or paste (50%). It is normally dissolved in a small amount of styrene before being added to the polyester.
- *Tertiary butyl peroxyoctoate*—This is used alone or in combination with other peroxides for heat-cured systems.

Promoters—Promoters accelerate the curing process by reacting with the peroxide compounds. PROMOTERS MUST BE KEPT OUT OF CONTACT WITH CONCENTRATED PEROXIDES. OTHERWISE VIOLENT REACTIONS WILL RESULT.

The principal promoters are:

- *Cobalt (or other metallic) soaps* such as cobalt naphthenate, cobalt octoate, manganese naphthenate, and so on.
- *Dimethylamine.* This may be used alone or with cobalt in a double-promoted BPO catalyst system.
- Other compounds such as *lauryl mercaptan.*

Exotherm—The reaction of styrene and polyester generates heat to form a solid polymer (exothermic reaction). It is mass-sensitive (a large mass contains more exothermic energy per unit of surface than does a small mass). On very thin sections, curing may be very slow; in large masses, the exothermic heat can cause cracking, or even charring and fire. In hand layup with about 30% glass content, the maximum thickness that can be produced in a single step, with a normal catalyst system, is about ³⁄₈ in.; thicker sections should be made in multiple layers.

Fillers and glass fibers act to soak up some of the exothermic heat. The peak exotherm temperature for a given system may be compared to other systems by the "SPI Exotherm Test." Details may be obtained from the Soc. of the Plastics Industry (New York City).

Abbreviations

BMC	Bulk molding compounds
BPO	Benzoyl peroxide
CPVC	Chlorinated PVC
DMA	Dimethylamine
FRP	Fiberglass-reinforced plastic
PVC	Polyvinyl chloride
RPM	Reinforced plastic mortar
RTM	Resin transfer molding
SMC	Sheet molding compound

Gel time—In the first stage of cure, the polyester becomes a gelatine-like material that will no longer flow as a liquid. The time to reach the gel stage is important, for it establishes the working time available for "wet out" (the wetting of the glass reinforcement) and fabrication. The variables affecting the gel time (sometimes called "pot life") are: polyester formulation (catalyst, promoter and inhibitor levels); ambient temperature; and mass of material.

A useful rule-of-thumb for processes where the resin pot is continuously replenished is: The usage rate of material must be twice the mass of material in the pot, divided by the pot life.

Example: A resin pot (wet-out bath, holding pot) contains 15 lb of resin that will gel in 30 min. If the usage rate (fresh resin being added continuously or at frequent intervals) is at least 30 lb per ½ h the resin pot will not gel, and a continuous operation can be sustained—assuming the fresh resin is mixed thoroughly with the old resin.

The same rule-of-thumb can be used to size the resin pot or to establish a usable gel time for continuous layup, filament winding, or pultrusion fabrication.

Temperature versus time—The ambient temperature is more important for establishing a complete cure than is the time involved. The maximum Barcol hardness (45–50) is achieved only with heat cure; room-temperature curing yields a Barcol hardness of 25–30.

Heat curing—If heat curing can be used (it is sometimes not feasible for large structures) the promoter content can be reduced or eliminated. Some producers prefer to use a low level of promoter with a heat cure. The promoted resin will serve to complete the cure (over a period of several days) in the event of an oven malfunction.

Inhibition—Curing is inhibited by exposure of the polyester resin to air. Some resins (notably the bisphenol A type) are seriously affected by this phenomenon. If a film cannot be placed over the part, a resin that contains wax should be used. The resin manufacturer's literature will note the resins containing wax, as well as the amount and technique to be used.

Moisture also inhibits the curing of polyester resins. So do many sulfur-containing compounds such as rubber or neoprene.

Shrinkage

Shrinkage of a polyester part is caused by heat expansion and contraction, or resin shrinkage during polymerization. The shrinkage of resin can create several difficulties: internal stresses, possibly causing cracks; the part sticking in the mold; and warping (from uneven internal stress).

Shrinkage can be reduced by: higher filler and/or glass content; reduced styrene content; reduced exotherm (lower curing temperature); thermoplastic additives. In a cylindrical part, curing from the inside out will reduce the shrinkage stresses, as compared with curing from the outside in.

Safety

In this day of increasing government regulations, safety precautions are not only desirable, but mandatory. The use of safety shields, respirators and other devices must be considered for each operation.

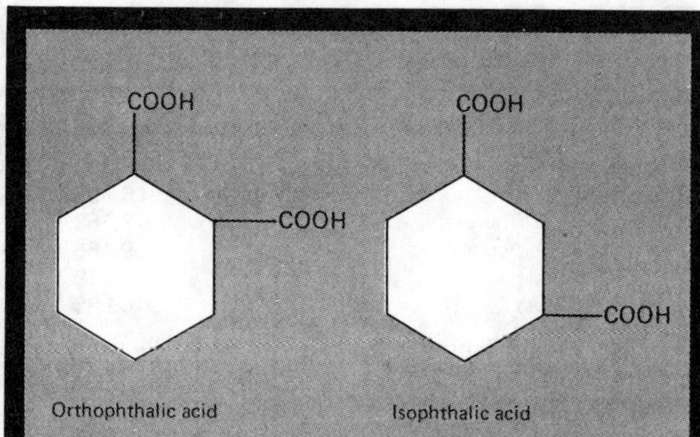

Structure of phthalic acid isomers Fig. 1

Resin manufacturers' literature will ordinarily give guidelines as to what to expect with toxic chemicals, but regulations about levels change frequently, so it is essential to keep up to date. The following subjects should be considered: styrene fumes; resin flammability; peroxide handling; peroxide/promoter/resin storage; amines (can cause skin problems); and exotherm.

Isophthalic and orthophthalic polyesters

The essential difference between these lies in the aromatic benzene-type ring of the phthalic anhydride molecule. The chemical formula of an isophthalic compound is the same as that of an orthophthalate, but the molecular arrangement is different (see Fig. 1).

Orthophthalic polyesters (often called general-purpose polyesters) are not as resistant to chemical attack as are isophthalic resins. Thus, the isophthalic polyesters are more often used in corrosion-resistance applications.

Bisphenol A polyester resins

Bisphenol A (a benzene-like compound) is reacted with propylene oxide to form a glycol (an alcohol compound). This glycol is reacted with fumaric acid to form the resin bisphenol A fumarate. Most producers sell this latter compound in a flake form or as a solution in styrene.

Early studies of this resin disclosed that, although it had exceptional corrosion resistance, it was brittle. A modified version having high elongation was developed. The optimum resin was found to contain 10% of the flexibilized resin mixed with the standard bisphenol A.

In the corrosion-resistant-equipment market, the bisphenol A fumarate has a well-established reputation, and it is widely specified.

Vinyl ester resin

Vinyl ester is a reaction product of an epoxy resin with a polyester (such as methyl methacrylate). It shares many of the attributes of polyesters—acid resistance, fast curing with peroxide catalysts, and dilution with styrene. In addition, the vinyl ester retains the high strength and elongation aspects of an epoxy, along with some of the epoxy's resistance to solvent attack. Since there are many types of epoxies, there are several vinyl esters now on the market, such as high-temperature versions, fire-retardant types, and resins having high toughness.

Vinyl ester is the only polyester competing in the high-pressure-pipe market, long dominated by epoxy resins. It has demonstrated excellent cyclic pressure performance.

Additives

Additives are substances blended into the polyester resin to provide a special property. Fillers and pigments are not included in this section.

Fire-retardant additives—A considerable measure of fire retardancy can be achieved with relatively high (near 50%) levels of hydrated alumina, $Al(OH)_3$. This combination increases the viscosity; ordinarily, it is used only in compression molding.

Antimony oxide (at a 5–15% level) is used to enhance the nonflammability of halogenated polyester resins. This additive is only effective in the presence of chlorine or bromine compounds.

Thixotropy—In hand-layup work, resin drainout is sometimes a problem. To combat this, a colloidal silica (made from silicone compound decomposition) is often used as a thixotropic agent to increase resin viscosity.

Conductivity—In some applications (such as explosive-vapor ducts, or chimney liners), static charges will build up in the nonconductive FRP. To reduce the danger of electrostatic sparks, conductive layers are added to the laminate. The conductivity can be achieved by additions of certain types of graphite, carbon black, graphite fibers or metal-coated glass fibers.

Low profile—To reduce shrinkage in molded articles, a small percentage of a thermoplastic powder can be added. Several types are used, such as polyethylene or polystyrene.

Resin thickening—Certain compounds (notably calcium or magnesium oxides) when added to polyester cause a gelation reaction that does not affect the double bonds and does not create any appreciable cross-linking. These compounds are used to produce sheet or bulk molding compounds for high- or moderate-pressure molding. The gelled material is tackfree and can be handled; when heated to 200°F or more, the compound reverts to the low-viscosity state and flows readily.

Fillers and pigments

Fillers—The primary purposes for using fillers are to reduce cost and reduce exothermic heat. The principal fillers are: calcium carbonate (ground limestone), talc, aluminum silicate (kaolin clay) and silica (sand). While calcium carbonate is favored for many noncorrosive applications, it cannot be used in corrosion-resistant equipment. Clay is the filler favored for most molded corrosion-resistance applications.

Reinforced-plastic mortar (RPM) systems (usually in filament-wound pipe or tanks) consist of approximately 30% resin, 30% fiberglass and 40% sand. The sand is usually 50 to 100 mesh, with all fines removed. Coarse-filler systems (such as used in RPM) can be loaded to a higher level than fine-filler ones such as clay or talc. Needless to say, in corrosion equipment, the filler must resist the environment expected in service. Some fillers are treated with silane compounds to enhance shear strength (and chemical resistance) at the resin-filler interface.

Pigments—Pigments are used for color, but more importantly (in corrosion applications) for ultraviolet resistance. It has been conclusively shown that polyesters (and most other organics) deteriorate from weathering, but the damage is essentially a surface phenomenon. Even after several years of exposure, with glass fibers showing all over the surface, strength properties are retained. In corrosion-resistant equipment, it is desirable to retain the resin-rich surface layer. A pigmented resin or surface coating (gel coat) can effectively block ultraviolet rays and keep the resin-rich surface. Many types of pigments are available: Two that are commonly used are carbon (black) and titanium dioxide (white).

Gel coats and surface coatings

As mentioned before, gel coats are often used to protect surfaces from weathering. A gel coat is a highly pigmented polyester. The name stems from the method of application; the coating is applied to the mold. When the coating reaches a gel state, the remainder of the laminate is applied.

In corrosion-resistant equipment, most specifications require an inner resin-rich layer, using a C glass veil or a polyester surface mat (mats will be discussed in a later section). The purpose of the resin-rich layer is to provide protection for the glass fibers comprising the structural portion of the laminate.

Post-fabrication coatings (paints, gel coats or surface mats) may be used to provide extra corrosion resistance in exposed areas. The nature of the coating may be quite different from the body of the laminate. A pigmented urethane or silicone coating, for example, can provide better ultraviolet resistance than an ordinary polyester gel coat.

Special resins

Fire-retardant resins

The principal means of fire-retarding a polyester (or epoxy) resin is to replace hydrogen atoms in the molecule with halogen (chlorine or bromine) atoms. Thus, chlorendic anhydride replaces phthalic anhydride in a polyester, and brominated compounds may be used for an epoxy resin.

The fire retardancy of halogenated resins is enhanced by the addition of a small percentage (3–5%) of antimony oxide (Sb_2O_3). This filler does not add fire retardancy to plain resins, only to the halogenated variety.

Although not in general use, furan resins have considerable promise for the corrosion-resistant-equipment market. Their corrosion resistance to acids and alkalis is outstanding; resistance to solvents is also superior to that of most other resins. In addition, furan resins are low-smoke organic compounds and are inherently fire-retardant.

The furfuryl alcohol molecule has a five-sided ring with a methyl alcohol molecule attached at one point. In the presence of a strong acid, a methyl linkage bonds the rings together with the OH groups, forming water. Acid type and acid level are critical. A post-cure is necessary to eliminate water byproducts.

Because of the high exothermic temperature, the water from the condensation reaction, and the brittle nature of the resin, the danger of cracking is high. In addition, furan resins have a short working time (10–30 min). For these reasons, most fabricators do not like to work with them.

Epoxy resins

Epoxies are not a single class of resins but rather describe a family of resin types. The common factor is the epoxy linkage: an oxygen molecule linked to two carbon atoms in the organic chain. This reactive group can join to an amine or an acid to make a strong and durable bond.

Since the epoxy and hardener join together in the polymerization reaction, the choice of hardener can have a strong effect on the properties of the polymer. The cross-linking action evolves no byproducts, and the exothermic temperature is low.

Epoxies are the strongest of the common polymers, with tensile strengths of 12,000–15,000 psi. Some formulations with post-curing have high heat resistance. Adhesion to most inorganics (and some organics) is excellent, and epoxies have been the first-choice resin for most high-pressure fiberglass pipe, especially when accompanied by high-temperature requirements.

Epoxies are attacked more readily by strong acids than are polyesters, but alkali resistance and solvent resistance are superior.

The most common epoxy is made from bisphenol A; other types are epoxy novolac (a reacted combination with phenolic) and cycloaliphatics. Hardeners include aliphatic amines and polyamides for room-temperature curing (mostly used in adhesives). Aromatic amines and acid anhydrides are ordinarily used for filament winding, molding or other heat-cured-resin processes. Several latent hardeners are used for making epoxy pre-impregnated fabrics.

Thermoplastic/thermoset combinations

Combination or hybrid structures have been made from thermoplastic/thermoset combinations for enhanced corrosion resistance. The thermoplastics normally used include polypropylene, polyvinyl chloride (PVC), chlorinated polyvinyl chloride (CPVC) and FEP (fluorinated ethylene-propylene) Teflon.

One procedure is to *heat-bond a fiberglass cloth to a thermoplastic sheet.* The side of the cloth away from the thermoplastic is left dry to provide a bonding surface for the thermosetting resin.

This combination of cloth and thermoplastic may be cut, heat-softened, formed and welded into the proper shape and dimensions (such as a tank, scrubber, venturi or other piece of equipment). After forming, the thermosetting plastic and fiberglass structure is built up on the glass-cloth side of the structure (usually on the outside). This buildup is usually accomplished by hand layup, but filament winding, sprayup and compression molding can be used.

Fiberglass-reinforcement prices* **Table II**

Type of reinforcement	Price ($/lb)
Unsplit rovings (forming packages)	0.53
Unsplit rovings (conventional)	0.56
Split rovings for chopping	0.61
Chopped-strand mat	0.75 (1½ to 3 oz), 0.81 (1 oz)
Continuous-strand mat	0.77 (1½ to 3 oz), 0.79 (1 oz)
Woven roving	0.78 (24 oz), 0.82 (18 oz)
Chopped strands	0.58
Milled fibers	0.58
Surface mat	2.21/100 sq ft (0.010 in. thick) 3.31/100 sq ft (0.015 in. thick) 4.45/100 sq ft (0.020 in. thick)
Glass flake	0.87 (1/8 in.) 0.89 (1/32, 1/64 in.)

*Published list prices from Owens-Corning Fiberglas Corp. (May 1979), for delivered truckload or carload.

Another approach is the *adhesive bonded thermoplastic/thermoset.* Filament-wound pipe is being made by overwrapping PVC or CPVC (the only polymers used at this time) with resins that can bond tightly to the thermoplastic substrate. In one instance, an epoxy resin is used, which bonds tightly to the thermoplastic. In another case, a specially modified polyester is used as a primer coat, then a standard polyester completes the overwrap.

The most serious problem with thermoplastic/thermoset hybrids arises from a thermal mismatch. The thermoplastic liner has an expansion rate several times higher than the reinforced thermoset's. Low temperatures can create stresses at the interface. In addition, thermal cycling can cause cracks to develop in some structures.

The current cost of the hybrid structures is about double that of a reinforced thermoset, without the thermoplastic liner. In some severe corrosion-service applications, however, the hybrid is cost-effective.

Reinforcements

The history of reinforced plastics dates back many years. High-pressure NEMA (National Electrical Manufacturers Assn.) laminates of phenolic, epoxy, melamine and other resins, some dating back to the early 1900s, are made with paper, cotton, linen and glass cloth.

In the present era of reinforced plastics for the corrosion market, glass-fiber reinforcements are used almost exclusively. Many types of reinforcements have been developed in continuous strands, woven fibers and

Glass-filament diameter **Table III**

Letter designation	Avg. dia. range, in.
A	0.00006-0.00010
B	0.00011-0.00015
C	0.00016-0.00020
D	0.00021-0.00025
E	0.00026-0.00030
F	0.00031-0.00035
G*	0.00036-0.00040
H	0.00041-0.00045
J	0.00046-0.00050
K*	0.00051-0.00055
L	0.00061-0.00065
M*	0.00066-0.00070
N	0.00066-0.00070
P	0.00071-0.00075
Q	0.00076-0.00080
R	0.00081-0.00085
S	0.00086-0.00090
T*	0.00091-0.00095
U	0.00096-0.00100
V	0.00101-0.00105
W	0.00106-0.00110
X	0.00111-0.00115
Y	0.00116-0.00120
Z	0.00121-0.00125

*Fiber diameters most commonly used in reinforcements

chopped fibers. The level of reinforcements ranges from less than 10% to above 80% by weight in some filament-wound structures.

Types of glass fibers

Four basic glass compositions are presently used in plastics reinforcement:

The lowest-cost type is **A (alkali) glass,** similar in composition to ordinary window or bottle glass. Several firms make mats from A glass, but its use is not widespread.

The most common reinforcement is **E glass,** a boro-aluminosilicate containing normally small percentages of calcia (CaO) and magnesia (MgO). The alkali metal (sodium and potassium) content is kept very low. The product's combination of good water resistance and reasonable cost has not been excelled in its 40-year existence.

C glass, a calcium aluminosilicate, has exceptional acid resistance and is useful in surface mats, glass flakes for coatings, and yarns for acid-resistant cloth. C glass has poor water resistance and has not been used as a major reinforcement.

S glass is used in high-strength aerospace applications. It is several times costlier than E glass. Although there are no known applications of S glass in the corrosion market, it has exceptional resistance to acids (approximately equal to C glass), and excellent resistance to water.

Approximate glass-fiber-reinforcement prices are shown in Table II.

Fiberglass nomenclature

Fiber filament diameter and weight are standardized and related by letter designations, as shown in Tables III and IV.

Glass fibers are grouped into "strands" with a specific number of fibers, which may be in the hundreds, or even thousands. Strands are combined into "rovings," which contain a specific number of strands, usually between 8 and 120. The weight of fiberglass is referred to as "yield," which is the number of yards of strand (or roving) that weighs one pound. Notice that yield is *inversely* proportional to weight. A heavy strand will have a low yield, since it doesn't require a great length to make up a full pound weight (see Table IV).

In order to find the yield of a roving, therefore, it is necessary to *divide* the strand yield by the number of strands in the roving. For example, a roving with 8 strands of K filament with 408 fibers per strand will have a yield of 3,700 ÷ 8 = 463 yd/lb.

Conversely, the filament diameter can sometimes be determined from the roving yield (which is printed on the package), and the number of strands in the roving. For example, if a roving has 16 strands and its yield is 225 yd/lb, the yield of each strand is 16 × 226 = 3,600 yd/lb. Since there are only two standard filaments with this approximate yield (see G-816 and K-408 in Table IV), the roving probably consists of one of these two filaments.

Relation of strand size and yield (Common standard products) **Table IV**

Filament designation	No. of fibers in strand	Approximate strand yield, yd/lb
G	204	15,000
	408	7,500
	816	3,600
K	204	7,500
	408	3,700
	816	1,800
M	2,080	675
	2,080	450
	4,080	225
T	2,080	225
	4,080	113

Fiberglass sizes

In the parlance of textile-fiber producers, a "size" is a material applied to the fibers in the forming process. With only a few exceptions, sizes are emulsions of organic compounds in water (certain S glass sizes are not). For reinforcement glass fibers (rovings or mats), the size applied in the forming process contains elements for compatibility with the end-process and the matrix resin. Ordinarily, sizes are used for general-purpose compatibility with epoxy or polyester resins, but special-purpose sizes have been developed for individual uses (such as for filament-wound vinyl-ester pipe).

Sizes have three components: coupling agents, film formers, lubricants and emulsifiers. Here follows a brief discussion of each component.

Coupling agents—The purpose of the coupling agent is to provide a bridge between the fiberglass surface and the polymeric matrix binder. Since the chemistry of polyesters is quite different from epoxies, an optimum coupling agent for epoxy applications is quite different from the optimum for polyesters. Coupling agents may interact unfavorably with film formers and with the fabrication process, so numerous compounds have been used to find the best combinations.

The earliest coupling agents (still used in some applications) were organic chromium compounds. Most of the current coupling agents are rather complicated silicone compounds, called silanes; recently, certain titanate compounds have been developed for some applications.

Experience has shown that epoxy resins have greater tolerance for size variability than do polyester resins. Polyester-compatible reinforcements (rovings or mats) can be used with an epoxy resin; it is not advisable to use an epoxy-compatible roving with a polyester resin—a very low-strength product is likely.

Film formers—Film formers give a leathery toughness to the glass fibers while holding them together for subsequent processing. If there were no film formers, the filaments would "balloon" due to static electricity and would have a harsh feel.

Compounds used for film formers include solid and liquid epoxies and polyesters, polyvinyl acetate and other thermoplastic emulsions.

The film former can make a great difference in the strength of the fabricated product, although its contribution is not so great as that of the coupling agent. In certain sensitive tests, such as fatigue performance, the type of film former is important. In some instances, interaction effects have been found between the film former, coupling agent, and the type of matrix resin employed in a structure.

Lubricants and emulsifiers—Generally speaking, lubricants are water-soluble oils (poly alcohols, for example) that help decrease fiberglass damage during fabrication into rovings or during other processing methods. Often, the size emulsifiers (detergent-type compounds) also serve as lubricating elements.

Forms of reinforcement, and uses

Rovings—Two types of rovings are available:

- *An unsplit type,* used as a continuous reinforcement in filament winding, pultrusion, or woven roving. Fig. 2a shows a typical roving package.
- *A split-strand roving,* which is used for chopping into lengths of $\frac{1}{8}$ in. to 2 in. One type is factory-chopped and sold in $\frac{1}{8}$-in. or $\frac{1}{4}$-in. lengths as chopped strands for mixing into bulk molding compounds (BMC). The split-strand rovings are also used in sprayup, in making chopped-strand mat, or in preforms (a contoured mat).

In the past few years, a newer type of continuous-strand product has become available—a thick-walled cylindrical forming package, which can be dried and wrapped immediately after forming. This product is normally sold at a slightly lower price than conventional roving.

Chopped-strand mat—Chopped-strand mat is made by chopping split-strand fibers from forming cakes (or rovings) in 2-in. lengths, gathering the fibers on a conveyor chain, coating the mat with a binder, drying and

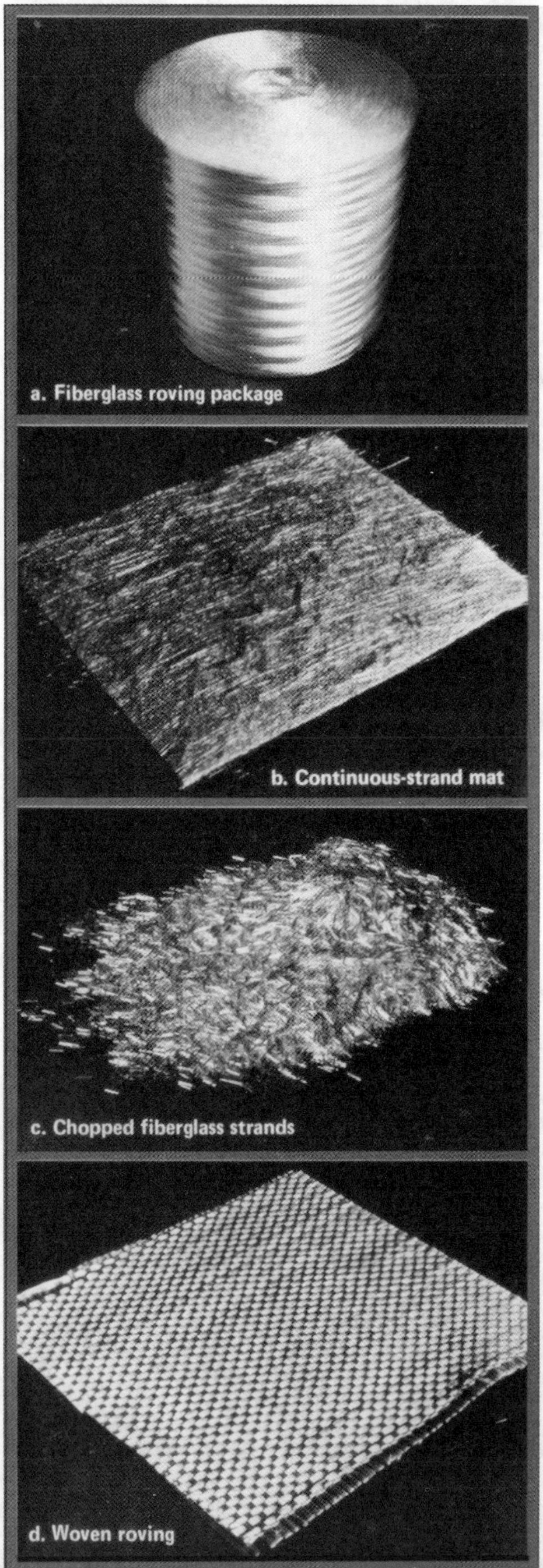

Fiberglass forms for composite fabrication Fig. 2

(Owens-Corning Fiberglas Corp.)

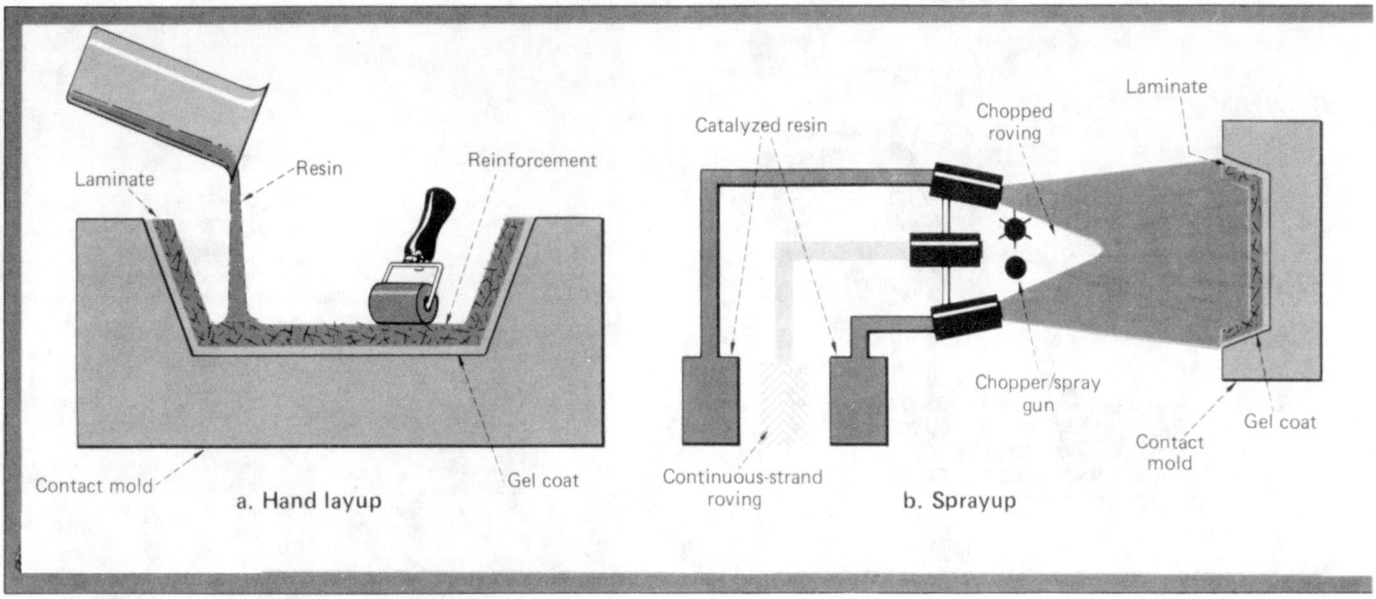

Schematic diagrams of FRP processing methods **Fig. 3a, b**

heat-setting the binder, and rolling up the mat. Two types of binders are used: soluble and insoluble (in polyester resin).

Chopped-strand mat is available in a number of widths (up to 76 in. wide); weights range from $\frac{3}{4}$ to 3 oz/ft^2. The principal application is in hand layup for tanks, pipes and other structures. Other applications are in translucent panels, filament-wound tanks and molded structures.

Continuous-strand mat—Continuous-strand mat (Fig. 2b) is made in the U.S. only by Owens-Corning Fiberglas (an imported French mat is sold by Certain-Teed). It is made in a patented process by swirling fibers (directly from the bushing) onto a conveyor chain. A binder agent is added just as in chopped-strand mat. The mat is available in widths up to 76 in. and weights from $\frac{3}{4}$ to 3 oz/ft^2.

Continuous-strand mat has greater tensile strength, but greater springback in the wet state; it is seldom used in hand layup because it will not conform as easily as chopped-strand mat. The high strength makes continuous-strand mat useful in compression molding and in pultrusion.

Chopped strands—Chopped fiberglass strands (Fig. 2c) have a high percentage ($1\frac{1}{2}$–2%) of an insoluble binder. Fiber lengths are $\frac{1}{8}$ to $\frac{1}{2}$ in. The strands are sold for mixing with polyester or epoxy resins and fillers to make bulk molding compounds (sometimes called dough molding compounds, or premixed compounds). The high content of insoluble binder is necessary for the strands to withstand the mixing process. If individual filaments are mixed in a compound, they tend to ball up and separate from the mixture. The compounds are used in compression molding, transfer molding and injection molding.

Milled fibers—Fiberglass strands are ground up by a hammer mill to produce milled fibers. Four sizes are manufactured (depending on the hole size in the hammer mill screen): $\frac{1}{32}$, $\frac{1}{16}$, $\frac{1}{8}$ and $\frac{1}{4}$ in.

Milled fibers have been used in a number of molding

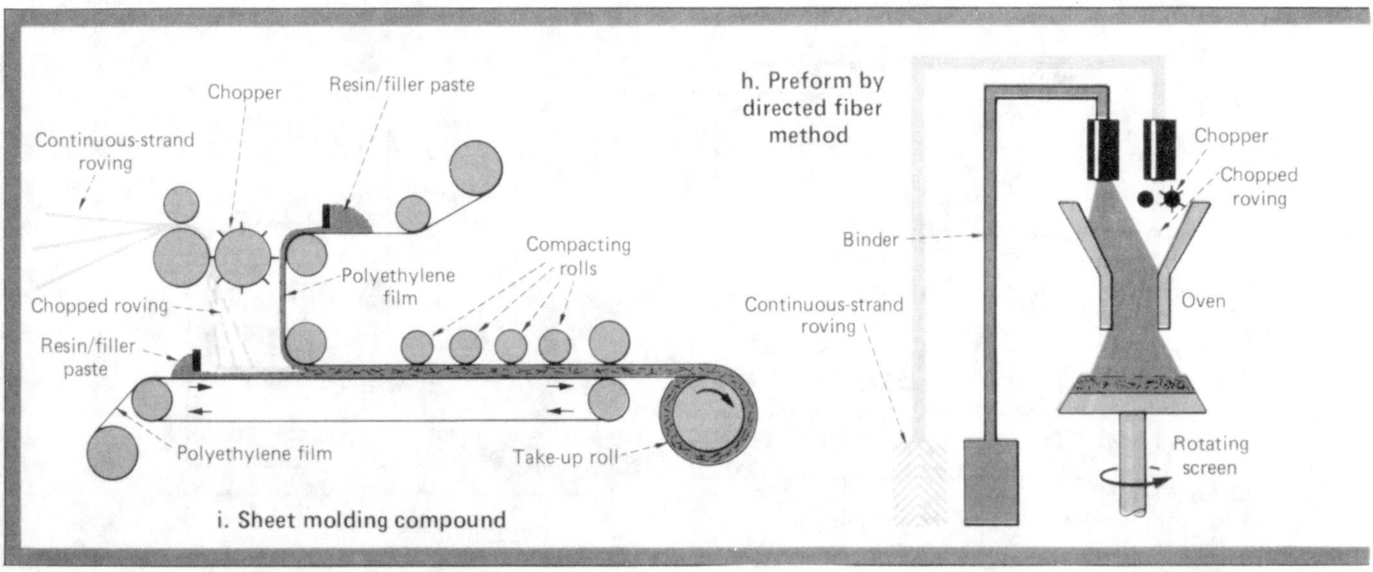

Schematic diagrams of FRP processing methods **Fig. 3h, i**

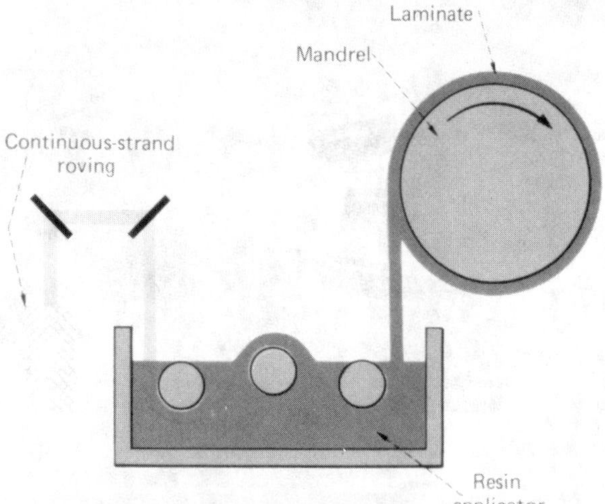

c. Filament winding

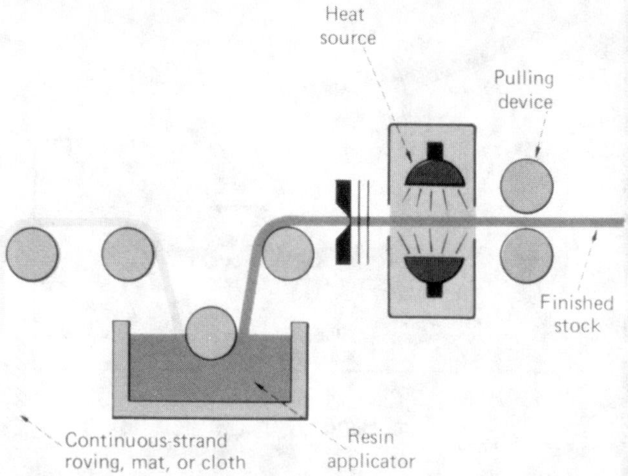

d. Continuous pultrusion

compound formulations. Recently, these fibers have found a potentially large market in reinforcing molded urethanes for automotive applications.

Woven rovings—A woven roving is a heavy cloth (Fig. 2d) normally made from 218-yield rovings in an approximately square weave pattern. Two common weights of fabric are: 24 oz/yd^2 (5 x 4 with 218 yield) and 18 oz/yd^2 (5 x 4 with 300 yield) where 5 x 4 indicates 5 strands/in. in the warp direction and 4 strands/in. in the fill direction.

Other weights have been developed for special purposes, and other warp and fill ratios, besides 5 x 4, have found particular applications. The 24-oz fabric is by far the most popular. Woven roving is made only with polyester-compatible sizes in widths up to 10 ft, usually with the same rovings sold for filament winding or pultrusions.

The greatest use of woven roving is to provide a structural backbone for hand-layup (contact-molded) structures. The usual construction method (discussed in the

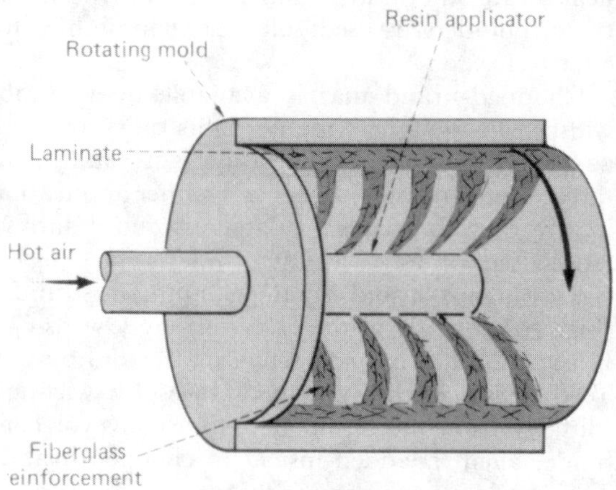

e. Centrifugal casting

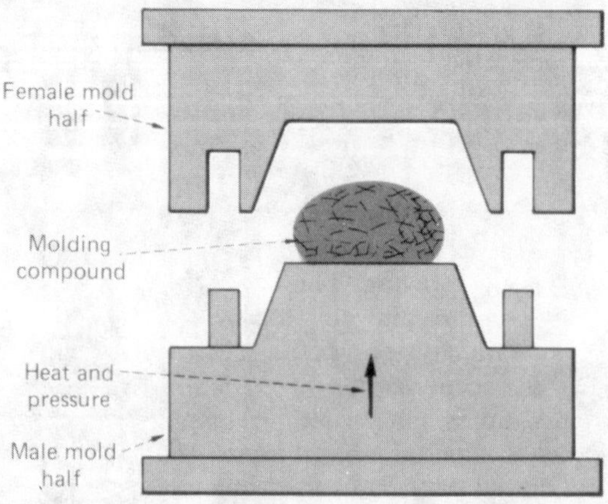

g. Compression molding

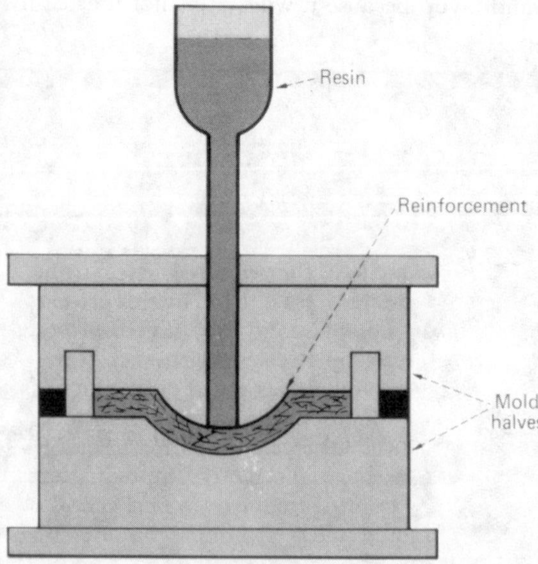

f. Resin transfer molding

Schematic diagrams of FRP processing methods

Owens-Corning Fiberglas Corp.

Sprayup chopping operation Fig. 4

Filament-winding a large-diameter chemical storage tank onsite Fig. 5

process section following) is to interweave woven rovings with mat layers.

Processing methods

Contact molding

Contact molding is defined as a zero-pressure molding method in which only one side "contacts" the mold surface. There are two principal techniques—hand layup and sprayup. Both have higher labor costs but lower tooling costs than other molding processes.

Hand layup—Mat or mat-and-woven-roving layers are wetted out with resin (normally polyester, but sometimes epoxy) and laid up on a male or female mold (Fig. 3a). Wetting should take place from bottom to top, to help eliminate air bubbles. Resin is poured onto the working surface, and mat is laid on top of the resin. The best practice uses alternate layers of mat and woven roving. Often a gel coat is applied to the mold surface prior to layup of the reinforcement. Strong and excellent structures have been made by this process. It is unbeatable for one-of-a-kind or limited-production structures. There is no restriction on size, from very small to very large. A wide variety of resins can be used.

A degree of automation (an impregnating machine) has been introduced in recent years for assembly-line structures such as boats. Formerly, impregnation was accomplished by pouring resin from a bucket and working it in with a hand roller. The impregnator consists of two motor-driven rolls with a pool of resin between the rolls. The mat and woven roving pass between the rolls and the resin is squeezed into the glass reinforcement. The rolls can wet out 50 ft or more of laminate per minute. Molds for hand layup are often made from FRP. Coated plaster, cast metal, sheet metal, wood and silicone rubber also are used.

Sprayup—By chopping the glass in a mechanism that also "throws" the chopped fibers (and at the same time sprays polyester resin) a structure can be built up layer by layer in a mat-like style (Fig. 3b, 4). Following deposition of fiber and resin, it is necessary to roll-out the structure (compacting the resin and glass) with a special rolling wheel. Sprayup is a fast process, with rates of 20–30 lb/min of product possible. Like hand layup, the chopped glass and resin can be applied over a gel-coated mold.

A typical structure is the one-piece shower stall or tub-shower combination.

A number of chopping and spraying guns are on the market. There are two basic types of spray guns—one with internally mixed resin and the other with externally mixed resin. The internal-mix system has the reputation for achieving a more thorough mixing, but the external mix avoids the buildup of cured resin on the gun and the necessity for immediately flushing the mixer after spraying. There are also two types of spray nozzles—airless and air atomized. Because of the need for reducing styrene vapors, the airless gun has gained favor in recent years.

Sprayup is often used in combination with other processes, especially hand layup and filament winding. Automated, tape-controlled machines have been developed that produce complex shapes, such as end-caps for tanks, large-diameter pipe fittings, and other structures that are required in quantity.

Filament winding

This process is most important in producing low-cost, high-strength, corrosion-resistant equipment (Fig. 3c, 5). Filament-wound shapes combine the highest-strength fiberglass composite with the lowest-cost raw materials and a highly automated process.

Structures normally have only one smooth side (as in contact molding); recently, some have been made that incorporate a molding step after winding to obtain a smooth exterior and provide exterior attachments (bosses, undercuts, odd curvatures).

The filament-winding process had its beginning in the aerospace industry for producing structures with a high strength/weight ratio. Most solid-fueled rocket motors are still made by the filament-winding process (see photo, p. 96). In the rocket motor development, a great deal of glass sizing research, stress analysis techniques, and resins development became available for adaptation to commercial structures.

Table V illustrates some advantages and limitations of filament winding for making pipe, tanks, ducts or other corrosion-resistant structures.

Features of filament winding	Table V
Advantages	
Automatable	
High-strength products	
Versatile in size range	
Low-cost raw materials	
Room-temperature or heat curing	
High production rates possible	
Strength orientation possible	
Limitations	
Reverse curves difficult	
Low-angle (parallel to principal axis) winding difficult	
Complex (double-curvature) shapes difficult	
Air-pollution problems	

Winding-angle effects—There are two basic types of filament winding—helical and biaxial.

Consider a cylindrical structure. In *helical winding,* a constant angle (traditionally measured from the rotational axis), alternately plus and minus, is maintained. Each pair of angles constitutes one layer, even though the layer may be constructed in steps. The required hoop/axial stress ratio is accomplished by winding at an optimum angle. For example, it has been mathematically shown (and confirmed by experiment) that a winding angle of 54¾ deg will result in a 2/1 hoop-to-axial strength ratio. This is the optimum for an internal pressure requirement with no restraints (such as in a closed-end cylindrical tank).

In *biaxial winding,* the required strength ratio is accomplished by combinations of 0-deg and 90-deg windings. The total thicknesses of the 0-deg and the 90-deg layers determine the strength ratio (assuming equal glass content in each layer). For example, the 2/1 hoop-to-axial strength ratio would consist of twice as many windings at 90 deg as those at 0 deg.

Materials—The principal reinforcement is fiberglass roving; aramid yarns (an aromatic polyamide fiber) and graphite rovings have also been employed. In aerospace structures, preimpregnated epoxy tapes are often used; the tapes are made from fiberglass, aramid, graphite, and boron fibers.

Polyester resins account for probably more than half of the total business in filament winding, principally in tanks and large-diameter pipe. For small-diameter, high-pressure pipe, epoxy is the predominant resin, although vinyl ester is used by some producers. Other resins used include furans (for corrosion resistance) and polyimides (for high-temperature performance).

Impregnation baths—There are many techniques (in wet-filament winding, which is the most common form) for wetting the strands prior to emplacement on the mandrel. The purpose of the impregnation bath is to wet the fiberglass strands (or other reinforcement) and remove excess resin. The excess resin can be squeezed out at the mandrel by the winding tension and resulting pressure, but it is economically preferable to remove as much resin as possible at the bath.

The dip bath system is the most common. The glass strands are guided over and under bars submerged in the resin. For ease in threading, it is desirable to have all the top bars (the ones that the strands pass under) fastened to side-rails so that all the bars can be lifted out of the bath simultaneously.

Rollers can be used to reduce the tension, but "roll wraps" can occur if strands break; this will eventually shut down the equipment. Some designers have used immersed rollers with the strands passing over the top of the rollers. This is effective, but results in less "working" of the glass to obtain complete wet-out.

The usual method of removing the excess resin is by passing the wet strands between a metal bar and a rubber (or urethane) bar, which are pressed together. Other methods incorporate slots, dies or rollers.

Winding machines—The most common machine has gears and roller chain drives to link the mandrel and carriage. Other types include cam-driven and electronic-drive machines. The latter is a recent development that incorporates a variable-speed drive continuously adjusted to maintain the desired speed ratio between mandrel and carriage.

Many fabricators have designed and built their own equipment, and there are a number of special-purpose machines. One machine that provides a continuous winding operation for large-diameter pipe has a cantilevered steel-ribbon mandrel that advances as it rotates, and disassembles at the free end. The endless steel ribbon returns through the hollow core to the beginning.

Cost factors—The high production rates possible on filament-winding equipment (linear speeds of 300–500 ft/min are common in medium to large sections), plus highly automated handling equipment, have reduced costs in making filament-wound products. Recent surveys [*2*] have shown that polyester (and even epoxy) filament-wound pipe can compete favorably against plain carbon steel in some circumstances. Recent price increases for steel may make filament-wound fiberglass even more competitive.

Active sections in ASTM committee D-20.23 — Table VI

Section No.	Title
D-20.23.10	Reinforced thermosetting resin pipe
D-20.23.11	Reinforced plastic mortar pipe
D-20.23.12	Mechanical test methods
D-20.23.13	Chemical and environmental test methods
D-20.23.14	Pipe fittings, flanges and joints
D-20.23.15	Tanks and process equipment
D-20.23.16	Underground tanks and structures
D-20.23.17	Fume handling equipment
D-20.23.18	Reinforced thermosetting resin materials of construction

Future trends in filament winding—

- Combinations with pultrusion for more-automated systems.
- Automated handling of raw materials and products, with greater use of fillers (such as currently used in making RPM pipe).
- Combinations with thermoplastics.

Pultrusion

Pultrusion is a truly continuous process (Fig. 3d); with an automated cutoff saw, a pultrusion unit can run with virtually no attention, except for occasional checking for glass breakouts and resin levels.

Ordinarily, two layers of mat encapsulate each roving layer, with rovings spaced about four per inch of width. The cured thickness of a single layer of mat and rovings is about 0.040 in.

Materials and process—Pultrusion resins are predominantly polyester, but epoxy is used in a few special-purpose applications. The range of polyester types is from lowest-cost orthophthalic to premium grades such as vinyl ester or bisphenol A. The limit on fillers is the resin viscosity; at high filler loadings, the rovings will not wet adequately. Pigments are often added for coloration and ultraviolet resistance. An internal mold release is a must.

Resin impregnation baths are usually 3 to 6 ft long. In a way, this is strange, since filament winders (whose speeds are 50 to 100 times higher) normally use much shorter resin baths. In filament winding, much of the wet-out occurs on the mandrel and tensile forces tend to squeeze out air. In pultrusion, as soon as the rovings reach the die cavity, any remaining air is trapped in the laminate, which at this point is cured by the heat source.

Surface mats are sometimes added to provide a resin-rich appearance. Fiberglass surface mats have no strength in the wet state. They should be fed into the structure as close to the curing die as possible. Polyester surface mats (unwoven cloth) have greater strength and will survive the trip through the impregnation bath; unfortunately, they are more expensive than glass.

The hot dies for curing are normally 2 to 4 ft. long and should have as high a polish as possible. They are made from hardened steel, polished, and then hard-chrome-plated. Ordinarily, the die is split longitudinally. Surface smoothness is critical.

The relationships of glass, resin, filler and stripping-die area are mathematically solvable on a volume basis.

The total amount of roving can be calculated from the relationship:

$$Y_{tot} = 1 + W_g(\rho_m/\rho_g - 1)/36\ W_g\rho_m A$$

where Y_{tot} = total yield of glass rovings in yd/lb; W_g = weight fraction of glass in the composite; ρ_m = density of resin and filler mixture in lb/cu in.; ρ_g = intrinsic fiberglass density in lb/cu in.; A = curing-die area in sq in.

From this, the number of roving packages can be determined from:

$$N = Y_g/Y_{tot}$$

where N = number of packages of fiberglass; Y_g = yield of individual package in yd/lb.

Centrifugal casting

The centrifugal casting process for making fiberglass pipe is one of the oldest FRP processes (Fig. 3e). A cylindrical mold is spun about its long axis. A mat or mat-and-cloth reinforcement is laid in the mold, resin is poured in, and the mold is turned. The laminate is compressed against the mold.

In a more recent development, sprayup glass is chopped and resin sprayed inside while the mold is spinning. In other operations, a low-viscosity premixed compound is dispensed into the rotating mold.

Centrifugal casting requires a modest investment, and produces parts having smooth surfaces and low void content. The process, which has shape limitations, is used for making tubes, pipes and tanks.

Resin transfer molding

This is a relatively new process, sometimes called resin injection molding, which appears to have a large potential for producing complex shapes, using woven roving, mats and surface mats, or gel coats. See Fig. 3f.

The process uses matched molds, male and female, made from reinforced plastic or metal. The dry reinforcements (mat, cloth, woven roving, etc.) are placed in the mold, which is then closed. Tight seals, usually O-rings, are necessary at all mold joints. A vacuum pulls some of the air out of the mold, and then resin, promoted and catalyzed for a pot life of about 30 min, is pumped in at the bottom. When resin flows out at the top of the mold, pumping is stopped. After the resin has hardened, the mold can be opened and the part removed.

In comparison to *hand layup*, RTM gives two molded sides, dimensional accuracy and faster production cycles—but requires higher capital investment.

In comparison to *compression molding*, the RTM process has a lower capital investment, but a higher labor factor and slower cycle, so that for runs of more than about 10,000 parts, compression molding is cheaper. Some fairly large parts have been built via RTM, and the process has considerable potential.

Compression molding

Many mass-produced parts for corrosion-resistant equipment are being made by one of the several forms of compression molding (see Fig. 3g). These include pipe flanges, pipe fittings, manholes and small tanks.

All of these parts have a complex shape and must be produced in large quantities. Compression molding is a process using two matched molds, male and female. The molds are preferably made from high-carbon steel hardened on the surface after machining. The mold faces are then polished to a high gloss, and chrome-plated for wear resistance. They usually will last for 10,000 parts before requiring stripping and replating. Molds may be made from FRP or filled epoxy, depending on: type of part and shape, mold material (and cost), number of parts to be produced, type of reinforcement and resin.

Many of the variations in compression molding depend on the type of material being molded. Several innovations in materials systems have occurred in recent years. The types of materials now being used are bulk molding compound, mats or preforms with liquid resins, and sheet molding compound.

Bulk molding compound—This material is sometimes called "dough molding compound" or "premix." Bulk molding compounds (BMC) are made from polyester or epoxy resin having relatively low glass contents (10–15%) and high filler contents (30–40%). The ingredients are charged into a heavy-duty mixer, such as a dough mixer, and blended. The fiberglass is either $\frac{1}{8}$- or $\frac{1}{4}$-in.-long chopped strand with a high amount of insoluble binder ($1\frac{1}{2}\%$ to 2%) to prevent it from filamentizing. Experience has shown that filamentized strands ball up and separate out in the mixing process. With higher glass contents, this separation also occurs.

The polyester or epoxy resin is catalyzed for high-temperature cure to achieve a long life at room temperature. Heat activates the catalyst system and allows the polymerization to occur quickly.

Bulk-molding-compound parts have moderate tensile strengths (around 15,000 psi) with better impact resistance than a filled resin. They are used in many low-stressed applications.

Mats or preforms with liquid resin—Continuous strand mats are often used with liquid resin in compression molding where the draw (depth) is not too great. In more-complex sections, a preform is used (Fig. 3h). This is made of chopped glass fibers (1–2 in.) blown against a screen shaped like the part. The screen has a vacuum pulled on the opposite side. After sufficient fiber buildup, a binder is sprayed onto the preform, and the matlike structure is dried.

The preform is placed in the mold in the same manner as the mat. The exact quantity of filled resin is poured over the preform, and the mold is closed. The high pressure of the press pushes the liquid resin into the fiberglass. When curing is complete, the part is strong and durable, with 20–25% glass content.

Sheet molding compound—A quantity of approximately 0.5%–1% magnesium or calcium oxide will cause polyester resins to thicken to a viscosity of about 18–20 million cP. This is a level at which the material is essentially non-tacky and has a leather-like stiffness. If the thickened polyester is heated, the resin becomes liquid again and will cure, under the influence of a catalyst, in a normal manner.

Principal relevant ASTM standards — Table VII

Std. No.	Title
ANSI/ASTM C581-74	Chemical resistance of thermosetting resins used in glass fiber reinforced structures
ANSI/ASTM D1599-74	Short-time rupture strength for plastic pipe, tubing, and fittings
ANSI/ASTM D2143-76	Standard test method for cyclic pressure strength of reinforced thermosetting resin pipe
ANSI/ASTM D2562-70	Standard recommended practice for classifying visual defects in parts molded from reinforced thermosetting plastics
ANSI/ASTM D2584-72	Standard test method for ignition loss of cured reinforced resins
ASTM D2992-71	Standard method for obtaining hydrostatic design basis for reinforced thermosetting resin pipe and fittings
ASTM D2996-71	Standard specification for filament wound reinforced thermosetting resin pipe
ASTM D3299-74	Standard specification for filament wound glass fiber reinforced polyester chemical resistant tanks
ASTM D3681-78	Test for determining chemical resistance of reinforced thermosetting resin pipe in a deflected condition

The sheet molding compound (SMC) is made by continuously laying down liquid resin, mixed with fillers, thickening additives and a catalyst, on a traveling polyethylene film (Fig. 3i). Glass fibers (2 in. long) are chopped and sprinkled onto the resin layer. A second layer of resin is laid down onto another film that is pressed down on top of the glass fibers.

The sandwich of films, glass and resin passes under rollers, or a pair of chain belts, that massage the layers and work the resin into the fiber bundles. The SMC is then rolled up and set aside for "maturation," a period of several hours during which thickening occurs. At the end of maturation, the sheet can be cut into sections to make a weighed charge for the molding process. The film can be stripped off since the SMC is non-tacky.

The charge can be varied in both weight and shape, as required for the part. While the hot mold halves are open, the charge is placed in the mold. Emplacement may be manual or mechanical. The mold is closed and the charge flows to fill the mold cavity. Pressure may reach 1,500–2,000 psi from the hydraulic press before and during the curing cycle, which lasts 1 to 2 min.

A variation of sheet molding compound, SMC-II, uses an adjusted chemical system to thicken the resin to a viscosity level approximately half as high as the standard SMC. This resin is still non-tacky, but it will flow to fill the mold at a lower pressure. This lower pressure allows the use of reinforced plastic or aluminum molds that are much lower in cost than the hardened steel molds required for standard SMC.

The properties of the SMC parts are equivalent to preform parts (20,000–30,000 psi tensile at about 30% glass content) but with flow properties, reproducibility and cost similar to BMC parts.

Mold releases

Mold releases are used in one of two ways, or a combination of both—internal, mixed in the resin, or external, applied to the mold. The common *internal* mold releases are: lecithin; zinc, calcium or other stearates; organic phosphate compounds. *External* mold releases include: fluorocarbons (dispersions of a solid fluorocarbon, in a volatile solvent); silicone oils; waxes.

The silicone oils are good releases but cannot be used if any secondary bonding is planned. Some filament winders use fluorocarbon; some prefer waxes or silicones. In pultrusion, only an internal release can be used; in compression molding, it is common to use both internal and external mold releases.

Miscellaneous molding methods

There are also a number of minor molding methods that are infrequently used in the manufacture of corrosion-resistant structures. Among these:

Autoclave—A laminate shaped over a mold, usually male, is placed in a heated pressure tank and cured. This process is used mostly in the aerospace industry for cloth/epoxy laminates.

Cold molding—This is a form of compression molding that uses a promoted polyester resin, and requires no heat for curing. It is slower than normal hot-compression molding. The investment required and the product obtained are similar to those for resin transfer molding.

Vacuum-bag molding—This is similar to the autoclave molding system, but uses a vacuum bag. It involves less investment and is a lower-pressure system.

Injection molding—This is a high-investment, mass-production process mostly used for small thermoplastic items, both unreinforced and with up to 40% glass fibers. The nature of the process involves high-pressure pumping and movement of materials through small orifices; the glass fibers are reduced to a very short length, probably less than $\frac{1}{32}$ in.

In recent years, techniques have been developed for molding thermoset resins, primarily reinforced ones. Because of high investment and high mold cost, the process is limited to large production runs.

Rotational molding—This process slowly rotates a closed, split mold on two axes. A dry powder or a liquid tumbles inside the heated mold and gradually builds up a shell. The process is limited to hollow shapes but can produce closed-end structures without the use of a core mold. Thermoplastic resins, reinforced and unreinforced, are most commonly used, but some liquid thermosets have been molded by this process.

Other processes—These include: cold stamping of reinforced thermoplastic sheets; fiberglass/polyester-backed thermoplastic sheet (vacuum-formed); encapsulation; and continuous lamination.

Standards and specifications

The production of corrosion-resistant composites is a rapidly growing business, and there have been numerous failures due to misapplication, poor craftsmanship, or underdesign.

The most active standards group by far is ASTM Committee D-20.23 "Reinforced Plastic Piping Systems and Chemical Process Equipment." There are nine sections, dealing with various products and test methods. See Table VI.

At present, the various sections of D-20.23 are working on approximately thirty different draft standards or revisions to standards. These include specifications, test methods or recommended practices for:

Underground petroleum-storage tanks.
Contact-pressure-molded chemical-resistant tanks.
Contact-pressure-molded ducts.
Corrosion resistance of molded fittings.
Manholes.
Underground chemical tanks.
Sewer and industrial pressure pipe.

Some of the principal ASTM standards that pertain directly or indirectly to chemical equipment made from composite materials are listed in Table VII.

Four other standards should be mentioned in connection with corrosion-resistant equipment:

- U.S. Dept. of Commerce NBS PS15-69, "Custom contact-molded reinforced-polyester chemical-resistant process equipment."
- British Standard BS 4994: 1973 "Specification for vessels and tanks in reinforced plastic."
- British Standard BS 5480: Part 1: 1977 "Specification for glass fiber reinforced plastics (GRP) pipes and fittings for water supply or sewerage."
- Amer. Pet. Inst. Std. 5 LR.

References

1. Mallinson, J. E., private communication, May 1978.
2. Marshall, S. P., and Brandt, J. L., Installed Cost of Piping, *Chem. Eng.*, Aug. 23, 1971, pp. 68–82.

The author

J. Albert Rolston, P.E., is a consulting engineer, Box 62, Granville, OH 43023, telephone 614-366-6117. He received a B.S. and an M.S. in chemical engineering at North Carolina State U. Mr. Rolston spent 12 years at Owens-Corning Fiberglas Technical Center and has a total of 19 years' experience working with fiberglass composites. He is a licensed engineer in Ohio and Virginia, a corrosion specialist for the National Assn. of Corrosion Engineers, a fellow of AIChE and a member of the Amer. Soc. for Testing and Materials, the Soc. for the Advancement of Material and Process Eng., and the Soc. of the Plastics Industry.

Using fiberglass-reinforced plastics

Process equipment for corrosive environments is increasingly being made of fiber-reinforced plastics. This article explains what FRP is, discusses corrosion resistance and other properties important in selecting a material of construction, examines where FRP can and cannot be used, and provides some general guidance for selecting such plastics.

Robert C. Talbot, *Ashland Chemical Co.*

☐ Fiberglass-reinforced plastic (FRP) has been used for various types of equipment in the chemical process industries (CPI) since the early 1950s, and its use continues to grow. Process vessels of all shapes and sizes, columns, scrubbers, hoppers, hoods, ducts, fans, stacks, piping, filter-press frames, pumps, pump bases, valve bodies, elevator buckets, heat-exchanger shells and tube sheets, mixer drive shafts, mist-eliminator blades, floor toppings, and vessel and tank lining systems are just a few examples of equipment made of FRP.

The chief reason for the popularity of these materials is their excellent resistance to corrosion, as well as their ability, when the appropriate additives have been incorporated, to resist heat and/or fire. This article will discuss these properties and other important factors that determine a particular material's applicability under certain conditions. But first, we will discuss a few of the basics of FRP.

What is FRP?

FRP stands for, most commonly, *fiberglass-reinforced plastic*, although since the introduction of other fiber reinforcements, FRP is also used to mean simply *fiber-reinforced plastic*. Terms used interchangeably with FRP are *reinforced thermoset plastic* (RTP) [which should not be confused with *reinforced thermoplastic*], *reinforced thermoset resin* (RTR), and *glass-reinforced plastic* (GRP). *Laminate* refers to the material that is manufactured from thermosetting resin and reinforcement fibers at the time of equipment fabrication.

FRP is a composite material consisting of a thermosetting polymer, often a type of polyester, reinforced with glass (or other) fibers. Reinforcements and processing methods are discussed in detail in Ref. [*1*].

There are six types of FRP resins: isophthalic, orthophthalic, chlorendic, bisphenol A fumarate, furan, and vinyl ester. Each is described briefly below. Their properties are compared in Table I, and some actual applications in corrosive environments are summarized in Table II.

Isophthalic resins

Isophthalic resins are non-fire-retardant, either rigid or flexible unsaturated polyester resins based on isophthalic acid and glycols of various types.

These resins are used for moderate corrosion resistance applications up to 180°F. At temperatures not exceeding 180°F, they generally exhibit excellent resistance to water, weak acids and alkalis, and good resistance to solvents and petroleum products such as gasoline and oil. The flexible isophthalics exhibit a lesser degree of chemical resistance than the rigid isophthalics of higher molecular weight.

Orthophthalic resins

The difference between orthophthalic and isophthalic resins is in the position of the two –COOH groups in the phthalic acid molecule: they are on adjacent carbons (e.g., in the *ortho-* position) in orthophthalic acid, and are separated by one carbon (e.g., in the *meta-* position) in isophthalic acid. Orthophthalic resins (often called general-purpose polyesters or ortho resins) provide only a minimum of corrosion resistance, and thus are not usually used in instances where corrosion resistance is needed.

Chlorendic resins

Chlorendic resins are unsaturated halogenated polyester resins based on HET (hexachlorocyclopentadiene) acid or chlorendic anhydride reacted with a stable glycol (such as neopentyl glycol).

They are particularly well suited for use at elevated temperatures, up to 350°F, because the chlorendic backbone is highly resistant to distortion by heat.

Originally published October 29, 1984

They are able to handle aggressive, highly oxidizing environments, concentrated acids and some solvents very well, but are poor in alkaline service.

Chlorendic resins can be formulated to achieve Class I fire rating (based on the standard test, discussed later). The only material with a better rating (of 0) is asbestos.

Bisphenol A fumarates

These are unsaturated, rigid polyesters made by reacting bisphenol A with propylene oxide to form a glycol, then reacting the glycol with fumaric acid to produce the resin.

Bisphenol A fumarate resins exhibit excellent corrosion resistance to both aggressive acid and alkali environments up to 250°F, but are not suitable in strong oxidizing conditions or solvent applications. They are substantially better than the isophthalics in severe corrosion applications.

Fire-retardant varieties of bisphenol A resins are available.

Vinyl esters

Vinyl ester resins are methacrylated epoxies that are very similar to polyester, although they are classified separately. They offer excellent physical strength and, in general, much better impact strength than rigid polyester resins. While the standard vinyl ester resins are limited to 200–225°F in most applications, other versions with higher-density crosslinking are suitable for up to 250°F.

These resins exhibit excellent resistance to acids, alkalis, hypochlorites and many solvents. They are preferred for filament winding, especially for machine-made piping.

Brominated versions of vinyl esters are available for fire-retardant applications.

Furan resins

Furan resins are based on a furan polymer derivative of furfuryl alcohol. They exhibit excellent resistance to strong alkalis and acids containing chlorinated organics, and are superior to polyesters and vinyl esters in solvent resistance. Furan laminates are good up to about 250°F for many corrosive applications. However, the furan material is not suitable for oxidizing chemicals, and should not be used for chromic or nitric acids, peroxides, or hypochlorites. Most furans offer some degree of fire retardance.

Generally, the furan resins are considered to be the best materials for all-around corrosion resistance. Unfortunately, they are one of the most difficult resins to work with and cure at room temperatures. Only qualified fabricators—those with proven performance—should be used for furan applications.

Evaluating corrosion resistance

As shown in Table I, FRPs do not resist all environments, nor do they respond equally to specific corrosives. Unless a successful installation exists to serve as a

Comparison of properties of various types of FRP — **Table I**

	Isophthalic	Orthophthalic	Chlorendic	Bisphenol A fumarate	Furan	Vinyl ester	Carbon steel	Stainless steel
Corrosion resistance								
Acids	B	C	A	A	A	B	C	B
Alkalis	B	C	C	A	A	A	B	B
Peroxides	C	C	A	B	C	B	C	C
Hypochlorites	C	C	A	B	C	B	C	C
Solvents	B	C	B	B	A	B	A	A
Flame retardance	C	C	A	C	B	C	A	A
Structural strength	A	B	A	A	A	A	A	A
Thermal insulation	A	A	A	A	A	A	C	C

A = High, B = Moderate, C = Low

Summary of actual applications of FRP in corrosive service — **Table II**

Equipment	Corrosive environment	Type of resin
Venturi cooler	160°F boiler fluegas	C
Stacks, chimney liners 500 ft and higher	Fluegas at 350-375°F, with peaks to 500-700°F	C
	Lined stacks for 310°F (continuous) fluegas	C,B
Scrubbers (Cross-flow; ionizing; cyclone; co-current; countercurrent; spray; packed tower; vertical; parallel flow; horizontal)	300-350°F bagasse fluegas, trash burner fluegas	C
	Acidic inert gas at 120-140°F	C
	Coal dust containing fluoride up to 350°F	C
	340°F fluegas and scrubbed carbonate liquor	C
Circulating tanks and piping	Acidic fluids up to 180°F	C
Mist eliminators, grating support beams	Sulfurous acid to 350°F and thiosorbic lime	C
	Coke quenching to 200°F	F
Fans and blowers	350°F fluegas	C
Electrostatic precipitators vibrator housings	H_2S and SO_2 fumes at 280-360°F	C
Piping with abrasion-resistant lining	Scrubbing-liquor slurry	C
Desulfurizing units	H_2S, SO_2 and citric acid	C
	H_2S, SO_2 and monoethanolamine to 270°F	F
Incinerator duct, stack, scrubbers	Trash, garbage, chemicals (no damage after short-duration exposures to 600°F	C
Smelter duct, stack scrubbers	Copper, lead, zinc, to 280°F	C

C = Chlorendic, B = Bisphenol A, F = Furan

guide, testing and evaluation must be done on an application-by-application basis. This is the more common situation.

Because of the many variables that can affect chemical resistance, it is often somewhat difficult to evaluate the results of chemical resistance tests and to select an economically serviceable material. This section describes the parameters that must be evaluated, factors that affect performance, some testing methods, and how to evaluate test results.

Chemical attack

Chemical attack on FRP is a complicated phenomenon that can occur in a number of different ways, which may be broadly classified as follows:

- Resin attack. This includes: (1) degradation and disintegration of a physical nature due to absorption, permeation or solvent action, which, under mild conditions, may manifest itself as reversible or irreversible swelling; (2) oxidation; (3) hydrolysis; (4) radiation, including nuclear; (5) thermal attack; or (6) combinations of these.
- Glass attack. Certain environments might not attack the resin itself, but upon absorption will attack the glass reinforcement. Fluorides, alkalis, hot water and its vapor, and hydrochloric acid under certain conditions, are typical examples of glass-attacking agents.
- Glass-resin interface attack. The glass-resin interface can be attacked by many substances, even though the resin itself might be reasonably resistant to attack. Wicking (degradation of the reinforcement due to an inadequate resin cover) and delamination are typical effects of this type of attack.

As a result of exposure to an environment, the FRP can respond in a number of different ways. It might not be affected at all. Or it might be embrittled, softened, discolored, charred, crazed (developing a network of fine cracks), weakened, delaminated, dissolved or blistered. Or it might exhibit any combination of these effects.

Although all FRP materials will exhibit attack in fundamentally the same manner in a particular environment, certain types—chlorendic, bisphenol A, and vinyl ester—are significantly more resistant. This is due to their somewhat unique structures, which have a built-in steric protection of ester groups. However, knowledge of the chemical or molecular structure of such resins *does not* eliminate the need for actual testing to determine suitability for a particular application. For example, certain vinyl esters are reasonably resistant to alkaline environments under some conditions; others are only fairly so.

Temperature and concentration effects

Changing the temperature of an environment will influence the response of FRP laminates. An increase in temperature will accelerate the rate of attack, in some instances to the point where degradation can be extremely rapid.

However, FRP laminates are heterogeneous and, in contrast to homogeneous metals and alloys, do not react at a predictable rate when exposed to a corrosive environment. The rule of thumb (often used for metals and alloys) that the corrosion rate will double for each 10°F

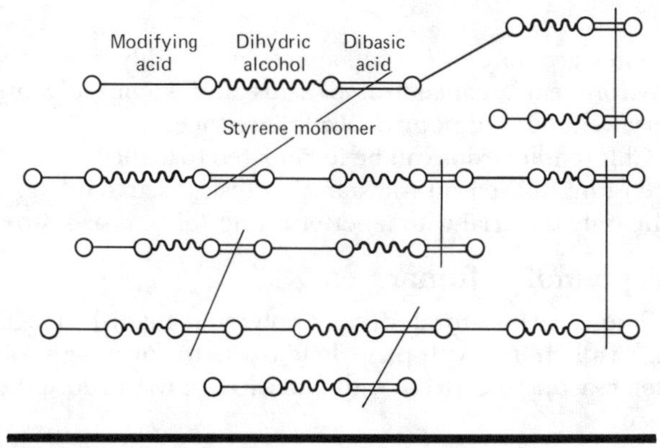

Complete crosslinking in polyester resin **Fig. 1**

rise in temperature is not applicable to FRP materials. Temperature effects cannot be extrapolated reliably to predict material response at different temperatures. An FRP material may be severely attacked by a substance at 180°F but possess excellent resistance up to 150°F.

A decrease in temperature will almost always reduce the degree of attack. However, phase separation or precipitation of one of the components of the environment as a result of the drop in temperature can completely alter the mode of attack.

There are similar problems in attempting to predict response based on concentration of the environment. A decrease in concentration generally results in decreased attack, but not in all instances, particularly when the material is exposed to corrosive organics. An increase in concentration will not necessarily accelerate attack on FRP—for example, 10% sodium hydroxide is *more* aggressive than a 50% solution.

Impurities or very small amounts of a substance can also significantly influence the response of FRP laminates and can often accelerate attack. Hydrochloric acid, for example, is not the same as hydrochloric acid contaminated with small amounts of chlorobenzene or other organics. In addition, certain conditions may cause a minor component to separate out of solution into a concentrated layer or phase and result in significantly increased attack in that region.

For reliable corrosion-resistance results, FRP must be tested under controlled, known conditions and under conditions that very closely simulate the actual environment in which the material would be used.

Laminate cure

Cure plays one of the more important parts in the chemical resistance developed by FRP materials.

Theoretically, the curing reaction should go to completion at room temperature, with all double bonds reacted, to achieve the complete cure or crosslinking shown schematically for polyester in Fig. 1. However, complete crosslinkage is rarely achieved at ambient temperature, with only about 95% of the unsaturation typically being converted. This accounts for reduced chemical resistance and, in general, for the inability of a resin to achieve full property potential. On the other hand,

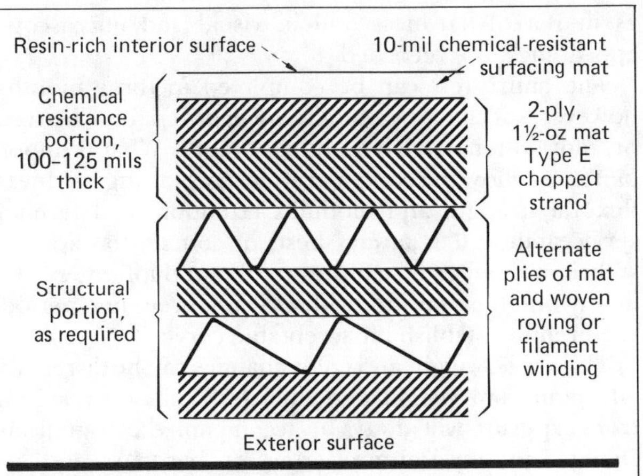

Typical laminate construction to provide chemical resistance **Fig. 2**

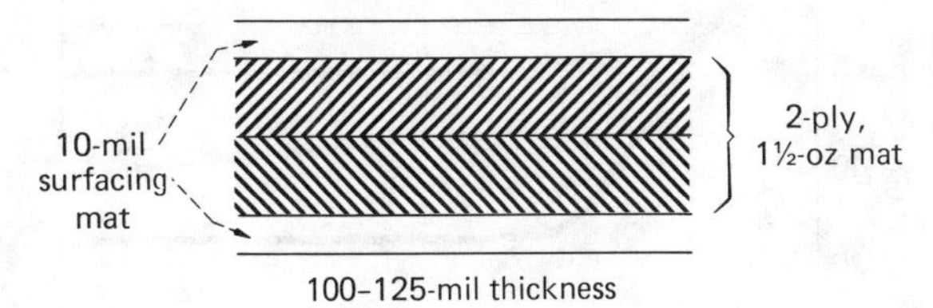

Standard laminate construction for chemical resistance testing **Fig. 3**

maximum chemical resistance and certain other improvements can be expected if an elevated temperature is used to more fully react the resin.

It is extremely important that, for FRP materials to perform at full potential, as complete a cure as possible be achieved.

This also indicates that candidate materials must be tested at essentially the same level of cure. A comparison cannot be made of one material cured at room temperature with another cured at an elevated temperature, because an elevated-temperature cure is not representative of most industrial-equipment fabrication techniques, where room temperatures are used almost exclusively.

The degree of cure can be determined by various methods, the simplest of which is the Barcol hardness test, or ASTM Method D-2583 [*2*]. A hardness value, on a scale of 0 – 100, is obtained by measuring the resistance to penetration of a sharp steel point under a spring load with an instrument called the Barcol Impressor. Standards and specifications generally accept a Barcol hardness value of at least 90, which indicates that full property potential, including sufficient (though not necessarily maximum) chemical resistance, has been achieved.

Some FRP laminate surfaces that are exposed to air will not cure well, resulting in reduced chemical resistance. Barcol hardness readings may not detect this condition in all cases, so an acetone sensitivity test should also be used, to check surfaces. If the surface feels tacky or soft after the application of acetone, it is not properly cured, and a post-cure at elevated temperature (up to 250°F) may be required before exposure of the material to a corrosive environment.

Laminate construction

Fabrication techniques also contribute to corrosion resistance.

Equipment for chemical-resistance applications must have a resin-rich surface containing a chemical-resistant surfacing veil or mat, whose purpose is to isolate the fiberglass reinforcement from attack that would result in wicking and blistering. The corrosion barrier should be about 125 mils thick and fabricated with two or three plies of Type E chopped strand mat. The structural portion of the laminate (the outer shell of the wall of a vessel or the interior of an I-beam) is high in glass fiber content; for the maximum strength/thickness ratio, chopped glass mat, cloth, woven roving, or, for very high strength, filament-wound reinforcement, can be used.

Fig. 2 illustrates the typical laminate construction required for chemical-resistant equipment. Fig. 3 shows the standard laminate currently used for chemical resistance tests by ASTM Method C-581 [*3*]; this simulates the interior chemical-resistant barrier surface of the actual construction in Fig. 2.

Candidate materials must be tested at the same nominal thickness and construction. While this is a relatively simple concept, there are some ¼-in.-thick laminates in the field that were submitted for comparative testing with the standard ⅛-in.-thick specimens, as well as laminates with two or more surfacing veils. The response of the thicker laminate or the one with two veils will always be better than that of the standard laminate. Curves F and G in Fig. 4 point out very well how this can be misleading. The thicker laminate (Curve F) appears to be significantly better than the thinner one (Curve G), even though they are constructed in the same way and of the same material. Castings are not representative of an actual fabrication and should not be used to assess chemical resistance.

Test methods

The ideal method for determining the response of FRP materials to a corrosive environment is to expose one 10-mil-surfaced face of the standard laminate shown in Fig. 3 to the environment under stress, with the magnitude of the load and the direction of the force corresponding to what would be encountered in actual service. Unfortunately, although this method is used by a number of investigators, it is somewhat complex, and not practical for evaluating a large number of laminates or materials under field conditions.

Another excellent test method is to totally immerse the standard laminate in the environment, again under stress. This is more representative of actual service conditions, since stress and chemical exposure are simultaneous. However, this method is also rather difficult

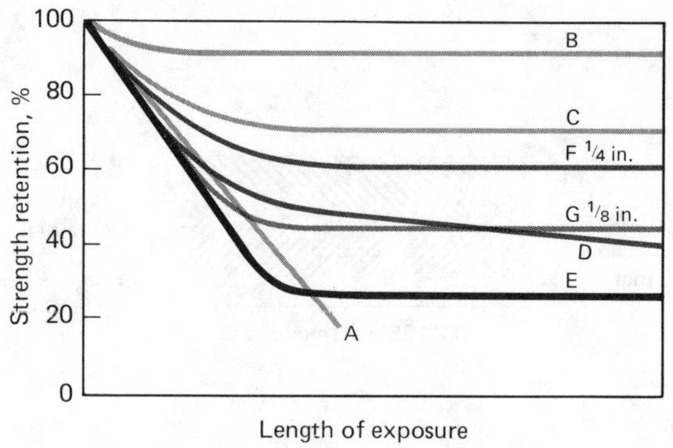

Flexural strength retention as a function of exposure **Fig. 4**

and is not practical either for in-the-field testing or for testing a large number of laminates.

A more practical method is to totally immerse standard laminates (as shown in Fig. 3) according to ASTM Method C-581 procedure. Although not as representative of actual conditions, it has been proven over the years and does successfully predict response and stability. It also provides an added safety factor, since it involves a significantly more severe exposure than actual service conditions, where only one surface would be exposed.

This total immersion method consists of the simultaneous exposure of a minimum of three test laminates to the environment, either in the laboratory under simulated conditions or under actual field conditions, in such a way that all surfaces are freely in contact with the test environment. A water condenser prevents loss of the solution; in other tests, the glass container would be insulated to control temperature. Certain solutions, e.g., sodium hypochlorite, should be changed frequently to ensure that the laminates are exposed to maximum concentration at all times.

Test laminates are usually removed after one, two and three months (at the minimum). After each exposure period, changes in flexural strength and modulus, as well as in Barcol hardness and in visual and microscopic appearance, are recorded.

Heat-resistance characteristics of FRP laminates and resin castings **Table III**

	Isophthalic	Vinyl ester	Bisphenol A fumarate	Chlorendic
Continuous operating temperature of FRP laminate system containing 65% glass, °F	275	250	300	350
Heat-deflection temperature of clear cast resin containing no glass reinforcement, °F	206	212	270	302

The entire test can be completed in three months; however, specimens generally are exposed for at least six or, more often, twelve months or more. What is important is to follow the course of attack by plotting hardness, flexural strength and modulus retention to determine serviceability. It is always best, of course, to expose a sufficient number of samples to obtain duplicate results in any one time period, with at least three time periods required to establish an acceptable curve.

Thickness, weight and color changes can be disregarded: Significant weight and thickness changes resulting from exposure will always be accompanied by significant changes in physical properties or hardness and are meaningless in themselves; changes in laminate color have also been found meaningless in corrosion-resistance testing.

In certain processes, it is important to know whether the FRP material will change the color of the solution and, particularly, whether discoloration persists with time. Tests to determine this must be run using the actual surface-area-to-volume ratio and expected residence time. It has been found that discoloration usually does not continue after the first exposure. Thus, a surface preconditioning, i.e., subjecting the surface to hot water or atmospheric steam for about four hours, or to the actual product for about four hours at the appropriate temperature, will often prevent discoloration.

Evaluation of test results

Evaluating the test results is the most difficult part of any material-selection program, and is much more complicated for nonhomogeneous structures such as FRP laminates than for metals. It is impossible to correlate property changes to a rate of deterioration, but experienced assessment of the property-change/time curves, such as Fig. 4 showing flexural-strength retention vs. months of exposure, can be used with a high degree of confidence to predict resistance and serviceability.

Rapid harmful reactions, represented by Curve A in Fig. 4, and those that are so slow as to produce only a minimum property change, Curves B and C, can be assessed quite easily. Obviously, Material A is unsuitable, while Materials B and C should, for most applications, be satisfactory for essentially an indefinite service life. However, if the laminate surface exhibits relatively bad blistering or crazing or is subject to other attack, the material's service life may not be satisfactory, even though Curves B and C indicate otherwise. In such cases, experience must serve as a guide.

Borderline cases, such as Curves D and E, often present problems. The flexural-strength retention of Material E levels off at a relatively low value over time. Depending on the material's retention of other properties, its visual appearance and its resistance in similar environments, it might still be economically serviceable if a greater-than-ordinary thickness were used to provide the typical 5–10 to 1 safety factor. An economic evaluation of competitive materials must be done before a decision can be made. For Material D, testing should continue, since strength retention did not level off over the length of the test. If the material continues to lose flexural strength at a rate of no more than about 0.5% per month, it can be used quite successfully, with an expected service life of from ten to twenty years.

Some investigators believe that the attack of FRP materials generally occurs at a decreasing rate approximated by a logarithmic function, and that if the retention data are plotted on log-log coordinates the resulting straight lines can be extended to predict service life, and property retention at some future time. However, such a log-log plot rarely results in a straight line. Also, because the lines are so close to each other and because of normal data fluctuations, it is quite difficult to confidently extend the lines to the five-, ten- or twenty-year time frame.

It should be emphasized that none of these property changes correlates with another or with degradation of the material, and that no one or two properties should be or can be used as a basis to accept or reject a material. Selection must be based on a combination of the test results. A conservative approach, rather than an overly optimistic one, is warranted, since it is better to err in that direction than to risk early failure.

Fire retardancy

All plastics are based on organic constituents. As such, they are inherently flammable and, once ignited, will burn profusely until completely consumed. However, there are several methods for making many of the thermoset resins fire-retardant.

In an estimated 20% of all chemical-resistant FRP applications, there is a need for some degree of fire retardancy. In addition, many of these applications are associated with higher operating temperatures—from 250 to 450°F—and require a material having high heat resistance as well. An example is a boiler duct handling wet SO_2 fluegas at 310°F.

Developing flame retardancy

Flame retardancy can be achieved by using numerous chemical additives, both organic and inorganic. Because most of these have a negative effect on mechanical and/or chemical-resistance properties, only the most widely used system for attaining optimum fire retardancy in a chemical-resistant FRP will be discussed here—a halogenated thermosetting resin system into which the inorganic synergist antimony oxide (Sb_2O_3) has been added.

This additive is referred to as a synergist because it enhances the natural fire retardancy of halogenated resins. Yet, if Sb_2O_3 were added to a nonhalogenated resin, it would have no effect on that laminate's burning characteristics.

The general practice is for the FRP equipment fabricator to add and mix powdered Sb_2O_3 into the liquid uncured halogenated resin, then add the curing catalyst to the resin/Sb_2O_3 mix. This is combined with the glass fibers to form the FRP laminate. Since the Sb_2O_3 is a white powder, it colors the laminate an opaque white, which is not necessarily a needed characteristic.

Fire retardancy tests and ratings

The most commonly used test for ignition and flammability is the National Fire Protection Assn.'s NFPA 255 [*4*], or ASTM Method E-84 [*5*]. It is frequently referred to as the "tunnel test," and its purpose is to determine the comparative burning characteristics of a material by evaluating the flame spread over its surface, which is exposed to a test fire in a forced-air chamber, or tunnel. A flame-spread classification (FSC), which considers both the ignition time and the distance the flame

Filament winding of a large-diameter FRP pipe

FRP flocculation feed tank, filament wound of vinyl ester resin, in wastewater treatment system

front advances down the tunnel during the 10-min duration of the test, is determined for the test material. The FSC is then compared with those for asbestos-cement board and select-grade red oak flooring, which have FSCs that have been established arbitrarily as 0 and 100, respectively.

An FSC of "25 or less" is generally required to meet most fire codes, although there are some notable exceptions. An FRP fabrication with an FSC of 35 might have a significantly lower cost than one with an FSC of 20; if the intended application can be safeguarded in another manner, such as with sprinklers or by being remotely located, the use of the FRP with the 35 FSC would probably be permitted (and considered safe).

Heat resistance

The various thermosetting resin technologies have different heat-resistance properties. Table III compares the continuous-operation temperature limits of four common laminates. Since the type and amount of reinforcement used also affect laminate performance under exposure to high temperature, a comparison of heat-deflection temperatures (HDTs) for the clear cast resins containing no reinforcement is made too.

The engineer designing equipment for high-temperature operation should also be aware of the reduction in mechanical properties of the laminate at higher temperatures, and incorporate this into the design for both normal and anticipated excursions.

An illustration of the need for both high heat resistance and fire retardancy is the use of FRP composites in power generation. Here, the more heat-resistant and acid-resistant chlorendic polyester resins are used in the construction of scrubbers, mist-eliminator blades, breeching ducts, and chimney liners in contact with hot, corrosive fossil-fuel gases.

Because of the low thermal conductivity of FRP laminates—about 1.5 Btu/h-ft^2-°F-in., compared with 150 Btu/h-ft^2-°F-in. for stainless steel—a low external temperature (for example, in the annulus of an FRP liner insert in a chimney) can result in a median wall temperature in a relatively thick laminate section that is considerably below that of the hot material (i.e., the gas inside the chimney liner).

Choosing a material

To compare FRP with other materials (e.g., metals and alloys) for use in CPI equipment, a fairly detailed analysis of the specific application must be conducted. While economics (as discussed briefly in the accompanying article) may well be the controlling influence, other factors also must be considered:

- All chemicals to which the equipment will be exposed—feedstocks, intermediates, products and byproducts, waste materials, and cleaning agents.
- Normal operating concentrations of chemicals, and maximum and minimum concentrations.
- The pH range of the system.
- Normal operating temperatures of the equipment, and maximum and minimum temperatures.
- Abrasion resistance and/or agitation requirements.
- Fire retardancy and heat resistance requirements.
- Thermal insulation requirements.

FRP vs. selected other materials

FRP offers corrosion resistance equal to that of carbon steel, to both internal and external exposed surfaces. However, carbon steel and other low-alloy metallics oxidize and require protective coatings to make them fully corrosion-resistant, whereas FRP does not. Carbon steel vessels require extensive surface preparation and application of corrosion-resistant liners, e.g. flake-glass-filled polyesters or vinyl esters and/or other materials, to provide adequate protection; FRP, on the other hand, is resistant throughout.

The strength properties of FRP have been discussed and compared with those of other materials, in the companion article to this one.

FRP can be fabricated into intricate one-piece shapes, requiring fewer joints than other materials, with more continuous, unobstructed surface area.

FRP is inherently nonconductive—an excellent electrical insulator, as well as a very good thermal insulator and a good noise barrier.

References

1. Rolston, J. A., Fiberglass composite materials and fabrication processes, *Chem. Eng.*, Jan. 28, 1980, pp. 96-110.
2. ASTM Method D-2583, Standard Test Method for Indentation Hardness of Rigid Plastics by Means of a Barcol Impressor, ASTM, Philadelphia, Pa., Feb. 27, 1981.
3. ASTM Method C-581, Standard Practice for Determining Chemical Resistance of Thermosettting Resins Used in Glass Fiber Reinforced Structures Intended for Liquid Service, ASTM, Philadelphia, Pa., Nov. 28, 1983.
4. NFPA 255, Method of Test of Surface Burning Characteristics of Building Materials, National Fire Protection Assn., Quincy, Mass., 1984.
5. ASTM Method E-84, Standard Test Method for Surface Burning Characteristics of Building Materials, ASTM, Philadelphia, Pa., June 26, 1981.

The author

Robert C. Talbot, Ashland Chemical Co., Polyester Div., P.O. Box 2219, Columbus, OH 43216, is responsible for marketing the company's corrosion-resistant and fire-retardant resins. He is a graduate of Erie County (N.Y.) Technical Institute, and joined Ashland in 1976 when it acquired the Hetron product group from Hooker Chemical Co. Mr. Talbot, a 30-yr veteran in the fiberglass-reinforced-plastics industry, is a member of the Natl. Assn. of Corrosion Engineers, the Soc. of the Plastics Industry and the Technical Assn. of the Pulp and Paper Industry.

FRP underground horizontal tanks for corrosive chemicals

There are many advantages to storing corrosive chemicals in underground tanks. Fiberglass-reinforced plastic (FRP), proven in the storage of flammable fuels, may be the ideal material of construction.

N. J. Kraus, Owens-Corning Fiberglas Corp.

☐ In considering the use of underground FRP tanks, the important questions are: Will the FRP withstand the corrosive service? Is it advantageous to have the tanks underground? What sort of service life can be expected? This article attempts to present some of the answers.

Substances that can be stored

Many thousands of FRP tanks have been installed underground for the storage of flammable fuel since their introduction in the early 1960s. They range in size from 550 to 50,000 gal capacity, and the list of petroleum products stored includes: gasoline—regular, premium and unleaded, as well as gasohol; diesel fuel; jet fuel (aviation); fuel oils for heating; waste oil; deicing fluids.

However, these tanks are being used on an ever-increasing basis for handling other corrosive chemical environments. By far, the great majority have been above-ground—vertical or horizontal.

Among the many corrosive chemical environments for which underground FRP tanks have been built: alum, ethylene glycol, ferric chloride, formic acid, hydrochloric acid, hydrogen peroxide, isopropyl alcohol, potassium permanganate, sodium chloride, sulfuric acid.

While virtually all chemicals that can be stored in aboveground FRP tanks can be handled in underground FRP tanks, the latter are particularly advantageous for environments having flash points under 200°F, due to National Fire Protection Assn. requirements (NFPA 30) for handling such environments.

Typical of the services of this type are: formaldehyde, methanol, isopropyl alcohol, Cellosolve acetate, naphtha, furniture-polish formulations, paint formulations.

Advantages of underground tanks

The *safety* feature afforded by underground tanks, combined with the *elimination of the need for diking* above-ground tanks, clearly makes underground tanks the preferred method of storage for many corrosive chemical environments.

This article was presented at the National Assn. of Corrosion Engineers 1979 Plastics Seminar, held in New Orleans, Oct./Nov. 1979.

There are additional features that provide real economic benefits:

First, there is the *temperature moderation* that is achieved in underground burial. The earth temperature will remain fairly constant at approximately 50-55°F once a depth of 3 or 4 ft is reached. When materials need to be maintained above their freezing points to prevent solidification or to prevent them from salting out, this can be accomplished in an underground tank with little or no heat input in some cases, depending upon the temperature level required. Additionally, the cost of insulating an aboveground tank can be eliminated.

Additional savings may be realized because the underground installation *minimizes or eliminates the cost of ladders, ladder lugs and catwalks* required in many aboveground installations.

Utilization of ground space or building space is another advantage. Plant buildings can often be used to better advantage than for sheltering tanks. And outdoor space utilization can often be improved by underground tank installation. Underground tanks, properly installed, can tolerate as much as 32,000-lb loaded vehicles driving back and forth over them, without structural damage, as evidenced by service-station tank installations.

In many applications, *the potential for gravity flow into the underground tank* is an advantage. FRP underground tanks are used for waste storage and treatment since the outflow from many waste lines can readily be gravity-fed into one tank and then pumped to the ultimate destination.

Another advantage is the safety of the FRP underground tank installation where dangerous corrosive materials are involved. Such tanks *do not have the vulnerability to accidental damage* that can occur in the plant from a variety of sources, or from high winds and wind-blown debris in exposed locations, or from vandalism.

Finally, the *size and capacity* to which FRP underground horizontal tanks can be built provides distinct advantages. Underground tanks of 48,000-gal capacity, or even larger, are not merely possibilities; they are proven by a successful history of installations. There are many support problems with aboveground tanks that are much less severe by several orders of magnitude with underground tanks where the backfill is used for support.

A single 50,000-gal FRP underground tank can be shipped, completely shop-fabricated, to the jobsite. In

FRP tank installed, Stony Ridge, Ohio, October 1964, for gasoline storage **Fig. 1.**

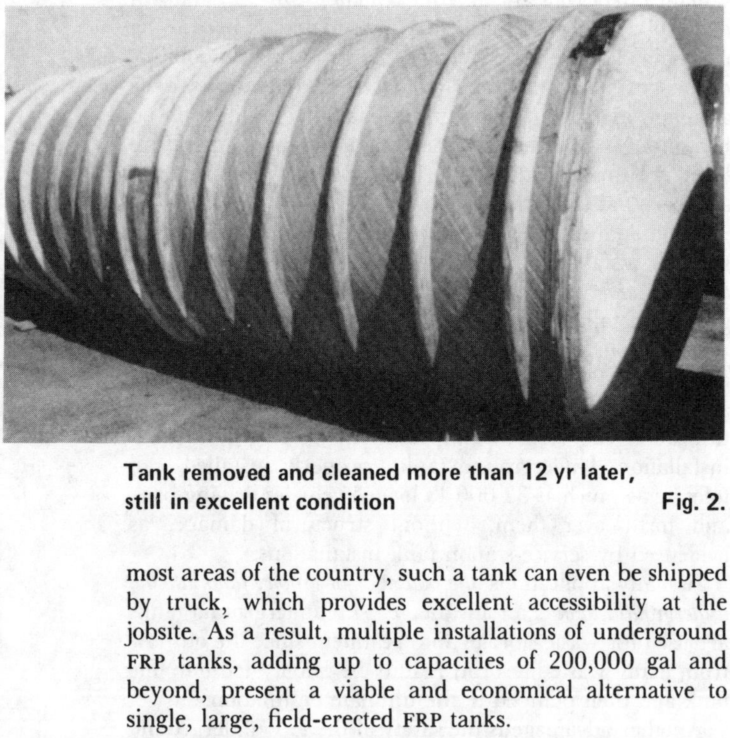

Tank removed and cleaned more than 12 yr later, still in excellent condition **Fig. 2.**

most areas of the country, such a tank can even be shipped by truck, which provides excellent accessibility at the jobsite. As a result, multiple installations of underground FRP tanks, adding up to capacities of 200,000 gal and beyond, present a viable and economical alternative to single, large, field-erected FRP tanks.

Underground service tests

Do FRP tanks really perform long-term underground?

In an attempt to confirm the engineering projections of long service life for underground FRP tanks, let us evaluate the performance of two tanks that had been in long-term service underground for the storage of gasoline. Both were removed from their installation sites not because of any problems but rather to test and to observe the physical condition of the tanks after extended periods of burial.

Tank 1, which was installed in Houston, Tex., in May 1964, was removed from service 6 yr and 10 mo later. Tank 2, which was installed near Stony Ridge, Ohio, in October 1964 (Fig. 1), was removed from service 12 yr and 9 mo after installation. The exterior surface of both tanks exhibited a white, somewhat chalky appearance—typical of an air-cured FRP surface exposed to moisture over long periods of time. This is only a surface effect and was easily scrubbed off with an abrasive cleaner, exposing a laminate of fine appearance (Fig. 2).

The interior surface of both tanks was in excellent condition, with no evidence of any cracking or crazing or any attack. Though a coating of dark-brown sediment covered much of the inner surface on most areas of the tank wall, this was easily washed off, revealing the original smooth surface to be essentially unchanged in color. Some of the secondary layups, presumably because they were rougher in texture than the molded tank surface, tended to retain more of the discoloration, but there was no evidence of delamination or of deterioration of the laminate.

Sections were then cut from both tanks, and physical-property tests were conducted to form a basis of comparison with the average mechanical properties of laminates made at the same approximate date as the two tanks being evaluated.

The results of these tests are shown in the table.

There is some experimental scatter in the results that shows the mechanical properties of the buried tanks to be in some cases higher, and in some cases lower than the test material. However, there is no major loss of mechanical properties, and all values exceed the design values.

The excellent condition of the laminate and its retention of mechanical properties after these extended periods of exposure demonstrate very effectively the long-term service that can be realized with FRP underground tanks.

While the results of similar tests on tanks exposed to other corrosive chemical environments over long periods are not available, these tests on the fuel tanks do suggest that laboratory corrosion test data, when favorable and if properly evaluated, provide a sound basis for predicting in-service performance. There is considerable evidence that FRP underground tanks resist external chemical attack under a wide range of conditions.

Similarly, we can utilize the shorter-term environmental test data that have been compiled in this industry to evaluate the chemical or mechanical effects of various internal corrosive environments. Such an evaluation would not differ from the evaluation of an FRP tank for aboveground service. The design value for mechanical properties is based on the values obtained with test coupons removed from the environment at various times during a period of, say, 12 mo. The plot of each property (such as flexural strength) versus time usually decreases and then levels off, and the value at which it levels off is taken as datum for the design value.

Materials of construction

Most of the discussion concerning the use of FRP tanks underground for storage of flammable fuels has involved comparisons of their performance and cost as related to metal tanks. However, the evaluation of underground FRP tanks for other corrosive chemical services involves the more complex comparison of the performance and cost of the FRP *underground* tank with those of *aboveground* tanks constructed of either FRP or metals. In the chemical process industries, the internal corrosion problems that were handled by the more traditional materials, such as

Mechanical properties of buried FRP tanks, compared with time zero

	Tank 1	Tank 2
Burial time	6 yr 10 mo	12 yr 9 mo
Flexural strength	33,500 psi	27,800 psi
Flexural modulus	1.83×10^6 psi	1.19×10^6 psi
Barcol hardness	48	35
Flexural strength retention	94%	106%
Flexural modulus retention	122%	91%
Barcol hardness retention	120%	95%

rubber-lined steel, stainless steel or one of the alloys, have presented a sufficient range of difficulties without introducing the external corrosion problems of underground installation. So there is very little precedent for installing underground metal tanks for corrosive chemical services.

However, with FRP tanks, the materials engineer can evaluate underground installation without concern about external corrosion. FRP tanks have convincingly demonstrated that they resist external corrosive effects of soil or groundwater attack. And they are impervious to attack due to galvanic action or to the effects of stray electrical currents from grounded equipment, factors affecting the performance of metal tanks underground.

FRP underground tanks for chemical service differ very little from their counterparts used for the storage of flammable fuel. The entire inner surface, including that of all secondary laminates used for the construction of the tank and its accessories, is comprised of a 20-mil layer, resin-rich and reinforced with "C"-glass surface veil. This is followed by an interior layer (approximately 75% resin) reinforced with randomly oriented chopped-glass fibers, usually to a combined thickness of 100 mil for the total corrosion liner.

The outer structural layer varies somewhat from fabricator to fabricator. Commonly, there is a chopped-glass-reinforced laminate, built up to the design thickness. In many cases, external reinforcing ribs are then used to achieve rigidity in the straight shell portion of the tank and to restrict post-installation deflection. The design of underground FRP tanks is based on partial use of the backfill for structural support.

While isophthalic resin has proved an excellent material of construction for flammable-fuel service, FRP underground tanks made for other applications will probably use other resins, such as polyester or vinylester, that are compatible with the internal corrosive environment. These resins are at least as resistant to external groundwater as is isophthalic resin.

Fiberglass composite materials and processes of construction are discussed in a previous report (*1*).

Other considerations

One of the obvious questions that will be raised about FRP underground tanks is: What about the possibility of *leakage of corrosive chemicals* into the groundwater table? If an aboveground tank develops a leak, this will generally be immediately detected. Of course, this is not possible underground. Careful monitoring of the liquid contents, i.e., inventory control, to determine that no losses have occurred, coupled with regular periodic inspections of the condition of the interior surface, are the best methods to maintain assurance of the integrity of the underground tank. There are also monitoring systems that can signal leakage of corrosive contents when chemical attack on a current-carrying wiring system is interrupted and a signaling device is activated.

Some consideration should be given to the *placement of accessories* on FRP underground tanks. The most practical approach is to place nozzles for vents, piping connections and manways on the top centerline of the tank shell. It would be well to avoid placing nozzles on the bottom centerline or on the tank ends, to eliminate the possibility of stresses on such points of attachment, which could damage the tank. Such placements are not of themselves impossible, and occasionally, for example, drain sumps are installed on the bottom centerline. In such cases, however, extremely careful hand-placement of the backfill beneath the tanks and around such projections is vital.

To obtain long-term corrosion-resistance properties and obviate problems of leakage caused by undue mechanical stresses, FRP underground tanks must be properly installed in the ground.

The basic advice for *good tank installation:*

- Retain the circular shape of the cylindrical wall.
- Provide continuous support.
- Allow minimal long-term settlement to avoid deformation.
- Maintain proper support when a high groundwater condition exists. This may include anchor straps in shallow burials.

As with all flexible conduits, underground FRP structures require the support of the surrounding soil to perform as desired (*2*). The premise on which flexible-conduit design theory is based is quite simple: Use the inherent structural strength of the soil in conjunction with the structure within it. There is a definite soil-structure interaction that occurs when a structure is buried.

Granular backfills provide better support, are much easier to place and are by far the best type of backfill to use when installing underground tanks.

References

1. Rolston, J. A., Fiberglass composite materials and fabrication processes, *Chem. Eng.*, Jan 28, 1980, pp. 96–110.
2. Pearson, L. E., Pinard, L. G., and Greenwood, M. E., Solving underground corrosion problems with Fiberglas reinforced plastics, Soc. of the Plastics Industry (SPI) West Technical Conference, 1978.

The author

Norbert J. Kraus is market manager—chemical process industries, Non-Corrosive Products Div. of Owens-Corning Fiberglas Corp., Fiberglas Tower, Toledo, OH 43659, 419-248-7501. He is responsible for expanded market penetration of FRP tanks into the chemical process industries. Mr. Kraus holds a Bachelor of Arts degree from St. Gregory College in Cincinnati, and has been associated with the FRP industry for 17 years. He has been active in the American Soc. for Testing and Materials and has been chairman of the corrosion resistant structures committee of the Soc. of the Plastics Industry since 1975.

Designing fiberglass-reinforced-plastic vessels for agitator service

Here is practical, step-by-step information on how to design fiberglass-reinforced-plastic tanks that have turbine agitators mounted on top.

J. B. Fasano, Chemineer, Inc., and T. M. Eberhart, Hercules Inc.

☐ This is an analysis of various aspects of mounting configurations, baffle theory, agitator loads and mechanical design criteria. It explains in some depth the theory of two of the most popular configurations—the top-center, nozzle-mounted agitator and the frame- or vessel-mounted agitator. Two examples are worked in complete detail, one for each of the two configurations. References are quoted for additional information.

Agitators

A turbine agitator is a mechanical device that produces motion in a fluid through the rotary action of impellers. The turbine impeller consists of fixed-angle blades attached to a hub driven by an agitator shaft. There are five major components: the prime mover, the drive, the shaft seal, the shaft, and the impeller.

Turbine impellers are characterized by the type of flow produced. Axial-flow turbines move the fluid parallel to the axis of the agitator shaft, while radial-flow types discharge the fluid from the impeller region perpendicular to the axis. The most common turbine is the 45°-pitched-blade one (see Fig. 1).

Mounting configuration

Top-entering mountings are shown in Fig. 2–6. In Fig. 2, the agitator is supported independent from the vessel. The beams must carry all agitator loads. This

Originally published May 5, 1980

Glossary of terms

A	Shear bond area of weld, in.^{2}
A_b	Projected baffle area, in.^{2}
D	Dia. of impeller, in.
E	Modulus of elasticity, psi
F_h	Hydraulic force, horizontal component, lb_f
F_T	Tangential force due to torque, lb_f
F_t	Turbine thrust, lb_f
F_p	Thrust due to pressure, lb_f
F_s	Fatigue stress limit, lb_f
F_w	Weight of shaft and turbines, lb_f
F_y	Vertical support reaction, lb_f
F'_y, F''_y	Components of F_y, lb_f
H_{pt}	Impeller power, hp
I_s	Moment of inertia, circumferential stiffeners, in.^{4}
J	Polar moment of inertia, in.^{4}
K	Coefficient of hydraulic force
L	Shaft projection, in.
L_b	Support beam, spacing, in.
M	Bending moment, lb-in.
N	Agitator shaft-speed, rpm
P_{bmax}	Baffle pressure, psi
R	Dished-head radius, in.
S	Head stresses, psi
S_1, S_2	Maximum radial stresses, psi
S_b, S_t	Gusset stresses, psi
S_c	Critical buckling stress, psi
S_F	Design stress for fatigue resistance, psi
S_{comp}	Compressive stress, psi
S_t	Shear stress in the gusset wall, psi
S_{ten}	Tensile stress, psi
T	Torsional load, psi
T_Q	Torque, lb-in.
W	Supported weight, lb_f
Z	Section modulus, in.^{3}
a	Nozzle, inside radius, in.
b	Nozzle flange, outside radius, in.
c	Gusset, radius at the centerline of handholes, in.
d	Dia., vessel, in.
d_h	Dia., handholes, in.
f	Natural frequency of vibration, cpm
f_t, f_y	Distributed support reactions, lb_f
h	Circumferential stiffener spacing, in.
k_1, k_2	Deflection coefficients
n_h	Number of handholes
q_1, q_2	Stress coefficients
r	Tank radius, in.
r_o	Radius of the base of a conical gusset, in.
t	Average shell thickness between stiffeners, in.
t_g	Thickness, conical gusset, in.
t_n	Thickness, nozzle neck, in.
w	Minimum weld width, in.
y, y_1, y_2, y_{ro}	Deflections, in.
α	Angular deflection, agitator shaft
γ	Correlation factor, local buckling stress
ρ	Laminate density, lb/in.^{3}

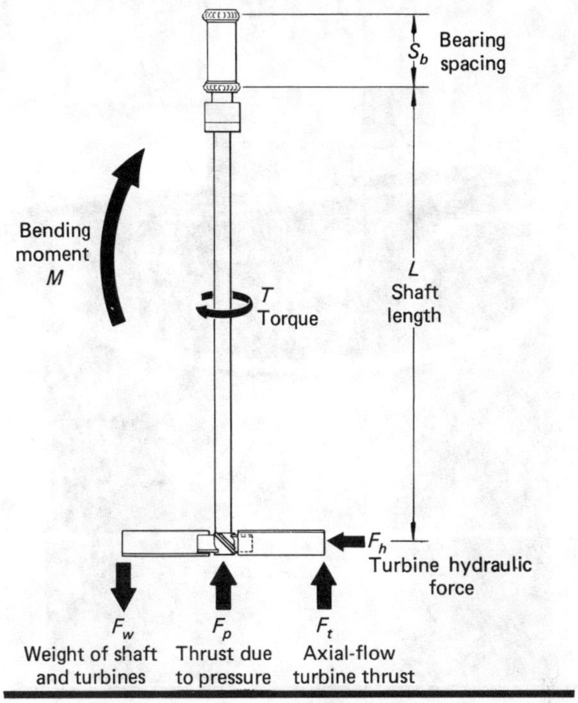

Forces and moments acting on the agitator **Fig. 1**

mounting is used where (a) an existing vessel does not have sufficient strength to take agitator loads, or where (b) an agitator for a new vessel is most economically supported in this manner. In many instances, large agitators for fiberglass-reinforced-plastic (FRP) tanks are most economically supported independent of the vessel. Sufficient structural engineering methods are available to permit design of independent mountings. For most applications, only the effect of fluid forces against the baffles and tank wall must be considered in designing fiberglass-reinforced plastic tanks with independent mountings.

In Fig. 3, the agitator is supported by a structure mounted on the vessel. In this case, the FRP vessel must be designed to handle all agitator loads. An auxiliary stuffing box or lip seal may be mounted on the nozzle of a closed-top vessel if a seal is required. A separate tank-supported seal cannot be used because of seal alignment requirements.

When a separate, tank-supported mechanical seal is needed, the configuration of Fig. 4 should be used. The prime mover and drive are mounted on an independently supported beam structure. The drive is connected to the shaft by means of a flexible gear-type coupling. The beam structure resists the torque load and supports the weight of the prime mover. All other loads must be handled by the nozzle and tank top. A pedestal assembly (mounted on the vessel nozzle) contains the mechanical seal. This configuration is often used with very large drives (greater than 10,000 lb) in order to isolate the mechanical loads from the vessel.

The configuration shown in Fig. 5 is the most common arrangement for smaller, closed-top tanks of FRP requiring a shaft seal. Typically, this will apply to FRP

Independent support (on open tank) Fig. 2

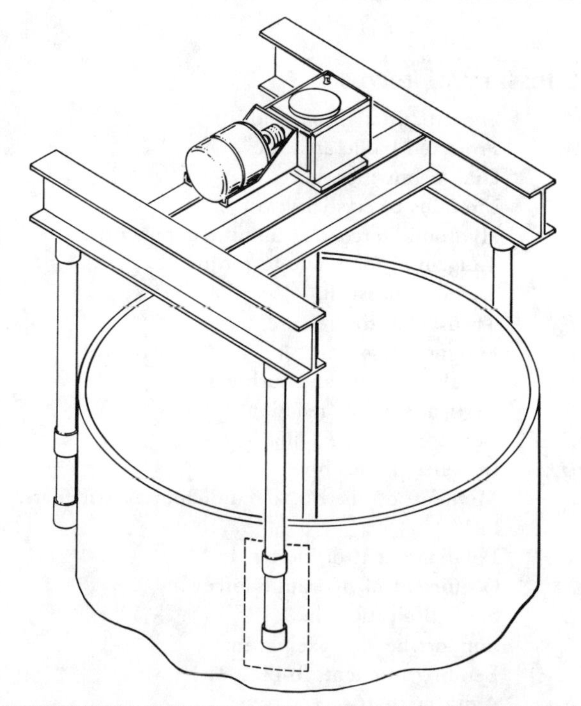

Frame-mounted agitator supported by tank wall Fig. 3

tanks 6 ft dia. and smaller. The pedestal is the transition piece between the agitator drive and the mounting flange. A lip seal, stuffing box, or mechanical seal can be mounted within the pedestal.

A turbine agitator mounted independent of the vessel on beams (Fig. 6) has a flexible bellows to connect the agitator flange to the vessel flange. This arrangement has the advantage of essentially removing all loads from the FRP vessel. Where mechanical seals are required, this option is less costly than the arrangement shown in Fig. 4. In addition, agitator vibration is isolated from the vessel. Any type of seal is compatible with this configuration. Of the mounting configurations discussed, the vessel-mounted support (Fig. 3) and the vessel-mounted pedestal (Fig. 5) are the only two commonly used mountings that require design considerations of the FRP vessel due to the agitator.

Further discussions in this paper are directed toward the design of FRP vessels for these latter two types of agitator mountings.

Agitator loads

It is necessary to understand agitator loads in order to relate them to the design of FRP vessels. Refer to Fig. 1 for the following discussion.

The forces and moments created by a fluid agitator are a result of the fluid motion produced by rotation of the turbine impeller. The power required to rotate a turbine impeller results in a torque that can be calculated from:

$$T_Q = 63{,}025 H_{pt}/N \qquad [1]$$

where H_{pt} is the impeller shaft horsepower, and N is the impeller shaft rotational speed in rpm. Torque as calculated by Eq. 1 implies an unchanging load.

Actual torque (or power draw) shows some variability, which is a function of the turbulent conditions within the agitated fluid. Usually, the larger the intensity of agitation, the larger the variability, up to a maximum of $\pm 15\%$.

Hydraulic forces acting on the impeller generate moments that act on the shaft. Because of the random nature of the forces and the rotation of the shaft, these bending moments often reverse during operation. The horizontal component of these hydraulic forces can be estimated from:

$$F_h = KT_Q/D \qquad [2]$$

where D is the impeller dia., in. K will vary between 0.3 and 2.1, depending upon impeller and vessel geometry and the level of agitation.

The bending moment, M, is the product of the hydraulic force, F_h, and the distance, L, from the turbine to the first support bearing in the agitator drive (the shaft length):

$$M = F_h L \qquad [3]$$

For multiple impellers, the M is cumulative.

Long, cantilevered shafts can produce very large bending moments. Such moments are quite often the largest loads to be accounted for in FRP vessel design. Most often, adequate design may be achieved by including only the static weight and the bending moment. Other loads do have an effect, but they are usually minor by comparison.

Tests using full-scale agitators have shown that off-center or angular mounting greatly increases dynamic

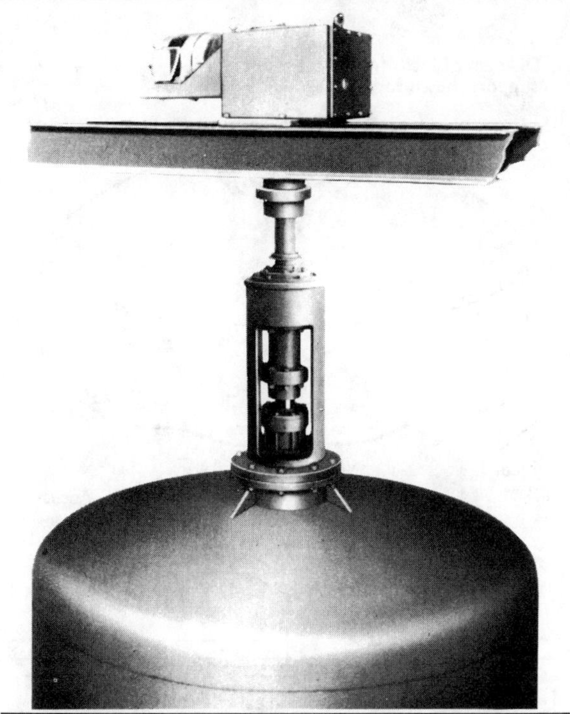

Independent support (tank-supported mechanical seal) Fig. 4

Vessel-mounted pedestal (shaft seal in pedestal) Fig. 5

bending moment. Because of this, the tolerances used during fabrication of the vessel can have more influence on design than any other single factor. It is important that mounting flanges be installed with extra care in order to ensure that the flange centerline coincides with the tank centerline and that the flange face is perpendicular to the axis of the vessel.

When mounting an agitator on a steel supporting frame, allowances should be made for horizontal alignment, and the unit should be shimmed so that the shaft will align with the axis of the vessel.

Intentional off-centering or angular orientation for the purpose of random agitation without the aid of internal baffling should be treated as a special design situation. When this type of installation is used, the designer should contact the agitator manufacturer for an estimate of the expected maximum shaft bending-moment.

The agitator shaft and drive are subjected to vertical forces. The principal vertical force is the weight of the shaft and impeller, F_w. In high-pressure applications, the upward thrust, F_p, acts on the bearings in the agitator drive. A pitched-blade or axial-flow impeller normally pumps downward and generates an upward thrust, F_t. As mentioned above, the forces F_p and F_t generally are not significant.

Agitator loadings on vessels

Irrespective of the type of mounting, the radial and tangential loads applied to the baffles and sidewall are essentially the same for identical impellers rotating at the same speed. For an axial-flow impeller, the baffle loading distribution is minimal at the top and increases to a maximum at the bottom. An assumption of triangular load distribution is conservative. The maximum pressure in this case would be twice the pressure in a uniform load, and, assuming that four baffles are used, can be found from:

$$P_{bmax} = \frac{34{,}377}{A_b}\frac{H_{pt}}{Nd} \quad [4]$$

where A_b is the projected baffle area, in.^{2} and d is the tank dia., in. Designing for this pressure acting on the projected area of the baffle ensures adequate design.

During agitation, a continuous radial load is experienced by the tank wall. The maximum pressure (near the lower tangent line) will generally be no larger than 0.02 psi. An agitator will also contribute a localized, randomly fluctuating pressure of as much as ± 1 psi.

With respect to the design of FRP tanks, only two types of agitator mountings will be discussed: the nozzle-supported agitator and the frame-mounted agitator supported by the tank sidewall. An example of the nozzle-supported agitator can be found in Fig. 5, and the frame-mounted agitator supported by the tank wall in Fig. 3.

When designing a support of any kind to hold an agitator, it is essential to limit deflection at the impeller to $\frac{1}{8}$ in./ft of shaft extension. With a nozzle-supported agitator, any flexure of the tank head, due to bending moment, will result in an angular deflection of the impeller shaft. This same type of deflection can occur on frame-mounted units supported by the sidewall. Frames should be designed to limit this angular deflection; the techniques and data for frame design are well known. The methods for designing FRP tanks to support frame-mounted units, however, are not well known and will be discussed in further sections.

Independent support (bellows connector) Fig. 6

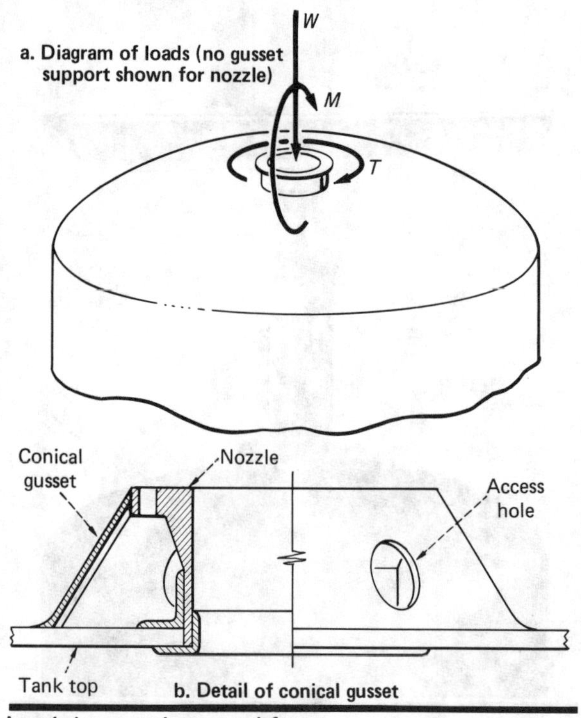

Loads imparted to vessel from nozzle-mounted agitator Fig. 7

Nozzle loadings for a nozzle-mounted agitator are depicted in Fig. 7a. These loadings result from: the bending moment (Eq. 3), the static weight, and the torque (Eq. 1).

The frame-mounted unit (assumed to be a rigid frame, as in Fig. 3) will transmit forces to the tank sidewall, as shown in Fig. 8. The force F_T is the tangential force generated by the agitator torque. The force F_y is the sum of the static-weight contribution plus the force generated by the bending moment. The reaction forces required by the vessel are shown also.

Baffles

Mixing in vessels by an agitator is most effectively accomplished through the use of baffles. There are various types of mounting positions and baffling that can be used to accomplish mixing. The most common arrangement is with the agitator shaft coincident with the centerline of the vessel, and four baffles attached to the tank wall, each having a width that is $1/12$ the tank dia. and being set off the tank by $1/72$ of the tank dia. The plane of the baffles should intersect the centerline of the vessel. This is termed "standard baffling." For FRP tanks, triangular or wedge-shaped baffles are the most cost-effective and provide essentially the same baffling as standard baffles. The design of triangular baffles will be considered later on.

Mechanical design criteria

In designing FRP vessels to handle agitator loads, there are primarily five criteria. All of these, where applicable, must be satisfied.

Tensile and compressive strength

Sufficient information exists to adequately define these in terms of the ultimate strengths for filament-wound as well as contact-molded vessels. Generally, the allowable limits, as a minimum, are those permitted by the National Bureau of Standards (NBS) Voluntary Product Standard 15-69 [*10*] for hand-layup construction and the American Soc. for Testing and Materials (ASTM) D3299-74 [*1*] for filament-wound construction.

Strain

A strain requirement is generally used to prevent liner failure and is usually limited to 0.001 in./in. The mechanical forces of agitation will cause strains higher than those occurring in static conditions.

Fatigue strength

Cyclic or repeated loading can result in fatigue failure at stress levels well below a material's ultimate static strength. Few conclusive fatigue-strength data have been published pertaining to corrosion-resistant fiberglass laminates. In addition, fatigue data are generally taken under ideal conditions—that is, no stress concentration, uniaxial stress condition, and so on. The number of variables usually associated with fatigue data, when combined with material variables (such as type and direction of reinforcement, reinforcement weight ratio, resin strength and fabrication quality), makes correlation of data relating to FRP very difficult.

Agitator loadings provide one of the most extreme variations of possible stress cycles. Knowing the maximum number of cycles to which the structure will be subjected, along with the stress ratio (max. compres-

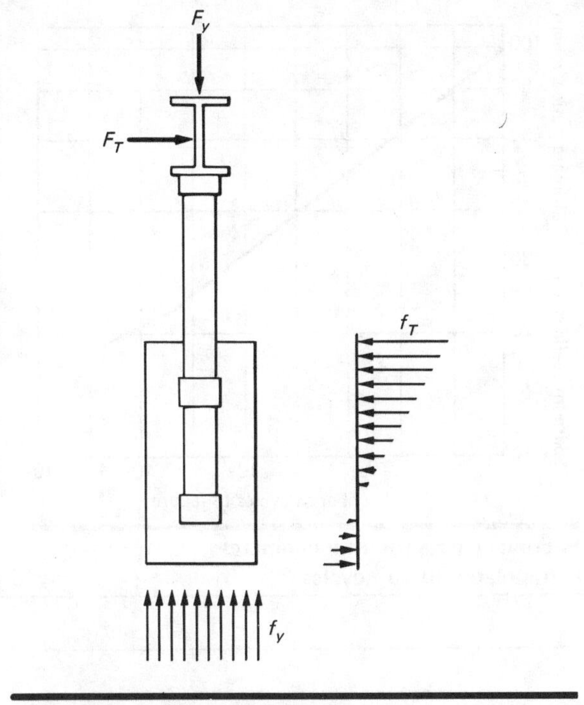

Load diagram for wall support of frame-mounted agitator **Fig. 8**

sion/max. tension), we can determine the fatigue strength from typical test data. When plotted to log-log scale (stress versus number of cycles, or SN), most fatigue data closely approximate two straight-line segments.

When a curve reaches a constant stress that is independent of the number of cycles of loading, the corresponding stress is referred to as the "fatigue limit" or "endurance limit." This usually occurs in the vicinity of a few million cycles of loading. All agitators will provide over 2 million cycles throughout their useful life. Therefore, when no endurance limit exists, maximum stresses must be determined from the SN curve for the maximum number of cycles expected over the useful life of the equipment.

Many materials do not have an endurance limit. FRP falls into this category. For such materials in agitator-applied service, maximum stress levels should be limited according to the anticipated number of cycles of applied load over the expected life of the equipment. An approximate indication of the SN behavior of FRP laminates is given in terms of the percentage of ultimate strength in Fig. 9. This figure represents a compilation of data from several references [*8,9,11*]. The data of Isham [*8,9*] extend approximately to 2×10^6 cycles. The rate of decline beyond this number of cycles has been shown to be much slower than for the initial part of the curve.

Because of the limited basis for predicting fatigue life of corrosion-resistant laminates, a safety factor of 2 is suggested for use with Fig. 9. Other factors—stress ratio, multiaxial stress conditions, elevated operating temperatures, exposure to corrosive environments, and so on—have a marked effect on the fatigue characteristics of laminates. This demands that a significant amount of engineering judgment be used for any particular application of general fatigue data.

The fatigue characteristics of steel used for supporting structures are generally much better known. Therefore, the design of the steel support will not be discussed here.

Local buckling

Frame-mounted agitators supported off the tank wall present concentrated compressive loads on the sidewall. The relative thickness of the wall requires that a localized buckling analysis be made. For FRP tank design, a safety factor of 5 is sufficient.

Deflection

The entire support system of an agitator should be made sufficiently strong to limit shaft deflection to less than or equal to $\frac{1}{8}$ in./ft of agitator shaft length. Unbalanced hydraulic forces are the primary cause of this deflection and should not be confused with the static deflection caused by weight. Supports that permit dynamic shaft deflections much larger than $\frac{1}{8}$ in./ft can severely shorten agitator life. If the criteria for shaft deflection are not controlling, deflection should be limited to $\frac{1}{2}\%$ of span.

Nozzle-mounted agitators

Head design

An investigation of stresses and deflections caused by agitator loads was made for FRP tanks with dished heads [*5*]. Here, for typical support reactions caused by agitation equipment, it was found that stresses, rather than static deflections, governed the design, and that the component of stress due to bending moment was from three to five times that due to the agitator's dead weight.

For the purposes of analysis, it was assumed that the area of support had a radius equal to the radius of the base of a 60-deg. conical gusset (see Fig. 7b). In all cases, nozzles used for agitator support should be properly gusseted. The effectiveness of 60-deg. plate gussets in evenly distributing stress to the head will be significantly less than for a comparably sized conical gusset (see Fig. 4 and 5).

As can be expected, stresses due to locally applied loads quickly die out as the radial distance from the point of load is increased. The point of investigation of stresses is taken as the location of the attachment of the gusset to the head, r_o. Because deflections and stresses vary directly with the magnitude of applied loads, coefficients dependent upon the geometry of the supporting structure can be defined in order to simplify the governing equations. By this procedure, deflections can be related to applied loads by:

$$y_1 = k_1 W \quad [5]$$

and

$$y_2 = k_2 M \quad [6]$$

where k_1 and k_2 are deflection coefficients. Similarly, maximum radial stress is given by:

$$S_1 = q_1 W \quad [7]$$

and

$$S_2 = q_2 M \quad [8]$$

where q_1 and q_2 are stress coefficients that take into account the combined effects of radial moments and membrane forces.

These deflection and stress coefficients are functions of r_o, R and t [*5,12*]. In addition, k_1 and k_2 are inversely proportional to the elastic modulus, E, which was assumed to have the PS 15-69 [*10*] value for the purpose of analysis. Fig. 10 to 13 give values of the coefficients for 4-ft to 8-ft-dia. tanks of varying head thickness with the load applied to a 6-in.-dia. conically gusseted nozzle. Fig. 14 through 17 show how each of the coefficients is affected by nozzle size.

Because of the complex nature of the interaction between head stiffness and the static and dynamic load characteristics of an agitator, it is very difficult to simplify the analysis of head stability. The area of concern becomes one of angular vibration, which will be highly dependent on the agitator's mass distribution. It has been found from test results that hydraulic forces due to fluid motion will be random in nature as long as the tank is uniform and the agitator centered and balanced. For agitators with impellers operating below the liquid interface, this effectively minimizes the cyclic nature of bending as long as the mounting nozzle is aligned properly and the agitator is balanced within reasonable tolerances. Again, proper installation is seen to be an important factor in nozzle-mounting design.

Practically speaking, nozzle-mounted agitators have not been utilized to a great extent on FRP tanks. This is due to the lack of understanding of how nozzle loads affect head design, the lack of fatigue data (particularly related to failure of the bond of the nozzle to the head), and because of the unpredictable nature of head instability. For these reasons, the use of nozzle-mounted agitators has been limited to small units mounted on small- to medium-sized tanks where the loads have been low enough to afford an extra degree of conservatism in design.

Because of the unpredictable nature of head instability, it is recommended that agitators with easily accessible change-gears be used for nozzle-mounted installations. Alternately, the head stiffness may require field modification in instances where dynamic deflection is found to be unacceptably high.

The two primary factors influencing top-head design of tanks with nozzle-mounted agitators are then local radial stress (due to dead weight and bending moment) and installation tolerances. Optimum head design may entail thickening of the head at the center to reduce local stresses or stiffening the top in order to raise the head's natural frequency to a safe operating level.

Conical-gusset design

Assuming that a conical gusset is provided and that the nozzle neck and gusset share in the resistance of the agitator's torque reaction, shear stress in the gusset wall near the attachment to the flange is given by:

$$S_t = Tb/J \qquad [9]$$

where $J = 2\pi(a^3t_n + b^3t_g)$. Normally, the gusset will be provided with two or four handholes for bolt access. The removal of material at the centerline of the handholes may be enough to make the location of the holes

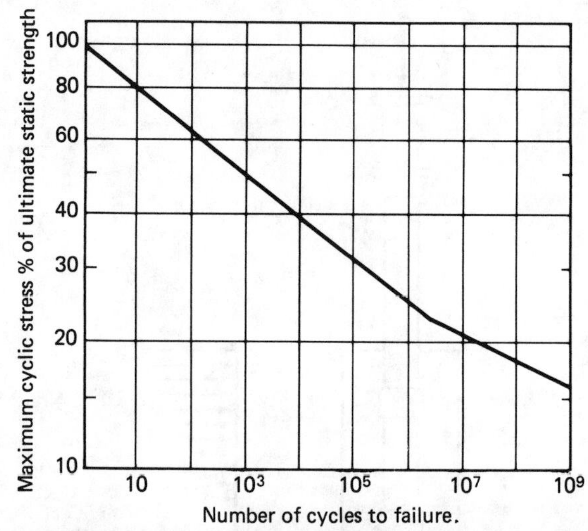

Endurance data for FRP laminates extrapolated to 10^9 cycles **Fig. 9**

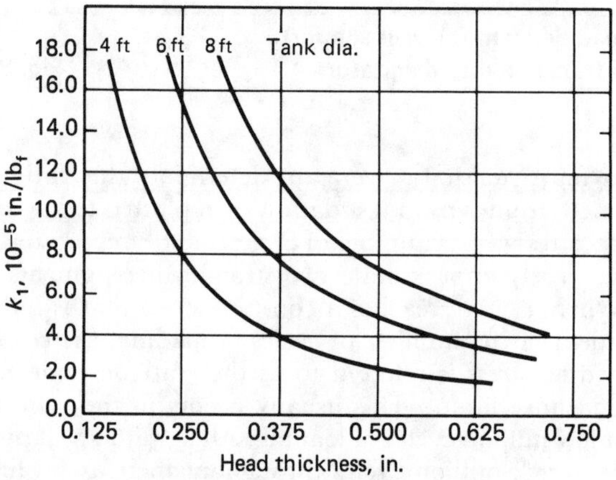

Deflection coefficient vs. head thickness for load applied to a 6-in.-dia., conically gusseted nozzle Fig. 10

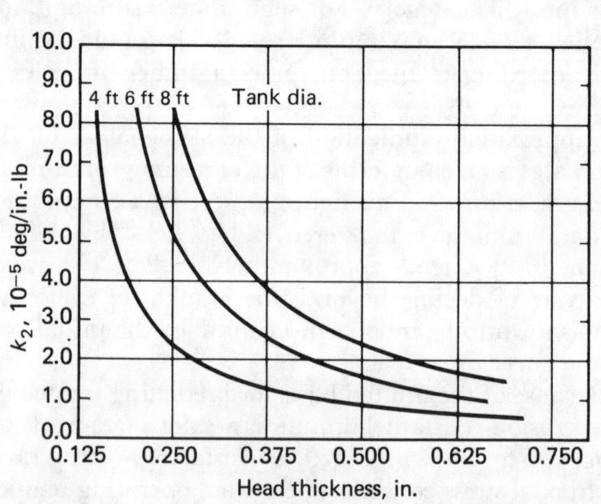

Deflection coefficient vs. head thickness for moment applied to 6-in.-dia., conically gusseted nozzle **Fig. 11**

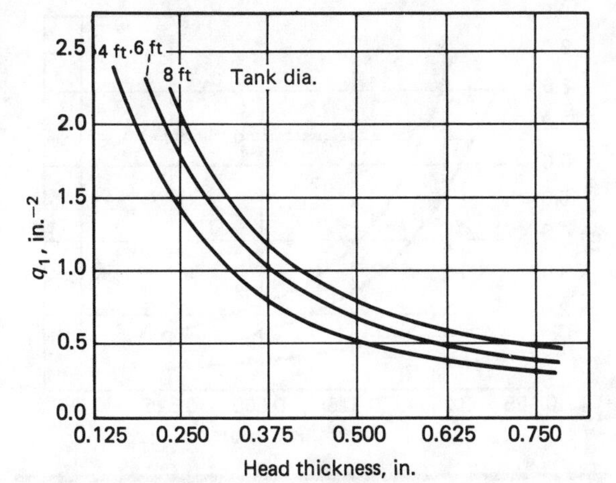

Stress coefficient vs. head thickness for load applied to a 6-in.-dia., conically gusseted nozzle **Fig. 12**

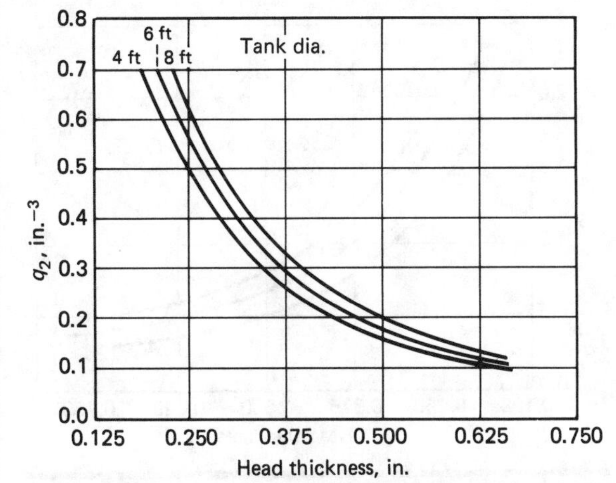

Stress coefficient vs. head thickness for moment applied to a 6-in.-dia., conically gusseted nozzle **Fig. 13**

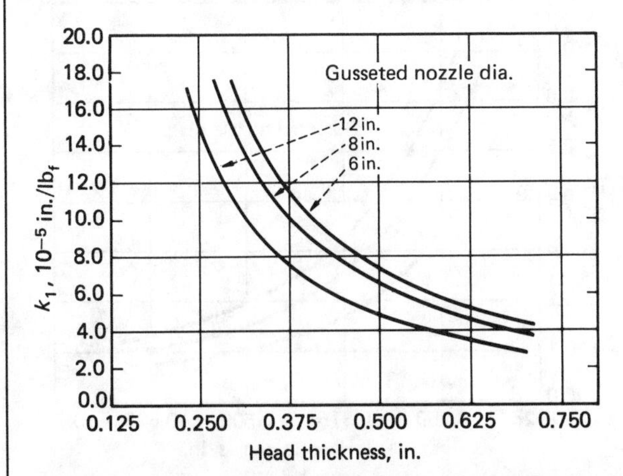

Deflection coefficient vs. head thickness for load applied to nozzles on an 8-ft-dia. tank **Fig. 14**

the governing stress area for design. An approximate relationship for J at the centerline of the handholes is:

$$J = 2\pi[a^3 t_n + c^3 t_g(1 - n_h d_h / 2\pi_c)] \quad [10]$$

Stress due to torsion can then be found by substituting c for b in Eq. 9 and solving as before.

Conical-gusset stress due to bending is given by:

$$S_b = M/Z \quad [11]$$

where $Z = \pi(a^2 t_n + b^2 t_g)$ at the attachment of the gusset to the flange. With four holes, the section modulus at the centerline of the holes is (approx.):

$$Z = \pi[a^2 t_n + c^2 t_g(1 - d_h/2\pi)] \quad [12]$$

Bending stress can then be calculated as before (see Eq. 11). Direct compressive stress due to agitator weight will be in addition to compressive stress due to bending.

It should be noted that the equations for bending and shear stresses contain no correction factors for the stress concentrations that will occur at the edge of the hole. While shear-stress concentration will be minimal, the stress concentration factor due to direct stress has been shown to be three at the edge of a round hole.

Because of the small diameter of the gusset, stress will not be limited by local buckling, but will be governed by the stress limit for static or cyclic load application.

Frame-mounted agitators

If it has been determined that the top head will require significant modification from a standard design, it may be economically feasible to mount the agitator on a steel supporting frame that is in turn attached to the tank at its periphery. Fairly large agitators (up to 20 hp, 5,000 lb on a 12-ft-dia. tank) have been successfully mounted by this method. The primary area of concern is the attachment of the supporting structure to the tank sidewall.

The most significant loading applied to the attachment is the vertical reaction, F_y. For a properly designed mount, this reaction is composed of roughly equivalent parts due to agitation plus supporting-structure dead weight and to the agitator's bending moment. If the center of gravity of the agitator coincides with the center of the support, then the component of force due to weight, F'_y, is given by:

$$F'_y = W/4 \quad [13]$$

The reaction due to bending moment will be greatest when the moment vector is parallel with the support beams. This condition results in equal reactions on each side of the support, but the reactions are of opposite sign (one upward and one downward). On the side of the support that bears the downward reaction, the vertical load due to bending moment, F''_y, will have the same sign as F'_y, and will create the worst-case loading on the support. The magnitude of F''_y (max) is:

$$F''_y(\text{max}) = M/2L_b \quad [14]$$

where L_b is equal to the beam support spacing, in. The vertical support reaction is then:

$$F_y = F'_y + F''_y \quad [15]$$

Once the magnitude of the vertical load has been determined, the local shell compressive-stress and the shear stress of the laminate attachment must be analyzed based on the actual shell thickness, type and size of support lug, and the method of lug attachment. The support attachment configuration must be designed based on the use of appropriate safety factors in conjunction with the material's compressive strength, shear strength, local buckling stress, and cyclic stress limits.

The compressive buckling of thin cylindrical shells has been shown to be governed by the following relationship:

$$S_c = \gamma\, 0.6Et/r \qquad [16]$$

where γ is a correlation factor (less than unity) that accounts for the difference between theoretical and experimental results. Values of γ can be conservatively approximated by $\gamma = [1.17 - 0.13 \ln(r/t)]$ for direct axial compression. This relationship is based on plots of the correlation factor, γ, for r/t values ranging from 10 to 400 [*2*].

The torque reaction will cause a lateral force, F_t, on each support. This will result in a linearly varying shear load applied on the support, f_t, as shown in Fig. 8. The magnitude of this stress is usually small as long as the projection of the support above the attachment is minimized. Lateral stress can, however, influence the type of laminate required for attachment of the support to the vessel sidewall.

Other types of attachments of agitator support frames to FRP tanks have been used successfully. These include beams mounted to flanges of open-top vessels by means of appropriate load-spreading baseplates (of steel construction) and similar types of support used with flat-top tanks.

General design

Sidewall stress and stability

As previously mentioned, sidewall pressures will be randomly applied due to agitation in the range of ±1 psi. The effect of localized pressures of this magnitude were analyzed [*3*]. The results of the investigation showed that neither local flexural stresses nor deflections are likely to be large enough to affect the design of the sidewall.

An investigation of the sidewall's dynamic stability was also made [*6*]. From this, it was found that the period of vibration of the shell can be in the range of frequencies normally used for shaft speed of turbine agitators. The natural frequency of the sideshell can be approximated by:

$$f = 581.6\, T/d^2 \sqrt{E/\rho} \qquad [17]$$

where the density of fiberglass, ρ, will normally range from 0.05 to 0.07 lb/in.^{3}, depending on glass-to-resin weight ratio and resin density. The second mode of vibration will occur at 2.83 times the frequency of the first.

In determining the free-ring frequency, stiffening effects of the end-constraints have been ignored. Although end-effects, nozzle reinforcement and attachments may increase the frequency of vibration, it is

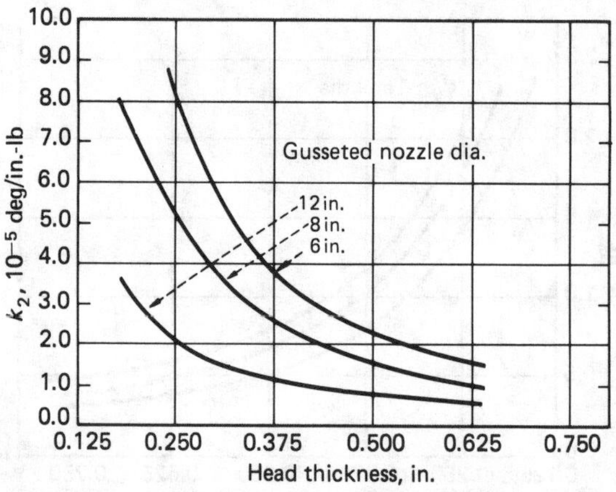

Deflection coefficient vs. head thickness for moment applied to nozzles on an 8-ft-dia. tank Fig. 15

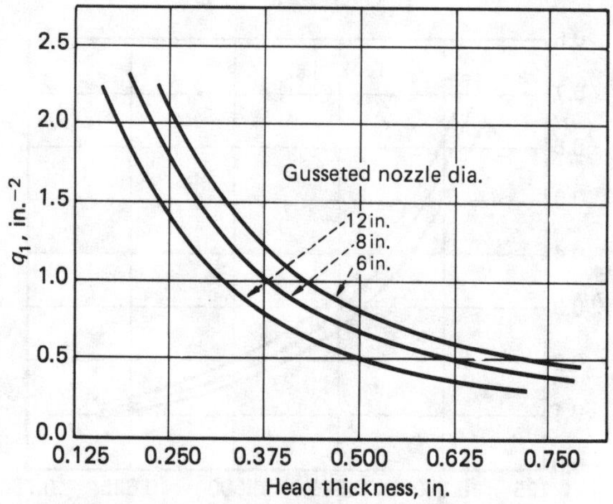

Stress coefficient vs. head thickness for load applied to nozzles on an 8-ft-dia. tank Fig. 16

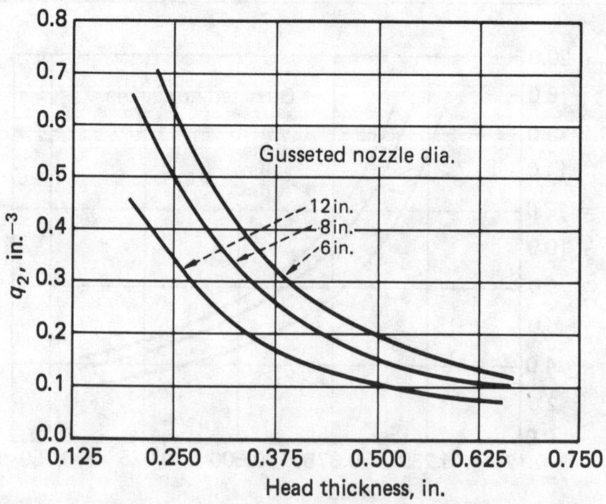

Stress coefficient vs. head thickness for moment applied to nozzles on an 8-ft-dia. tank Fig. 17

recommended that circumferential stiffeners be added if the operating speed of the agitator is greater than 80% of the theoretical frequency so calculated. If stiffeners are required, they should have a centerline spacing of from one half to two thirds that of the tank diameter. The required rib stiffness is:

$$EI_s = 1.98 \times 10^{-6} \rho\, th\, d^4 N^3/f \qquad [18]$$

where t is the average shell thickness between stiffeners, I_s is the moment of inertia of the circumferential rib stiffener, and h is the stiffener spacing in inches.

Baffles

The maximum distributed load applied to the face of a flat baffle has been given previously (Eq. 4). Over the normal range of agitator applications, this load will result in low stresses and deflections for wedge-shaped baffles. This is due to the nature of the continuous support of the baffle along the tank wall. The moment that is applied to the tank shell will not affect the sidewall's design under normal operating conditions.

The superiority of this baffle design over a flat-plate-type baffle is obvious. First, the flat-plate baffle is subject to significant bending stresses and deflections under load. And second, gusseted baffles apply a much higher sidewall moment at the location of the gussets, whereas the wedge-type baffle distributes this moment over the entire length of the baffle. If a gusseted-type baffle is used, the design of the attachment and sidewall should be based on an analysis of the stresses arising from locally applied moments on a cylindrical shell [4].

Design examples

Nozzle-mounted agitator

Consider a 3-hp agitator to be mounted on a 6-in.-dia. nozzle on a 6-ft-dia. tank. The agitator will have a 2-in.-dia. shaft with a single 33-in.-dia. turbine rotating at 84 rpm. The agitator weighs 580 lb and generates 2,250 lb-in. torque and 6,000 lb-in. bending moment when operating. The design chosen uses a ½-in.-thick ASME dished head and a 6-in. conically gusseted nozzle with ¼-in. nozzle neck and gusset thickness. Deflection and stress coefficients for the configuration chosen can be determined from Fig. 10 through 13 as:

$$k_1 = 5.0 \times 10^{-5} \text{ in./lb}_f$$
$$k_2 = 3.0 \times 10^{-5} \text{ deg/lb-in.}$$
$$q_1 = 0.70 \text{ in.}^{-2}$$
$$q_2 = 0.18 \text{ in.}^{-3}$$

The deflection analysis is as follows:

$$y_1 = k_1 W = 5.0 \times 10^{-5}(580) = 0.029 \text{ in.}$$
$$y_2 = k_2 M = 3.0 \times 10^{-5}(6{,}000) = 0.18 \text{ deg}$$
$$r_o = 9 \text{ in.}$$

Therefore, vertical deflection (at r_o) due to angular movement of the nozzle is:

$$y_{ro} = 9 \sin(0.18 \text{ deg}) = 0.028 \text{ in.}$$

Total deflection:

$$y = y_1 + y_{ro} = 0.057 \text{ in.}$$

Deflection should be limited to 0.5% of span. In fact, deflection = 0.057/72 × 100 = 0.08 (< 0.5%). Shaft deflection should be limited to ⅛ in./ft of extension. Shaft deflection/ft of extension = 12 sin(0.18 deg) = 0.037 in. (< ⅛ in.).

Head stresses are:

$$S_1 = q_1 W = 0.70(580) = 406 \text{ psi}$$
$$S_2 = q_2 M = 0.18(6{,}000) = 1{,}080 \text{ psi}$$
$$S = S_1 + S_2 = 1{,}486 \text{ psi}$$

Using a safety factor of 10 on the ultimate flexural stress of 22,000 psi (for ½-in.-thick hand-layup laminates) yields a static allowable stress of 2,200 psi. Referring to Fig. 9 yields a reduction in strength to 16% of the static ultimate for 10^9 cycles of loading. Since the agitator will reach 6.6×10^8 cycles after 15 yr of operation, the design stress for fatigue resistance should be:

$$S_F = 0.16(22{,}000)/2 = 1{,}760 \text{ psi}$$

where the suggested safety factor of 2 has been used with the Fig. 9 data. Therefore, the maximum stress for fatigue resistance governs the head design. The safety factor for fatigue design is then:

$$\text{S.F.} = 0.16(22{,}000)/1{,}486 = 2.4$$

which is greater than the value of 2 suggested for use with Fig. 9.

For the analysis of gusset stresses, J is given for the centerline of the holes in the conical gusset by Eq. 10 as $J = 286.8$ in.^{4}, where $c = 6.375$, $t_n = t_g = 0.25$ in., $a = 3.0$ in., and the gusset is assumed to have four 4-in.-dia. handholes for bolt access. Torsional shear stress is then:

$$S_t = T_c/J = (2{,}250)6.375/286.8 = 50 \text{ psi}$$

Similarly, using Eq. 12:

$$Z = 18.7 \text{ in.}^3$$
$$S_b = M/Z = 6{,}000/18.7 = 321 \text{ psi}$$

The direct compressive stress due to agitator weight is 15 psi, so that the combined direct compressive stress is $S_{comp} = 321 + 15 = 336$ psi. Adding a factor of 3 for stress concentration at the hole yields:

$$S = 336 \times 3 = 1{,}008 \text{ psi}$$

Again, the fatigue stress limit will govern the design. The factor for fatigue is given by:

$$\text{S.F.} = 0.16(18{,}000 \text{ est. comp. strength})/1{,}008 = 2.9$$

Tensile strength will be:

$$S_{ten.} = 321 - 15 = 306 \text{ psi}$$
$$S = 306 \times 3 = 918 \text{ psi}$$

and the resulting safety factor for tensile stress is:

$$\text{S.F.} = 0.16(12{,}000)/918 = 2.1$$

Frame-mounted agitator

A 7.5-hp agitator is to be mounted on a 12-ft-dia. tank using two W6 x 12 steel wide-flange beams. The tank has a 12-ft straight length. The agitator is to rotate a 36-in.-dia. turbine on a 144-in.-long shaft at 45 rpm. The agitator generates 10,500 lb-in. torque and a bending moment of 56,000 lb-in., and the agitator plus its support weighs 1,500 lb.

The agitator mounting design chosen attaches to an area on the tank where the shell thickness is $\frac{5}{16}$ in. The steel base plate of the support is $\frac{1}{4}$-in. thick by 6-in. wide. The weld has a minimum thickness of $\frac{5}{16}$ in. The average shell thickness is $\frac{3}{8}$ in.

From Eq. 13:

$$F_y' = W/4 = 1{,}400/4 = 350 \text{ lb}$$

Similarly from Eq. 14 and 15:

$$F_y'' = M/2L_b = 56{,}000/2(36) = 778 \text{ lb}$$

and:

$$F_y = F_y' + F_y'' = 1{,}128 \text{ lb}$$

Thus, the sidewall must withstand a 1,128-lb vertical load at each support location.

Eq. 16 can be used to predict the allowable compressive stress to prevent buckling:

$$S_c\ (t = \tfrac{5}{16} \text{ in.}) = 206 \text{ psi, where } E = 0.9 \times 10^6 \text{ psi}$$

$$S_c\ (t = \tfrac{1}{2} \text{ in.}) = 418 \text{ psi, where } E = 1.0 \times 10^6 \text{ psi}$$

and a safety factor of 5 has been used.

Therefore, the minimum weld width (to adequately distribute compressive load to the shell) is:

$$w = 1{,}128/0.3125(206) = 18.4$$

Using a reinforced thickness of $\frac{1}{2}$ in. at the edge of the base plate results in a stress of:

$$S = 1{,}128/0.50(6) = 376 \text{ psi}$$

Since 376 psi $< S_c\ (t = \frac{1}{2}$ in.), this weld thickness is adequate.

Assuming an allowable average shear-bond-stress of 100 psi, the minimum shear-bond-area of the weld below the baseplate is:

$$A = 1{,}128 \text{ psi}/100 \text{ psi} = 11.3 \text{ in.}^2$$

The compressive stress in the weld material at the edge of the plate will be governed by the fatigue stress limit:

$$F_s = 19{,}000 \text{ psi} \times 0.16 = 3{,}040 \text{ psi}$$

The support area is $\frac{1}{4} \times 6 = 1.5$ in.^{2}. Compressive stress is:

$$S = F_y/A = 1{,}128/1.5 = 752 \text{ psi}$$

Therefore, the safety factor for fatigue stress is:

$$\text{S.F.} = 3{,}040/752 = 4.0$$

This is adequate since it is larger than the safety factor of 2 that was suggested for use with the data of Fig. 9.

The natural frequency of the shell can be approximated by:

$$f = 581.6(0.375/144^2)\sqrt{1.8 \times 10^6/0.07} = 53 \text{ cpm}$$

where the flexural modulus of the composite (structural and liner) has been taken as 1.8×10^6 psi. Since the estimated natural frequency is greater than 80% of the operating rpm of the agitator, it is suggested that one rib be added to stiffen the tank shell. The stiffness required is given by Eq. 18 as:

$$EI_s = 1.98 \times 10^6(0.07)(0.375)(72)(45^3/53)(144)^4$$

$$EI_s = 2.8 \times 10^6$$

For a rib stiffener having a flexural modulus of 1×10^6 psi, the moment of inertia of the rib should equal or exceed 2.8 in.^{4}.

References

1. ASTM D3299-74, Standard Specification for Filament-Wound Glass-Fiber Reinforced Polyester Chemical-Resistant Tanks, Annual Book of ASTM Standards (Part 36), ASTM, Philadelphia, 1977.
2. Baker, E. H., Kovalevsky, L., and Rish, F. L., "Structural Analysis of Shells," McGraw-Hill, New York, 1972, pp. 229–235.
3. Bijlaard, P. P., Stresses from Radial Loads in Cylindrical Pressure Vessels, *Welding Journal Research Supplement,* 1954.
4. Bijlaard, P. P., Stresses from Radial Loads and External Moments in Cylindrical Pressure Vessels, *Welding Journal Research Supplement,* 1955.
5. Bijlaard, P. P., Computation of the Stresses from Local Loads in Spherical Pressure Vessels or Pressure Vessel Heads, *Welding Research Council Bull.* 34, 1957.
6. Den Hartog, J. P., "Mechanical Vibrations," 4th ed., McGraw-Hill, Inc., New York, 1956, pp. 165–166.
7. Gates, L. E., Hicks, R. W., and Dickey, D. S., Application Guidelines for Turbine Agitators, *Chem. Eng.*, Dec. 6, 1976, p. 165.
8. Isham, A. B., Design of Fiberglass-Reinforced Plastic Chemical Storage Tanks, *Proc. 21st Annual Tech. Conf.*, Reinforced Plastics Div., Soc. of the Plastics Industry, New York, 1966.
9. Isham, A. B., Influence of Chemical Exposure and Temperature on the Design Properties of Reinforced Plastic Chemical Storage Tanks, *Proc. of the 22nd Annual Tech. Conf.*, Reinforced Plastics Div., Soc. of the Plastics Industry, New York, 1967.
10. NBS PS 15-69, Custom Contact-Molded Reinforced-Polyester Chemical-Resistant Process Equipment, National Bureau of Standards, Voluntary Produce Standard, U.S. Govt. Printing Office, Washington, 1969.
11. Osgood, C. C., "Fatigue Design," John Wiley and Sons, Inc., New York, 1970, pp. 462–467.
12. Roark, R. J., "Formulas for Stress and Strain," 5th ed., McGraw-Hill, New York, 1975, p. 476, (case 2).

The authors

Julian B. Fasano is Manager, Process Engineering and Development, for Chemineer Agitators Div. of Chemineer, Inc., P.O. Box 1123, 5870 Poe Ave., Dayton, Ohio 45401, telephone 513-898-1111. He previously worked for Air Products and Chemicals, and PPG Industries. He received his B.S.Ch.E. and M.B.A. degrees from the University of Dayton, and an M.S.Ch.E. degree from Lehigh University. Mr. Fasano is a registered professional engineer in the state of Ohio and a member of AIChE.

Timothy M. Eberhart is a Senior Engineer at Hercules Inc.'s Allegany Ballistics Laboratory, P.O. Box 210, Cumberland, MD 21502, telephone 304-726-4500, where he is involved in the design of advanced-composite products. He received B.S. and M.S. degrees in Engineering Mechanics from Pennsylvania State University and has worked with General Electric Co. and Beetle Plastics, Inc., a Chemineer company. He is a member of the Soc. of Plastic Engineers and the American Soc. for Testing and Materials.

Abrasion of fiber-reinforced plastics in corrosive environments

Fiberglass-reinforced plastics often must handle abrasives under corrosive conditions. By considering particle kinetic energy and hardness, proper equipment and piping can be designed.

John H. Mallinson of J. H. Mallinson, PE & Associates

☐ The term "corbrasion" has been coined to describe the destruction of surfaces by the combined processes of corrosion and abrasion [1]. The two mechanisms are synergistic, and can result in catastrophic wear. An example involved a 24-in. cast-iron elbow that was installed in an acid evaporator line. Entrained crystals and acid—as a vapor stream—destroyed the elbow in two weeks.

Abrasion can be a problem for fiberglass-reinforced-plastic (FRP) piping and equipment, especially when it is used for ash handling and salt crystallization. FRP is commonly used for such purposes, yet little has been published on how to minimize abrasion.

Here, we will look at how FRP performs in abrasion service, based upon field experience. The object is to design and run a system that tolerates particles with an acceptable amount of wear.

We will first look at the effect of a particle's kinetic energy, then consider particle hardness. Also, the FRP system itself will be briefly discussed.

Particle size

The kinetic energy of a particle is a function of its mass and velocity. Mass can be translated into particle size, and this parameter will be looked at first.

Case histories show that most particles of 100 mesh or below in slurry systems do not pose a major abrasion threat. For example, many of the systems for handling fly ash or slurries perform excellently. However, as particle size increases, problems begin.

A mixture of fly ash and bottom ash is a real problem [2]. Clinkers, $\frac{1}{4}$ and $\frac{1}{2}$ in. in size, audibly roll along the bottom of a line. According to a study, the bulk of the heavy flow is confined to the bottom 60-deg arc of the pipe (30 deg off the vertical centerline). Wear is overwhelming in the bottom 40 deg as the heavy particles bounce and roll along the pipe.

Substantially all the wear occurred in an arc totaling 80 deg of the pipe. Only 22% of the pipe area bore the brunt of the abrasion, while 78% was untouched. The real heavy wear took place in a total arc of 40 deg—only 11% of the pipe's circumference.

Large-diameter (27 in.) FRP piping stands up against crystal-laden acidic vapors Fig. 1

This case study was on a commercially available abrasion-resistant (ABR) pipe. The total wall thickness was $\frac{3}{8}$ in. with a $\frac{1}{8}$-in. ABR liner. Flow of solids was 1,000 gal/min of 1–2% fly ash/bottom ash at 7 ft/s through an 8-in. line.

The particle size is, of course, of great importance in abrasion. In theory, the abrasive effect should be a function of the particle size cubed. (Volume is a function of the diameter cubed.) In practice, due to irregular shapes and definitions, a factor of particle size squared seems more likely.

Particle velocity

Since kinetic energy of a moving particle is directly proportional to the weight times the square of the velocity, the velocity of the particle assumes tremendous importance.

Particle velocity must be considered with respect to its impingement angle with the wall. If flow is parallel, even for high velocities, damage is not significant. If the particles hit the wall at an angle, then problems can occur.

FRP piping and equipment are suited for handling the softer types of materials

Mohs scale		Other examples
1 Talc	Very soft	Filter-press cake, crystal aggregates
2 Gypsum		Salt, crystalline salts, soft coal, graphite
	Soft	
3 Calcite		Marble, chalk, brimstone, barite
4 Fluorite		Limestone, soft phosphate
5 Apatite		Chromite, bauxite, hard phosphate
	Intermediate	
6 Feldspar		Hornblende
7 Quartz		Granite, sand
	Hard	
8 Topaz		Aluminum silicate, spinel
9 Corundum		Tungsten carbide, alumina, sapphire
10 Diamond	Very hard	——

■ For slurries, where particle flow is parallel to the wall: If the particle is not large or very hard, then conventionally designed piping systems will do well, even with liquid velocities to 12 ft/s. Also, tests have shown that velocities to 6 ft/s with standard FRP piping and fittings present no problem in long-term (10-yr) service for crystals and crystal aggregates.

■ In gases, where particles may be 0.2% by weight of the gas flows: In large (e.g., 27-in.-dia.) pipe, fittings, and vapor-conveying systems, velocities to 250 ft/s at 50°C have run for 12 yr in evaporators with elbows and straight runs (see Fig. 1). There is some checking and crazing of the resin surface, but no abrasion problems. However, those portions of the system subject to impact-type abrasion at the same vapor velocities show decided wear in a 4–6 yr period. Localized wear may reach $\frac{1}{8}$ in./yr on FRP systems.

An attempt to cover the high-wear areas in vapor flow with a ceramic-filled epoxy material was not successful, due to debonding of the material from the basic structure in 6–9 mo, even though great care was taken in installation.

In very-high-wear areas, corbrasion was minimized by imbedding a section of 16-gage Carpenter 20 or Hastelloy C sheet near the interior surface and laminating it on the vessel.

When designing an FRP system, care must be taken because conditions—and consequent wear—can vary tremendously, depending upon the particular locations.

For example, in a crystallizer booster, the velocity in the throat area was 1,150 ft/s at 180°F. Entrainment was estimated at less than 0.1%, but was definitely present. Some 15 of these units were in service for periods to 10 yr. The practical service life of a booster was typically 6 yr on a "run to destruction" basis. Wear in the throat area even at these velocities was virtually nil. The C glass was not worn through. C glass (or chemical glass) is a type of FRP that has high resistance to chemicals, particularly acids.

However, in the suction head, where there was turbulent flow, the temperature varied from 2 to 50°C. Severe erosion was evident. Here the flow changed from parallel to impact that was, in many cases, 90 deg to the wall. In 4 yr, erosion was severe, with perhaps $\frac{1}{4}$ in. worn away, or about $\frac{1}{16}$ in./yr. The calculated velocity in this area was about 160 ft/s, with flow directly impinging on the FRP.

In the above example, one booster nozzle was incorrectly installed so that it impinged the wall at an angle of about 10 deg. A hole was worn in the $\frac{3}{4}$-in. FRP wall in 6 mo.

When dealing with metallic corrosion, the engineer generally adds a corrosion allowance, in inches, to the thickness specified. For FRP, where abrasion is a factor, the engineer must likewise use an erosion allowance. It is particularly important where vacuum or pressure dictates the vessel design. Where flow is parallel to the wall, the allowance may be minimal.

Hardness of particles

Another basic consideration is hardness [*3,4*]. Commonly, material hardness is defined on the Mohs scale, with talc = 1 and diamond = 10 (see the table). The scale is linear to 9, but then sadly out of scale, as diamond should be about 42 rather than 10.

The Mohs scale is also a fairly good indication of abrasive qualities.

Quite fortunately, nearly all the crystalline salts are relatively soft, with a Mohs hardness of about 2. This permits FRP to handle this type of material very nicely as long as flow is parallel to the wall for slurries (up to 12 ft/s) and gases (up to 1,100 ft/s).

We are, of course, in trouble when the flow is changed to impingement. A localized slurry velocity of 30 ft/s directed against an FRP wall may wear through it in a few months. For gases, heavy wear can happen at 160 ft/s with entrained crystals. At 50 ft/s, soft crystals in a gas stream are not a serious problem.

Here are some examples of how crystals can cause problems in FRP piping and equipment:

1. Soft crystals comprising 0.1% by weight of gas streams impinging on an FRP wall can go through it at a rate of about $\frac{1}{16}$ in./yr at 250 ft/s, at 0–50°C. However, soft crystals in a gas stream at the velocity of sound parallel to an FRP wall created no real problem in 6 yr.
2. Small particles (100 mesh or finer) even at a Mohs hardness of 6–9 are not troublesome in slurries or suspensions in piping systems at velocities to 10 ft/s. However, larger particles ($\frac{1}{4}$–$\frac{1}{2}$ in.) are troublesome at slower velocities.
3. High-temperature (270°F) gas streams carrying about 0.1% solids at 1,100 ft/s and impinging on an FRP wall can wear a hole in it in 1–2 d.

Particle shape

Certainly, particle shape has a bearing. Well-formed spherical glossy particles such as fly ash present little difficulty. Most crystals with their soft geometric shapes can be handled. The problems are the hard, angular, irregular, large shapes such as bottom ash.

Still, even when we deal with soft, crystalline material, we may have a problem unless close attention is paid to system design. However, the real problems occur with ores, bottom ash, minerals, hard phosphates, and so on, where a large particle size and hard material

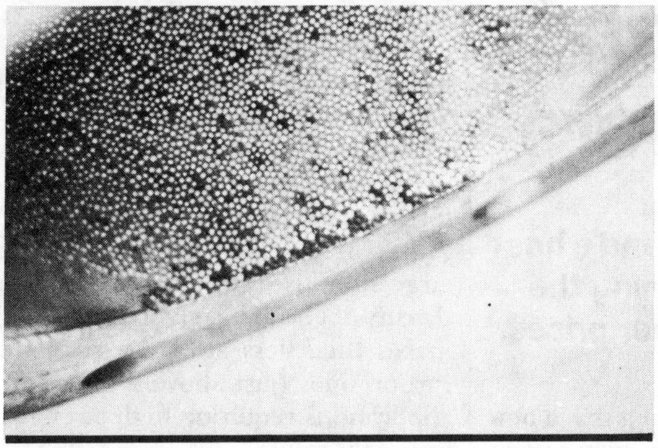

Spherical, abrasion-resistant filler inside an FRP pipe after 4-yr service Fig. 2

combine to cause difficulties. Virtually all of these have Mohs hardness numbers of 6–9.

Synthetic veiling

A veil (or surfacing mat) is employed with a reinforcing mat and fabrics to yield a smooth finish and serve as a high-resin barrier for corrosives. A reinforcing mat is made up of random, nonwoven fibers bonded with the resin.

Field tests have been run to determine the role of surface veiling—either C glass or polyester veiling—in heavily-abrasive corrosive streams. There did not appear to be any major difference in equipment performance solely attributable to veiling. In parallel flow, under cycling-temperature conditions, synthetic veiling can be a plus. Consider the use of double-veiling systems of C glass backed up by polyester veiling. For more on the role of veiling, refer to Ref. [*5,6*].

Abrupt turns

Almost as severe as direct impingement is a 180-deg reversal of flow in a vapor-handling unit such as a crystallizer booster. For example, the tube extends into the suction head. Vapor passing over the edge of the tube at 250–300 ft/s creates an area of heavy wear.

Laminate destruction by abrasion

The mechanism of severe abrasion, in which loss is $\frac{1}{16}$–$\frac{1}{4}$ in./yr in local areas, begins with the wearing away of the C glass layer. This is followed by the removal of the 100-mil corrosion barrier. Now we are into the woven roving. This is a heavy fabric made from continuous roving (a slightly twisted fiber strand). We then plow through successive layers of mat and roving until the vessel wall is breached.

The degree to which we have exposed the structural wall to chemical attack will depend on how corrosive the medium is. The use of additives to enhance the abrasion resistance of a laminate has much to recommend it, particularly as regards the corrosion barrier.

Effect of additives

When additives are used in laminates to enhance abrasion resistance, the physical properties of the laminates may be altered [*7*]. The tensile strength of a highly-corrosion-resistant laminate was studied as the loading of alumina was increased:

Additive, % by wt.	Tensile strength, lb/in.^{2}
0, control	12,000
10	11,400
20	8,000
30	7,000
55	4,000

Up to loadings of 10%, the effect is not significant. Above that, the effect becomes progressively severe. It can be compensated for by using heavier walls.

Tests have shown that increasing the additive loading from 10% to 40% does little to improve the abrasion resistance if the additive size is kept constant. However, changing the shape of the additive particle may have profound effects:

1. If we use a flakeglass-shaped particle, we get improvements in two areas:

- Less weight gained (vs. a control) in submersion tests. This means a better laminate with improved chemical resistance.
- Better abrasion performance, because we have layered the particles.

2. Still using the same additive compound, we can increase the size to 0.050–0.070 in. balls with a very heavy volumetric loading (50–60%) and greatly improve resistance to tremendously abrasive slurries [*2*]. Such a system becomes essentially Ceram Core pipe (see Fig. 2). Tests on this pipe for over 4 yr have been encouraging. As a result, it was installed in a fly ash/bottom ash line. In the same service, steel pipe would last about one year.

References

1. Wolf, E. J., and Heim, R., Thermoset Resin Monolithic Linings for Chemical and Abrasion Service, "Managing Corrosion with Plastics," Volume III, Natl. Assn. of Corrosion Engineers (Houston), 1977.
2. Bulletin on Ceram Core abrasion-resistant piping systems, A. O. Smith-Inland, Little Rock, Ark., 1978.
3. Perry, J. H., "Chemical Engineers' Handbook," 3rd ed., 1950, McGraw-Hill, New York., p. 1,114.
4. Perry, J. H., "Chemical Engineers' Handbook," 4th ed., 1963, McGraw-Hill, New York, Sect. 8, p. 7.
5. Bautista, T. O., The Role of Synthetic Veil in the Wear Factor of Corrosion Resistant Laminates, 35th Annual Technical Conference, Soc. of the Plastics Industry (SPI), New York., 1980.
6. Overholt, G. T., New Surfacing Fabrics for Corrosion Resistant Reinforced Plastics, 31st Annual Technical Conference, SPI, 1976.
7. Mallinson, J. J., Increasing the Abrasion Resistance of Reinforced Plastics, 31st Annual Technical Conference, SPI, 1976.

The author

John H. Mallinson is president of J. H. Mallinson PE and Associates, P.O. Box 1617, Front Royal, VA 22630. Tel: (703) 635-2731. The firm consults in fiberglass construction exposed to corrosive environments. Previously, Mallinson held management positions in engineering with several large chemical companies. He has had over 20 years' experience in the application of composites, and is the author of numerous papers in this field. He wrote "Chemical Plant Design with Reinforced Plastics," McGraw-Hill, 1969. He is a registered professional engineer in Virginia, and a member of AIChE, and the Natl. Assn. of Corrosion Engineers.

Growth of composites is bolstered by new fibers

Carbon fibers made from cheaper raw materials, and a host of new organic and inorganic fibers, are broadening the markets for plastic composites, and lowering fiber prices.

☐ The worldwide outlook for plastic composites—those materials whose strength is beefed up with glass, carbon or aromatic polyamide fibers (see box)—is beginning to brighten. Although the industry is still plagued with overcapacity, and growth rates for the 1980s will not match those of the palmy 1970s, the emergence of new markets in the chemical process industries (CPI) and the development of new high-performance fibers are drastically reducing fiber costs. And, at least in the U.S., a rebounding economy is giving new life to the traditional markets for composites.

The gradually diminishing price for fibers—from over \$300/lb in the late 1970s to about \$20 for carbon ones—has expanded use of composites in such key CPI applications as pipes, heat exchangers, compressors, valves, turbine blades and others. And experts are predicting that a new crop of fibers will be an even bigger booster for composites.

Carbon-fiber manufacturers, for example, are looking at cheaper feedstocks (coal pitch, asphaltene) to make products that perform as well as, but cost less than, those obtained from polyacrylonitrile (PAN) or petroleum pitch. Other firms are working on new organic and inorganic fibers—the latter made from silicon carbide and alumina, and exhibiting unprecedented high-temperature properties.

SOME CAVEATS—Not everything is rosy, however. Industry analysts say that composites, especially carbon-based ones, are still too pricey to displace metals. Indeed, though declining fiber costs are making composites more popular, their use still represents less than 3% of the total volume of materials (metals, alloys, plastics, composites) employed in CPI equipment. To compete with metals, they say, prices should come down further—say, from the current \$50–\$150/lb to \$10–\$50 for carbon-fiber composites.

Prices for the cheaper, glass-reinforced materials typically range from \$3–\$15/lb—an acceptable level for large-scale use. However, the higher density of glass fibers (about 3 g/cm³) makes them less attractive than the carbon ones (just above 2 g/cm³) in applications requiring high strength-to-weight ratios.

No major price reductions are expected in the plastic matrices, whose markets are saturated. And such advances as the automation of composite-fabrication processes are not seen as contributing much to the products' economic progress. Most producers feel that the real opportunity is in reducing further the price of fibers, which generally account for over 50% of the total volume of finished composites.

ANOTHER PITCH—Finding a cheaper raw material doesn't always guarantee a mass market, as Union Carbide Corp. has found out through the years. The company, which in 1974 developed technology to grow carbon fibers from petroleum pitch—a raw material that costs about a third as much as PAN—employed its new process to build a 500,000-lb/yr carbon-

Composites: How they're made today

Composites can be made by over a dozen methods. Ordinary fiberglass, for example, is manufactured by alternating layers of polyester resin and chopped glass fiber in a mold. High-performance composites are most often made with continuous fibers, generally by lamination, filament winding or pultrusion. Much of the current CPI interest centers on fabrication of thermoplastic composites with continuous fibers—a procedure that has already become common with thermosets.

Last year, ICI Americas Inc. (Wilmington, Del.) developed a process for making a thermoplastic/fiber composite called APC, for aromatic-polymer composite. The secret behind the process, says a company spokesperson, is the method by which the resin (polyetheretherketone, PEEK) is thoroughly impregnated with the continuous fibers.

APC is claimed to be superior to conventional epoxy/carbon composites, especially in areas where high strength-to-weight ratios are required. The company says that PEEK filaments can be woven to produce high-performance filters and belting for CPI use, while pump impellers for high-temperature and high-pressure applications have been molded from APC.

Another thermoplastic/continuous-fiber entry is a series of glass- or carbon-fiber sheets with a polyphenylene-sulfide (PPS, tradename Ryton) matrix, from Phillips Petroleum Co. Fiber reinforcement of PPS results in a fivefold to sevenfold increase in impact strength vs. conventional short-fiber-reinforced products. At the same time, says Don Brady, manager for thermoplastic composites, the products retain or exceed all PPS properties: high chemical, mechanical and flame resistance, and good thermal stability and electrical properties.

Typical applications being sought include baghouses for fluegas filtration, valve and pump components, housings, heat exchangers, pipe couplings and fittings, and compressor components. The price (at \$6–\$10/lb) is said to be comparable to that of stainless steel on a volumetric basis.

Originally published August 20, 1984

fiber facility in Greenville, S.C., largely to accommodate anticipated demand in the automobile industry. The hope was that heavy use of carbon fibers by automakers would reduce the cost from \$20 – \$26/lb to about \$6/lb. That market, however, failed to materialize.

Fortunately, Carbide has succeeded in specialized aerospace applications, in which pitch-based fibers with a tensile modulus of about 100 million psi sell for about \$1,200/lb. The firm has had to increase capacity fourfold at Greenville this year, say industry sources.

BRINGING DOWN COSTS—Researchers elsewhere continue to experiment with even cheaper raw materials for carbon fibers. Teh Fu Yen, a professor of environmental and civil engineering at the University of Southern California (Los Angeles), believes he has the key to making fibers for about \$5/lb. Yen is carrying out tests with asphaltene, a waste product of coal liquefaction.

In research supported by the National Science Foundation and by interested companies such as Bethlehem Steel Corp. (Bethlehem, Pa.), Yen is screening various raw materials by testing their ability to form a mesophase—a liquid-crystal state—upon heating in an inert atmosphere at 400 – 500°C. Those with the highest mesophase-forming ability are then "chemically modified to increase carbon-fiber growth," he says, declining to reveal further process details because of pending patents.

Work at General Motors Research Laboratories (Detroit) could also help break down the cost barrier even further. In 1979, GM researchers accidentally discovered that natural gas introduced inside a heated steel pipe diffuses through the metal, forming carbon fibers on the outside wall.

Since then, GM scientists have been trying to understand the basic mechanisms involved in growing the fibers in order to commercialize the technology. "It's possible that this will happen in a few years," notes Gary Tibbetts, staff research scientist, pointing out that similar methods are being studied both in Japan and in France. GM's goal is to reduce the cost of making carbon fibers to that of glass fibers currently used in autos.

The company's process for growing the fibers consists of two stages.

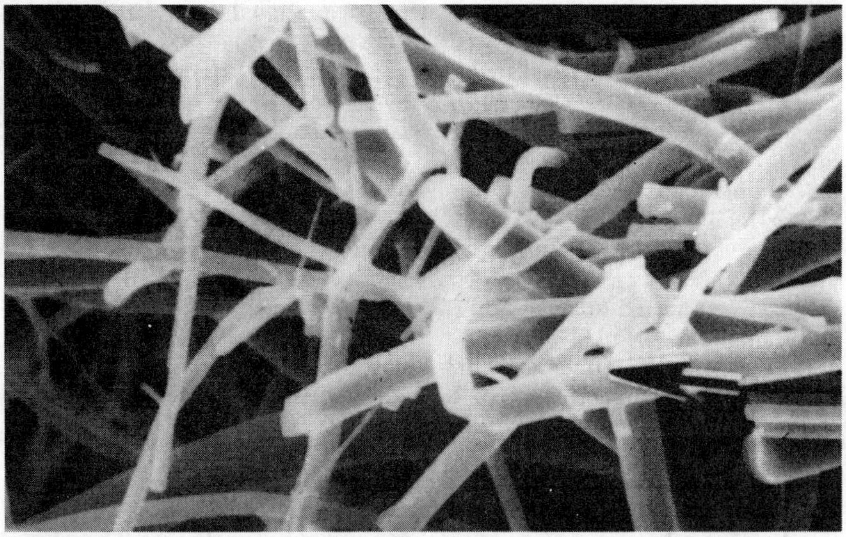

Lockheed Missiles and Space Co.

Network of Nextel and silica fibers. Arrow shows one of many places where the two ingredients fuse to form a rigid tridimensional structure

In the first, a steel tube is heated to about 1,000°C in a natural-gas atmosphere. At this temperature, the inside surface carburizes into tiny nuclei particles (a process called "metal dusting corrosion"). The nucleation sites then catalyze fiber growth by literally extruding carbonaceous filaments through the steel pipe. The rate of growth of the filaments is "surprisingly rapid," says Tibbetts.

In the second stage, the filaments—about 20 – 50 nanometers in dia.—are thickened by chemical-vapor deposition. Then, immersion in a higher concentration of natural gas forms layers of carbon and builds up fibers as much as 20 cm long and 10 micrometers in dia. These are heat-treated at 3,000°C to crystallize the carbon and form strands with high strength (3×10^9 Pa).

OVERSEAS ACTIVITY—Like GM, Japanese firms are optimistic about developing high-volume (and technically less demanding) markets for carbon fibers. Kureha Chemical Industries Co. (Tokyo), which besides Union Carbide is the only other major producer of carbon fibers from petroleum pitch, already has accomplished this to some extent: It sells chopped carbon fibers for less than \$6/lb, and composites for about \$75/lb. The low prices allow Kureha's clients to use the fibers as asbestos substitutes, and for gasket, seal and insulation applications. The tensile strength of the products, however, is the lowest among carbon fibers, varying from 60.2 – 73 kg/mm².

Kureha refuses to discuss its manufacturing process. However, Japanese government and industry sources say that the firm's knowhow centers primarily on the spinning of hot, molten isotropic petroleum pitch in a centrifuge. Kureha admits that its raw material — isotropic pitch — is different from Union Carbide's liquid-crystal pitch. The Carbide process has an extra step that involves heating isotropic pitch until a continuous mesophase — liquid crystal — forms. Carbide says that this step is the key to making its high-strength fibers.

Working with Sumitomo Metal Industries, Ltd. (Osaka), Kureha also has developed a process for making an even lower-grade, lower-price carbon fiber, using coal pitch as raw material. The product is targeted for use as a reinforcing material in cement. The companies are now studying whether demand from cement mills will warrant the building of a 1,000-m.t./yr carbon-fiber plant.

Mitsubishi Chemical Industries, Ltd. (Tokyo), is currently running a pilot plant in Kurosaki, Kyushu Island, which uses a byproduct from coal cokers—tar pitch—as raw material for making low-cost, high-strength carbon fibers. The firm is likely to begin construction of a demonstration plant by yearend. Although Mitsubishi is considering selling the material primarily in the automobile industry, it says that the tar-pitch route can also be used to produce material suitable for reinforcing cement.

MADE WITH SILICON—The Japanese also have a new silicon carbide fiber, developed by Nippon Carbon Co. (Tokyo), that is now being marketed in the U.S. by Dow Corning Corp. (Midland, Mich.). The prime advantage of the 10 – 15-micrometer fibers, called Nicalon, is their resistance to heat: The tensile strength of 200 kg/mm² is maintained at 1,000°C, dropping to about 130 kg/mm² at 1,200°C. In contrast, carbon fibers oxidize when heated to above 400°C.

Although the silicon fibers perform at their best in aluminum-matrix composites (primarily for aerospace applications), Dow Corning says Nicalon can be used just as easily in plastics matrices. The material, which can be woven or formed by conventional techniques, is claimed to form stronger bonds with most commercial resins than do carbon fibers.

Nicalon is made in a 12-ton/yr pilot plant at Yokohama, where it is spun from molten polycarbosilane, heated to oxidize the surface, and roasted at 1,100 – 1,500°C in an inert atmosphere to drive off nitrogen and methane. The fibers are then treated with various chemicals to permit bonding to metal or plastic matrices.

Although the material costs about $250/lb, a Dow Corning spokesman expects that "the price will come down considerably with large-scale production." Anticipated applications in the CPI include insulation, packings and gaskets; other probable uses include automobile- and aeroplane-engine parts.

General Electric Co. (Schenectady, N.Y.) also has started making experimental quantities of silicon carbide fibers. GE produces the materials by reacting carbon fibers with molten silicon at high temperatures. The material, called Silcomp, is currently being evaluated as a liner primarily for furnaces operating at temperatures up to 1,350°C.

The Specialty Materials Div. of Avco Corp. (Lowell, Mass.) also uses carbon fibers as the precursor for making silicon carbide fibers. In the Avco process, silicon carbide is produced by reacting chlorosilanes with propane at high temperatures. Carbide vapor is then deposited on the carbon substrate, forming fibers about 140 micrometers in diameter.

A new, lightweight fibrous tile used as a primary insulation on the underside of the space shuttle Discovery is reducing the weight of the spacecraft by about 900 lb. Manufactured by Lockheed Missiles & Space Co. (Sunnyvale, Calif.), the tiles are made by adding Nextel—a high-strength, large-diameter fiber containing a minute amount of boron—to pure silica fiber. During high-temperature sintering in a furnace to form tiles of the fibers, the boron welds the silica and Nextel into a rigid structure, resulting in tiles that are lighter but just as strong as those made from pure silica fibers.

ALUMINA FIBERS—Du Pont Co. also has come up with an inorganic fiber (Fiber FP) that, like Dow Corning's Nicalon, is being groomed for use in both metal- and plastic-matrix composites. It is a polycrystalline aluminum oxide produced as round filaments about 20 micrometers in dia. Each of these is coated with a thin layer of silica. The company says that even though the coating is less than 0.005 micrometer thick, it increases the tensile strength of the fiber by as much as 25%, compared with uncoated material.

According to Du Pont, aluminum or magnesium reinforced with 35 – 50% by volume of continuous FP filament will have three to four times the strength, four times the fatigue strength, and higher temperature capability (the fiber's melting point is 2,045°C) than unreinforced products. Because of such impressive properties, the material is currently being evaluated by Toyota Motor Co. in Japan for use in automobile piston rods.

A similar aluminum oxide fiber—named Saffil—is being marketed by ICI Americas Inc. (Wilmington, Del.) for use as insulation and for reinforcing metals. And last year, 3M (Minneapolis) started producing aluminum borosilicate fibers, called Nextel, that can be woven into fabric for use as conveyor belts, seals and reinforcements in high-temperature (up to 1,200°C) applications.

MORE ORGANIC FIBERS—Although Du Pont's aramid fiber Kevlar is not new, the company is now promoting one version (Kevlar 49) for use in plastic composites. A spokesman for the firm contends that the product matches carbon fibers in every respect, except perhaps in compression strength. This is why the firm is touting hybrids that mix Kevlar 49 with carbon fibers. In such combinations, the latter fibers contribute high modulus and compression strength, while Kevlar increases impact resistance and reduces overall cost.

American Kynol Co. (New York City) has begun to market a synthetic organic fiber that is highly aromatic, nonoriented and insoluble. The phenol-formaldehyde fibers are said to be easily blended into polypropylene, improving its heat- and flame-resistance, as well as strength. Fiber performance is said to be even better when a chlorinated hydrocarbon coupling agent is used during blending. Other advantages are low specific gravity (1.27), low abrasiveness, and nontoxicity. Among the minuses: low stiffness, and fading color.

Polyethylene fibers that are ten times stronger than steel fiber and twice as strong as carbon ones, have been developed by DSM (South Limburg, the Netherlands). Key to their strength comes from a gel-spinning process that enhances the polymer's ability to form covalent carbon-carbon bonds, and allows it to crystallize into long chains. The company has signed an agreement with Japan's Toyobo Co. (Osaka) to develop production knowhow; commercialization may come sometime in late 1985.

Meanwhile, Mitsui Petrochemical Industries says that it, too, has developed a polyethylene fiber, claimed to have a tensile strength of 30 tons/cm². The company is planning to build a pilot plant at Iwakuni in western Japan to produce test-market quantities of the material.

Several PAN-based carbon fibers have been introduced in the last year or so. Hercules, Inc. (Magna, Utah) has two new products—the Magnamite Series AS6 and IM6—that are said to be less brittle, with higher ultimate elongations (1.65% and 1.6%, respectively). The company's previous product, the AS4, had 1.5% ultimate elongation, and an approximately 20% lower tensile modulus.

A similar high-strength product called Celion ST was introduced last year by Celanese Specialty Operations (Chatham, N.J.). The product has 1.8% ultimate elongation, and a tensile strength of 610,000 psi—about 10,000 psi higher than Hercules' AS6, but lower than the IM6 by an equal amount. These PAN-fibers sell in the $20 – $30/lb range.

When and how to select plastics

Thermoplastics and thermosets can be used to fabricate corrosion-resistant piping and equipment. Selection largely depends upon strength and corrosion resistance of the material under consideration.

J. Albert Rolston, Consultant

☐ In a quiet evolution, various plastics and composites are finding increasing use in chemical-process-industries (CPI) equipment, and are replacing metals, ceramics and rubber as traditional materials. Still, according to engineers in several CPI firms, penetration of plastics into the CPI market is small, probably accounting for less than 5% of the total dollar volume of equipment. These engineers note that new plants generally order about twice as much new plastic equipment as older facilities do for replacements. In some applications (e.g., in handling high-purity chemicals), polymeric materials dominate; there seem to be no metals or ceramics used where minimum contamination is required. Here, we will look at some of the polymers and composites competing for recognition in the CPI equipment market.

The nomenclature of plastics can be confusing. Basically, there are two types of polymers: thermoplastic and thermoset.

Thermoplastics have a defined melting point; they are normally fabricated by softening or melting the plastic and forming it into a useful shape. The plastic is then cooled to retain the shape.

Thermoset polymers have cross-linked molecules and no definite melting point. They are made from liquid monomers (or solutions containing monomers), which are formed into the desired shape and polymerized by heat or chemical reaction. Thermosets are ordinarily used in combination with reinforcements (chiefly fiberglass). These composite materials are often called FRP (fiberglass reinforced plastics) or RTR (reinforced thermoset resins).

Strength properties of plastics

The strength properties of many thermoset composites compare favorably with those of carbon steel; however, the tensile and flexural moduli are usually much lower for plastics. Thermoplastics are much weaker than composites, and have correspondingly lower moduli.

In contrast to metals (at normal temperatures), plastics creep and have other problems that do not affect metals. Thus, specialized design techniques must be employed when plastics are specified. The use of these methods has hindered the growth of plastics in many areas, but especially in CPI plants where safety is paramount.

Table I [*1–3*] shows some physical properties for FRP composites. Some of these properties, or the way they are determined, may be unfamiliar. For example, with metals, tensile-strength tests are normally made on "dogbone"-shaped coupons. However, with some composites—such as filament-wound or pultruded shapes—the usual tests cannot be used, due to the extraordinary orientation of strength properties. For example, a tensile specimen with all of its fibers parallel will fail in shear (parallel to the fibers) when tested with a dogbone-shaped specimen.

The flexural strength of plastics may not be known by engineers who specify metals. Flexural strength and flexural modulus are found from bending tests—often the only way that such data can be obtained. The parallel-fiber specimen that could not be tested in tension, for example, could, however, easily be broken in a bending test to obtain flexural properties. But if the flexural specimen is too short (less than a 16:1 span [i.e., the distance between two support points]/depth ratio), it too may fail in shear along the neutral axis. It should be obvious that shear strength is important for many composites; failure in shear is a frequent mode of failure.

As a result of this and other complications, the standards for composites normally call for larger safety factors than do those for metals. For composites, the usual safety factor for tension is 10, while for shear or for buckling it is 5, based on short-term data. However, the situation is not as dismal as these factors indicate. Long-term strength data (for pipes or tanks, as an example) indicate that, for all plastics, failure is time-dependent. If the long-term strength (as predicted from standard test procedures) is used, the safety factor may be reduced to 2 or 3, a normal factor.

Table II [*4–7*] lists comparable properties for thermoplastics. The low values for tensile and flexural strength and tensile and flexural moduli are, in large part, responsible for the reluctance of many engineers to specify plastic piping and supported structures (such as lined-steel pipe) when a plastic's corrosion resistance is desired. For relatively small diameters (less than 12 in.),

Among plastics, thermoset composites are the strongest, and can compete with metals **Table I**

Materials and construction	Average reinforcing content, wt. %	Strength, 10^3 psi tensile	Strength, 10^3 psi flexural	Modulus, 10^6 psi tensile	Modulus, 10^6 psi flexural	Data source
Mat/polyester (hand layup)	25	9	16	1.0	0.7	[*1*]
Mat and woven roving/polyester (hand layup)	40	15	22	1.5	1.0	[*2*]
Mat and roving/polyester (pultrusion)						
Longitudinal direction	65	30	30	2.5	1.6	[*3*]
Transverse direction	—	7	10	0.8	0.2	

thermoplastic pressure-pipe can be used safely at most normal operating pressures and temperatures. In all, the widespread use of plastic piping in and out of the CPI attests to its excellent performance. But it is wise to be prudent in selection and design.

Standards

Despite the problems, there has been progress: Many design and test standards have been developed. For composites, specialized standards have been effected for CPI applications—more than are on the books for thermoplastics. There are numerous standards for FRP pipe and tanks; most of these have been written by ASTM, but other organizations (e.g., the Amer. Soc. of Mechanical Engineers and the Amer. Petroleum Institute) have also developed some. There are a number of standards for thermoplastic pipe, but for other applications, thermoplastics are not normally considered as structural members. These materials are usually supported by a strong material such as a metal, concrete, or thermoset (in the form of a composite).

For piping systems (thermoplastic or composite), the main standards are (using ASTM acronyms, i.e., RTR rather than FRP):

1. ASTM D1599, a test method for burst pressure for all types of plastic pipe.
2. ASTM D1598, for time-to-failure under constant internal pressure for all types of plastic pipe.
3. ASTM D2143, for the cyclic pressure-strength of reinforced thermosetting resin pipe (RTRP).
4. ASTM D2992, for defining the hydrostatic design basis for RTRP.
5. ASTM D2837, for defining the hydrostatic design basis for thermoplastic pipe.
6. ASTM D543, for measuring the resistance of plastics to chemical reagents in an unstressed condition.
7. ASTM C581, for measuring the chemical resistance of thermosetting resins used in RTR structures.
8. ASTM D3681, for measuring the chemical resistance of RTR in a deflected condition.
9. AWWA (Amer. Water Works Assn.) C950-81, a performance standard for selection and purchase of RTR pressure-pipe for water-transmission systems, in either buried or aboveground installations.
10. ASME Section X, a performance standard for fiberglass-composite pressure vessels.

There are numerous other ASTM standards, including specifications for polymers and cements, test methods for determining dimensions, and recommended practices for handling and installing pipe. Table III lists the principal ASTM standards concerning pipe, fittings and tanks, that apply to CPI installations.

Thermoplastics, not as strong as thermosets, often must be supported for large structures **Table II**

Material	Strength, 10^3 psi tensile	Strength, 10^3 psi flexural	Modulus, 10^6 psi tensile	Modulus, 10^6 psi flexural	Tensile elongation, %	Maximum service temperature*, °F
Polyvinyl chloride (PVC)	7	17	0.4	0.4	40	150
Chlorinated PVC (CPVC)	8	15	0.4	0.4	200	210
Polypropylene	5	6	0.4	0.2	150	200
Polybutylene	4	2.2	0.04	0.05	300	210
Polyvinylidene fluoride	6	—	0.2	0.2	300	300
Ethylene-chlorotrifluoro-ethylene	7	7	0.2	0.2	200	300
Fluorinated ethylene-propylene	3	—	—	0.1	300	over 300

*The maximum service temperature depends greatly upon environmental and stress.
Sources: From Ref. [*5-7*].

ASTM's principal standards for pipe, fittings and tanks in CPI applications **Table III**

Category key: G = general standard
T = thermoplastic standard
R = reinforced thermoset standard

ASTM No.	Category	Title
D543	G	Test method for resistance of plastics to chemical reagents
D638	G	Test method for tensile properties of plastics
D671	G	Test method for flexural fatigue of plastics by constant amplitude force
D790	G	Test methods for flexural properties of unreinforced and reinforced plastics and electrical insulating materials
D1598	G	Test method for time-to-failure of plastic pipe under constant internal pressure
D1599	G	Test method for short-time hydraulic failure pressure of plastic pipe, tubing and fittings
D2990	G	Test methods for tensile, compressive and flexural creep and creep-rupture of plastics
D1785	T	Specification for PVC plastic pipe, schedules 40, 80, and 120
D1794	T	Specification for rigid PVC compounds and CPVC compounds
D2657	T	Practice for heat-joining polyolefin pipe and fittings
D2837	T	Methods for obtaining hydrostatic design basis for thermoplastic pipe materials
D2855	T	Practice for making solvent-cemented joints with PVC pipe and fittings
D3139	T	Specification for joints for plastic pressure pipes using flexible elastomeric seals
D3915	T	Specification for PVC and related plastic pipe and fitting compounds
F645	T	Guide for selection, design and installation of thermoplastic water-pressure-piping systems
C581	R	Test method for chemical resistance of thermosetting resins used in glass-fiber-reinforced structures
C582	R	Specification for reinforced plastic laminates for self-supporting structures for use in a chemical environment
D2143	R	Test method for cyclic pressure strength of reinforced-thermosetting-plastic pipe
D2992	R	Method for obtaining hydrostatic design basis for reinforced-thermosetting-resin pipe and fittings
D2996	R	Specification for filament-wound reinforced-thermosetting-resin pipe
D3299	R	Specification for filament-wound glass-fiber-reinforced thermoset resin chemical-resistant tanks
D4097	R	Specification for contact-molded glass-fiber-reinforced thermoset resin chemical-resistant tanks
D4162	R	Specification for reinforced-thermosetting-resin sewer and industrial pressure pipe

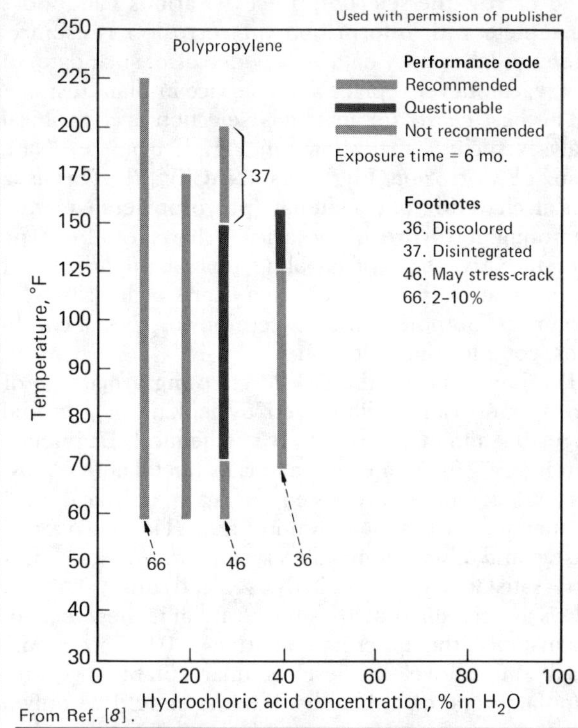

From Ref. [8].

This type of graphical analysis provides a useful way to evaluate corrosion data **Fig. 1**

Corrosion resistance of plastics

Plastics have made considerable inroads into the markets for traditional metal alloys (or lined steel and concrete) mainly because of their resistance in some extremely corrosive environments. In some instances, replacing equipment with plastic costs much less than the competing metal alloy would; in other cases, plastic may provide a longer service life at a lower cost.

The most striking example is handling halogen acids or compounds. Many of these severely corrode almost all stainless steels, but in most cases they can be handled with a low-cost plastic such as fiberglass-reinforced polyester or any of several thermoplastics. Another example: Chlorine bleaching agents, often used in paper mills, can corrode most alloys; much of the bleaching equipment is now made from plastics. Other instances abound for successful applications of plastics in the CPI.

Corrosion testing

Corrosion testing for any material is difficult, since many variables are involved, and these may interact unpredictably. The main ones are: chemical composition and concentration of the corrosive medium, temperature, stress (internal or imposed), and fluid motion.

Since these factors may vary over wide ranges, testing and selecting materials involves considerable effort. There are many sources for corrosion data; most plastics suppliers can provide this. But the variety of conditions makes it mandatory that a manufacturer's tables be used only as guidelines. Perhaps more than in any other area, experience dominates in determining the corrosion resistance.

To narrow the selection process, various handbooks are replete with information on corrosion resistance. Although laboratory data are needed also, such data, of course, can never replace a full-service in-plant test.

The best charts for materials selection are graphical analyses such as that shown in Fig. 1, from the Natl. Assn. of Corrosion Engineers (NACE) [*8*]. This is a typical chart for one material (polypropylene) in one environment (hydrochloric acid). Charts of this type offer the advantage of enabling quick comparisons of various materials. However, such charts ordinarily have numerous footnotes listing exceptions or special conditions, complicating selection.

The immensity of the task of choosing a material of construction can be illustrated by looking at a typical corrosion summary in "Perry's Chemical Engineers' Handbook" [*9*]. Numerous materials (metal alloys, plastics and ceramics) are listed for each chemical in a manner similar to that shown in Fig. 1. The behavior of plastics and other nonmetallics is categorized as: satisfactory, satisfactory for limited use, and unsatisfactory. Alloys are classified by the attack rate in inches/year. In this instance, the upper limit of attack is 0.05 in./yr. Any attack rate above this level is unacceptable. As with manufacturers' data, handbook tables should be only a starting point in any materials determination. But they can help to narrow down the choices. Beyond handbooks are texts on materials of construction. Some contain information on many types of materials; others give more-detailed data by narrowing the choices. An excellent survey volume is the NACE publication, "Corrosion Data Survey: Nonmetals Section," [*8*]. This book was published in 1975, but still contains valuable information about a wide variety of polymers and composites for a large number of chemicals, foods and miscellaneous formulations (such as insecticides, oils and waxes, medicines, etc.).

For more details, materials manufacturers, fabricators or consultants can offer guidance. Most textbooks are about five years or more behind current technology. Magazine articles or conference papers can provide useful and more-current design and background information. The most recent data, of course, are the least available, so selection becomes more difficult when one tries to locate up-to-date information.

Final tests and evaluations

When the choices are eventually narrowed down to a few, it is desirable, as already mentioned, to conduct final tests in the plant environment (or as close a simulation as possible). These tests should look at any expected stress conditions and at all of the process variables. Spool pieces (short, flanged pipe-sections) are often used to test new piping materials, for example. Tests should be run for as many months as possible before making the choice.

Economic evaluations should be made for the various competing materials. Often, it is cheaper to use a more expensive material; it yields a better return on investment. Table IV, from Ref. [*10*, Table 23-2, p. 23-13], shows a simplified analysis for comparing three materials, each having a different service life.

In making such an analysis, all variables must be

Plastic Piping Systems Inc.

In this chlor-alkali plant, PVC piping and valves are connected to FRP tanks **Fig. 2**

Plastic Piping Systems Inc.

For PVC, piping is a major use in chemical-process-industires plants **Fig. 3**

Plastic Piping Systems Inc.

Injection-molded plastic valves. Plastic valves are typically made in up to 4-in. sizes **Fig. 4**

Here is a typical investment analysis that is helpful in picking a material — **Table IV**

	Material A	Material B	Material C
Installed cost (investment)	$45,000	$65,000	$60,000
Estimated life, yr	4	6	10
Estimated maintenance rate, %	10	7	5
Annual replacement cost (installed cost ÷ life)	$11,250	$10,830	$6,000
Annual maintenance cost (installed cost X rate)	$4,500	$4,550	$3,000
Total annual cost	$15,750	$15,380	$9,000
Annual saving (compared with A)		$370	$6,750
Tax on saving @ 50%		$185	$3,375
Net annual saving		$185	$3,375
Return on investment (compared with A)		0.9%	22.5%

From Ref. [*10*]. Used with permission of publisher

included. For example, if equipment is expected to be obsolete in a short period (say five years), the equation may shift toward a lower-cost material with a shorter life. The cost of downtime, for both maintenance and replacement, should be factored into the analysis.

Often, building a plant for a new process or product precludes the more leisurely approach of testing a new material, and a more hurried decision must be made. Here, it is wise to rely upon those engineers with the widest range of materials experience. However, the designer often will unfortunately follow the advice of an equipment maker who has a good knowledge of the materials used in the past but may not be up-to-date on newer ones.

Corrosion data

Table V [*5, 11–14*] summarizes the corrosion resistance of several polymers and composites now being used in the CPI. Some of the materials are old reliables; others are more recent innovations. While only a few chemicals are listed, they are among the most aggressive reagents known.

Fluoropolymers usually have the best overall corrosion resistance. Unfortunately, they also are the most expensive. Many of the thermoplastics (e.g., polyvinyl chloride (PVC) and polypropylene, as well as the fluoropolymers) have better corrosion resistance than do the thermoset composites. FRP composites, for example, are not suitable for high concentrations of sulfuric acid, but chlorinated PVC can be used at moderate temperatures.

Some environments prove unsuitable for any of the normally used polymers. Hot oxidizing acids (such as strong nitric acid or oleum) cannot be handled by any common polymer; only fluorinated ethylene-propylene can survive them.

Similarly, there are some solvents that attack many polymers. These solvents must be looked at individually, but, in general, strongly polar ones are more difficult to handle. Some of the thermoset plastics (e.g., epoxy,

Tables such as this are useful for making preliminary evaluations of various plastics — **Table V**

Corrosion resistance of various polymer systems
Maximum recommended-use temperature, °F

Resin	Sulfuric acid 25%	Sulfuric acid 75%	Sulfuric acid 96%	Nitric acid 10%	Nitric acid 50%	Hydrochloric acid 10%	Hydrochloric acid 36%	Sodium hydroxide 5%	Sodium hydroxide 25%	Sodium hypochlorite 5¼%	Ferric chloride 40%	Monochloro-benzene	Distilled water
Thermoset polymer composites (fiberglass-reinforced)													
Isophthalic polyester	180	NR	NR	70	NR	180	125	120	NR	120	180	NR	160
Bisphenol A polyester	225	75	NR	150	NR	230	150	120	150	125	220	NR	200
Vinyl ester (standard)	210	120	NR	120	NR	210	150	150	120	150	210	NR	210
Vinyl ester (high-temperature)	220	100	NR	140	NR	220	150	150	120	150	220	75	220
Epoxy (special-cure)	NR	NR	NR	150	NR	150	NR	150	150	NR	205	200	225
Thermoplastic polymers (unreinforced)													
Polyvinyl chloride (PVC)	140	140	75	140	75	140	140	140	140	140	140	NR	140
Chlorinated PVC (CPVC)	180	180	120	180	75	180	180	180	180	180	200	NR	180
Polypropylene	150	NR	NR	180	NR	180	150	160	160	NR	150	NR	180
Polyvinylidene fluoride (PVDF)	240	175	120	240	120	240	240	NR	NR	NR	240	75	240
Ethlyene chlorotrifluoro-ethylene (ECTFE)	240	120	NT	240	120	240	240	240	240	NT	240	NR	240
Fluorinated ethylene-propylene (FEP)*	200	200	200	200	200	200	200	200	200	200	200	200	200

*Highest temperature for available data.
From Ref. [*5,11-14*].

NR = not recommended at any temperature
NT = not tested

J. Albert Rolston

Filament-wound FRP epoxy pipes that are used in a uranium solvent-extraction process **Fig. 5**

Heil Process Equipment, Xerxes Corp.

A venturi scrubber that removes dust and fumes in a superphosphate facility **Fig. 6**

furan or phenolic resins) are best overall for use with strong solvents. Monochlorobenzene, a typically aggressive aromatic solvent, can be handled by a reinforced epoxy structure, as noted in Table V.

Applications

There are almost no practical limits for applications of plastics in the CPI, except for excessively high temperatures (over about 500°F), at which almost any organic molecule degrades rapidly. In general, thermoplastics are used in smaller structures such as pipe, fittings, valves and pumps; the reinforced thermosets are used more for larger structures such as tanks, scrubbers and fans. Several photographs will show typical applications.

Fig. 2 is a chlor-alkali plant in which PVC pipe, fittings and valves are connected to FRP tanks. As previously mentioned, plastic structures are being used successfully in all but the highest temperature extremes of CPI environments. The lower temperature conditions account for nearly all (75–85%) of the plastic used.

For thermoplastics, piping accounts for the largest percentage of applications by far. PVC is used the most. Fig. 3 shows PVC pipe, again in a chlor-alkali plant.

Fig. 4 shows two diaphragm valves, injection-molded. Plastic valves are typically made in sizes up to 4 in.

There are many applications in which the pressure ratings of FRP pipe are believed to be superior to thermoplastic. Fig. 5 illustrates several filament-wound FRP epoxy pipes (2 and 6 in. dia.) used in a uranium solvent-extraction process for transporting a mixture of dilute sulfuric acid, kerosene and various amines.

The combination of a thermoplastic lining with an FRP shell is used primarily on tanks, scrubbers (see Fig. 6) and other custom-made products, although a few fabricators produce filament-wound pipe with a thermoplastic lining.

References

1. ASTM standard ASTM D4097, Contact-Molded Glass-Fiber-Reinforced Thermoset Resin Chemical-Resistant Tanks, 1982, ASTM, Philadelphia.
2. Soc. of the Plastics Industry/Materials Technology Inst. of the Chem. Proc. Industry, Quality Assurance Report, 1980, SPI, New York.
3. Morrison Molded Fiber Glass Co., "Extren Engineering Manual," 1971, Bristol, Va.
4. Data from R. & G. Sloane Mfg. Co., Sun Valley, Calif.
5. Data from Plastic Piping Systems Inc., Cranford, N.J.
6. Agranoff, Joan (ed.), "Modern Plastics Encyclopedia, 1983-1984," Oct. 1983. Vol. 60, No. 10A, McGraw-Hill, Inc., New York.
7. Mruk, S. A., Thermoplastics Piping—A Review, from Natl. Assn. of Corrosion Engineers (NACE) publication, "Managing Corrosion with Plastics," Vol. IV, Houston, Tex., 1979.
8. NACE, "Corrosion Data Survey: Nonmetals Section," Houston, Tex., 1975.
9. Perry, Robert H., and Green, Don W. (eds.), "Perry's Chemical Engineers' Handbook," 6th ed., Table 23-2, p. 23-16, McGraw-Hill Book Co., New York, 1984.
10. Perry, Robert H., and Chilton, Cecil H. (eds.), "Chemical Engineers' Handbook," 5th ed., Table 23-2, p. 23-13, McGraw-Hill Book Co., New York, 1973.
11. Data from Ashland Chemical Co., Columbus, Ohio.
12. Data from Dow Chemical Co., Midland, Mich.
13. Kidd, J. A., The Use of Dual Laminates in the Chemical Industry, from NACE publication, "Managing Corrosion with Plastics," Vol. IV, Houston, Tex., 1979.
14. Day, F. B., and Mills, C. C., Application of Plastics for Corrosion Control, from NACE publication "Managing Corrosion with Plastics," Vol. V, Houston, Tex., 1983.

The author

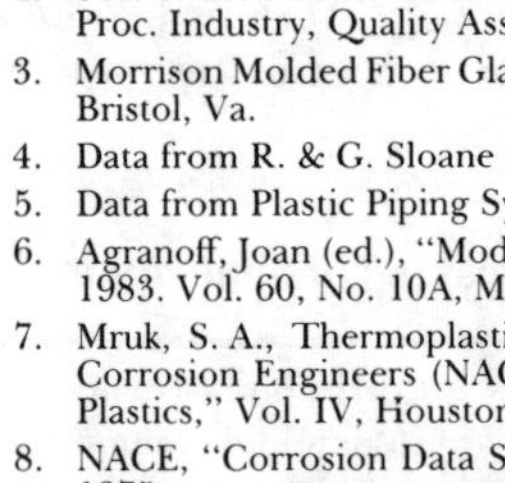

J. Albert Rolston, P.E., is a consulting engineer, P.O. Box 62, Granville, OH 43023. Tel: (614) 366-6117. He received a B.S. and M.S. in chemical engineering at North Carolina State University. He spent 12 years at Owens-Corning Fiberglas Technical Center and has well over 25 years' experience working with fiberglass composites. He is a licensed engineer in Ohio and Virginia, a corrosion specialist for NACE, a fellow of AIChE, and a member of ASTM and the Soc. of the Plastics Industry.

Beat corrosion with rubber hose

The uses of various rubber hose polymers are discussed here. Their physical properties and their compatibilities with six hundred fifty different chemicals are detailed in the tables.

Raymond Gallagher, *The Gates Rubber Co.*

☐ In most cases, hose is destroyed by a combination of corrosion and erosion. Corrosion starts rapidly, and when the stream has a certain velocity, the tube can be eroded, leaving a fresh surface exposed to further corrosion. Any pressure or stress can also accelerate the corrosion rate.

Advantages of rubber

Rubber has a number of advantages in a corrosive environment:

- Being neither a metal nor an electrical conductor, it is not subject to electrolytic corrosion.
- It can be compounded to resist almost any corrosive solution being handled.
- Because rubber hose covers metal, intergranular-, stress- and crevice-corrosion, and corrosion fatigue are avoided.
- It can flex, expand and contract without losing its integrity.
- It resists abrasion, and therefore erosion is minimized.
- It is light in weight.

Rubber linings are generally useful for most chemicals and concentrations but have a temperature limitation of around 200°F as an accepted maximum, depending on polymer type, and solution type and strength.

Common types of rubber linings

Soft natural rubber is a relatively low-cost hose material for both corrosion and abrasion protection. Ranging from 40 to 50 durometer (a numerical value measuring the hardness of rubber—0 durometer being softest and 100 being hardest), the natural rubbers exhibit superior abrasion-resistance when compared to other hose tube materials. Soft natural rubber also offers excellent protection from many corrosive materials. It is available in red, black or tan (color contamination may be a consideration). The upper temperature limit is 150-180°F.

The physical and chemical properties of the various other rubber polymers are detailed in Tables I and II.

Hose selection and use

Consider the following six points when selecting a hose assembly to handle chemicals:

- Correct size—both inside diameter and length.
- Adequate rated working-pressure.
- A tube stock designed to handle the conveyed products.
- A cover stock to resist the environment.
- Adequate reinforcement to withstand external abuse.
- A coupling that resists the chemical to be conveyed.

Care. Use the proper hose and couplings as recommended in Tables I and II. Never use a hose for higher chemical concentrations or temperatures than indicated. Do not exceed recommended rated working-pressure or allowable bend radius. Keep the hose out of working and traffic areas where it may be struck or damaged by vehicles or falling objects. Excessive end pull will quickly ruin such

Chemical resistance of hose polymer types **Table I**

	Hose polymer types														Coupling material				
Chemical	D_3 D_2 D	H	P	M	A_2 A	C	C_2	C_3	Q	K	W	Y	Z	V	Iron or carbon steel	304 SS	316 SS	Alum-inum	Brass
A																			
Absorption oil	X	X	X	2	2	1	2	X	–	2	–	–	–	1	–	–	–	–	1
Acetaldehyde	2	1	1	X	X	X	X	X	–	1	–	1	2	–	1	1	1	1	1
Acetal	–	–	–	–	–	–	–	–	–	2	–	–	–	–	–	–	–	–	1
Acetamide	–	2	2	–	–	2	–	X	–	–	–	1	–	X	–	–	–	–	–
Acetic acid (10%)	X	2	–	2	2	X	X	X	X	2	2	2	–	X	X	2	2	2	X
Acetic acid (25%)	X	2	2	2	2	X	X	X	X	2	–	2	X	X	X	2	2	2	X
Acetic acid (50%)	X	2	X	–	2	X	X	X	X	2	X	2	X	X	X	2	2	2	X
Acetic acid, glacial	–	–	–	X	X	X	X	X	–	–	X	–	X	X	–	2	2	–	–
Acetic acid, anhydride	X	2	–	2	X	X	X	X	X	2	X	–	X	X	X	2	2	2	X
Acetic oxide	X	2	–	2	X	X	X	X	X	–	–	1	X	–	1	1	1	2	–
Acetone	X	2	2	X	X	X	X	X	X	2	X	2	1	X	1	1	1	1	1
Acetonitrile	2	2	2	2	2	X	X	X	–	–	–	–	–	–	–	–	–	–	–
Acetophenone	X	1	1	–	X	X	X	X	–	2	–	–	–	X	–	–	–	–	–
Acetylene	Use acetylene welding hose														–	1	1	1	2
Acrolein (inhibited)	–	2	–	–	–	–	–	–	–	–	–	–	–	–	–	–	–	–	–
Acrylonitrile	2	X	X	X	X	X	X	X	–	2	–	1	–	X	1	1	1	–	–
Aero-safe 2300	–	2	1	–	–	X	–	–	–	–	–	X	–	–	1	–	–	1	1
Aeroshell type 1A,1AC,4	–	X	X	–	–	1	–	–	–	–	–	1	–	–	–	–	–	–	–
Air, 150°F	2	1	1	1	1	1	1	1	2	2	1	1	1	1	1	1	1	1	1
Air, 180°F	X	1	1	2	2	2	2	2	X	X	1	1	2	1	1	1	1	1	1
Air, 200°F	X	1	1	X	X	X	X	X	X	X	2	1	X	1	1	1	1	1	1
Aircraft hydraulic oil AA	–	X	X	–	–	1	–	–	–	–	–	1	–	–	1	1	1	1	1
Alkazene	X	X	X	X	X	X	X	X	–	–	–	–	–	2	–	–	–	–	–
Alkylaryl sulfonates	–	–	–	X	–	–	–	–	–	–	–	–	–	2	–	–	–	–	–
Allyl chloride	X	–	–	–	X	X	X	X	–	1	–	–	–	–	–	1	1	–	–
Aluminum acetate	1	2	–	X	2	2	2	2	–	–	–	–	–	X	–	1	1	–	X
Aluminum bromide	1	1	1	1	1	1	1	1	–	–	–	–	–	1	X	2	2	–	X
Aluminum chloride	1	1	1	1	1	1	1	1	–	1	–	1	–	1	X	2	2	X	X
Aluminum fluoride	1	1	1	1	1	1	1	1	–	1	–	–	–	X	X	2	2	2	X
Aluminum hydroxide	1	1	1	1	1	1	1	1	1	1	–	1	–	–	–	1	1	–	1
Aluminum nitrate	1	1	1	1	1	1	1	1	1	1	–	–	–	X	X	1	1	2	–
Aluminum salts	1	1	1	1	1	1	1	1	1	1	–	1	–	–	–	2	2	2	–
Aluminum sulfate	2	1	1	1	1	1	1	2	1	1	–	1	–	–	X	X	2	X	X
Alums (ammonium or potassium)	1	1	1	1	1	1	1	1	1	1	–	1	–	–	X	2	2	X	X
Amines (mixed)	2	2	2	–	2	2	–	–	–	–	–	–	–	–	–	1	–	X	X
Ammonia (anhydrous)	Use anhydrous ammonia hose only														1	1	1	–	–
Ammonia (aqueous)	1	1	1	1	1	1	1	1	1	1	–	2	1	–	–	1	1	–	X
Ammonium bisulfate (50%)	–	–	–	–	–	–	–	–	–	1	–	–	–	–	–	–	–	–	–
Ammonium carbonate	–	2	2	–	2	X	X	X	–	1	1	1	–	–	1	1	1	–	–
Ammonium chloride	1	1	1	1	1	1	1	1	1	1	–	1	X	–	–	2	2	–	X
Ammonium hydroxide	2	2	1	1	2	2	2	–	–	1	2	2	X	1	2	1	1	–	X
Ammonium metaphosphate	2	1	1	2	2	2	2	2	2	1	–	–	–	–	1	1	1	X	–
Ammonium nitrate (fertiilizer)	1	1	1	1	1	1	1	1	1	1	1§	–	–	–	1	1	1	2	X
Ammonium nitrite	X	–	–	–	2	X	–	–	–	–	–	–	–	–	–	1	1	–	–
Ammonium persulfate	2	2	2	2	–	X	X	X	–	1	–	–	–	–	–	1	1	X	–
Ammonium phosphate (mono- di-, tri-basic)	1	1	1	1	1	1	1	2	–	1	–	–	–	–	X	2	1	X	–
Ammonium sulfate	1	1	1	1	1	2	2	2	2	1	–	–	–	–	1	1	1	X	X
Ammonium thiocyanate	1	–	–	1	1	1	–	–	–	1	–	–	–	–	1	1	1	–	–
Amyl acetate	X	2	2	X	X	X	X	X	–	1	X	2	X	X	X	1	1	X	1
Amyl alcohol	2	2	2	2	2	2	2	2	2	1	2	1	–	1	1	1	1	1	1
Amyl borate	X	X	X	–	2	2	–	–	–	–	–	–	–	–	–	–	–	–	–
Amyl chloride	X	X	X	X	X	–	X	X	X	1	–	–	–	2	–	1	1	–	–
Amyl chloronaphthalene	X	X	X	X	X	X	X	X	–	–	–	–	–	2	–	1	1	–	–
Amyl naphthalene	X	X	X	X	X	X	X	X	–	–	–	–	–	2	–	1	1	–	–
Amyl phenol	–	–	–	–	–	–	–	–	–	2	–	–	–	–	–	1	1	–	–
Aniline	X	2	2	X	X	X	X	X	–	2	X	X	–	2	2	1	1	2	X
Aniline dyes	X	2	2	X	X	X	X	X	–	1	–	–	–	2	X	1	1	–	–
Aniline hydrochloride	2	2	2	–	X	2	–	–	–	–	–	–	–	–	–	X	X	–	X
Animal fats	X	X	2	2	2	2	1	–	2	1	–	2	–	1	1	1	1	1	–
Anti freeze (glycol base)	1	1	1	1	1	1	1	1	1	1	1	1	1	1	1	1	1	1	1
Antimony chloride (50%)	–	2	–	–	–	–	–	–	–	1	–	X	X	2	X	X	X	–	–
Aqua regia	X	X	X	X	X	X	X	X	X	2	X	X	X	2	–	X	X	–	–
Arco A.T.F. Dexron	–	X	X	–	–	1	–	–	–	–	–	1	–	–	–	–	–	–	–
Arsenic acid	X	2	2	–	–	–	–	–	–	1	–	–	–	–	2	–	1	2	–
Askarel (transformer oil)**	X	X	X	X	X	X	X	X	X	2	–	1	–	1	1	1	1	–	1
Asphalt (under 180°F)†	X	X	–	X	X	2	2	X	–	2	1	–	–	1	1	1	1	–	2
Asphalt (cut back)	X	X	–	X	X	2	2	X	–	2	–	–	–	1	1	1	1	–	1

Chemical resistance of hose polymer types—continued **Table I**

A—continued

	Hose polymer types														Coupling material				
Chemical	D_3 D_2 D	H	P	M	A_2 A	C	C_2	C_3	Q	K	W	Y	Z	V	Iron or carbon steel	304 SS	316 SS	Aluminum	Brass
ASTM Oil No. 1	X	X	X	2	1	1	2	2	2	1	1	1	1	1	1	1	1	1	1
ASTM Oil No. 2	X	X	X	–	–	1	–	–	–	1	1	1	1	1	1	1	1	1	1
ASTM Oil No. 3	X	X	X	X	X	1	X	X	X	1	1	1	1	1	1	1	1	1	1
ASTM Reference fuel A	X	X	X	1	1	1	1	2	–	1	1	1	1	1	1	1	1	1	1
ASTM Reference fuel B	X	X	X	X	2	1	2	X	X	2	2	1	–	1	1	1	1	1	1
ASTM Reference fuel C	X	X	X	X	X	2	X	X	X	X	X	2	–	1	1	1	1	–	1
ATF Special (automatic transmission fluid)	–	X	X	–	–	1	–	–	–	–	–	1	–	–	–	–	–	–	–

**Contamination of fluid may be a problem. †Above this temperature use hot asphalt hose §(5%)

B

	Hose polymer types														Coupling material				
Chemical	D_3 D_2 D	H	P	M	A_2 A	C	C_2	C_3	Q	K	W	Y	Z	V	Iron or carbon steel	304 SS	316 SS	Aluminum	Brass
Baltic Types 100, 150, 200, 300 500	–	X	X	–	–	1	–	–	–	–	–	1	–	–	–	–	–	–	–
Banvel (agricultural spray)	–	–	–	–	–	–	–	–	–	1	–	–	1	–	–	–	1	–	–
Bardol B	X	X	X	X	X	X	X	X	–	–	–	–	–	2	1	1	1	–	–
Barium carbonate	1	1	1	1	1	1	1	1	1	1	–	–	–	–	2	1	1	–	1
Barium chloride	1	1	1	1	1	1	1	1	1	1	–	2	X	–	X	1	1	–	2
Barium hydroxide	X	1	–	1	1	1	1	1	–	1	–	–	–	–	2	1	1	–	–
Barium sulfate	X	–	–	2	–	–	–	–	–	1	–	2	2	1	1	1	1	–	2
Barium sulfide	2	1	1	1	1	1	1	1	1	1	–	–	–	–	X	1	1	–	X
Beer*	1	–	–	–	2	–	–	–	–	–	–	–	–	–	2	1	1	1	1
Beet sugar liquors	1	1	1	1	X	1	1	1	1	1	–	–	–	1	1	1	1	1	–
Bellows 80-20 hydraulic oil	–	X	X	–	–	1	–	–	–	–	–	1	–	–	–	–	–	–	–
Benzaldehyde	X	2	2	X	X	X	X	X	X	1	X	2	–	X	1	–	–	1	–
Benzene (Benzol)	X	X	X	X	X	X	X	X	X	2	X	2	1	2	1	1	1	1	1
Benzenesulfonic acid (10%)	X	2	–	2	X	–	–	–	–	1	–	–	–	2	X	–	2	X	–
Benzine (petroleum ether)	X	X	X	–	X	2	–	–	–	2	–	–	–	2	1	1	1	1	1
Benzophenone	–	–	–	–	–	–	–	–	–	–	–	–	–	1	–	–	–	–	–
Benzyl alcohol	X	2	2	X	X	X	X	X	–	–	X	2	–	1	1	1	1	–	–
Benzyl benzoate	–	2	2	–	–	–	–	–	–	–	–	–	–	2	1	1	1	–	–
Benzyl chloride	–	–	–	–	X	X	–	–	–	–	–	–	–	2	1	–	–	–	–
Bismuth carbonate	X	–	–	2	X	–	–	–	–	1	–	–	–	2	1	1	1	–	–
Bitumastic	X	X	X	X	X	2	2	–	X	–	–	–	–	2	1	1	1	–	1
Black sulfate liquor	2	2	2	2	2	2	–	–	2	1	–	–	–	1	1	1	1	–	–
Blast furnace gas	X	X	X	X	X	X	–	–	–	1	–	–	–	1	1	1	1	–	1
Borax	2	2	2	1	2	2	–	–	2	1	–	–	–	1	2	1	1	–	2
Bordeaux mixture	2	2	–	–	2	2	2	2	–	1	–	–	–	2	–	1	1	–	–
Boric acid	1	1	1	1	1	1	1	1	1	1	–	1	X	1	X	2	1	1	X
Brake fluid (petroleum base)	X	X	X	X	2	1	2	X	–	1	–	1	–	–	1	1	1	–	1
Brake fluid (synthetic base)	1	1	1	X	X	X	X	X	2	–	–	1	–	X	1	1	1	–	1
Bromine	No hose available														–	–	–	–	–
Bromochloromethane	X	X	X	X	X	X	X	X	X	2	–	X	–	2	1	1	1	–	1
Bunker oil	X	X	X	X	2	2	–	–	–	–	–	2	–	1	1	1	1	1	1
Butadiene (monomer)	X	–	–	X	2	2	–	–	–	1	–	–	–	2	–	1	1	–	1
Butane (gas)	Use LPG hose only														1	1	1	1	1
Butane (liquid)	Use LPG hose only														1	1	1	1	1
Butter oil	X	–	–	–	2	–	2	–	–	–	–	2	–	–	1	1	1	1	1
Butyric acid	2	2	2	X	X	–	–	–	–	–	–	–	X	–	X	1	1	1	2
Butyl acetate	X	2	–	X	X	X	X	X	–	2	–	2	–	X	2	1	1	1	1
Butyl alcohol (butanol)	1	1	1	2	2	X	X	X	2	1	–	1	1	X	1	1	1	1	1
Butyl amine	X	X	–	X	X	X	X	X	X	–	–	–	–	X	1	1	1	1	1
Butyl "Carbitol"	X	2	2	–	2	2	–	X	–	–	–	–	–	1	1	1	1	1	1
Butyl ether	X	2	–	–	2	2	–	X	X	1	–	–	–	X	1	1	1	1	1
Butyl mercaptan (tertiary)	–	–	–	–	–	–	–	–	–	–	–	–	–	2	–	1	1	–	–
Butyl stearate	X	X	X	–	X	X	–	–	–	1	–	–	–	–	1	1	1	1	1
Butyraldehyde	X	X	X	X	2	–	–	–	–	–	–	–	–	X	–	–	–	–	1

*Use food-grade hose only.

C

	Hose polymer types														Coupling material				
Chemical	D_3 D_2 D	H	P	M	A_2 A	C	C_2	C_3	Q	K	W	Y	Z	V	Iron or carbon steel	304 SS	316 SS	Aluminum	Brass
Cake alum	1	1	1	1	1	1	1	1	1	1	–	–	–	–	2	2	1	–	–
Calcine liquor	–	1	1	–	–	1	2	2	–	–	–	–	–	1	1	1	1	2	–
Calcium acetate	2	1	1	X	X	X	X	X	–	–	–	–	–	X	1	1	1	1	1
Calcium bisulfate	1	1	1	1	1	1	–	–	–	1	–	–	–	1	–	2	1	–	X
Calcium bisulfite	2	1	–	1	1	1	2	–	–	–	–	–	–	1	–	1	1	–	–
Calcium carbonate	1	1	1	1	1	1	1	1	1	1	–	1	1	1	1	1	1	1	1
Calcium chlorate	2	2	2	1	1	1	–	–	X	1	–	–	–	–	–	2	1	–	–

Chemical resistance of hose polymer types—continued **Table I**

Chemical	Hose polymer types														Coupling material				
	D_3 D_2 D	H	P	M	A_2 A	C	C_2	C_3	Q	K	W	Y	Z	V	Iron or carbon steel	304 SS	316 SS	Aluminum	Brass
C—continued																			
Calcium chloride	1	1	1	1	1	1	1	1	1	1	1	1	X	1	X	2	1	–	2
Calcium hydroxide	1	1	–	1	1	2	2	–	2	1	–	–	–	X	X	X	1	–	2
Calcium hypochlorite (5%)	X	2	–	2	X	–	–	–	2	2	–	–	–	–	–	X	2	X	X
Calcium hypochlorite (15%)	X	2	–	2	X	–	–	–	X	2	–	–	X	–	–	X	2	X	X
Calcium nitrate	1	1	1	1	1	1	1	1	1	1	–	–	–	1	1	1	1	1	1
Calcium silicate	2	2	–	2	–	2	–	–	–	–	–	–	–	1	1	1	1	1	1
Calcium sulfate	–	1	1	1	1	1	1	1	–	1	–	–	–	1	1	1	1	–	1
Calcium sulfide	2	1	–	1	2	1	–	–	2	–	–	–	–	2	1	1	1	2	–
Caliche liquors*	2	1	–	1	–	1	–	–	–	1	–	–	–	–	–	1	1	–	–
Cane sugar liquors	2	2	2	1	1	1	2	–	–	1	–	–	–	–	1	1	1	1	2
Carbamates	X	X	X	X	X	X	X	X	–	–	–	–	–	2	–	–	–	–	–
Carbolic acid (phenol)	X	2	2	X	X	X	X	X	X	2	X	–	X	2	X	1	1	2	X
Carbon disulfide	X	X	X	X	X	2	–	X	X	2	–	1	–	1	2	1	1	2	2
Carbon dioxide (dry)	1	1	1	1	1	1	1	1	1	1	–	1	–	1	1	1	1	1	1
Carbon monoxide (hot, 150°F)	X	X	X	1	2	2	2	X	X	2	–	–	–	1	1	1	1	1	1
Carbon tetrachloride	X	X	X	X	X	X	X	X	X	X	X	X	1	1	X	2	2	2	2
Carbonic acid	1	1	1	1	1	1	–	–	1	1	1	–	–	–	X	1	1	2	X
Carter motor oil	–	X	X	–	–	1	–	–	–	–	–	1	–	–	1	1	1	1	1
Castor oil	X	2	–	2	2	2	–	–	–	1	1	1	–	2	1	1	1	1	1
Caustic potash (30%) (potassium hydroxide)	2	2	2	1	2	X	X	X	X	2	–	X	X	2	1	1	1	X	X
Caustic soda (50%)	1	2	–	1	2	X	X	X	X	2	–	X	–	X	2	1	1	X	X
Cellosolve acetate	X	2	2	X	X	X	X	X	–	1	–	2	–	X	1	1	1	–	–
Cellosolve, butyl	X	2	2	X	X	X	X	X	–	2	–	–	–	–	1	1	1	–	–
Cellugard, Cellugard 200	–	1	1	–	–	1	–	–	–	–	–	1	–	–	1	1	1	1	1
Cellulube Types 90, 150, 220, 300, 550, 1000, 220A, ST 220, A60	X	1	1	X	X	X	X	X	X	1	–	X	–	1	1	1	1	1	1
China-wood oil (tung oil)	X	–	X	2	X	2	2	X	–	1	–	–	–	–	1	1	1	1	1
Chlorine water (25% chlorine)	2	X	2	2	X	X	X	X	–	1	1	2	–	–	–	X	X	–	–
Chlorine gas (wet)						No hose available									X	X	X	X	X
Chlorine gas (dry)						No hose available									2	X	X	–	2
Chlorine trifluoride						No hose available									X	–	–	X	–
Chloracetic acid	X	X	X	2	2	X	X	X	–	1	–	X	–	–	X	X	X	–	2
Chlorobenzene	X	X	X	X	X	X	X	X	X	–	–	X	–	1	1	1	1	1	1
Chloroform	X	X	X	X	X	X	X	X	X	X	–	X	2	1	1	1	1	1	1
O-Chloronaphthalene	X	X	X	X	X	X	X	X	X	–	–	–	–	–	1	1	1	–	1
Chlorosulfonic acid						No hose available									–	–	–	–	–
Chlorothene	–	–	–	–	–	X	X	X	–	–	–	–	–	2	–	1	1	–	1
Chlorotoluene	X	X	X	X	X	X	X	X	X	–	–	–	–	2	1	1	1	1	1
Chromic acid (10%)	X	X	–	2	X	X	X	X	X	1	–	X	X	1	X	X	2	X	X
Chromic acid (25%)	X	X	–	2	X	X	X	X	X	2	–	X	X	1	X	X	2	X	X
Chromic acid (50%)	X	X	–	2	X	X	X	X	X	X	–	X	X	1	X	X	2	X	X
Citgo FR fluids	–	1	1	–	–	X	–	–	–	–	–	1	–	–	–	–	–	–	–
Citgo glycol FR – 20 XD	–	1	1	–	–	1	–	–	–	–	–	1	–	–	1	1	1	1	1
Citgo pacemaker invert FR fluid	–	X	X	–	–	1	–	–	–	–	–	1	–	–	–	–	–	–	–
Citgo tractor hydraulic fluid	–	X	X	–	–	1	–	–	–	–	–	1	–	–	1	1	1	–	1
Citric acid	2	2	2	1	1	X	–	–	–	1	1△	1	X	1	X	X	1	1	X
Clorox	2	2	–	2	–	–	–	–	–	2	–	–	1	–	–	2	1	–	–
Cobalt nickel plating solution	–	–	–	–	–	–	–	–	–	1	–	–	X	–	–	–	2	–	–
Cocoa butter	X	–	–	–	2	2	2	X	–	–	–	–	–	–	1	1	1	–	–
Cod liver oil	X	2	2	X	X	X	X	X	–	1	–	–	–	1	1	1	1	1	1
Coke oven gas	X	X	X	2	X	X	X	X	–	1	–	–	–	1	1	1	1	2	–
Condor Types 1000, 1002, 1004, 1006, 1008, 1010, 1012, 1014, 1016	–	X	X	–	–	2	–	–	–	–	–	2	–	–	–	–	–	–	–
Copper arsenate (cupric arsenate)	2	–	–	2	–	–	–	–	–	1	–	–	–	1	1	1	1	–	–
Copper chloride (cupric chloride)	2	2	–	2	2	2	2	–	–	1	–	–	X	1	X	X	1	–	X
Copper cyanide (cupric cyanide)	2	2	2	2	2	2	–	–	–	1	–	–	–	1	–	1	1	–	X
Copper nitrate (cupric nitrate)	2	1	1	1	1	1	1	1	–	1	–	–	–	1	X	1	1	–	X
Copper sulfate (cupric sulfate)	2	2	1	1	1	1	1	1	1	1	–	1	–	1	X	1	1	X	X
Cordage oil	–	–	–	–	–	1	2	2	–	–	–	–	–	1	–	–	–	–	–
Corn oil	X	2	2	X	X	2	2	X	–	1	2	–	–	1	1	1	1	1	1
Corn syrup	2	2	2	2	2	2	2	–	2	2	–	–	–	2	1	1	1	1	–
Cottonseed oil	–	–	2	2	1	2	1	2	X	1	1	1	–	1	1	1	1	1	1
Cottonseed oil (170°F)	X	–	2	2	2	2	1	X	X	X	2	2	–	1	1	1	1	1	1
Creosote (wood or coal tar)	X	2	–	X	X	2	X	X	–	2	X	–	–	1	2	1	1	1	X
Cresol (cresylic acid)	X	2	–	X	X	X	X	X	X	2	–	X	X	1	2	1	1	1	–
Cresol 85%, xylene 5%, DDT 10%	–	–	–	–	–	–	–	–	–	2	–	X	–	1	2	1	1	–	–

Chemical resistance of hose polymer types—continued **Table I**

C—continued

Chemical	Hose polymer types														Coupling material				
	D_3 D_2 D	H	P	M	A_2 A	C	C_2	C_3	Q	K	W	Y	Z	V	Iron or carbon steel	304 SS	316 SS	Aluminum	Brass
Crude wax	–	2	–	–	–	2	–	–	–	2	–	–	–	1	1	1	1	–	1
Cryolite (10%)	–	2	–	–	1	2	2	–	–	1	–	–	–	–	–	1	1	–	–
Cutting oil	X	X	X	X	2	1	2	X	–	2	–	–	–	1	1	1	1	–	1
Cyclohexane	X	X	X	X	X	2	–	X	X	1	–	1	–	1	1	1	1	–	1
Cyclohexanone	X	X	X	X	X	X	X	X	X	1	X	–	–	X	–	1	1	2	–
Cymene	X	X	X	X	X	X	X	X	X	2	–	–	–	2	–	–	–	–	1

*Fertilizers only. △(20%)

D

Chemical	D_3 D_2 D	H	P	M	A_2 A	C	C_2	C_3	Q	K	W	Y	Z	V	Iron or carbon steel	304 SS	316 SS	Aluminum	Brass
Dasco FR150, FR200, FR200B, FR310	–	1	1	–	–	1	–	–	–	–	–	1	–	–	–	–	–	–	–
Dasco IFR	–	–	–	–	–	1	–	–	–	–	–	1	–	–	1	–	–	1	1
DC200, DC510, DC550, DC560	–	–	–	–	–	1	–	–	–	–	–	1	–	–	–	1	1	1	1
Decalin	X	X	–	X	–	2	–	–	–	1	–	–	–	2	–	–	–	–	1
Deionized water	No hose available														–	–	–	–	–
Dectol R&O oils	–	X	X	–	–	1	–	–	–	–	–	1	–	–	–	–	–	–	–
Denatured alcohol	1	1	1	1	1	1	1	1	1	1	–	–	–	2	1	1	1	1	1
Developing solutions (hypos)	2	2	–	2	2	–	–	–	–	–	–	–	–	–	–	1	1	–	–
Dexron	–	X	X	–	–	1	–	–	–	–	–	1	–	–	–	–	–	–	–
Diacetone	X	2	2	X	X	X	X	X	–	1	–	–	–	X	1	1	1	–	1
Diacetone alcohol	2	2	–	2	–	X	X	X	–	1	–	–	–	X	1	1	1	1	1
Dibenzyl ether	X	2	2	X	X	X	X	X	X	–	–	–	–	X	1	1	1	1	1
Dibutyl ether	X	2	–	X	X	X	X	X	–	1	–	–	–	X	1	1	1	1	1
Dibutyl phthalate	X	2	1	X	X	X	X	X	–	1	–	1	–	2	1	1	1	1	1
Dibutyl sebacate	X	2	X	–	X	X	X	X	–	1	X	–	–	1	–	–	–	–	1
Dichlorobenzene	X	X	X	X	X	X	X	X	X	X	–	X	–	1	–	1	1	–	1
Diesel oil	X	X	X	X	2	1	2	2	–	1	–	1	1	1	1	1	1	1	1
Diethylamine	2	2	2	X	2	2	–	–	–	1	–	–	–	–	–	1	1	–	1
Diethylbenzene	X	X	X	–	–	–	–	X	–	–	–	X	–	1	–	–	–	–	–
Diethyl ether	X	2	2	–	X	–	–	–	–	1	–	–	–	X	1	1	1	1	1
Diethyl phthalate	X	2	–	–	–	X	X	X	–	1	–	–	–	–	–	1	1	–	1
Diethyl sebacate	X	2	–	X	X	X	X	X	–	1	–	1	–	2	–	1	1	–	1
Diethylene glycol	1	1	1	1	1	1	1	1	1	1	–	1	1	1	1	1	1	1	1
Di-Isobutyl ketone	X	2	2	X	X	X	X	X	–	1	–	–	–	X	–	1	1	–	1
Diisobutyl ketone	X	X	–	X	X	2	–	–	–	1	–	–	–	1	–	1	1	–	1
Diisobutylene	X	2	2	X	X	X	X	X	–	1	–	–	–	X	–	1	1	–	1
Diisopropyl ketone	X	2	X	X	X	X	X	X	–	1	–	–	–	–	–	–	–	–	1
Dimethyl ether	X	2	X	X	X	X	X	X	X	1	–	–	–	X	1	1	1	1	1
Dimethyl formamide	X	X	–	–	X	X	X	X	–	2	–	2	–	X	1	1	1	–	–
Dimethyl phthalate	X	2	2	X	X	X	X	X	–	1	–	–	–	1	–	–	–	–	1
Dioctyl phthalate	X	X	X	X	X	X	X	X	X	1	2	1	–	1	1	1	1	1	1
Dioctyl sebacate	X	2	–	X	X	X	X	X	–	1	1	–	–	2	–	–	–	–	–
Dioxane (diethylene ether)	X	2	2	–	X	X	X	X	X	1	–	2	–	X	1	1	1	1	1
Dipentene	X	–	–	–	–	X	–	–	–	–	–	–	–	2	1	1	1	1	1
Dirco oils	–	X	X	–	–	1	–	–	–	–	–	1	–	–	1	1	1	1	1
Dowtherm A	X	X	X	X	X	X	X	X	X	1	X	–	–	1	1	1	1	1	1
Dowtherm E	–	–	–	–	–	–	–	–	–	1	–	–	–	–	1	1	1	1	1
Duro FR–HD	–	X	X	–	–	1	–	–	–	–	–	X	–	–	–	–	–	–	–
Duro oils	–	X	X	–	–	1	–	–	–	–	–	1	–	–	1	1	1	1	1
DP47,200 fluid (Dow)	–	–	–	–	–	1	–	–	–	–	–	1	–	–	–	–	–	–	–

E

Chemical	D_3 D_2 D	H	P	M	A_2 A	C	C_2	C_3	Q	K	W	Y	Z	V	Iron or carbon steel	304 SS	316 SS	Aluminum	Brass
Energol HL 68	–	X	X	–	–	1	–	–	–	–	–	1	–	–	–	–	–	–	–
Energol HLPC 68	–	X	X	–	–	1	–	–	–	–	–	1	–	–	–	–	–	–	–
EP Hydraulic oils (Chevron)	–	X	X	–	–	1	–	–	–	–	–	1	–	–	1	1	1	1	1
Epichlorohydrin	–	X	X	–	–	–	–	–	–	2	–	X	–	X	1	–	–	–	–
Ethanolamine	2	1	2	X	2	2	–	2	–	1	–	1	–	X	1	1	1	–	1
Ethyl acetate	X	2	2	X	X	X	X	X	X	1	X	2	1	X	1	1	1	1	1
Ethyl acetoacetate	X	2	2	X	X	X	X	X	X	–	–	–	–	X	1	1	1	1	1
Ethyl acrylate	X	X	X	X	–	X	X	X	–	2	–	–	–	X	1	1	1	–	–
Ethyl alcohol (ethanol)	1	1	1	1	1	1	1	1	2	1	2	1	1	X	1	1	1	1	2
Ethyl amine	X	2	–	X	X	X	X	X	–	2	–	–	–	X	–	1	1	–	1
Ethyl benzene	X	X	X	X	X	X	X	X	–	2	–	–	–	1	1	1	1	–	1

Chemical resistance of hose polymer types—continued **Table I**

Chemical	Hose polymer types														Coupling material				
	D_3 D_2 D	H	P	M	A_2 A	C	C_2	C_3	Q	K	W	Y	Z	V	Iron or carbon steel	304 SS	316 SS	Aluminum	Brass
E—continued																			
Ethyl butyrate	X	2	–	–	X	X	X	X	–	1	–	–	–	–	–	1	1	1	–
Ethyl cellulose	–	–	–	–	–	–	–	–	–	1	–	–	–	–	1	1	1	–	1
Ethyl chloride	X	X	X	X	X	X	X	X	–	–	–	X	–	2	2	1	1	1	2
Ethyl ether	X	2	X	X	X	X	X	X	X	2	2	–	–	X	2	1	1	1	1
Ethyl mercaptan	X	X	X	X	X	X	X	X	X	–	–	–	–	2	2	–	–	–	–
Ethyl oxalate	2	2	2	–	X	X	X	X	–	–	–	–	–	–	–	–	–	–	–
Ethyl pentachlorobenzene	X	X	X	X	X	X	X	X	X	–	–	–	–	2	2	1	1	–	1
Ethyl silicate	2	–	–	–	1	1	2	2	–	–	–	–	–	–	1	1	1	1	1
Ethyl bromide (di)	X	X	X	X	X	X	X	X	X	2	–	–	–	1	–	1	1	–	1
Ethylene chloride (di)	X	X	X	X	X	X	X	X	–	X	X	X	1	2	2	1	1	1	2
Ethylene chlorohydrin	–	2	2	–	X	X	X	X	–	–	–	–	–	–	–	–	–	–	–
Ethylene diamine	2	2	2	X	2	2	–	–	–	–	–	X	–	–	–	–	–	–	1
Ethylene glycol	1	1	1	1	1	1	1	1	1	1	1	1	1	1	2	1	1	1	1
F																			
Factovis 52	–	X	X	–	–	1	–	–	–	–	–	1	–	–	1	1	1	1	1
Fatty acids	–	–	2	X	2	2	–	–	X	2	–	–	–	2	–	1	1	1	–
Ferric chloride	–	1	–	2	–	–	–	–	–	1	–	2	–	2	X	X	X	X	X
Ferric nitrate	2	2	–	2	2	–	–	–	–	1	–	–	–	–	X	1	1	–	–
Ferric sulfate	2	2	2	2	2	2	2	2	–	1	–	–	–	–	X	1	1	X	X
Ferrous chloride	–	1	–	2	–	–	–	–	–	1	–	–	–	–	X	1	2	–	2
Ferrous nitrate	–	2	2	2	2	2	–	–	2	–	–	–	–	–	–	1	1	–	–
Ferrous sulfate	–	2	2	2	2	2	–	–	2	1	–	–	–	–	X	1	1	–	2
Fire-resistant hydra-fluid (Texaco)	–	X	X	–	–	1	–	–	–	–	–	1	–	–	1	1	1	1	1
Fluoboric acid (65%)	2	–	–	2	2	–	–	–	–	1	–	–	–	X	–	1	1	–	–
Fluorine (liquid)						No hose available									–	–	–	–	–
Fluosilicic acid (50%)	–	X	–	2	2	X	X	X	–	1	–	X	X	X	–	–	–	1	–
Formaldehyde (37%)	X	2	1	2	2	2	–	–	–	2	2	2	–	1	–	1	1	1	1
Formic acid	X	2	–	2	1	–	–	–	–	1	–	X	X	X	X	2	1	–	2
FR Fluid D	–	X	X	–	–	1	–	–	–	–	–	1	–	–	–	–	–	–	–
FR Hydraulic fluid	–	–	–	–	–	1	–	–	–	–	–	1	–	–	–	–	–	–	–
FRM (Code 6SS22)	–	X	X	–	–	1	–	–	–	–	–	1	–	–	–	–	–	–	–
Freon 114						Special hose required									X	1	1	–	–
Freon 12						Special hose required									X	1	1	–	–
Fuel oil	X	X	X	X	2	1	1	2	–	1	–	1	–	1	2	2	2	1	1
Fumaric acid	2	–	–	–	–	–	–	–	–	–	–	–	–	–	–	1	1	–	–
Furan (furfuran)	X	X	X	–	X	X	X	X	X	–	–	–	–	–	1	1	1	1	1
Furfural (ant oil)	X	X	X	2	2	X	X	X	–	1	X	2	–	X	2	1	1	1	1
Fusel oil	–	–	–	–	–	–	–	–	–	1	–	–	–	–	–	–	–	–	–
Fyrguard 150, 200	–	1	1	–	–	1	–	–	–	–	–	1	–	–	1	1	1	1	1
Fyrquel 90, 150, 220, 300, 550, 1000, 15R&O, 220R&O, 550R&O	–	1	1	–	–	X	–	–	–	–	–	1	–	–	1	–	–	1	–
Firtec 290, MF	–	–	–	–	–	–	–	–	–	–	–	–	–	–	–	–	–	–	–
G																			
Gallic acid	2	2	–	–	X	X	X	X	–	1	–	–	X	–	X	1	1	–	–
Gasoline (standard)	X	X	X	X	1	1	1	2	X	1	1	1	–	1	2	1	1	1	1
Gasoline (premium)	X	X	X	X	2	1	2	2	X	1	1	1	–	1	2	1	1	1	1
Gasoline (unleaded up to 50% aromatics)	X	X	X	X	X	1	X	X	X	2	2	–	–	1	2	1	1	1	1
Glauber's salt	1	–	–	–	–	–	–	–	–	–	–	–	–	–	1	1	1	–	–
Glucose	1	1	1	1	1	1	1	1	1	1	1	–	–	1	1	1	1	1	1
Glue	X	X	X	1	2	2	2	–	–	1	–	1	2	1	2	1	1	1	X
Glycerine (glycerol)	1	1	1	1	1	1	1	1	1	1	1	1	–	1	2	1	1	1	1
Glycols	1	1	1	1	1	1	1	1	–	1	1	1	1	1	1	1	1	1	1
Glycol FR fluids	–	1	1	–	–	1	–	–	–	–	–	1	–	–	1	1	1	1	1
Grease	X	X	X	2	2	1	2	2	2	1	–	1	–	1	1	1	1	1	1
Green sulfate liquor	1	1	1	1	1	2	2	2	–	1	–	–	–	–	1	1	1	–	–
Gulf FR fluid G-200	–	1	1	–	–	1	–	–	–	–	–	1	–	–	1	1	1	1	1
Gulf FR fluid P-37, P-40, P-43 P-45, P-47	–	1	1	–	–	X	–	–	–	–	–	1	–	–	–	–	–	–	–

Chemical resistance of hose polymer types—continued **Table I**

Chemical	Hose polymer types														Coupling material				
	D_3 D_2 D	H	P	M	A_2 A	C	C_2	C_3	Q	K	W	Y	Z	V	Iron or carbon steel	304 SS	316 SS	Aluminum	Brass
H																			
Halowax oil	X	X	X	X	X	X	X	X	–	–	–	–	–	2	–	–	–	–	–
Harness oil	–	–	–	–	–	1	–	–	–	1	–	–	–	–	1	1	1	–	–
Heptane	X	X	X	X	2	1	2	2	–	2	–	1	–	1	1	1	1	–	1
n-Hexaldehyde	X	1	2	–	2	X	X	X	–	1	–	–	–	X	1	1	1	1	1
Hexane	X	X	X	1	1	1	2	2	–	1	–	1	–	1	1	1	1	–	1
Hexene	X	X	X	–	2	2	–	–	–	–	–	–	–	–	1	1	1	–	1
Hexyl alcohol (hexanol)	2	–	X	X	2	1	2	X	–	1	–	–	–	1	1	1	1	1	2
Houghto-Safe 271, 416, 520, 525, 616, 620, 625, 640	–	1	1	–	–	1	–	–	–	–	–	1	–	–	1	1	1	1	1
Houghto-Safe 1010, 1055, 1115, 1120, 1130	–	1	1	–	–	X	–	–	–	–	–	–	–	–	1	1	1	1	1
Houghto-Safe 5046, 5046W	–	X	X	–	–	1	–	–	–	–	–	1	–	–	1	1	1	1	1
Hy-Chock oil	–	–	–	–	–	1	–	–	–	–	–	1	–	–	–	–	–	–	–
Hydrafluid AZR&O, A, B, AA, C	–	X	X	–	–	1	–	–	–	–	–	1	–	–	–	–	–	–	–
Hydrafluid 760 (Texaco & Houghton)	–	X	X	–	–	1	–	–	–	–	–	1	–	–	1	1	1	1	–
Hydrasol A.	–	X	X	–	–	1	–	–	–	–	–	1	–	–	–	–	–	–	–
Hydraulic oils (Shell)	–	X	X	–	–	1	–	–	–	–	–	1	–	–	1	1	1	1	1
Hydraulic fluid HF-31	–	X	X	–	–	–	–	–	–	–	–	–	–	–	–	–	–	–	–
Hydraulic fluid HF-18, HF-20	–	1	1	–	–	1	–	–	–	–	–	1	–	–	–	–	–	–	–
Hydraulic safety fluid 200 & 300 (Texaco)	–	1	1	–	–	1	–	–	–	–	–	1	–	–	1	1	1	1	1
Hydraulic fluid (standard petroleum oils)	X	X	X	2	2	1	2	–	–	1	–	1	–	1	1	1	1	1	1
Hydraulic fluid (phosphate ester base)	–	1	1	–	X	X	X	X	–	1	–	1	1	1	1	1	1	–	–
Hydraulic fluid (water glycol base)	2	1	–	–	1	1	–	–	–	1	–	1	–	1	1	1	1	1	1
Hydrazine	X	2	2	X	X	X	X	X	X	–	–	X	–	X	–	–	–	–	–
Hydrobromic acid (37%)	2	2	–	2	X	X	–	–	–	2	–	X	X	–	–	–	–	X	–
Hydrochloric acid (15%)	2	2†	2	2	X	X	–	–	–	2	1	X	X	1	X	X	X	X	X
Hydrochloric acid (37%)	2	2†	X	2	X	–	–	–	X	2	X	X	X	1	X	X	X	X	X
Hydrocyanic acid (20%)	2	–	–	2	X	X	X	X	–	2	–	–	–	2	X	1	1	1	X
Hydrocyanic acid (98%)							No hose available								X	1	1	1	X
Hydro-drive oil (Houghton)	–	X	X	–	–	1	–	–	–	–	–	1	–	–	–	–	–	–	–
Hydrofluoric acid (10%)	X	2	–	2	2	X	X	X	–	2	–	X	X	1	X	X	X	X	X
Hydrofluoric acid (20%)	X	2	–	2	2	X	X	X	X	2	–	X	X	1	X	X	X	X	X
Hydrofluoric acid (48%)	X	X	–	X	2	X	X	X	X	2	–	X	X	1	X	X	X	X	X
Hydrofluoric acid (70%)	X	–	–	–	–	X	X	X	–	2	–	X	X	1	X	X	X	X	X
Hydrofluoric acid (conc.)	X	X	–	–	X	X	X	X	X	2	–	X	X	1	X	X	X	X	X
Hydrofluosilicic acid (50%)	–	X	–	2	–	X	X	X	–	1	–	–	X	X	–	X	X	X	X
Hydrogen (gas)							Use special hose								1	1	1	1	1
Hydrogen peroxide (10%)	X	X	–	2	X	X	–	–	–	1	–	–	X	1	X	2	1	1	X
Hydrogen peroxide (30%)	X	X	–	2	X	X	–	–	–	2	–	–	X	1	X	2	1	1	X
Hydrogen peroxide (70%)	X	–	–	–	X	X	–	–	–	–	–	–	X	1	X	2	1	1	X
Hydrogen sulfide (gas)							No hose available								X	2	1	2	X
Hydrolubric oil (Houghton)	–	X	X	–	–	2	–	–	–	–	–	1	–	–	–	–	–	–	–
Hydroquinone	X	X	X	X	X	–	–	–	–	–	–	–	–	2	–	1	1	–	–
Hykil No. 6 (33%); Water (67%)	–	–	–	–	–	2	–	–	–	1	–	–	–	–	1	–	–	–	–
Hypochlorous acid	X	X	X	2	X	X	–	–	–	–	–	–	–	2	–	–	–	–	–

†If under 0.035% free chlorine.

Chemical	D_3 D_2 D	H	P	M	A_2 A	C	C_2	C_3	Q	K	W	Y	Z	V	Iron or carbon steel	304 SS	316 SS	Aluminum	Brass
I																			
Imol, Imol S150, S220, S300, S500	–	X	X	–	–	1	–	–	–	–	–	1	–	–	1	1	1	1	–
Industron 53.	–	X	X	–	–	1	–	–	–	–	–	1	–	–	–	–	–	–	–
Ink (printers')	X	X	X	–	–	–	–	X	–	2	–	1	–	X	2	2	1	–	2
Ink oil	–	–	–	–	–	2	–	–	–	2	–	–	–	–	1	1	1	–	1
Insulating oil (transformer)*	X	X	X	X	2	1	–	–	–	1	–	1	–	1	1	1	1	–	1
Iodine	X	X	–	2	X	X	–	–	–	1	–	X	–	2	X	X	X	–	–
Irus fluid 902, 905	–	X	X	–	–	1	–	–	–	–	–	1	–	–	1	1	1	1	1
Isobutane							Use LPG hose only								1	1	1	2	1
Isobutyl alcohol	2	2	1	1	2	2	2	2	2	1	–	–	–	–	1	1	1	1	2
Iso octane	X	X	X	1	1	1	2	X	–	1	1	1	–	1	1	1	1	2	1
Iso octyl thioglycolate	–	–	2	–	–	–	–	–	–	–	–	–	–	1	–	–	–	–	–
Isopropyl acetate	X	2	X	X	X	X	X	X	–	1	–	–	–	X	1	1	1	1	1

Chemical resistance of hose polymer types—continued **Table I**

Chemical	Hose polymer types														Coupling material				
	D_3 D_2 D	H	P	M	A_2 A	C	C_2	C_3	Q	K	W	Y	Z	V	Iron or carbon steel	304 SS	316 SS	Aluminum	Brass
I—continued																			
Isopropyl alcohol	2	2	2	2	2	2	2	2	2	1	–	2	–	–	1	1	1.	1	2
Isopropyl ether	X	2	X	X	X	X	X	X	–	–	–	1	–	X	1	1	1	1	1
Isocyanate (toluene-2,4-diisocyanate)	–	2	2	–	–	–	–	–	–	–	–	–	–	2	–	–	–	–	–

*Petroleum type – Oil contamination could become a problem.

Chemical	D_3 D_2 D	H	P	M	A_2 A	C	C_2	C_3	Q	K	W	Y	Z	V	Iron or carbon steel	304 SS	316 SS	Aluminum	Brass
J																			
Jet Fuel** JP-3	X	X	X	X	2	1	2	–	X	–	–	1	–	1	2	1	1	2	1
Jet Fuel** JP-4	X	X	X	X	2	1	–	–	X	–	1	1	–	1	2	1	1	2	1
Jet Fuel** JP-5	X	X	X	X	X	1	–	–	X	–	1	1	–	1	2	1	1	2	1
Jet Fuel** JP-6	X	X	X	X	X	1	–	–	X	–	1	–	–	1	2	1	1	2	1
Jet Fuel** JP-X	X	X	X	X	–	1	–	–	X	–	–	–	–	–	2	1	1	2	1

**Use aircraft fueling hose only for fueling operations

Chemical	D_3 D_2 D	H	P	M	A_2 A	C	C_2	C_3	Q	K	W	Y	Z	V	Iron or carbon steel	304 SS	316 SS	Aluminum	Brass
K																			
Kerosene	X	X	X	X	X	2	2	X	–	2	1	2	–	1	1	1	1	1	1
L																			
Lacquers	X	X	–	X	X	X	X	X	X	1	–	–	1	X	X	X	1	1	1
Lacquer solvents	X	X	–	X	X	X	X	X	X	1	–	–	1	X	X	X	1	1	1
Lactic acid	2	–	–	1	1	X	X	X	–	1	–	2	–	1	X	2	1	X	2
Lactol	–	–	–	–	2	2	2	–	–	1	–	–	–	–	1	1	1	–	1
Lard†	X	X	X	X	2	–	–	–	–	–	–	–	–	–	1	1	1	1	X
Lard oil	–	–	–	–	2	–	–	–	–	–	1	–	–	–	1	1	1	1	–
Lasso (agricultural spray)	–	–	1	–	–	–	–	–	–	1	–	–	1	–	–	1	1	–	–
Latex	2	–	–	–	–	–	–	–	–	–	–	–	–	–	1	1	1	–	1
Lead acetate	2	2	2	X	X	X	–	–	–	1	–	2	–	X	2	1	1	–	1
Lead arsenate	2	–	–	2	2	2	–	–	–	1	–	–	–	–	1	1	1	–	–
Lead nitrate	2	2	–	–	2	2	–	–	–	1	–	–	–	–	1	1	1	–	–
Lead sulfamate	2	–	2	2	2	–	–	–	–	–	–	–	–	–	–	–	–	–	–
Lead sulfate	2	1	–	1	1	1	–	–	–	1	–	–	–	–	1	1	1	–	–
Lecithin	–	–	–	–	2	X	–	–	–	–	–	–	–	–	–	1	1	–	–
Ligroin (petroleum ether)	X	X	X	X	X	1	–	–	–	2	–	–	–	1	2	1	1	–	–
Lime (chlorinated)*	–	2	–	–	–	–	–	–	–	–	–	–	–	–	–	–	2	–	–
Lime bleach	2	2	2	X	X	2	–	–	–	–	–	–	–	–	X	2	1	–	–
Lindane (agricultural spray)	–	–	–	–	–	–	–	–	1	–	–	–	1	–	–	1	1	–	–
Linoleic acid	–	X	X	–	X	2	–	–	–	–	–	–	–	–	–	–	–	–	–
Linseed oil (boiled)	X	–	2	1	2	2	1	–	–	1	–	1	–	1	2	1	1	1	2
Liquid soap	2	2	2	–	–	–	–	–	–	2	–	–	–	–	1	1	1	1	1
Lubricating oils	X	X	X	2	2	1	2	2	2	1	1	1	1	1	1	1	1	1	1

*If free chlorine does not exceed 20%.

Chemical	D_3 D_2 D	H	P	M	A_2 A	C	C_2	C_3	Q	K	W	Y	Z	V	Iron or carbon steel	304 SS	316 SS	Aluminum	Brass
M																			
Magnesium carbonate	1	1	1	1	1	1	1	1	1	1	–	1	–	–	1	1	1	–	–
Magnesium chloride	1	1	–	1	1	1	1	2	–	1	–	1	–	–	X	2	1	X	2
Magnesium hydroxide	1	1	–	1	2	2	–	–	–	1	–	1	–	–	1	1	1	X	–
Magnesium nitrate	1	2	2	1	2	2	2	–	–	1	–	–	–	–	1	1	1	X	1
Magnesium sulfate	1	2	2	1	2	2	2	–	2	1	–	–	–	–	2	1	1	–	1
Magnus – light	–	X	X	–	–	1	–	–	–	–	–	1	–	–	–	–	–	–	–
Magnus – medium	–	X	X	–	–	1	–	–	–	–	–	1	–	–	–	–	–	–	–
Malathion (agricultural spray)	–	–	2	–	–	–	–	–	–	1	–	–	1	–	–	1	1	–	–
Malic acid	2	X	–	2	X	2	–	–	–	–	–	1	–	2	2	2	1	–	–
Maxmul (Pennzoil hydraulic fluid)	X	X	X	–	2	1	–	–	–	2	–	1	–	1	1	–	1	–	–
Mercuric chloride	2	2	2	1	1	2	2	–	–	1	–	1	–	–	X	1	1	X	X
Mercuric cyanide	2	2	2	1	1	2	2	–	–	1	–	–	–	–	–	–	–	X	–
Mercurous nitrate	2	2	2	1	1	2	2	–	–	1	–	–	–	–	1	1	1	X	–

Chemical resistance of hose polymer types—continued **Table I**

M—continued

	Hose polymer types														Coupling material				
Chemical	D_3 D_2 D	H	P	M	A_2 A	C	C_2	C_3	Q	K	W	Y	Z	V	Iron or carbon steel	304 SS	316 SS	Aluminum	Brass
Mercury	2	2	–	1	1	2	2	–	–	1	–	1	–	–	1	1	1	X	X
Mercury vapor	2	2	–	1	1	2	2	–	–	1	–	–	–	–	1	1	1	–	–
Mesityl oxide	X	2	2	X	X	X	X	X	X	–	–	–	–	X	1	1	1	1	1
Methyl acetate	X	2	2	X	X	X	X	X	–	2	–	–	–	X	1	1	1	1	1
Methyl acrylate	X	X	X	X	X	X	X	X	–	–	–	–	–	X	1	1	1	1	1
Methyl alcohol (methanol)	1	1	1	1	1	1	1	1	1	1	2	1	1	X	1	1	1	1	2
Methyl amine (25% aqueous sol'n)	2	2	2	–	2	X	–	X	–	2	–	–	–	–	1	1	1	–	–
Methyl amyl carbinol	–	–	–	–	–	–	–	–	–	1	–	–	–	–	1	1	1	–	–
Methyl bromide	X	X	X	X	X	X	–	X	–	–	–	–	–	1	1	1	1	–	1
Methyl butyl ketone (MBK)	X	2	2	X	X	X	X	X	X	–	–	–	–	X	1	1	1	1	1
Methyl cellosolve	X	X	X	X	–	–	–	–	–	–	–	–	–	X	–	–	–	–	–
Methyl chloride	X	X	–	X	X	X	X	X	–	2	–	–	–	2	1	1	1	–	1
Methyl ethyl ketone (MEK)	X	2	2	X	X	X	X	X	X	2	X	2	2	X	1	1	1	1	1
Methyl formate	X	2	2	X	2	X	X	X	–	–	–	–	–	–	1	1	1	1	1
Methyl isobutyl ketone (MIBK)	X	2	–	X	X	X	X	X	–	2	X	–	1	X	1	1	1	1	1
Methyl isopropyl ketone	X	2	–	X	X	X	X	X	–	2	–	–	–	X	1	1	1	1	1
Methyl methacrylate	X	X	–	2	X	X	X	X	–	2	–	–	–	X	1	1	1	–	–
Methyl salicylate	–	2	X	–	X	X	X	X	–	–	–	–	–	–	1	1	1	1	1
Methylene chloride	X	X	X	X	X	X	X	X	X	2	–	X	–	2	1	1	1	–	1
Methylene dichloride	X	X	X	X	X	X	X	X	X	–	–	X	–	2	1	1	1	–	1
Milk†	1	–	–	–	–	–	–	–	–	–	–	–	–	–	X	1	1	1	X
Mineral oil	X	X	X	1	1	1	2	2	2	1	–	1	–	1	1	1	1	2	1
Mineral spirits	–	X	X	X	–	1	2	2	–	1	–	–	–	–	1	1	1	2	1
Miners oil	–	–	–	–	–	–	–	–	–	1	–	–	–	–	1	1	1	–	–
Mobile hydraulic oils	–	X	X	–	–	1	–	–	–	–	–	1	–	–	1	1	1	1	–
Mobilmet S122	–	X	X	–	–	1	–	–	–	–	–	1	–	–	–	–	–	–	–
Molasses	2	–	–	–	2	2	–	–	–	1	–	–	–	–	2	1	1	2	X
Monochlorobenzene	X	X	X	X	X	X	X	X	X	X	X	X	–	1	1	1	1	–	1
Monoethanolamine	–	2	–	X	–	2	–	–	–	2	–	–	–	X	1	1	1	–	1
Mould oil	–	–	–	–	–	–	–	–	–	1	–	–	–	–	1	1	1	–	–
Muriatic acid (hydrochloric)	2	2	X	2	X	–	–	–	X	2	X	X	X	1	X	X	X	X	X
Mustard	1	1	–	1	1	–	–	–	–	–	–	1	–	–	X	1	1	–	–

†Use food-grade hose only.

N

	Hose polymer types														Coupling material				
Chemical	D_3 D_2 D	H	P	M	A_2 A	C	C_2	C_3	Q	K	W	Y	Z	V	Iron or carbon steel	304 SS	316 SS	Aluminum	Brass
Naphtha (low aromatic content**)	X	X	X	X	X	2	–	X	–	1	2	2	–	1	2	1	1	–	1
Naphthaline†	X	X	X	X	X	X	X	X	–	2	–	2	–	1	1	1	1	–	1
Natural gas	Use special hose														1	1	1	–	2
Neutral oil	X	X	X	–	2	2	2	–	–	1	–	–	–	1	1	1	1	–	1
Nickel acetate	2	1	–	–	–	–	–	–	–	–	–	–	–	X	1	1	1	1	1
Nickel chloride	2	2	2	2	2	2	2	2	2	1	–	–	–	1	X	2	2	X	X
Nickel nitrate	2	2	2	2	2	2	2	2	2	1	–	–	–	–	–	–	2	X	–
Nickel plating solution	2	–	–	2	–	2	–	–	–	1	–	–	–	–	–	1	1	–	–
Nickel sulfate	2	2	2	2	2	2	2	2	2	1	–	–	–	1	–	2	1	X	X
Nicotine salts	–	–	–	–	–	–	–	–	–	1	–	–	–	–	1	X	2	–	–
Niter cake (sodium bisulfate)	1	1	1	1	1	1	1	1	1	–	1§	–	–	1	X	2	1	–	–
Nitric acid (10%)	X	2	–	2	X	X	X	X	X	2	–	X	X	1	X	2	2	–	X
Nitric acid (25%)	2	2	–	2	X	X	X	X	X	2	X	X	X	1	X	2	2	–	X
Nitric acid (40%)	X	X	X	X	X	X	X	X	X	2	X	X	X	1	X	2	2	–	X
Nitric acid (60%)	X	X	–	X	X	X	X	X	X	2	X	X	X	2	X	2	2	–	X
Nitric acid (red fuming)	X	X	X	X	X	X	X	X	X	X	X	X	X	2	X	2	2	2	X
Nitrobenzene	X	X	X	X	X	X	X	X	–	2	X	2	–	2	1	1	1	1	1
Nitroethane	2	2	2	2	X	X	X	X	–	–	–	–	–	–	–	1	1	–	1
Nitrogen	1	1	1	1	1	1	1	1	1	1	1	1	1	1	1	1	1	1	1
Nitrogen tetroxide	No hose available														–	–	2	–	–
Nitromethane	2	2	2	–	X	X	X	X	–	–	–	2	1	–	–	1	1	–	1
Nitropropane	X	2	2	–	X	X	X	X	–	1	–	–	1	X	–	1	1	–	1
Nitrosyl chloride	–	–	–	–	–	–	–	–	–	1	–	–	–	–	–	1	1	–	–
Nyvac FR fluid	–	1	1	–	–	1	–	–	–	–	–	1	–	–	1	1	1	1	1
Nyvac FR200 fluid	–	1	1	–	–	1	–	–	–	–	–	1	–	–	1	1	1	1	1
Nyvac 20(WG), 30(WG)	–	1	1	–	–	1	–	–	–	–	–	1	–	–	1	1	1	1	1
n-Octane	X	X	X	X	–	2	–	X	–	2	–	2	–	2	1	1	1	–	1

**Special naphthas may require special hose. §(5%)

Chemical resistance of hose polymer types—continued **Table I**

Chemical	Hose polymer types														Coupling material				
	D_3 D_2 D	H	P	M	A_2 A	C	C_2	C_3	Q	K	W	Y	Z	V	Iron or carbon steel	304 SS	316 SS	Aluminum	Brass
O																			
Octyl alcohol	2	–	X	–	2	2	2	2	–	–	–	–	–	1	1	1	1	1	2
Oil (SAE)	X	X	X	2	2	1	2	2	2	1	1	1	1	1	1	1	1	1	1
Oleic acid	X	2	2	2	2	2	2	X	–	1	–	1	–	2	2	2	1	1	2
Oleum (fuming sulfuric acid)	X	X	X	X	X	X	X	X	X	X	X	X	X	X	–	–	1	–	–
Olive oil	X	2	2	X	X	2	2	X	–	–	–	2	–	1	2	1	1	1	2
Oxalic acid	X	2	2	2	X	X	X	X	–	2	1†	–	X	2	X	2	1	2	X
Oxygen	Use 2B or 16B hose														1	1	1	1	1
Ozone	X	2	1	2	2	X	2	X	2	2	2	2	–	2	1	1	1	1	1

†Solid at room temperature. Use solvent to make liquid.

Chemical	D_3 D_2 D	H	P	M	A_2 A	C	C_2	C_3	Q	K	W	Y	Z	V	Iron or carbon steel	304 SS	316 SS	Aluminum	Brass
P																			
Pacemaker Types 150T, 300T, 500T (Citgo)	–	X	X	–	–	1	–	–	–	–	–	1	–	–	–	–	–	–	–
Paint	–	–	–	–	–	–	–	–	–	–	–	–	1	–	–	1	1	1	1
Palm oil	X	2	–	2	2	1	2	–	–	1	–	1	–	–	1	1	1	1	1
Palmitic acid	X	2	2	X	2	2	–	X	–	1	2	1	–	1	1	2	1	1	X
Paraffin (petroleum)	X	X	X	X	2	1	2	–	–	–	–	1	–	–	2	1	1	–	1
Paraformaldehyde	X	–	–	2	2	2	2	–	–	2	–	–	–	–	–	1	1	1	–
Peanut oil	–	X	–	–	2	1	–	–	–	–	–	1	–	–	1	1	1	1	1
Pennant motor oils	–	X	X	–	–	1	–	–	–	–	–	1	–	–	1	1	1	1	1
Perchloric acid	2	2	–	2	2	–	–	–	–	2	–	X	X	–	–	2	1	–	–
Perchloroethylene	X	X	X	X	X	X	X	X	X	2	X	X	2	1	1	1	1	–	1
Pentasol	2	2	2	2	2	2	2	2	2	2	–	–	–	1	1	1	1	1	1
Petroleum oils	X	X	X	2	2	1	2	2	2	1	1	1	–	1	1	1	1	1	1
Phenol (carbolic acid)	X	2	2	X	X	X	X	X	X	2	X	X	X	1	X	1	1	–	X
Phorone	–	2	–	–	–	–	–	–	–	–	–	–	–	–	1	1	1	–	1
Phosphoric acid (50%)	2	2	–	1	2	2	–	–	2	1	–	–	X	1	X	1	1	X	2
Phosphoric acid (85%)	2	2	2	1	2	X	–	–	–	2	–	–	X	–	X	2	2	X	X
Picric acid (water solution)	2	2	–	2	2	2	–	–	X	2	X	–	X	1	X	1	1	X	X
Pine oil	X	X	X	X	–	2	–	X	–	1	–	–	–	2	1	1	1	–	–
Pinene	X	X	X	–	–	2	–	–	–	–	–	–	–	1	1	1	1	1	1
Piperazine hydrochloride solution (34%)	–	–	–	–	–	2	–	–	–	–	–	–	–	–	–	–	–	–	–
Plating solution (chrome)	–	2	2	–	–	–	–	–	–	–	–	–	X	2	–	X	X	–	–
Polyester resin	–	–	–	–	–	–	–	–	–	–	–	–	2	2	–	–	–	–	–
Polyurethane (to 125°F)	–	2	2	–	–	–	–	–	–	–	–	–	–	2	–	–	–	–	–
Potassium acetate	2	2	2	–	–	–	–	–	–	–	–	2	–	X	–	1	1	–	–
Potassium carbonate	1	1	1	1	1	1	1	1	1	1	–	1	1	1	2	1	1	–	X
Potassium chlorate	2	2	2	2	2	2	–	–	–	2	–	–	–	–	1	1	1	–	–
Potasium chloride	1	1	1	1	1	1	1	1	1	1	1†	1	1	1	2	2	1	X	X
Potassium cyanide	2	2	2	2	2	2	2	2	2	2	–	–	–	–	2	1	1	–	X
Potassium dichromate	X	2	–	1	2	–	–	–	–	1	–	–	–	1	–	1	1	–	–
Potassium hydroxide (30%) (caustic potash)	2	2	2	1	2	X	X	X	X	2	–	X	X	2	X	1	1	X	X
Potassium nitrate	1	1	1	1	1	1	1	1	1	1	1†	–	–	1	X	1	1	–	2
Potassium permanganate	X	–	–	–	X	–	–	–	–	1	–	–	X	2	2	1	1	–	–
Potassium sulfate	2	1	1	2	1	1	1	1	1	1	–	1	–	–	2	1	1	1	2
Potassium sulfite	2	2	2	2	2	2	2	2	2	2	–	–	–	–	1	1	1	–	–
Powerlube (Carter)	–	X	X	–	–	1	–	–	–	–	–	1	–	–	1	1	1	1	1
Primatol A, S, P (agricultural spray)	X	X	2	X	–	X	–	X	–	2	–	–	–	–	–	–	–	–	–
Propane gas	Use LPG hose only														1	1	1	–	1
Propionic acid	2	2	2	2	X	X	X	X	–	1	–	–	–	–	1	1	1	–	–
Propyl acetate	–	2	–	–	X	X	X	X	–	1	–	–	–	X	1	1	1	1	1
Propyl alcohol (propanol)	2	2	2	2	2	2	2	2	2	1	–	1	–	2	1	1	1	1	2
Purina insecticide	–	2	2	–	X	X	–	X	–	–	–	–	–	2	–	–	–	–	2
Puropale RX oils	–	X	X	–	–	1	–	–	–	–	–	1	–	–	–	–	–	–	–
Pydraul F-9	X	2	2	X	X	X	X	X	–	2	–	2	2	1	1	1	1	–	–
Pydraul 50E	–	2	2	–	–	–	–	–	–	–	–	1	2	2	1	1	1	–	–
Pydraul 150	X	2	2	X	X	X	X	X	X	2	–	2	2	1	1	1	1	1	1
Pydraul A-200	X	X	X	X	X	X	X	X	X	2	–	X	2	1	1	1	1	–	–
Pydraul 280	X	2	2	X	X	X	X	X	X	2	–	–	2	2	1	1	1	–	–
Pydraul 312	X	X	X	–	X	X	X	X	–	2	–	1	–	1	1	1	1	–	–
Pydraul 540	X	X	X	X	X	X	X	X	X	2	–	X	–	1	1	1	1	–	–
Pydraul 625	X	2	2	X	X	X	X	X	X	2	–	2	2	1	1	1	1	–	–
Pydraul 10-E, 29-E-LT, 30E, 60, 65E, 115E	–	2	2	–	–	X	–	–	–	–	–	X	–	–	1	1	1	1	1

Chemical resistance of hose polymer types—continued **Table I**

Chemical	Hose polymer types														Coupling material				
	D_3 D_2 D	H	P	M	A_2 A	C	C_2	C_3	Q	K	W	Y	Z	V	Iron or carbon steel	304 SS	316 SS	Aluminum	Brass
P—continued																			
Pydraul 135	–	1	–	–	–	X	–	–	–	–	–	2	1	1	1	1	1	–	–
Pyrene (carbon tetrachloride)	X	X	X	X	X	X	X	X	X	X	–	X	–	1	X	2	2	X	2
Pyridine (50%)	–	–	–	X	X	–	–	–	–	2	X	X	–	X	–	1	1	1	1
Pyrogard 51, 53, 55	–	2	2	–	–	X	–	–	–	–	–	2	–	–	1	1	1	–	–
Pyrogard 160, 230, 630	–	–	–	–	–	–	–	–	–	–	–	–	–	2	1	1	1	–	–
Pyrogard C, D	–	X	X	–	–	1	–	–	–	–	–	1	–	–	1	1	1	1	1
Pyronal (transformer oil)	–	–	–	–	–	–	–	–	–	–	X	–	–	1	–	–	–	–	–
R																			
Ramrod (agricultural spray)	–	–	–	–	–	–	–	–	–	1	–	–	1	–	–	–	–	–	–
Rando oils	–	X	X	–	–	1	–	–	–	–	–	1	–	–	–	–	–	–	–
Rape-seed oil	–	2	2	X	–	–	–	–	–	–	–	–	–	–	–	–	–	–	–
Red oil (commercial oleic acid)	X	2	2	2	2	2	2	X	–	1	–	1	–	2	2	2	1	1	2
Refined wax (petroleum)	X	–	–	–	–	1	2	2	–	–	–	–	–	1	1	1	1	–	1
Regal oils, R&O	–	X	X	–	–	1	–	–	–	–	–	1	–	–	1	1	1	1	1
Rubilene oils	–	X	X	–	–	1	–	–	–	–	–	1	–	–	–	–	–	–	–
S																			
Safetytex 215	–	1	1	–	–	X	–	–	–	–	–	X	–	–	–	–	–	–	–
Salicylic acid	2	2	2	–	–	X	–	–	–	–	–	X	–	2	–	1	1	2	–
Salt water (sea water)	1	1	1	2	2	2	2	2	1	1	–	1	1	1	2	1	1	–	X
Santosafe W-G15, W-G20, W-G30	–	1	1	–	–	1	–	–	–	–	–	1	–	–	1	1	1	1	1
SCC 7204 (Stauffer)	–	–	–	–	–	–	–	–	–	–	–	–	–	–	–	–	–	–	–
Sevin	–	–	2	–	–	–	–	–	–	2	–	–	–	–	–	–	–	–	–
Sewage	X	–	–	2	2	2	2	–	–	1	–	2	1	–	X	1	1	2	1
SFR Fluid B (Shell)	–	2	2	–	–	X	–	–	–	–	–	–	–	–	–	–	–	–	–
SFR Fluid C (Shell)	–	2	2	–	–	X	–	–	–	–	–	–	–	–	–	–	–	–	–
Silicone grease	–	–	–	2	2	2	2	–	–	2	–	1	–	2	1	1	1	–	1
Silicone oils	–	–	–	2	2	2	2	–	–	2	X	1	–	2	1	1	1	–	1
Silver nitrate	1	1	1	1	1	1	1	1	1	1	–	1	–	1	2	1	1	1	2
Skydrol 500 A & 7000	X	2	1	X	X	X	X	X	X	1	X	2	2	X	1	1	1	1	–
Soap oil	–	–	–	2	2	2	–	–	–	1	–	–	–	–	1	1	1	–	–
Soap solutions	X	1	1	1	X	1	2	2	2	1	–	1	–	1	1	1	1	1	1
Soda ash (sodium carbonate)	1	1	1	1	1	1	1	1	1	1	1	1	1	1	1	1	1	X	2
Sodium acetate	2	2	2	X	X	X	–	–	–	1	–	1	1	X	1	1	1	1	1
Sodium bicarbonate	1	1	1	1	1	1	1	1	1	1	1	1	1	1	2	1	1	–	2
Sodium bisulfate (niter cake)	1	1	1	1	1	1	1	1	1	1	1†	1	1	1	X	1	1	X	X
Sodium bisulfite	1	1	1	1	1	1	1	1	1	1	1†	1	–	1	1	1	1	–	–
Sodium borate	1	1	1	1	1	1	1	1	1	1	–	1	–	1	1	1	1	–	–
Sodium chloride	1	1	1	1	1	1	1	1	1	1	1	1	1	1	2	2	1	X	X
Sodium cyanide	1	1	1	1	1	1	1	1	1	1	–	–	–	1	2	1	1	X	X
Sodium fluoride (70%)	2	2	2	–	–	–	–	–	–	2	–	–	–	–	–	–	2	–	–
Sodium hydroxide (40%)	1	2	2	1	1	2	2	–	2	1	2	2	X	2	2	1	1	X	X
Sodium hydroxide (50%, 115°F)	1	–	–	1	2	X	X	X	X	2	X	–	X	X	2	2	2	X	X
Sodium hydroxide (50%, 180°F)	X	–	–	2	–	–	–	–	–	X	X	–	X	X	X	2	2	X	X
Sodium hydroxide (60%)	2	–	–	2	2	X	X	X	X	–	2	–	X	X	X	2	2	X	X
Sodium hypochlorite (5%)	X	–	1	1	–	X	X	X	–	2	1	–	–	1	X	X	2	X	X
Sodium hypochlorite (20%)	X	–	1	1	X	X	X	X	–	2	–	–	–	X	X	X	2	X	X
Sodium metaphosphate	2	2	2	2	2	2	–	–	2	1	–	–	–	2	X	1	1	1	X
Sodium nitrate	X	2	2	2	X	X	–	–	2	1	–	1	1	–	1	2	2	2	2
Sodium perborate	X	2	2	X	X	X	–	–	–	1	–	–	–	–	X	1	1	1	X
Sodium peroxide	–	1	–	1	1	–	–	–	–	1	–	–	X	1	X	1	1	1	X
Sodium phosphates	2	2	2	–	X	–	–	–	–	1	–	–	–	–	–	1	1	X	X
Sodium sulfite	2	2	2	2	2	2	2	2	2	1	–	–	–	–	1	1	1	–	–
Sodium thiosulfate (HPO)	1	1	–	1	1	1	1	1	1	1	–	2	–	–	X	1	1	2	X
Sodium tripolyphosphate (STPP)	–	2	–	–	–	–	–	–	–	–	–	–	–	X	–	1	1	X	X
Solnus oils	–	X	X	–	–	1	–	–	–	–	–	1	–	–	–	–	–	–	–
Solvac 1535G	–	X	X	–	–	1	–	–	–	–	–	1	–	–	–	–	–	–	–
Soybean oil	X	2	–	2	2	2	–	–	–	1	1	–	–	1	1	1	1	–	–
Spent acid	X	–	–	2	–	–	–	–	–	–	–	–	–	2	–	1	1	–	–
Stannic chloride	2	X	–	X	X	2	2	X	2	1	–	–	X	1	X	–	–	–	X

†(5%)

Chemical resistance of hose polymer types—continued **Table I**

Chemical	Hose polymer types														Coupling material				
	D_3 D_2 D	H	P	M	A_2 A	C	C_2	C_3	Q	K	W	Y	Z	V	Iron or carbon steel	304 SS	316 SS	Aluminum	Brass

S—continued

Chemical	D_3 D_2 D	H	P	M	A_2 A	C	C_2	C_3	Q	K	W	Y	Z	V	Iron or carbon steel	304 SS	316 SS	Aluminum	Brass
Stanoil No. 15, 18, 25, 31, 35, 51	–	X	X	–	–	1	–	–	–	–	–	1	–	–	–	–	–	–	–
Stauffer Jet 1	–	–	–	–	–	–	–	–	–	–	–	–	–	–	–	–	–	–	–
Stauffer Jet 2	–	–	–	–	–	–	–	–	–	–	–	–	–	–	–	–	–	–	–
Staysol FR	–	X	X	–	–	1	–	–	–	–	–	1	–	–	–	–	–	–	–
Steam*	Use steam hose only														1	1	1	–	2
Stearic acid	2	2	–	2	2	2	2	–	–	1	–	–	–	–	X	2	1	X	X
Stoddard solvent	X	X	X	–	–	–	2	–	–	2	–	–	–	1	2	1	1	–	1
Styrene (monomer)	X	X	X	–	–	X	–	X	–	2	–	X	2	2	2	–	2	–	2
Sucrose solutions	1	1	–	1	1	1	–	–	–	1	–	–	–	–	1	1	1	–	–
Sulfamic acid (10%)	X	–	–	1	–	–	–	–	–	1	–	–	–	1	–	–	–	–	–
Sulfamic acid (10%, 170°F)	X	–	–	2	–	–	–	–	–	X	–	–	–	2	–	–	–	–	–
Sulfur (200°F)	Use molten sulfur hose														2	2	1	2	X
Sulfur chloride	X	X	X	2	X	X	–	–	–	2	–	–	–	1	X	X	2	–	X
Sulfur dioxide (dry)	X	X	2	2	X	X	–	–	–	2	–	–	X	1	2	1	1	1	1
Sulfur trioxide (dry)	X	X	–	X	X	X	X	X	–	1	–	–	–	1	2	2	2	2	–
Sulfuric acid (10%)	1	2	2	1	1	2	2	2	2	1	1	X	X	1	X	X	2	X	X
Sulfuric acid (30%)	2	2	–	1	1	–	–	–	–	1	–	X	X	1	X	X	2	X	X
Sulfuric acid (50%)	X	X	–	1	2	X	X	X	X	1	X	X	X	1	X	X	X	X	X
Sulfuric acid (75%)	X	X	X	2	X	X	X	X	X	1	X	X	X	1	X	X	X	X	X
Sulfuric acid (93%)	X	X	X	X	X	X	X	X	X	–	X	X	X	1	2	X	2	X	X
Sulfuric acid (98%)	X	X	X	X	X	X	X	X	X	X	X	X	X	2	2	X	2	X	X
Sulfuric acid, fuming	X	X	X	X	X	X	X	X	X	X	X	X	X	2	2	–	1	–	X
Sulfurous acid (10%)	X	2	–	1	–	X	–	–	X	1	–	–	X	1	X	2	1	X	X
Sulfurous acid (75%)	X	X	–	1	X	X	–	–	X	1	–	–	X	1	X	X	2	X	X
Sun R&O oils	–	X	X	–	–	1	–	–	–	–	–	1	–	–	1	1	1	1	1
Sunsafe F	–	X	X	–	–	1	–	–	–	–	–	1	–	–	1	1	1	1	–
Suntac HP oils	–	X	X	–	–	1	–	–	–	–	–	1	–	–	1	–	1	1	–
Suntac WR oils	–	X	X	–	–	1	–	–	–	–	–	1	–	–	1	–	1	1	–
Sunvis oils 700, 800, 900	–	X	X	–	–	1	–	–	–	–	–	1	–	–	1	1	1	–	–
Super hydraulic oils (Conoco)	–	X	X	–	–	1	–	–	–	–	–	1	–	–	1	1	1	1	–
303 fluid (Conoco)	–	X	X	–	–	1	–	–	–	–	–	1	–	–	1	1	1	–	–
Synthetic oil (Citgo)	–	X	X	–	–	–	–	–	–	–	–	–	–	–	1	1	1	–	–
Syrup	1	–	–	–	2	–	–	–	2	1	–	–	–	–	–	1	1	–	–

†(5%) *Consult industrial hose Catalog 39993 for selecting the correct steam hose.

T

Chemical	D_3 D_2 D	H	P	M	A_2 A	C	C_2	C_3	Q	K	W	Y	Z	V	Iron or carbon steel	304 SS	316 SS	Aluminum	Brass
Tall oil (to 150°F)	X	X	X	X	2	2	2	–	–	–	–	–	–	1	–	X	2	–	–
Tallow	–	2	2	-	2	2	–	–	–	1	–	1	–	–	2	2	2	1	2
Tannic acid (10%)	2	X	–	2	2	X	–	–	–	1	–	–	–	1	2	1	1	2	X
Tar (bituminous)	X	X	X	–	2	2	2	X	–	–	–	2	–	1	1	1	1	1	2
Tartaric acid	2	X	–	1	2	2	–	–	–	1	1△	1	–	1	–	2	2	2	–
Tellus oils	–	X	X	–	–	1	–	–	–	–	–	1	–	–	1	1	1	1	1
Tenol oils	–	X	X	–	–	1	–	–	–	–	–	1	–	–	1	1	1	–	–
Tergitol	–	–	–	–	–	–	–	–	–	2	–	–	–	–	2	1	1	–	2
Terpineol	X	X	–	–	–	–	–	X	–	–	–	1	–	–	–	–	–	–	–
Terresstic	–	X	X	–	–	1	–	–	–	–	–	1	–	–	1	1	1	–	–
Tetraethyl lead (TEL)	–	–	–	–	–	–	–	–	–	2	–	–	–	–	–	–	–	–	–
Tetrahydrofuran (THF)	–	X	2	–	–	–	–	–	–	1	–	2	–	X	–	–	–	–	–
Titanium tetrachloride	–	X	X	–	X	X	X	X	–	–	–	–	–	2	1	2	2	X	X
Toluene (toluol)	X	X	X	X	X	X	X	X	X	2	X	2	1	1	1	1	1	1	1
Toluene diisocyanate	–	2	2	–	–	–	–	–	–	–	–	2	–	2	–	–	–	–	–
Transformer oil* (petroleum type)	X	X	X	X	2	1	–	–	–	1	–	1	–	1	1	1	1	1	1
Transformer oil* (askarel types)	X	X	X	X	X	X	X	X	X	2	–	–	–	1	–	–	1	–	–
Transmission fluid (Type A)	X	X	X	–	X	1	2	2	–	–	2	–	–	1	1	1	1	–	1
Tributoxyethyl phosphate	X	2	2	X	–	X	X	X	–	–	–	–	–	–	1	–	–	X	–
Tributyl phosphate	X	X	X	X	X	X	X	X	–	–	–	–	–	X	1	–	–	X	–
Trichloroethylene	X	X	X	X	X	X	X	X	X	X	X	X	2	1	X	–	1	X	1
Tricresyl phosphate	X	2	1	X	X	X	X	X	–	–	X	–	1	1	1	–	2	X	–
Triethanolamine (TEA)	2	2	1	2	2	2	–	–	–	1	–	X	–	X	–	1	1	–	1
Tripolyphosphate (STPP)	–	2	–	–	–	–	–	–	–	2	–	–	–	–	–	2	1	X	–
Tung oil	X	2	–	2	2	2	–	–	–	2	–	X	–	1	1	1	1	1	1
Turpentine	X	X	X	X	X	2	–	–	–	X	2	–	–	1	–	1	1	1	2
Tycol Avalon 50, 57, 60	–	X	X	–	–	1	–	–	–	–	–	–	–	–	1	1	1	–	–
Tycol A Turbio 37, 50, 58, 60	–	X	X	–	–	1	–	–	–	–	–	–	–	–	1	1	1	–	–

△(20%) *Oil contamination may be a problem.

Chemical resistance of hose polymer types—continued **Table I**

Chemical	Hose polymer types														Coupling material				
	D_3 D_2 D	H	P	M	A_2 A	C	C_2	C_3	Q	K	W	Y	Z	V	Iron or carbon steel	304 SS	316 SS	Aluminum	Brass
U																			
Ucon M1	–	1	1	–	–	1	–	–	–	–	–	1	–	–	1	1	1	1	1
Ucon hydrolube types 150CP, 200CP 275CP, 300CP, 550CP, 900CP, 150DB, 275DB, 150LT, 200LT, 275LT, 300LT, 200NM, 300NM	–	1	1	–	–	1	–	–	–	–	–	1	–	–	1	1	1	1	1
Union C-2 fluid	–	X	X	–	–	1	–	–	–	–	–	1	–	–	1	1	1	–	–
Union CP oil	–	X	X	–	–	1	–	–	–	–	–	1	–	–	1	1	1	1	1
Union ATF Dexron	–	X	X	–	–	1	–	–	–	–	–	1	–	–	1	1	1	–	–
Union ATF Type F	–	X	X	–	–	1	–	–	–	–	–	1	–	–	1	1	1	1	1
Union hydraulic oil AW	–	X	X	–	–	1	–	–	–	–	–	1	–	–	1	1	1	1	1
Union hydraulic tractor fluid	–	X	X	–	–	1	–	–	–	–	–	1	–	–	1	1	1	1	1
Urea solution	1	2	–	1	1	2	–	–	2	1	–	1	–	–	1	1	1	–	–
V																			
Varnish	X	X	X	X	X	X	X	X	–	2	–	–	1	2	2	1	1	–	2
Vegetable oils**	X	X	–	–	2	–	–	–	–	–	–	2	–	–	1	1	1	1	–
Versilube F-44, F-50	2	2	2	2	2	2	2	2	2	–	–	1	–	1	1	1	1	1	1
Vinegar	2	2	–	X	2	X	X	X	–	1	2	–	–	1	X	2	1	X	X
Vinyl acetate	X	2	X	X	X	X	X	X	–	1	–	–	–	X	–	1	2	1	2
Vinyl chloride (monomer)	X	X	X	X	X	X	X	X	X	2	–	1	–	2	2	1	1	2	X
Vinyl fluoride	–	–	–	–	–	–	–	–	–	–	–	–	–	1	–	–	–	–	–
Vitrea oils	–	X	X	–	–	1	–	–	–	–	–	1	–	–	1	1	1	–	–
W																			
Water	1	1	1	1	1	1	1	1	1	1	1	1	1	1	2	1	1	1	1
Wines**	1	–	–	–	2	–	–	–	–	–	–	1	–	–	X	2	1	–	X
X																			
Xylene	X	X	X	X	X	X	X	X	X	2	X	2	1	2	2	2	2	1	–
Z																			
Zeric	–	X	X	–	–	1	–	–	–	–	–	1	–	–	–	–	–	–	–
Zinc acetate	2	2	2	X	X	X	–	–	–	–	–	–	–	X	1	1	1	1	1
Zinc chloride solutions	2	2	–	1	1	1	1	2	–	1	–	2	–	1	X	2	1	X	X
Zinc sulfate solutions	X	2	2	2	2	2	2	2	–	1	–	–	–	–	X	2	1	X	X

**Use food-grade hose only.

Hose polymer types

- D – Natural rubber (and also styrene butadiene)
- D_2 – The same as D, but white, for food products
- D_3 – The same as D, but tan
- H – Butyl
- P – Ethylene propylene
- M – Hypalon
- A – Neoprene
- A_2 – Neoprene, tan, for food products
- C – Buna-N
- C_2 – A blend of Buna-N and plastic
- C_3 – Modification of Buna-N, with some properties of neoprene
- Q – Tufflex
- K – Gatron
- W – Epichlorohydrin
- Y – Polyester elastomer (Hytrel, by Du Pont)
- Z – Nylon
- V – Fluoro-elastomer

Resistance to chemicals

- "1" – Excellent resistance. The fluid is expected to have minor or no affect on the polymer.
- "2" – Good resistance. This hose polymer should give reasonably satisfactory service.
- "X" – Not recommended. This polymer is unsatisfactory for the chemical and should not be used.
- "–" – (Dash) Insufficient data is available for the material. Testing is advised.

Physical characteristics of major hose stock types **Table II**

Hose compound / Strength and resistance	Natural rubber (and styrene butadiene*)	Butyl (IIR)	Ethylene propylene (EPDM)	Hypalon (CSM)	Neoprene (CR)	Buna N (NBR)	Tufflex†	Gatron†	Fluoro-elastomer (FPM)	Epi-chloro-hydrin	Polyester elastomer	Nylon
Tensile strength	Excellent	Fair to good	Good	Good	Good	Fair to good	Good	Good	Fair	Good	Good	Good
Tearing	Good to excellent	Good	Good	Fair	Good	Fair to good	Good	Fair	Fair	Good	Good	Good
Abrasion	Excellent	Fair to good	Good	Good	Good to excellent	Fair to good	Good	Fair	Good	Fair to good	Good	Good
Flame	Poor	Poor	Poor	Good	Very good	Poor	Fair to good	Poor	Good	Poor	Poor	Good
Petroleum, oil and commercial gasoline	Poor	Poor	Poor	Good	Good	Good to excellent	Fair	Excellent	Excellent	Excellent	Good	Excellent
Gas permeation	Fair	Out-standing	Fair to good	Good to excellent	Good	Good	Good	Good	Good	Good	Good	Excellent
Weathering	Poor	Excellent	Excellent	Very good	Good to excellent	Good** to poor	Excellent	Excellent	Excellent	Very good	Good	Excellent
Ozone	Poor	Excellent	Out-standing	Very good	Good to excellent	Good** to poor	Excellent	Excellent	Excellent	Very good	Good	Excellent
Heat	Poor	Excellent	Excellent	Very good	Good	Good	Fair	Fair	Out-standing	Excellent	Excellent	Good
Low temperature	Good	Very good	Good to excellent	Poor	Fair to good	Poor to fair	Excellent	Fair to good	Good	Excellent	Excellent	Excellent
General chemicals	Good	Good	Good	Good	Good	Fair to good	Good	Excellent	Excellent	Fair to good	Good	Excellent

*Styrene butadiene polymer (SBR) has properties very similar to natural rubber.
**Good to poor depending on requirements and compounding.
†Trademarks of Gates Rubber Co.

hoses, which can cost anywhere from \$3 to \$45 per foot. If the hose is suspended, it should be well supported with minimum stress on the couplings.

Inspection. All acid-chemical hoses should be inspected on a regular basis:

Daily or after each usage, check the hose and coupling assembly for evidence of external damage.

Every 30 to 90 days, depending on usage, pressure-test the hose and coupling assembly.

Any hose assembly showing evidence of external damage during inspection, or leakage during pressure testing, should be removed from service. Inspections must be made by one person who is responsible for this function. Daily inspections should be made of each hose in service. Lay out hose full length in a clean, dry area. Examine the hose cover for blistering, excessive abrasion or cuts.

Cuts that damage or expose the reinforcement are cause for replacing the hose. Abrasion that exposes the reinforcement is also cause for hose replacement. A blistered or loose outer cover, likewise. Also check out hose couplings, and check the hose cover for 18 in. from the coupling; if soft spots are found, replace the hose and re-couple. Any evidence of coupling slippage is also cause for hose replacement or recoupling.

Storage. New hose should be stored in an area protected from direct contact with the weather. It should be kept in the original shipping carton, preferably on pallets. Heavy objects should not be placed on the hose. Between periods of usage, flush hose, and coil where it cannot be damaged. Take caution to store hoses away from areas where they will be exposed to high temperature or concentrations of ozone.

Physical and chemical properties

The ratings in the tables are intended as guides only. They are compiled from the best data available, but, where possible, tests should be performed under actual application conditions. Ratings are based on a temperature of 70° F and 100% concentration or saturated solutions unless otherwise noted. Properties are subject to a considerable amount of control through compounding, so characteristics have been generalized. When available, the ASTM designation (from D141B) has been given in parenthesis.

The author

Ray Gallagher is a senior product development engineer in the general hose technical department of The Gates Rubber Co., 999 South Broadway, Denver, CO 80217, telephone (303) 744-4041. He is responsible for the design, building and testing, cost analysis, and production capability functions of Gates industrial hoses. He has 28 years experience in the rubber manufacturing industry and is a former U.S. representative to the International Standards Organization Committee on Hose; past member of the Rubber Div. of the Amer. Chemical Soc.; and former member of the Technical Committee of the Rubber Manufacturers Association.

Choosing plastic pumps

Plastics—especially fluoropolymers—have exceptional corrosion resistance to a wide range of chemicals, and thus are well suited for making pumps.

Edward A. Margus, *Vanton Pump & Equipment Corp.*

☐ Industrial plastic pumps are often specified when corrosion is a problem. Pumps can be made from a wide variety of engineering plastics, and one of them can usually handle a corrosive fluid in question. Plastics are also used for erosion—many plastics offer erosion resistance superior to metals. And when zero contamination of product streams is needed, plastic is specified. Many plastics perform excellently when subjected to thermal, hydraulic and mechanical shock.

Some engineers dismiss plastic pumps because their mechanical properties are lower than those of common structural metals such as low-carbon steel. However, pump designers can engineer units that take into account the plastic's properties. These pumps are often thought to be not rugged enough, but this is not so. In fact, except for fiberglass-reinforced plastic, the plastics in pumps are unfilled, unmodified, unplasticized, virgin resins, which are high-quality to meet industrial needs.

However, plastic pumps cannot be used under all conditions. Working temperatures cannot be much above 225°F. Above such temperatures, loss of mechanical properties and stress-cracking corrosion may hinder performance. Also, there are limits on flowrate and head. This varies with each manufacturer's line, but for one line the limits are 1,000 gpm and 190 ft.

Plastic pumps used in plants are often centrifugals. Three types are used: horizontal, horizontal self-priming, and sump. Fig. 1 shows a typical horizontal type. The impeller, casing and heavy-sectioned sleeving over the shaft are made of engineering plastics.

Plastic as structural material

To better understand how plastics are used in constructing pumps, we will discuss how plastics behave under load. Not only are these materials weaker than metals, but they behave differently.

Fig. 2 shows a typical stress-strain curve for a low-carbon steel. Up to the proportional limit, stress is proportional to strain; the graph is linear. Designers keep stresses below the proportional limit. Here, the metal springs back to shape once the load is removed.

Plastics do not have an elastic region, and, thus, do not show linear behavior (see Fig. 3). Consequently, there is

Typical plastic horizontal centrifugal pump used to handle corrosive chemicals **Fig. 1**

no fixed relationship between stress and strain, even at low loads. Also, for plastics:

- When the load is released the material will not snap back into its original position. Permanent deflection takes place.
- Under sustained load, the material creeps. Steel does not.
- Temperature greatly affects behavior. For steel, its behavior will not change much below 500°F.

However, as noted, designers have found ways to cope with such behavior in constructing pumps. Much data exists on the behavior of engineering plastics. Nonetheless, owing to the nonlinear behavior of plastics, numbers cannot be plugged into formulas to accomplish a design, as with metals. Therefore, only highly qualified designers have the expertise to create workable pumps.

Materials of construction

Several types of tough, industrial plastics are used to manufacture pumps:

- *Vinyls*—The most common are polyvinyl chloride (PVC) and chlorinated PVC (CPVC). PVC has good chemical resistance and is an excellent choice for service temperatures to 130°F. CPVC has an advantage over PVC in that it withstands temperatures to 225°F.

Service temperatures for pumps should not be confused with those listed in tables by plastics manufacturers. Maximum temperatures listed by manufacturers are those at which the plastic still retains its basic shape but may have lost its mechanical properties. Pump makers are more conservative and only list temperatures at which the plastic can carry the load imposed on it.

- *Polypropylene*—This plastic has one of the lowest densities of the widely produced engineering plastics (its

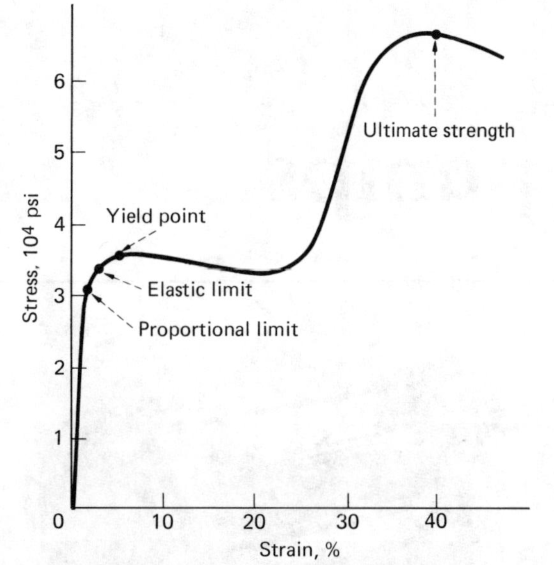

For low-carbon steel and other metals, designers keep stresses below the proportional limit **Fig. 2**

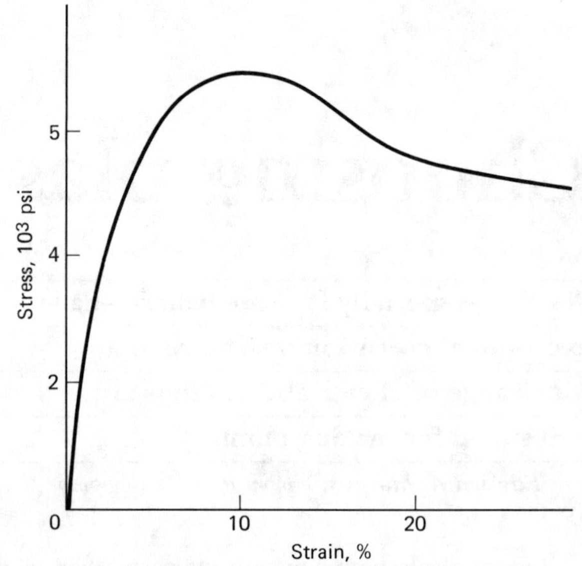

The stress-strain curve for a plastic has no elastic region and no linear behavior **Fig. 3**

specific gravity is 0.90), and a relatively good stiffness. Thus, it has a good strength-to-weight ratio. Polypropylene offers broad chemical resistance to acids, caustics and solvents. It is subject to attack by strong oxidizing agents.

■ *Fluoroplastics*—These plastics have excellent corrosion resistance. Polytetrafluoroethylene (PTFE) is perhaps the most inert compound known. Polyvinylidene fluoride (PVDF) has better mechanical properties than PTFE; it is stronger, stiffer, and less subject to elongation under load. It withstands temperatures to 300°F, and has excellent chemical resistance, even to strong oxidizers.

Two modified PTFEs, created to serve for special tasks and well suited for use in pumps, are now covered.

FTFE offers some improved properties over the virgin resin

	PTFE	FTFE
Compressive creep, %		
78° F, 2,000 psi, 24 h	16	7
500° F, 600 psi, 24 h	30	10.5
Hardness, Rockwell R	15–17	27–29
Compressive stress (73° F, 1% strain)	600	1,200
Lubricated dynamic wear factor, K*	20,000 x 10^{-10}	2–5 x 10^{-10}
Operating temperature, °F	–400 to +500° F	–400 to +500° F
Chemical resistance	Almost inert	Almost inert

*Special company test, where $K = R/(PVt)$
K is a relative number; the higher its value, the faster the wear rate, $(in^3)(min)/(lb)(ft)(h)$
R is radial wear, in.
P is pressure, psi
V is velocity, ft/min
t is time, h

Improved PTFE

1. A proprietary filled PTFE (FTFE)—Rulon, made by Dixon Industries Corp., Flow Control Product Unit (Bristol, R.I.)—designed as a superior material for sleeve bearings. PTFE is not a suitable bearing material unless it is fully contained to prevent cold flow. Properties of PTFE and FTFE are compared in the table.

2. Ethylene-chlorotrifluoroethylene (ECTFE) copolymer—Halar, made by Allied Corp., Plastics Div. (Morristown, N.J.)—which was designed to overcome PTFE's drawbacks in strength and fabricability. (Although PTFE is a thermoplastic, it does not melt enough to be injection-molded. It must be formed by the same methods used for sintering metals, then machined into shape.)

ECTFE was formulated to be an excellent structural material, as well as have excellent coating properties and abrasion resistance. Also, ECTFE was designed to be formed using conventional molding methods. Mechanical properties of PTFE, ECTFE and the filled fluoropolymer are compared in Fig. 4. The properties of PVDF do not appear in the figure, but they will be mentioned.

■ The ultimate tensile strength (Fig. 4a) is the stress at rupture under tension. For FTFE, the manufacturer did not provide a number since sleeve bearings are not loaded under tension. Still, the value for FTFE is lower than for the other two polymers. For PVDF, the value is about 4,500 psi.

■ The flexural (bending) modulus (Fig. 4b) measures stiffness or resistance to bending, such as a load imposed on a cantilever beam. Since no plastic has a stiffness near that of steel, the value of this modulus is critical. It is important because pump impellers and casings are subject to bending stresses (as well as to shear, tension and compression). The modulus for PVDF is the same as that of ECTFE. The modulus for ECTFE is about three times those for PTFE and FTFE.

■ Creep (Fig. 4c) is a measure of stretching under a

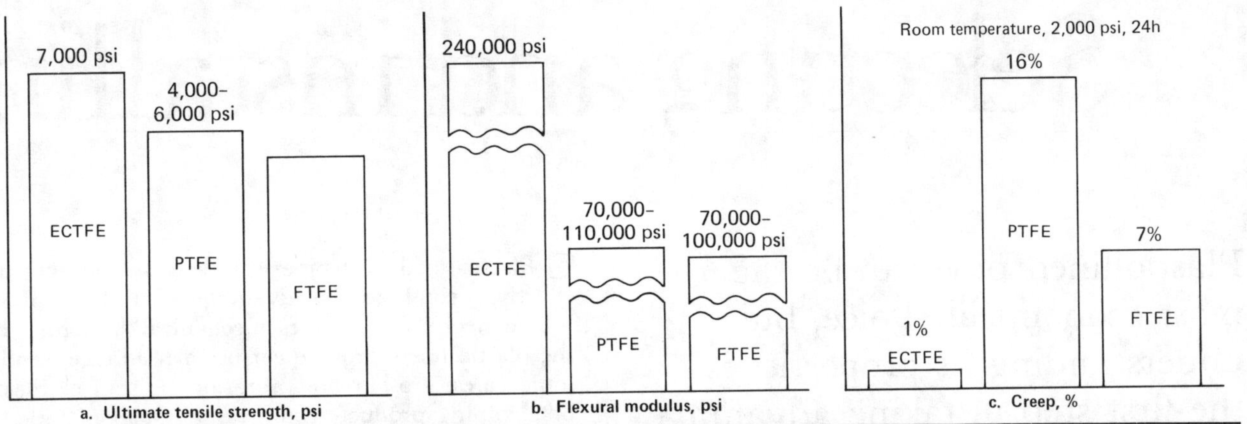

Mechanical properties of some fluoroplastics that are used in making industrial pumps **Fig. 4**

load as a function of time. This is a serious problem for highly loaded structural parts, and ECTFE has a low level of creep—only 1%. (For PVDF, this is about 2%.) While 1% would horrify a bridge designer, a manufacturer of plastic pumps can work around such a figure. Moderate creep is not a serious problem with sleeve bearings if they are properly contained—there is simply no place for the material to go. Pump designers try to work at maximum stress levels below 2,000 psi.

From Fig. 4, ECTFE (with PVDF close behind) looks like a superior material, and FTFE does not. However, when chemical resistance is considered, FTFE looks much better. Fig. 5 shows that ECTFE has, in general, a lower temperature rating than do PTFE and FTFE.

The temperature ranges in Fig. 5 may be high for use in pumps. For gaskets, where the plastic is completely supported and under little stress, temperatures may be realistic. But for use in structures, such as impellers and casings, one pump maker limits the working temperature to 225°F, even for the best plastics.

Further, the chemical resistance of these materials varies. As noted, virgin PTFE is almost inert. In developing FTFE, the manufacturer tried to retain this chemical resistance, and the two plastics have about the same corrosion resistance. ECTFE's chemical resistance is excellent, but it is not quite as inert as PTFE. ECTFE does, however, resist almost every chemical found in industrial service to 225°F. Only a few chemicals can attack this material; among them are dioxane at room temperature. PVDF also offers excellent resistance.

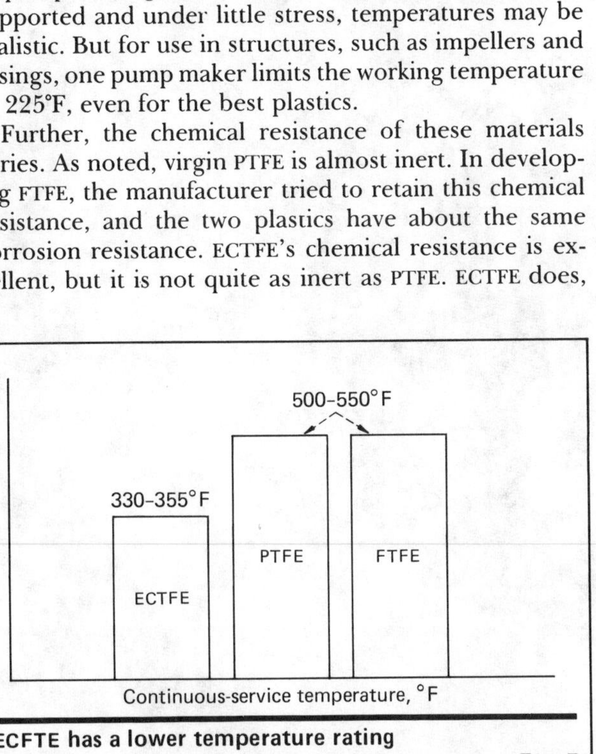

ECFTE has a lower temperature rating than do these two other fluoropolymers **Fig. 5**

ECTFE vs. PVDF

Since both of these are top candidates for plastic pumps in severe service, a brief comparison of their properties is presented:

Chemical resistance—Both offer excellent resistance here, altough neither is as inert as pure PTFE. Particularly at elevated temperatures and high concentrations of corrodents, ECTFE has an edge over PVDF in resisting oxidizing acids and hydroxides. In general, ECTFE may have a slight advantage in terms of working temperatures. PVDF, on the other hand, resists bromine better. Both plastics show some permeability to liquid bromine, but PVDF's absorption rate is about one-tenth that of ECTFE.

Mechanical properties—PVDF is somewhat stiffer under tensile and bending loads, and has slightly higher tensile strength. However, ECTFE has better impact resistance. The coefficient of thermal expansion is about the same for the two fluoroplastics.

Formability—ECTFE may be slightly more versatile, and has the added advantage of producing an exceptionally smooth surface during injection molding. This contributes to its high resistance to abrasion and erosion.

The author

Edward A. Margus is vice-president of engineering for Vanton Pump & Equipment Corp., 201 Sweetland Ave., Hillside, NJ 07205; tel: (201) 688-4216. With the company since 1951, he and other staff members have pioneered the use of industrial-grade plastics for chemical pumps. He serves on the boards of directors of media and natural-resources groups. He has held numerous positions in the machine-tool and tooling industries. Margus holds B.S. and M.S. degrees in mechanical engineering from New Jersey Institute of Technology.

Selecting and installing

Plastic-lined pipe can be the most economical choice, but understanding its properties is the first step in taking advantage of its unique features.

Germán O. Castro, *Dow Chemical U.S.A.*

☐ Plastic-lined steel pipe is a designed product that effectively combines the advantages of steel and plastic.

The steel shell provides mechanical strength while the plastic liners provide corrosion resistance—and in some cases, erosion and abrasion resistance. Plastic-lined piping products comprise a system of pipe, fittings, and valves (see Fig. 1 and 2) specifically designed for handling corrosive liquids and gases commonly encountered in the chemical, petrochemical, oil, pulp and paper, and mining industries. They are also applicable where product purity is a concern, as in the pharmaceutical, water-handling and food-processing industries.

Originally published March 22, 1982

plastic-lined pipe

Summary of applications

Since the introduction of plastic-lined piping in 1946, the number and types of applications have steadily grown. Here is a brief summary of major uses for plastic-lined piping and the industries where they occur.

Chemical and petrochemical industries—There are a multitude of products being handled in the manufacture of organic and inorganic chemicals, pharmaceuticals, plastic resins, and agricultural products. Plastic-lined pipe is used for handling raw materials, in-process and finished-product streams. Liners have been tested with hundreds of chemicals, and recommendations are available from the manufacturers of these products.

Oil and gas industry—Oxygen, carbon dioxide, and hydrogen sulfide in the presence of water cause electrochemical reactions that create pitting and crevice-corrosion in both carbon steel and stainless steel. Also, chlorides, besides causing stress-corrosion cracking, penetrate the oxide coating on stainless steel, causing the pitting problems usually encountered in the oil and gas industry. Natural gases (whether sweet or sour), fresh or salt water, and wet hydrogen sulfide can be handled with plastic-lined steel pipe, as can the treating chemicals commonly used, such as hydrochloric acid, corrosion inhibitors, aromatic solvents, carbon disulfide, toluene and xylene.

Pulp and paper industry—Plastic-lined pipe is used to handle caustic, sodium sulfate, sodium sulfide, sodium silicate, sodium peroxide and sodium phosphates. In bleaching areas, recycling of these agents increases their concentration and hence their corrosiveness. This in turn can result in caustic embrittlement in stainless and carbon steels, and corrosion in carbon steel by sulfuric acid or premature failure of stainless steel due to hypochlorite and chlorine dioxides.

High-purity process streams—There are numerous instances where it is absolutely essential to maintain the purity of a process stream. Some examples are: water-treatment facilities, including skid-mounted units for municipal and laboratory waste-disposal systems; electric-power plants, where contamination of deionized water must be avoided; and in food processing, where contamination also cannot be tolerated.

Plastic-lined pipe vs. alternatives

The performance of plastic-lined piping makes it a logical choice in many situations where conditions render alternative piping systems marginal, if not unsatisfactory. These include installations where plastic pipe failure due to physical impact could create a safety hazard or in applications where the need to field-join fiber-reinforced plastics makes fabrication quality-control and installation difficult. Another instance is the possibility of glass-lined pipe being attacked by hydrofluoric acid or alkaline materials, or where rubber and other plastics could be attacked by solvents.

Table I lists the most widely used plastic liners. Even though we have indicated their maximum temperature limits, the aggressiveness of some chemicals can reduce these upper limits and must be taken into consideration. This will be discussed in more detail along with other critical factors that need to be recognized in selecting an adequate liner and installing lined systems in the field.

The following section suggests the steps needed to determine the best piping system for a particular use. Some of the critical factors may be recognized as not too different from those required for other piping systems. These suggestions are not intended to be all-inclusive. (Refer to Fig. 3 for an outline of these steps.)

Selecting the lining

Chemical service—The first step is a full analysis of the chemicals involved. Is the fluid being conveyed a single

Cutaway of lined spool showing perforated collar and picked pipe to hold lining in place Fig. 1

product or a composite of several products? This analysis should contain the expected concentrations of the major components as well as of trace components. Small quantities, even at parts per million, of a noncompatible fluid can damage a liner.

Most chemical-resistance charts are based on a one-component system, not a mixture of chemicals. Therefore, it is important that all components be considered separately for compatibility with the liner. It is advisable that someone experienced in making judgments regarding the aggressiveness of mixtures analyze the process conditions to make accurate projections as to material suitability. This is one of the many services available from some plastic-lined-pipe manufacturers.

Table II outlines the ASTM standards for the plastics used as liners.

Physical state of the fluid—The state needs to be defined in order to determine the life and the performance of the liner; is it a gas, or a liquid?

Temperature—Temperature in many instances is given as "ambient" when recommendations are requested. Ambient could be $-20°F$ in Michigan during the winter or 75°F in Texas during the same period. Consideration must be given to the effect on the process temperature due to external circumstances (sun effects for example) which may increase the fluid temperature above the maximum recommended for that liner. The heat of reaction, if encountered, must also be considered. If this heat develops, it could also increase the temperature above that recommended for a liner.

Therefore, the operating temperature (its maximum as well as minimum ranges) and the frequency of temperature cycles should be considered in evaluating those conditions that might cause departures from the design operating temperature.

Flowrate—Fluid flowrate and velocity are factors for consideration in selecting the pipe size.

Recommended flow velocities for liquids should be between 4–6 ft/s, while for gases they should be 33–35 ft/s through 3-in. sizes, and up to 83 ft/s in larger diameters. Note that these velocities should be reduced when abrasive particles are present in liquid systems or moisture is present in gas streams.

These velocities may be used to approximate the line size for pressure-drop calculations. The final line size should provide an economical balance between pressure drop and reasonable velocity.

Handling slurries—For slurry handling, the particle size, or mesh, should be evaluated as well as the solids concentration. With such applications, it is advisable that the manufacturer be contacted to obtain test results and data needed for evaluation. These factors will

A sampling of the many sizes and shapes of fittings designed for lined-pipe systems Fig. 2

Recommended temperatures for various liners — Table I

Liner materials		Recommended temperature ratings °F	°C
PVDC	Saran* polyvinylidene chloride	0-175	(−18-79)
PP	Polypropylene	0-225	(−18-107)
PVDF	Kynar† polyvinylidene fluoride	0-275	(−18-135)
PTFE	Polytetrafluoroethylene	−20-500	(−29-260)
FEP	Fluorinated ethylene-propylene	−20-300	(−29-149)
PFA	Perfluoroalkoxy	−20-500	(−29-260)

*Trademark of The Dow Chemical Co.
†Trademark of Pennwalt Corp.

determine the erosion and abrasion parameters that need to be considered.

Tracing—Heat tracing is also an important concern. The temperature of the tracer—whether the source is electric, steam, or Dowtherm* fluids—needs to be monitored in order to minimize temperature excursion. If localized overheating occurs, it may weaken the liner and eventually cause its failure.

*Trademark of The Dow Chemical Co.

Steam cleaning—The method and technique of cleaning the system should also be considered. Where steam is used, care should be exercised to avoid using steam having a temperature higher than the maximum recommended for the liner.

Vacuum

Vacuum effects are another important consideration that cannot be overlooked in the selection of plastic-lined pipe.

Here is the textbook formula* to determine the collapse pressure for unrestrained, uniform, long tubes, supported at their ends:

$$P_{cr} = (2E/(1 - \mu^2))(t^3/d^3)$$

Where:

P_{cr} = collapse pressure (vacuum), psi; E = modulus of elasticity, psi; μ = Poisson's ratio; t = wall thickness of the liner; and d = O.D. of the liner.

This formula indicates that the vacuum needed to collapse a tube decreases as the diameter of the tube increases. As the tube thickness increases, the vacuum needed to collapse the tube becomes higher.

Therefore, it is easy to see that thick plastic liners

*Seely, Fred B., and Smith, James O., "Advanced Mechanics of Materials," Chapter 21 (Buckling of Cylindrical Tubes Under Uniform External Pressure), Wiley, New York, 1952.

PLEASE FILL IN THE BLANKS: ONE FOR EACH PIPE LINE

PIPING:
EXPECTED PIPING SIZE________
EXPECTED LENGTH________

PROJECT NAME________
PROJECT NO. ________ DATE________
COMPANY NAME________
ENGINEER________ PHONE________

EXPECTED QUANTITY OF FITTINGS:
90° ELL______ 45°______ TEE______ CROSS______
OTHER TYPE OF FITTINGS________ QUANTITY________
TYPE OF VALVES________ QUANTITY________

FLUID STREAM: SINGLE PRODUCT?________ OR MIXTURE?________

Composition[A]	Liquid (Yes, No)	Gas (Yes, No)	Concentration % or ppm	Temperature °F (°C)[B]			Pressure psig (Kg/cm^2)	Vacuum—inches[C] of Hg absolute	Flow gpm (liters/sec.)	Velocity ft./sec. (m/sec.)	Specific gravity	Particle size[E] Mesh – % slurry	Steam cleaning?			Tracing[D]			
				Normal	Minimum	Maximum							No	Yes: Steam temperature	Yes: Steam pressure	No	Yes: Method	Yes: Tracing fluid	Yes: Temperature of tracing

Notes: A DO NOT USE BRAND OR GENERIC NAMES – USE THE CORRECT CHEMICAL NAME
B CONSIDER EFFECT OF SUN, HEAT, AND/OR COLD WEATHER
C CHECK YOUR UNITS OF MEASUREMENT
D CHECK MANUFACTURER'S WRITTEN SPECS
E IF AVAILABLE, SUPPLY HARDNESS OF PARTICLE, MESH SIZE, PERCENT SLURRY

Technical aspects to consider when selecting plastic-lined piping products — Fig. 3

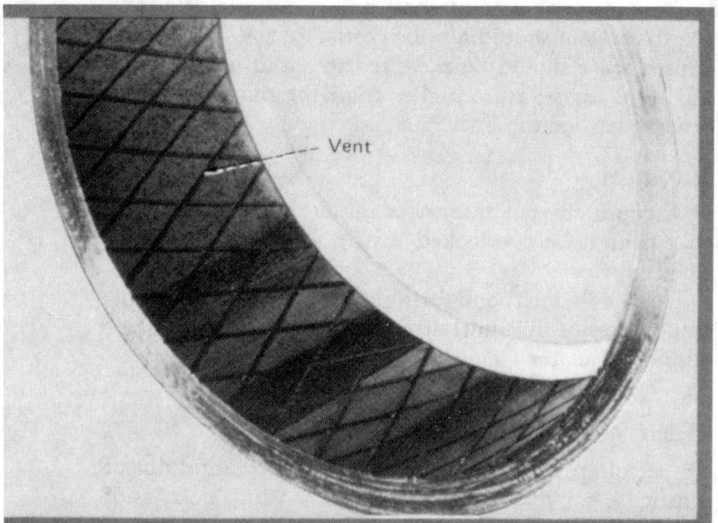

Grooves in steel shell provide a venting path Fig. 4

offer real performance benefits in piping applications involving vacuum.

It is important to note that this formula does not apply to lined fittings or valves, because of the irregularities of shape encountered on these items. Also, there is some question as to how precisely it applies to plastic-lined piping products, since the formula does not account for the restraint provided by the steel shell. Some research is being done in this area; meanwhile, ratings of these items are being determined by qualification tests in the appropriate ASTM standards.

Where possible, the manufacturer should be consulted for recommendations and limitations under vacuum conditions. Problems have been encountered in the field where vacuum develops due to hydrostatic leg effects, or when vacuum develops due to condensation of vapors. The problem of vacuum collapse becomes more severe at elevated temperatures because the modulus of the material is reduced.

Care should be exercised to analyze the system for possible vacuum exposure, even if vacuum is *not* normally expected. A simple matter of shutting down a system or trapping condensibles between two closed valves can lead to liner collapse when the vapors condense and partial vacuum results.

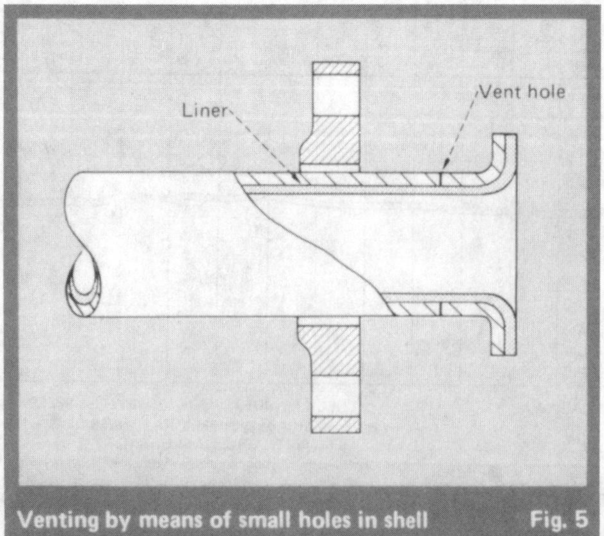

Venting by means of small holes in shell Fig. 5

Permeation

The fluorocarbon resins, being inert to most chemicals and solvents, provide excellent corrosion protection. These resins (PTFE, FEP, and PFA), even though chemically inert, are somewhat susceptible to absorption and permeation.

The vapors of some chemicals have the ability to slowly penetrate and eventually pass through the plastic. However, it is important to note that several factors influence the rate of permeation: a) molecular size of the permeant, b) operating temperature, and c) internal pressure.

Due to the possibility of permeation, it is important that the space between the liner and the steel shell be vented to allow escape of minute quantities of permeant vapors; and to reduce this permeation, a thick liner should be used because the permeation rate is inversely proportional to the thickness of the liner.

Two major venting techniques are in use. One consists of a series of right- and left-handed spiral grooves on the inner steel surface, providing a positive venting path. The small quantities of vapors pass freely through the spiral grooves to the end of the pipe section where they may be safely vented away behind a plated metal reinforcement collar (see Fig. 4).

A second technique involves the making of small holes in the steel shell, allowing the permeants to escape (see Fig. 5). Basic design and methods of fabrication of plastic-lined piping and components can have a major effect on the variables that affect system performance. For this reason, techniques in manufacturing and technical differences among manufacturers should be evaluated before selection of a system.

The steel shell

The plastic should not be expected to function in a structural capacity. This function is carried out by the steel shell. The selection of the metallurgy will depend on the pressure/temperature rating of the material and the corrosiveness of the atmosphere in which the system is being installed.

Piping—The carbon steel shell that houses the pipe liner is usually made of one of the following materials:

ASTM A53, Grades A and B	Pipe, steel, black and hot-dipped zinc-coated, welded and seamless.
ASTM A106, Grades A and B	Seamless carbon-steel pipe for high-temperature service.
ASTM A587	Electric-welded, low-carbon-steel pipe for the chemical process industries.

Castings and forgings—In general, the industry follows the recommendations given by ANSI and ASTM codes and standards for steel, ductile iron, and cast iron forgings and castings. These two voluntary associations provide guidelines for dimensions, material thickness, and classifications according to pressure and temperature ratings—as in the case of ANSI codes for Class 125, 150, and 300. ASTM standards relate to the chemical composition, method of heat treatment, mechanical properties, and methods of testing.

Table III outlines some of both the ANSI and ASTM standards most commonly used, classified according to metallurgy.

Installed-cost comparison

Cost, obviously, is an essential consideration in selecting the most appropriate liners. Material cost alone, however, is not sufficient for an adequate evaluation. The real basis for this is the total cost (combination of material and installation costs). In addition, all costs pertaining to maintenance and product life need to be evaluated as well.

To assist this decision process, a comprehensive report on installed costs of various systems has been published by CHEMICAL ENGINEERING, Nov. 2, 1981, pp. 97–102, based on a study by Dow Chemical U.S.A. entitled "Installed Cost of Corrosion Resistant Pipe." The report covers a variety of metallurgies and plastic-lined/steel systems, detailing both material and labor costs.

Field fabrication

Techniques for field fabrication of plastic-lined piping have progressed to the point that today it is a very simple task.

Basically, only a few steps are required:

A. Cut pipe to proper length.
B. Determine the steel cutback in order to expose the liner.
C. Remove the cut steel ring.
D. Thread the steel pipe.
E. Install flanges.
F. Flare plastic over flanges.

Although this is a simplified outline, all manufacturers can provide full details for field forming and fabrication. They have developed the tools and techniques for ease and excellence in quality of field-fabricated joints (Fig. 6).

Field-installation guidelines

In general, plastic-lined steel piping systems are treated according to standard practices for steel piping. However, the following additional practices should be observed:

Installation—Piping and equipment should be installed so as to avoid vacuum due either to hydrostatic-leg effects or to condensation of vapors. This applies particularly to large-diameter piping systems. If such a situation is encountered, use venting techniques and consult the piping manufacturer for vacuum ratings and recommendations.

Assembly—When assembling, always use a full complement of clean, new machine bolts. Apply bolt lubricant to the full length of the threads. Use manufacturer's recommended bolt torques for each liner type and pipe size. Take care to avoid over-torquing, which can cause damage to the plastic sealing surfaces.

Tighten the flange bolts with a torque wrench, using a "criss cross" pattern that alternately tightens the bolts

ASTM standards for plastic liners — Table II

Standard	ASTM No.
Standard specification for	
Poly(vinylidene chloride) (PVDC), Saran	F 599
Propylene and polypropylene (PP)	F 492
Poly(vinylidene fluoride) (PVDF), Kynar	F 491
Polytetrafluoroethylene (PTFE)	F 423
Perfluoro (ethylene-propylene copolymer) (FEP)	F 546

Standards for forgings and castings — Table III

Steel	ANSI No.	ASTM No.
Steel Flanges and Flanged Fittings	B16.5	
Steel Valves	B16.34	
Carbon Steel Castings Suitable for Fusion Welding for High Temperature Service		A216 (Gr. WCB)
Forgings, Carbon Steel for Piping Components		A105
Piping Fittings of Wrought Carbon Steel and Alloy Steel for Moderate and Elevated Temperatures		A234
Ductile iron		
Ductile Iron Pipe Flanges and Flanged Fittings	B16.42	
Ferritic Ductile Iron Pressure Retaining Castings for Use at Elevated Temperature		A395
Cast iron		
Cast Iron Pipe and Flanges and Flanged Fittings	B16.1	
Gray Iron Castings for Valves, Flanges, and Pipe Fittings		A126

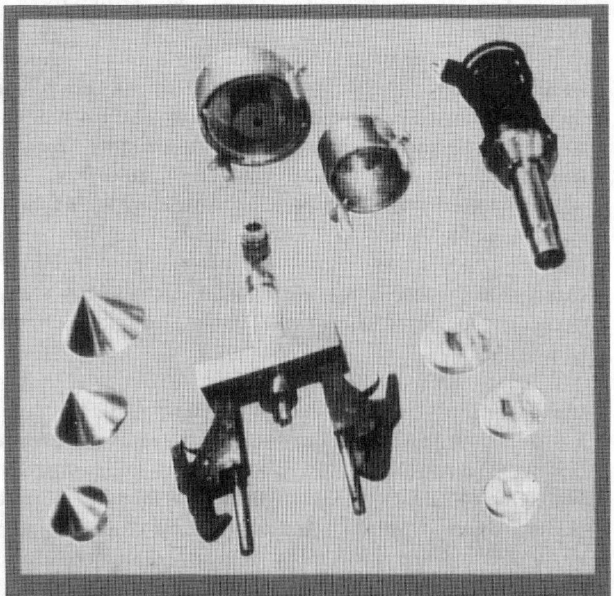

Some tools for field fabrication Fig. 6

located 180° apart. After the first thermal cycle, all bolts should be checked and retorqued if necessary.

In bolting to metals, glass, ceramics, etc., ring spacers or gaskets should be used only according to written specifications from the manufacturers. Gaskets, in general, are not required since the flared plastic face serves this purpose.

Hanger spacing—Hanger spacing for plastic-lined piping is essentially the same as recommended for steel piping. Special attention should be given to the steel wall-thickness (Schedule 40, 30, 20, or 10) because this variable will determine the free span. Supports such as spring-loaded hangers, clevis-type hangers with adjustable rods, guided support shoes, adjustable pipe stanchions, and pipe-roll support may be considered, and should allow freedom of movement resulting from temperature changes.

In addition, for plastic-lined pipe it is recommended that each spool be supported near the flange connection for maximum protection against excessive deflection. Additional support is recommended where flow changes direction and in areas of high load concentrations such as clusters of valves or fittings.

Thermal expansion—Consideration for thermal expansion must be given to the system, and allowances made for expansion and contraction. Where deemed necessary by the engineer or designer, stress analysis should be done in order to avoid overstressing of piping, fittings, and gasket faces. Thermal expansion can be compensated for by using expansion loops or expansion joints.

Handling precautions—To obtain maximum performance, it is important that the plastic on the end-joints be protected from damage during handling and installation. The protective end-cover should be left in place until the pipe is ready for installation. In removing these covers, care should be exercised so as not to damage the plastic sealing faces.

Never make welds on plastic-lined pipe or use it as ground for electric welders, since the heat generated could severely damage the liner. Do not flame-cut plastic-lined pipe.

Painting precautions—Sandblasting and painting should be done with care to protect the raised plastic face. Make sure that protective end-caps are in place at all times if sandblasting is performed before installation. Direct the blast away from the flange face.

If the exterior of the pipe is to be treated with a heat-curable protective coating, exercise caution during the heating process. Never apply heat in excess of the liner's rated maximum temperature limit. In systems where venting is performed by drilled holes, be careful not to plug them.

Heat tracing—Handling materials that have high freezing points or high viscosity often requires the use of a pipe tracing system. Four systems that have been successfully used in tracing plastic-lined pipe are: hot water, heat-transfer fluids such as Dowtherm, electrical resistance, and steam. Any heat-tracing system used with plastic-lined pipe must be applied in accordance with the manufacturer's written recommendation.

Among the factors to be considered in selection of tracing methods are: the temperature rating of the plastic liner for the intended chemical, service, climate, length of heating season, temperature range to be maintained, length of tracer run, initial costs, operating costs, and need for future expansion.

Steam tracing—Steam tracing of Saran, PVDC and polypropylene-lined pipe is not recommended, because it requires much more care in design and maintenance.

However, the tracing of PVDF, FEP, PFA and PTFE-lined pipe with steam may be done successfully, keeping in mind the maximum temperature limitations for these liners.

Electrical tracing—Tracing lined pipe with electrical heaters has been successful in many applications, particularly those in which continuous heating is not needed. Such tracing is available in strip or cable form.

In electrical tracing systems, it is especially important that the temperature controller sensing-device be placed in direct contact with the heating strip or cable, to ensure that the maximum operating temperature of the pipe lining is not exceeded. As with the fluid tracing systems, flanges, fittings, valves, etc., should be protected from excessive heat losses. Refer to manufacturer's literature for specific recommendations.

Liquid tracing systems—Hot water and heat-transfer fluids are generally preferred when there is a need for closer control of operating temperatures or to hold them within the temperature limitations of plastic liners. The choice of heat-transfer fluids will depend on the temperature range.

Insulation—The techniques used in insulating steel pipe can also be used with plastic-lined steel pipe. But, care should be exercised to ensure the integrity of the venting system.

References

1. Atkinson, Harvey E., Fluoroplastic Linings for Corrosives Service, E. I. du Pont de Nemours & Co. Reprinted from *Chem. Eng.*, Dec. 25, 1972, p. 76.
2. Ayres, J. M., Keys to Success in Corrosive Fluid Handling With PLP, The Dow Chemical Co., Bay City, Mich.
3. Kassouf, M., Plastic-Lined Piping Systems for Corrosion Service, Current Developments in Design and Utilization of Plastic-Lined Pipe, The Dow Chemical Co., Bay City, Mich. Presented at the 16th Annual Liberty Bell Corrosion Course, Philadelphia, Pa., 1978.
4. Corrosion Control in Petroleum Production, National Assn. of Corrosion Engineers, Houston, Tex., 1979.
5. Plastic-Lined Piping Systems, Dow Chemical U.S.A., Midland, Mich.
6. Marshall, Stephen P., Design Concepts for Plastic-Lined Steel Piping and Their Impact on Performance, The Dow Chemical Co., Bay City, Mich. Presented at Corrosion 73, National Assn. of Corrosion Engineers, Anaheim, Calif., Mar. 22, 1973.
7. Teflon in Chemical Service, E. I. du Pont de Nemours & Co., Wilmington, Del.
8. *J. of Teflon*, No. 27 and No. 41, E. I. du Pont de Nemours & Co., Wilmington, Del.

The author

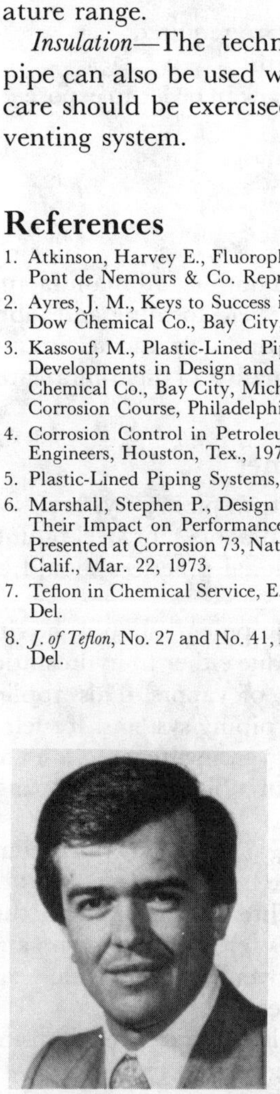

Germán O. Castro is a Project Leader for the Functional Products and Systems Dept.'s R&D organization of Dow Chemical U.S.A., Midland, MI 48640, where he provides technical service and development support for Dow's plastic-lined piping products. He holds a B.S. in mechanical engineering from Florida International University and a master's degree in business administration from Central Michigan University. He is a member of the American Soc. for Testing and Materials and Subgroup F of the ASME/ANSI B31.3 code committee for chemical plant and petroleum refinery piping.

Specifying plastic-lined pipe

Here is a rundown on the kinds of plastic-lined pipe and fittings available, and procedures for specifying a project.

Ralph E. Johnsen, The Heyward-Robinson Co.

☐ With the development of improved materials, plastic-lined pipe is used with increasing frequency and for an increasing range of corrosive flow media. Specifying plastic-lined pipe can be complex. Seven parameters must be considered in making the appropriate selection: Linings, operating pressure (or vacuum), temperature, availability of piping components, cost, installation (field or shop fabrication), and available standards.

Linings

Polypropylene—This plastic, derived from propylene, is resistant to corrosives such as caustics and phosphoric and sulfuric acids.

PVDF (polyvinylidene fluoride)—This is a partially fluorinated plastic that is sold by Pennwalt Corp. under the tradename Kynar. It has excellent resistance to chlorine, bromine and other aggressive chemicals and acids.

FEP (fluorinated ethylene propylene copolymer), PTFE (polytetrafluoroethylene) and PFA (perfluoralkoxy)—These linings are completely fluorinated and therefore have equal resistance to corrodents. They are resistant to almost all corrodents except: fluorine percursors (oxygen difluoride and chlorine trifluoride); elemental fluorine, and molten alkali metals. All three are manufactured by Du Pont and marketed by them as "Teflon." PTFE is also manufactured by I.C.I. and sold as "Fluon" and by Allied Chemical Corp. as "Halar." Specific corrosion data on these plastics are available from the manufacturers.

Each material is effective with a range of corrosive media. Other materials used to line steel pipe include Saran (Dow), CPVC—chlorinated polyvinylchloride (Celanese), and PVC.

Liner thickness can affect performance. The "industry standard" is the minimum allowable thickness specified by ASTM. It is adequate for many corrodents when used as a liner in positive-pressure service. When vacuum conditions are present, thicker linings are advisable. Table I gives the availability of both "industry standard" and "heavy wall" linings.

Kynar Reg. T.M.—Pennwalt Corp.

How pipe is lined

Although many lining materials exist, there are four primary methods of placing the liner within the pipe, varying from one manufacturer to another:

- Pulling the liner into an oversized steel tube, and swaging the steel over the liner. Depending upon the liner material, the interior of the tube can contain tiny picks or flutes to lock the liner in place (used by Dow).
- Pulling an oversized liner through a swaging die into a standard steel pipe. The pipe is submerged in a stabilizing water bath to relax the liner, which decreases liner length while increasing its diameter, causing an interference fit inside the steel pipe (used by Peabody/Doré).
- Wrapping PTFE tape around a mandrel and placing it into a sintering oven. Controlled heat cycles fuse the PTFE into a solid tube for insertion into standard steel pipe (used by Plastic Omnium).
- Pulling an oversized liner through hot sizing dies, cooling it and placing it into a standard steel pipe. Heating in a controlled oven relaxes the liner, causing it to expand inside the pipe for a tight fit. A pipe to be fabricated in the field is treated similarly, except that different heat controls create a liner with a slightly smaller outside diameter than the shop-fabricated liner. The field liner is slid into the pipe (used by Resistoflex).

In practice, the method of manufacture does not seem to significantly affect pipe performance, and each producer makes good-quality pipe.

Operating pressure (or vacuum)

Lined pipe and fittings are available with 125-lb, 150-lb and 300-lb ratings. Because these are not full ANSI ratings, the suitability of the pipe must be verified by the manufacturer.

(Text continues on p. 281)

Originally published March 22, 1982

Liner thickness and pipe lengths available in various materials from four major manufacturers — **Table I**

Manufacturer	Dow																	
Size, in., nominal	1		1½		2		3		4		6		8		10		12	
	Std.	Hvy.	Std.	Hvy.	Std.	Hvy.	Std.	Hvy.	Std.	Hvy.	Std.	Hvy.	Std.	Hvy.	Std.	Hvy.	Std.	Hvy.
PP Lining thickness, in.		0.150		0.160		0.172		0.175		0.207		0.218		0.218	0.180		0.180	
PP Maximum length, ft.		20		20		20		20		20		20		20	10		10	
KYN Lining thickness, in.		0.150		0.160		0.172		0.175		0.207		0.218		0.218				
KYN Maximum length, ft.		20		20		20		20		20		20		20				
TFE Lining thickness, in.		0.120		0.120		0.120		0.120		0.120		0.145		0.150		0.188		0.219
TFE Maximum length, ft.		10		10		10		10		10		10		10		10		10
Minimum length, in.	2 7/16		2 11/16		3 1/16		3 7/16		3 9/16		3 15/16		4 3/16		5		5	

Manufacturer	Peabody/Doré																	
Size, in., nominal	1		1½		2		3				6		8		10		12 – 16[a]	
	Std.	Hvy.	Std.	Hvy.	Std.	Hvy.	Std.	Hvy.	Std.	Hvy.	Std.	Hvy.	Std.	Hvy.	Std.	Hvy.	Std.	Hvy.
PP Lining thickness, in.		0.125		0.160		0.172		0.175		0.207		0.240		0.285		0.285		0.285
PP Maximum length, ft.		20		20		20		20		20		20		20		10[b]		10
KYN Lining thickness, in.	0.075	0.150	0.075	0.160	0.080	0.172	0.090	0.175	0.095	0.207	0.130	0.218	0.130	0.218				
KYN Maximum length, ft.	20	20	20	20	20	20	20	20	20	20	20	20						
FEP Lining thickness, in.	0.075	0.100	0.075	0.120	0.080	0.130	0.090	0.150	0.095	0.160	0.110	0.160	0.130	0.180	0.140	0.180	0.140	0.200
FEP Maximum length, ft.	20	20	20	20	20	20	20	20	20	20	20	20	20	20	10[c]	10[c]	10[c]	10[c]
TFE Lining thickness, in.		0.093		0.093		0.125		0.125		0.125		d		d				
TFE Maximum length, ft.		20		20		20		20		20								
PFA Lining thickness, in.	0.085		0.085		0.085		0.090		0.095		0.130		0.130					
PFA Maximum length, ft.	20		20		20		20		20		20		20					
Minimum length, in.	3 1/4		3 1/2		4		4 1/2		4 1/2		5		5 1/2		6		6 1/4	

Manufacturer	Plastic Omnium																	
Size, in., nominal	1		1½		2		3		4		6		8		10–16		18-20	24-30
	Std.	Hvy.	Std.	Hvy.	Std.	Hvy.	Std.	Hvy.	Std.	Hvy.	Std.	Hvy.	Std.	Hvy.	Std.	Hvy.	Std.	Hvy.
TFE Lining thickness, in		0.115		0.115		0.115		0.120		0.120		0.120		0.140		0.160	0.200	0.250
TFE Maximum length, ft.		10		10		10		10		10		10		10		10	10	10
Minimum length, in.	3 3/16		3 11/16		3 15/16		4 3/16		4 7/16		5 1/2		6		e	e		

Manufacturer	Resistoflex																	
Size, in., nominal	1		1½		2		3		4		6		8		10		12	
	Std.	Hvy.	Std.	Hvy.	Std.	Hvy.	Std.	Hvy.	Std.	Hvy.	Std.	Hvy.	Std.	Hvy.	Std.	Hvy.	Std.	Hvy.
PP Lining thickness, in.		0.150		0.160		0.175		0.175		0.210		0.220		0.220		0.250		
PP Maximum length, ft.		20		20		20		20		20		20		20		10		
KYN Lining thickness, in.		0.100		0.100		0.100		0.100		0.125		0.145		0.160		0.220		
KYN Maximum length, ft.		20		20		20		20		20		20		20		10		
FEP Lining thickness, in.	0.075		0.075		0.080		0.090		0.095		0.130		0.130		0.160			
FEP Maximum length, ft.	20		20		20		20		20		20		10		10			
TFE Lining thickness, in.	0.062	0.125	0.062	0.125	0.062	0.125	0.073	0.125	0.125	0.125	0.140	0.195	0.160	0.250	0.185		0.185	
TFE Maximum length, ft.	15	15	15	15	15	15	15	15	15	15	10	10	10	10	10		10	
Minimum length, ft.	3		3		3 1/4		3 9/16		4 1/16		5 1/2		6 1/2		8 1/2		8 1/2	

Empty boxes indicate unavailable item.

[a] Available through 24 in., not shown in catalog.

[b] 20 ft. available at 150° F.

[c] 20 ft. available at 225° F

[d] On application.

[e] For 10-in. dia. and larger.

Size, in.	Min. length, in.	Size, in.	Min. length, in.
10	6 3/4	18	No min. set as standard
12	8	20	
14	10 1/2	24	
16	10 7/8	30	

Vacuum ratings: Pipe and some fittings (see notes) will take full vacuum to temperatures (°F) shown, unless limited by corrodents **Table II**

Manufacturer	Dow[a]																	
Size, in., nominal	1		1½		2		3		4		6		8		10		12	
	Std.	Hvy.	Std.	Hvy.	Std.	Hvy.	Std.	Hvy.	Std.	Hvy.	Std.	Hvy.	Std.	Hvy.	Std.	Hvy.	Std.	Hvy.
Polypro		225°		225°		225°		225°		225°		225°		225°	NR		NR	
Kynar		275°		275°		275°		275°		275°		275°		275°				
TFE		500°		500°		500°		500°							NR		NR	

Manufacturer	Peabody/Doré[b]																	
Size, in., nominal	1		1½		2		3		4		6		8		10		12–16	
	Std.	Hvy.	Std.	Hvy.	Std.	Hvy.	Std.	Hvy.	Std.	Hvy.	Std.	Hvy.	Std.	Hvy.	Std.	Hvy.	Std.	Hvy.
Polypro		225°		225°		225°		225°		225°		225°		225°		100°		100°
Kynar	275°	275°	250°	275°	225°	275°	200°	275°	175°	275°	150°	275°	100°	275°				
FEP	250°	300°	250°	300°	225°	275°	200°	250°	200°	250°	200°	200°	100°	100°	NR	NR	NR	NR
TFE		500°		500°		500°		500°		275°								
PGA PFA	300°		250°		250°		200°		200°		200°		100°					

Manufacturer	Plastic Omnium[c]																	
Size, in., nominal	1		1½		2		3		4		6		8		10		12–30	
	Std.	Hvy.	Std.	Hvy.	Std.	Hvy.	Std.	Hvy.	Std.	Hvy.	Std.	Hvy.	Std.	Hvy.	Std.	Hvy.	Std.	Hvy.
TFE		450°		450°		450°		450°		300°		70°		NR		NR		NR

Manufacturer	Resistoflex[d]																	
Size, in., nominal	1		1½		2		3		4		6		8		10		12	
	Std.	Hvy.	Std.	Hvy.	Std.	Hvy.	Std.	Hvy.	Std.	Hvy.	Std.	Hvy.	Std.	Hvy.	Std.	Hvy.	Std.	Hvy.
Polypro		200°		200°		200°		200°		200°		200°		200°		200°		
Kynar		275°		275°		275°		275°		275°		180°		180°		70°		
FEP	300°		300°		300°		70°		70°		70°		NR		NR			
TFE	450°	450°	450°	450°	450°	450°	450°	450°	450°	450°	70°	450°	NR	350°	NR		NR	

R = Not recommended for vacuum service.

a. Dow ratings shown are for pipe and fittings. For 4, 6 and 8 in. TFE consult manufacturer.

b. Peabody/Doré ratings shown are for pipe and fittings.

c. Plastic Omnium ratings shown are for pipe and fittings through 3 in. size. For 4 and 6 in. sizes, rating is for pipe only. For larger pipe and fittings consult manufacturer.

d. Resistoflex ratings shown are for pipe and fittings through 4 in. size, and for pipe only for sizes 6 in. and larger. Consult manufacturer for ratings of larger fittings.

Available fittings lined with plastic, from four leading manufacturers — Table III

Item	Dow			Peabody/Doré				Resistoflex†				Plastic Omnium
	Polypro	Kynar	TFE	Polypro	Kynar	FEP	TFE/PFA	Polypro	Kynar	FEP	TFE/PFA	TFE
90° elbow	1 – 10	1 – 8	1 – 12	1 – 16	1 – 8	1 – 16	1 – 6	1 – 10	1 – 10	1 – 10	1 – 12	1 – 30
90° reducing ell	1 – 10	1 – 8	1 – 10	1 – 12	1 – 8	1 – 8	1 – 6					
45° elbow	1 – 10	1 – 8	1 – 10	1 – 16	1 – 8	1 – 16	1 – 6	1 – 10	1 – 10	1 – 10	1 – 12	1 – 30
Tee	1 – 10	1 – 8	1 – 12	1 – 16	1 – 8	1 – 16	1 – 6	1 – 10	1 – 10	1 – 10	1 – 10	1 – 30
Reducing tee	1 – 10	1 – 8	1 – 8	1 – 12	1 – 8	1 – 14	1 – 6	1 – 8	1 – 8	1 – 8	1 – 8	1 – 30*
Cross	1 – 8	1 – 8	1 – 8	1 – 12	1 – 8	1 – 12	1 – 6	1 – 10	1 – 10	1 – 10	1 – 10	1 – 12
Lateral	1 – 12	1 – 8		1 – 12	1 – 4	1 – 8	1 – 6	1 – 8	1 – 8	1 – 8	1 – 8	
Concentric reducer	1 – 12	1 – 8	1 – 12	1 – 12	1 – 8	1 – 12	1 – 6	1 – 10	1 – 10	1 – 10	1 – 10	1 – 14
Eccentric reducer	1 – 12	1 – 8	1 – 8*	1 – 12	1 – 8	1 – 12	1 – 6	1 – 10	1 – 10	1 – 10	1 – 10	1 – 14
Reducing flange	1 – 12	1 – 8	1 – 12	1 – 12	1 – 8	1 – 16	1 – 6	1 – 20	1 – 20	1 – 20	1 – 20	1 – 30
Blind flange	1 – 12	1 – 8	1 – 12	1 – 16	1 – 8		1 – 6	1 – 12	1 – 12	1 – 12	1 – 12	1 – 30
Spacer	1 – 12	1 – 8	1 – 12	1 – 16	1 – 8		1 – 6	1 – 10	1 – 10	1 – 10	1 – 12	1 – 30
Tapered spacer	1 – 12	1 – 8	1 – 12	1 – 16	1 – 8		1 – 6	1 – 8	1 – 8	1 – 8	1 – 8	1 – 30*
Distance piece	1 – 12	1 – 8	1 – 12	1 – 12	1 – 8	1 – 8	1 – 6	1 – 10	1 – 10	1 – 10	1 – 12	1 – 30*
Instrument tee	1 – 8	1 – 8	1 – 12	1 – 12	1 – 8	1 – 16	1 – 6	1 – 10	1 – 10		1 – 10	1 – 30
Strainer	1 – 12	1 – 8		1 – 8	1 – 8	1 – 8	1 – 6					

*Not shown in catalog, but are available.

†Resistoflex makes 45- and 90-deg long radius bends in ½ and ¾-in. sizes, lined with TFE.

General notes: Blank boxes indicate unavailable items; all sizes shown are nominal pipe sizes in inches; reducing fitting sizes are limited—consult manufacturer; sizes 1¼, 2½ and 5 in. are normally not available.

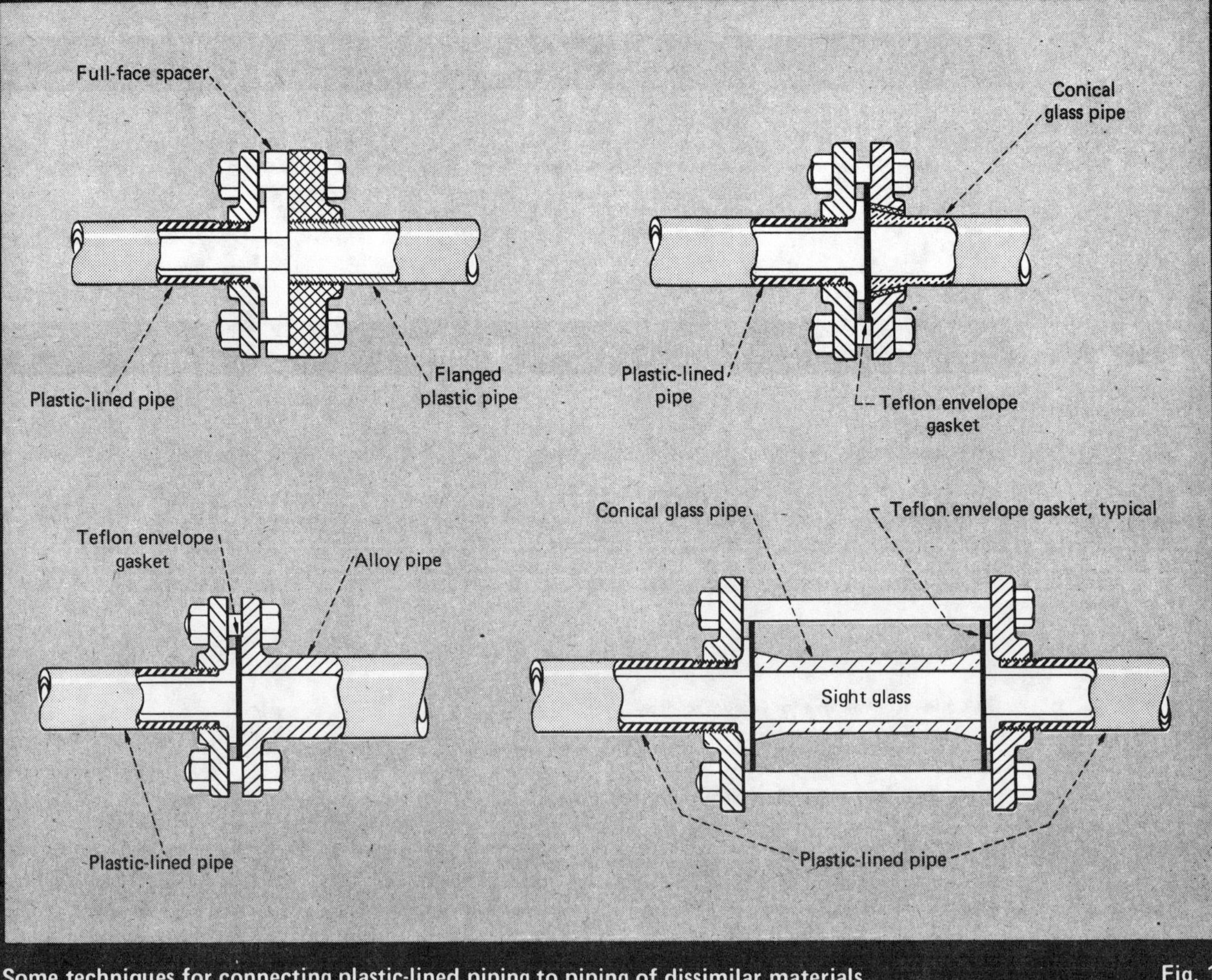

Some techniques for connecting plastic-lined piping to piping of dissimilar materials — Fig. 1

Valves suitable for use with various types of lined pipe and fittings — **Table IV**

Valve type	Manufacturer	Material	Size range, in.	Pressure, psi	Temperature, °F	Full vacuum at °F
Ball	WKM	FEP	1 – 10	150	300	200
Butterfly	Tufline	FEP	4 – 12	150	300	200
		PFA	4 – 12	150	400	200
	Durco	TFE	3 – 12	150	300	75
	Garlock 100	TFE	2 – 42	150	400	400
	Garlock 200	TFE	2 – 8	150	350	350
Check	Peabody/Doré	PP	1 – 8	150	225	225
		Kynar	1 – 8	150	275	275
		FEP	1 – 8	150	300	300
		PFA	1 – 8	150	400	400
	Dow	PP	1 – 8	150	225	225*
		Kynar	1 – 6	150	275	275*
		TFE	1 – 8	150	500	450*
	Tufline	FEP	1 – 3	150	300	250
		PFA	1 – 3	150	400	350
	150 psi	FEP	4 – 8	150	300	250
	150 psi	PFA	4 – 8	150	400	350
	300 psi	FEP	4 – 8	300	300	250
	300 psi	PFA	4 – 8	300	400	350
Clamp	Resistoflex	TFE	1 – 4	150	350	400
			6 & 8	150	350	350
Diaphragm	Grinnell	TFE	½ – 2	150	300	250
		(R2 Diaphragm)	2½	150	300	225
			3	150	300	175
			4	150	300	150
			5	150	300	100
	Hills-McCanna	FEP	1 – 4	150	400	250†
		Kynar	½ – 4	150	300	250*
		PP	½ – 4	150	225	225*
	Dow	PP	1 – 4	150	225	225*
			6	125	225	225*
			8	100	225	225*
		Kynar	1 – 4	150	275	275*
			6	125	275	275*
			8	100	275	275*
Plug (2 way)	Tufline	FEP	½ – 8	150	300	250
		PFA	½ – 8	150	400	360
	300 psi	PFA	½ – 8	300	350	360
	Durco	TFE	1 – 8	150	300	300
	Dow	PP	1 – 2½	150	225	225
		PP	2 – 3	150	225	225
		Kynar	2 – 3	150	275	275
		TFE	1 – 6	150	450	450*
	Peabody/Doré	PP	1 – 6	150	225	225
		Kynar	1 – 6	150	275	275
		FEP	1 – 6	150	300	300
		PFA	1 – 6	150	400	400
Plug (3 way)	Tufline	FEP	1 – 4	150	300	250
		PFA	1 – 4	150	400	350

*Vacuum ratings are for sizes 1 – 3 in.

†Recommend elastomer diaphragm

Vacuum service introduces additional complexity for it may require the writing of two separate specifications, one for vacuum service and the other for non-vacuum service. If a significant amount of non-vacuum pipe is needed, two specifications should be written—however this increases the possibility of misapplying the pipe in the field. If one spec is used for both pressure services, it may limit the competition among suppliers from the standpoints of lining materials and method of application. Regardless of the actual operating pressure in vacuum service, the lined pipe should be designed to withstand full vacuum at the operating temperature. Here, too, the suitability of the pipe should be verified by the supplier. Vacuum ratings published in manufacturers' catalogs generally apply to pipe (Table II). Fittings have poorer ratings in most cases. The manufacturer must verify ratings for valves, expansion joints, etc.

Temperature

The primary factor affecting the selection of plastic-lined piping is temperature. In fact, the pressure rating of a lined pipe is invalid if it is not accompanied by a

Miscellaneous items for use with lined pipe — Table V

Item	Material	Size range, in.
Expansion joint	TFE	½ – 36
Hose	TFE	1 – 8
Dip tube	TFE	1 X ½ – 4 X 3
Sparger	TFE	2 X 1 – 6 X 3
Vacuum breaker	PP, Kynar, FED, PFA	3
Flush bottom valve	PP, Kynar, FEP, PFA	3 – 8
Sight glass	PP, Kynar, PFA	1 – 8
Mixing tee	PP, Kynar, FEP, PFA	2 – 6
Strainer	PP, Kynar	150 & 300 psi, 1 – 8
Jacketed pipe	TFE	1¼ X ½ – 12 X 10
Thermowell	TFE	1 X ½ – 2 X 3
Loading arm	TFE	1 – 4
Spectacle blind	PP, Kynar, PFA	1 – 12
Floor drain	PP	2 – 6
Through-tank connector	PP, Kynar	1½ – 8
Support ell	TFE	1 – 4
300 X 150 psi reducing flange	TFE	1 X ½ – 6 X 4

Piping-material cost ratios — Table VI

Carbon steel (Sch. 40)	0.86
304 S. S. (Sch. 5)	1.41
Saran lined	1.42
Polypropylene lined	1.50
316 S. S. (Sch. 5)	1.61
PVDF (Kynar)	2.01
Armored glass	2.38
FEP Lined	2.45
Alloy 20 (Sch. 5)	2.50
PTFE Lined	2.55
PFA Lined	2.68
Nickel 200 (Sch. 5)	3.23
Glass-lined steel	3.29
Hastelloy C–276 (Sch. 5)	4.25

Taken from a 1978 Dow study. Cost ratio based on a shop-fabricated complex system 4 in. in dia. and 500 ft. in length.

maximum temperature for its use. The maximum temperatures of the more common materials are:

Polypropylene	225°F
PVDF	275°F
FEP	300°F
PTFE/PFA	500°F

The corrosion resistance of polypropylene and PVDF to certain chemicals varies with temperature above 70°F. FEP, PTFE and PFA resist corrosion equally throughout their effective temperature range. Manufacturers' corrosion charts show the variation of resistance to specific corrodents. Reference to at least two different manufacturers' charts or consultation with the manufacturer's technical department for clarification prior to specifying a material is mandatory.

Even within its temperature limitation, a lining is physically weakened at elevated temperature. This can seriously diminish the integrity of the lining under vacuum service. Consult the manufacturer to determine the magnitude of this effect.

Availability of piping components

Many components are available to complete a piping system. Tables III and IV show the lined fittings and the valves available from a number of manufacturers. Other items made for use in lined piping systems are shown in Table V.

In general, the minimum size for all plastic-lined pipe is 1 in., however, ½-in., and ¾-in. PTFE lined pipe are available. Maximum size varies with liner material and manufacturer, as does liner thickness (Table I).

Cost

After the piping materials for a specific service are determined, their costs must be compared. Dow Chemical Company has prepared a comprehensive comparison of the installed cost of 26 corrosion-resistant piping materials. Table VI gives the cost ratios of 14 commonly used piping materials.*

Installation

The installation of plastic-lined pipe is similar to that of flanged carbon steel pipe, but some special precautions must be employed. The flared plastic ends must be protected until the piece is installed. Bolt torques recommended by the manufacturer to guard against liner damage must not be exceeded. Gaskets are not used in bolting two plastic-lined items to each other; however, manufacturer specifications must be followed for gaskets or spacers to join dissimilar materials (Fig. 1).

Plastic-lined pipe can be fabricated either in the shop or in the field. Shop fabrication, the most economical method, is recommended when equipment arrangement is firm and detailed piping drawings are available. In the shop, pipe is cut to exact length, flanges are attached, the liner is installed and its ends are flared. Each item can be "piece marked" to conform to piping drawings showing the location of each piece for field identification.

Field fabrication should be used when equipment locations and piping lengths may not be firm at the time the pipe is purchased. Procedures for field flaring are specified by each manufacturer.

A compromise is available that allows shop fabrication when equipment arrangement is not firm. The line is completely fabricated in the shop with the exception of the closure piece, which is supplied for field fabrication to exact length.

Specification for a project

A piping specification is used by the process engineer to develop piping and instrumentation drawings, by the designer to lay out the piping run and produce detailed drawings, by the purchasing agent for procurement, and by the field engineer to ensure proper installation. Thus the specification must include: the maximum pressure and temperature ratings; list of flow media; the

*Barrett, Orrin H., Installed Cost of Corrosion-Resistant Piping, *Chem. Eng.,* Nov. 2, 1981, pp. 97–102. See also *Chem. Eng.,* Nov. 20, 1978, p. 138.

Job No.: ____________

Date: ____________

INITIAL PIPING SPEC SERVICE INDEX

By Mat'l Department — By Process Department

SPEC DESIGNATION	SERVICE & CODE	MAX. OPER.		SIZE RANGE	MAT'L	REMARKS
		PSIG	°F			

Initial piping-spec service index provides information needed for proper specification of materials **Fig. 2**

complete description of pipe, fittings, valves and other piping items, and installation details. To be certain that a complete specification is achieved, the following procedure has proven useful.

1. The process engineer specifies maximum operating pressure and temperature and the flow media. An "Initial Piping Index" can be useful in supplying this information (Fig. 2). At this point it is necessary to ascertain that all possible media are included, that the operating conditions (pressure, temperature and vacuum) are not minimized or exaggerated, and that the size range is all-inclusive and accurate.
2. Completely describe all pipe and components. Reference the applicable ASTM standards to which all material and tests must conform. Include the size range required. Note the limitations of individual components—e.g., does a valve, hose, expansion joint, or sight glass have a pressure or temperature rating lower than that of the specification?
3. Describe the type of valve required for each function, and reference a particular manufacturer's model and an "approved equal." Do not specify two different types of valves for the same function in the same size range. This is confusing and provides a chance for error.
4. To assist the designer in laying out the pipe and in preparing detailed drawings, provide information giving the maximum and minimun lengths available, the methods of making branch connections, and the connections to dissimilar piping materials. Include instructions regarding the use of reducing tees, instrument tees, gaskets and spacers.

Most frequently, piping components fail because their design capabilities have been exceeded. Know what is required of the components you specify and do not expect more from them than what they have been designed for. Reliable manufacturers of plastic-lined piping will not recommend the use of their products for services and conditions not previously tested.

One last caution: do not write requirements into the specification that exclude otherwise qualified suppliers. This may result in your having only a single source and the loss of the benefits of competitive bidding.

Note

The data contained in this report were primarily obtained from the representatives of Dow, Peabody/Doré, Plastic Omnium and Resistoflex. It is not the intent of this paper to recommend or eliminate any manufacturer of plastic-lined piping as a potential supplier. Other sources for the plastic-lined steel pipe considered are:

Carborundum Corporation—polypropylene, PVDF, FEP, PTFE, PFA lined steel pipe

Ethylene Corporation—TFE lined steel pipe

Fusibond Corporation—polypropylene and PVDF lined steel pipe

The data shown in the tables are to be used only as a guide and should be verified with the manufacturer prior to application.

The author

Ralph E. Johnsen was Chief Materials Engineer for The Heyward-Robinson Co. at the time of writing this article. He is now N.Y. Regional Sales Manager for Herman Goldner Co. of Philadelphia, P.O. Box 917, Bronxville, NY 10708; telephone: (215) 365-5400. He received his degree in mechanical engineering from Westchester Community College, and is a member of the National Assn. of Corrosion Engineers.

Section VII
REFRACTORIES, INSULATION AND CERAMICS

Saving heat energy in refractory-lined equipment

High-temperature equipment operating under severe conditions must use the least amount of energy in order to economically process many materials.

James E. Neal and *Roger S. Clark, Johns-Manville Corp.*

☐ Refractories are vital to processes using great heat, such as those found in the chemical, hydrocarbon, metallurgical and ceramic industries. Originally, refractories were seen merely as a means for containing heat within a given space, but they are now considered critical to the success of an efficient energy-conservation program and a cost-efficient manufacturing process. The proper use of refractories also promotes safety and health for the workers and creates a more comfortable environment overall.

In our discussion, we will provide information on the types and properties of refractories, refractory selection, installation and/or construction, heat-flow theory, and economics.

A basic approach

The traditional approach to insulation is to only consider the amount of heat escaping through the insulation when we should instead be thinking of the amount of heat contained or restricted from flowing.

If we look at insulation in a resistive sense, we cannot help but be impressed by its performance. For example, let us examine:

1. The radiative heat transfer between two radiantly black surfaces, each 1 ft^2. One surface is at 100°F, and the other at the temperature, T, as indicated by the abscissa in Fig. 1. This heat transfer does not include convective and conductive heat flow of the air, both of which are small by comparison.

2. The heat transfer resulting from placing 1 in. of insulation in the form of a refractory fiber having a density of 10 lb/ft^3 between the two plates.

There will be quite a difference in heat transfer for each condition. For instance, suppose we have a hot-face temperature of 2,000°F. In this situation, and with no insulation, we find a radiant heat flow of over 60,000 Btu/(h)(ft^2) from Fig. 1. With insulation, the total heat flow from the hot face to the cold face is about 1,500 Btu/(h)(ft^2). Over 58,000 Btu/(h)(ft^2) has been retained. This is a ratio of 40 to 1.

Obviously, it would be economically as well as practically impossible to operate this system without the insulation. The capacity of the power source that would be necessary to maintain the temperature of 2,000°F would have to be increased 40 to 50 times if insulation were not used.

Insulation-selection criteria

An insulation allowing no heat flow would be ideal, of course. Several insulation systems approach this goal, but they must operate in a vacuum. In most applications, this condition is impractical to realize. Insulation selection is not limited to resistance to heat flow. Other criteria that may be involved are: temperature limitations, thermal-shock resistance, coefficient of expansion, strength, hardness, compressibility, specific heat, erosion resistance, chemical resistance, environmental resistance, space limitations, ease of application, and economics for the various refractory materials.

The choice is generally a compromise governed to a great extent by the cost of insulation and its installation. However, the increasing cost of energy introduces a new parameter, life-cycle cost. This is based on the initial cost plus maintenance and operating costs (in-

Originally published May 4, 1981

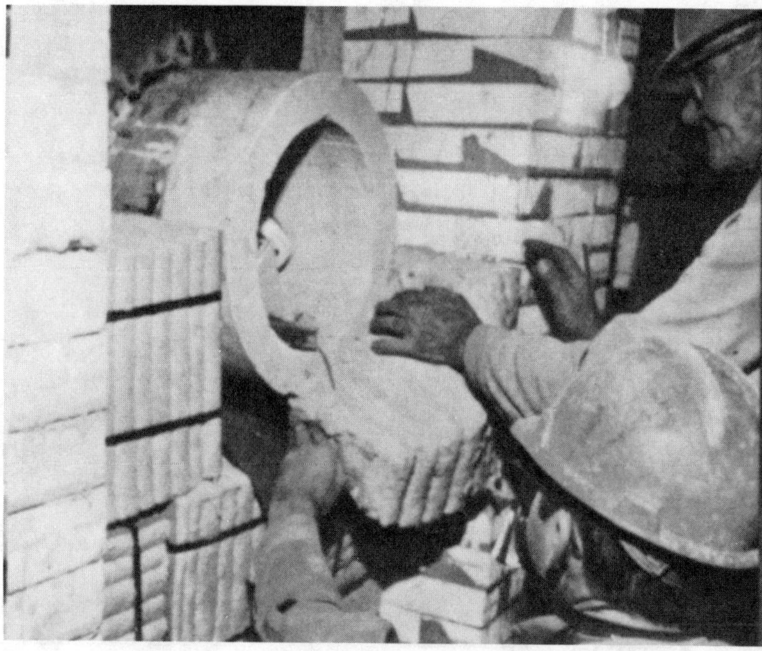

cluding energy cost) for a reasonable life expectancy of the system, rather than the lowest initial cost.

Compromise begins with the initial conception of the refractory or insulating material. The components of the refractory/insulating system are critical parts in this selection.

Types of refractories

The essential properties of the available refractories are shown in Table I. The basic advantages and disadvantages for each material are:

Dense castables, for use in temperatures to 3,300°F, are modern monolithic versions of the age-old fired-clay brick. Their insulating qualities are marginal. However, they have lower thermal conductivity than fired brick, and provide greater strength and lower first cost than true insulating firebrick (IFB).

True insulating firebrick is available to 3,300°F as are insulating castables. In general, these materials weigh substantially less than dense refractories, have much better insulating qualities (being two to four times more resistant to heat transmission), but are not as strong. In most applications, the thicknesses called for by nonstructural considerations provide ample strength.

Refractory fiber products such as felts and blankets are available up to 3,000°F. They have three major advantages: excellent insulating values, ease of application, and low heat storage. Low heat storage is of great advantage in cyclical heat applications, such as for a periodic kiln. Heat absorbed by the walls of such a kiln, and then lost when the kiln cools, can exceed the amount of heat actually used for production.

Calcium silicate, mineral-wool, fiberglass boards and blankets, all at the lower fringe of refractory-temperature ranges, are often used because of their combination of qualities. For example, calcium silicate usually offers structural-strength and high thermal-insulating values, and is not affected by moisture or humidity. Mineral wool and fiberglass generally combine the properties of light weight, heat resistance, low conductivity and high sound absorption.

Castables

Castables, also called refractory concretes, are available in dense and insulating versions. Castables are supplied dry, and are mixed with water before installation. They are installed by pouring, trowelling, and pneumatic gunning, or occasionally, by ramming. As a result, castables give a smooth, practically jointless construction to monolithic hearths, walls and roofs of furnaces and kilns.

Applications of castables in industries or processes using heat vary from aluminum furnaces or reformers for ammonia plants, to zinc furnaces. The growing use of castables stems directly from their advantages. Castables save money during installation because they go into place faster and easier. Mixing, conveying and placement of the unfired castable mixture are often mechanized. Furthermore, castables save time by eliminating the cutting of brick to fit a furnace or kiln, and eliminate the need for inventories of special brick shapes.

Gunning castables into place requires more skill than pouring them, and uses more material, but gunning

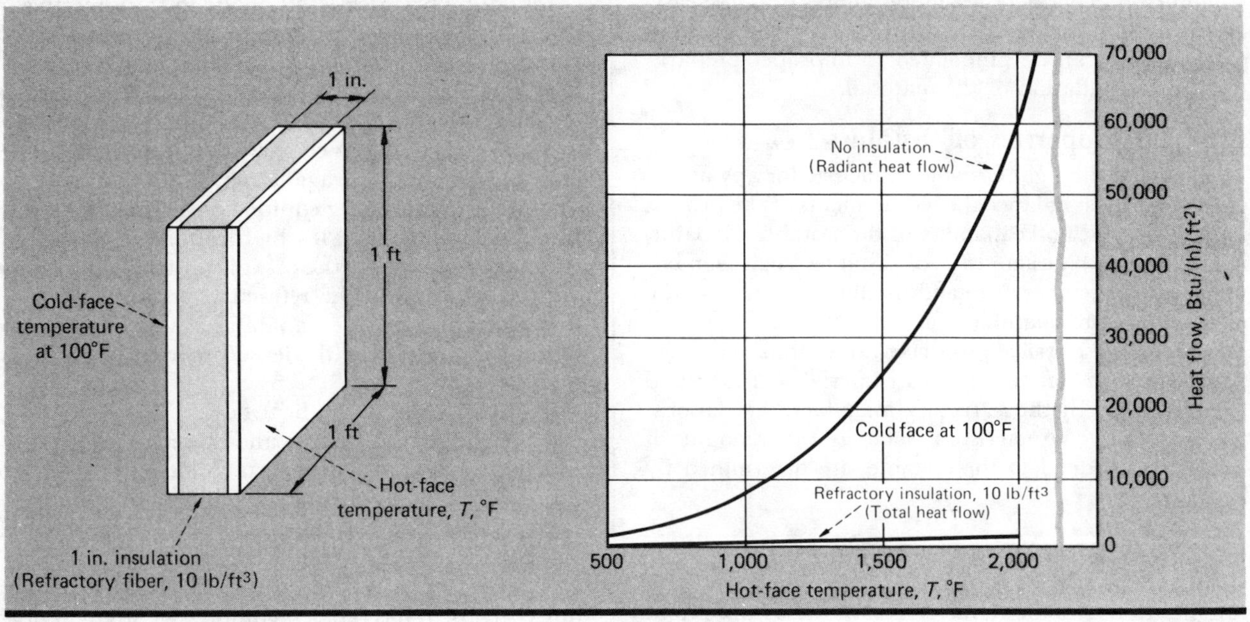

Refractory insulation restricts heat loss **Fig. 1**

applies more material in less time than any other method. It also makes possible the installation of castables to horizontal, vertical or overhead structures without using forms.

Castables also have a number of performance advantages. First, dense castables have low permeability, partly because of the nature of the material and partly because the structure has few joints. As a result, castables are usually the best lining for vessels or chambers operating at other than atmospheric pressures. Second, most castable refractories have good volume stability within the specified temperature range, and many have good resistance to impact and mechanical abuse.

Installation practices for castables may cause premature failure and uneconomic life of the refractory. The reason is that the user or installer of such castables performs most of the ceramic manufacturing processes. The manufacturer of castables simply grinds and mixes raw materials together, then bags and ships them. The user adds water (whose amount and quality are critical to complete the mixing operation), and fires the emplaced castable to the necessary refractory quality.

The amount of water required for casting is probably the most commonly neglected variable associated with castable installation. Some castables are more sensitive to water requirements than others. Too much or too little water can produce poor physical characteristics, such as poor strength or poor abrasion resistance.

Refractories for use in fired heaters **Table I**

	Mineral-wool block	Calcium-silicate block	Haydite, vermiculite	Insulating castables	Dense castables	Insulating firebrick	Dense firebrick	Refractory-fiber boards and shapes	Refractory-fiber blankets	Refractory-fiber modules
Maxium operating temperature, °F	1,500	2,000	1,800	2,800	3,300	3,200	3,300	2,600	2,600	2,600
Thermal insulation	1	2	3	3	4	3	4	2	1	1
Heat storage	1	1	2	3	4	2	4	1	1	1
Thermal-shock resistance	4	4	2	2	1	2	1	1	1	1
Chemical resistance	4	4	4	2	1	2	1	3	3	3
Erosion resistance	4	3	3	2	1	2	1	3	4	3
Strength	4	3	2	2	1	2	1	3	4	2
Resiliency	2	3	4	4	4	4	4	2	1	1
Acoustical insulation	2	3	3	3	4	3	4	2	1	2
Installed cost	Low	Medium	Low	Medium	Medium	Medium	Medium	High	Medium	High

Code: 1 - Excellent, 2 - Good, 3 - Fair, 4 - Poor.

Unfortunately, there are more failures of dense-castable installations than there should be. In practically all cases, failure can be pinpointed to improper preparation and installation of the material.

Physical properties of castables

To choose the right castable refractory for any application, we must balance the requirements of the process against the cost and capability of the castable. Capabilities and other characteristics of castables, such as maximum service temperature and modulus of rupture, are obtainable from manufacturers' data sheets. Test procedures for these physical properties are established by the American Soc. for Testing and Materials (ASTM), and are conducted by the refractory-manufacturers' laboratories. Other characteristics, such as the amount of water to be added to the castable, are determined by actual tests.

Chemical properties of castables

In refractory compositions, the alumina-silica ratio roughly indicates the refractoriness of the composition: the higher the alumina (Al_2O_3) content, the higher the refractoriness. Lime (CaO) in the composition appears as a result of the binder, and also contributes to the ultimate refractoriness.

Iron oxide (Fe_2O_3) becomes important in refractory compositions when the process being carried out in the refractory-lined vessel involves a reducing atmosphere. For example, carbon monoxide in the vessel's atmosphere at certain temperatures can react with the iron oxide to deposit carbon within the lining. If the reaction continues, the carbon will cause the lining to crack. Likewise, a hydrogen atmosphere will reduce Fe_2O_3.

Thermal expansion of castables

Unlike brick, castables are not fired before installation. Only after prolonged exposure to heat do castables acquire the reversible expansion and contraction characteristics found in fixed ceramics. Because of this, castables show the results of complex forces working on them during the initial heatup. The fired aggregate expands according to its chemical composition, the cement phase shrinks as it loses the water of hydration, and the castable itself expands or contracts as it sinters and mineralogical reactions occur.

Most of what we know about the thermal expansion of refractories is based on brick. This knowledge is difficult to transfer to castables. From an application standpoint, two factors are important:

- Fireclay castables do not normally require an expansion allowance.
- Certain dense, highly refractory castables have substantial reversible thermal expansion, up to 2,000°F or more. Castables of this type may require an expansion allowance in massive installations.

Insulating firebrick

The objective of any insulation manufacturing process is to create a product of great porosity. After all, only dead-air spaces insulate. The surrounding material is useless in the insulating sense, and it is there only to provide structural strength.

Insulating firebrick (IFB), made to this objective, reduces heat loss by two thermal mechanisms. As compared with dense refractories, IFB has high resistance to heat flow and, being lightweight, stores very little heat. Initially, IFB was applied to cyclically operated equipment, allowing energy conservation from reduced heat flow and low heat storage. Later, IFB was used in continuous heat-processing equipment such as tunnel kilns. In such equipment, IFB's high insulating ability conserves fuel and lowers the cost of construction by reducing the thickness of the refractory system.

Let us compare the capabilities and limitations of insulating firebrick with dense firebrick. We will find that IFB is:

- Highly resistant to heat flow.
- Very light in weight, and stores much less heat. The light weight expedites installation and bricklaying. Moreover, the supporting structure and foundations can be lighter and less expensive. This weight reduction can be as much as 80%.
- Structurally self-supporting at elevated temperatures because it has the best compressive strength of any insulating refractory. IFB is compatible with dense firebrick to firm up the whole construction.
- Machined to finished shape. It has tighter tolerance than dense fireclay brick. This provides a tighter refractory construction, less heat loss through the joints, and faster installation.
- Low in impurities that can adversely affect refractory performance—for example, a furnace atmosphere having hydrogen as one of its constituents.
- Available in a large variety of shapes, both as supplied by the manufacturer and as made from slabs without mortared joints.
- Readily cut or sawed on the job.

On the other hand, we find IFB to be:

- Typically lower in strength than dense fireclay brick. This will influence the design of large structures.
- Prone to hot-load deformation failure from heavy loads at high temperatures.
- Generally not good for abrasion resistance. For example, it should not be used in a flue that would have high velocities and/or abrasive matter present.
- More prone to thermal and mechanical spalling than most dense refractories. This prohibits its use in some very abusive applications. For example, the slab-and-billet reheat furnace found in rolling mills.
- Incompatible with a chemical environment that will flux an alumina-silica refractory body.

Shapes and sizes

IFB is available in almost all the standard refractory shapes in addition to special shapes such as arch, key-wedge, and circle brick, and in nonstandard thicknesses (obtained by mortaring two or more pieces).

The classification by ASTM (C 155-70), as shown in Table II, groups the brick in accordance with bulk density (a good indication of thermal conductivity) and the behavior of the reheat change test.

The reheat change test for each group is conducted at a temperature 50°F below the normal maximum recommended service temperature. For example, Group 23 brick (ordinarily referred to as 2,300°F brick, and recommended for this temperature) is tested at 2,250°F. A maximum of 2% linear shrinkage is allowed.

If precise operating conditions are known, fairly exact factors of safety can be determined. This is seldom the case. Hence, a rule of thumb has been developed by refractory users that a 200°F safety factor will be allowed between the furnace design temperature and published temperature-use limits.

For example, a Group 23 brick would not be used for a furnace having a design temperature above 2,100°F. This rule has evolved through long experience. Its rationale is the difficulty of operating heating equipment at a precise design temperature for long periods of time. During that period, we can expect temperature variations within the furnace, and it is almost inevitable that an upset condition will occur where the design temperature will be exceeded.

Excellent information on construction methods is published by refractory manufacturers for their products, and is equally applicable to IFB. These handbooks include design shortcuts, tables of brick shapes, calculation methods, and many details for building and supporting refractory structures.

Brick equivalents

Because refractory brick differs in size from products such as building brick, the refractories industry has developed a term to describe what is meant by a "brick." This term is a "brick equivalent" (BE), and the brick having the dimensions of 9 in. by 4½ in. by 2½ in. is defined as one brick equivalent. Thus, a brick having dimensions of 9 in. × 4½ in. × 3 in. is 1.20 BEs.

To illustrate the concept of BE, let us consider a furnace wall that might require 1,200 BEs. The wall could be built from twelve hundred 9 × 4½ × 2½ bricks, or one thousand 9 × 4½ × 3 bricks. Or, it could be constructed from four hundred 11¼ × 9 × 3 bricks because this brick contains 3.00 BEs.

The idea of brick equivalents is useful in estimating the brick requirements. Calculations can be done first in brick equivalents, and then converted into the exact number of pieces of each size desired. The information in Table III is helpful in estimating quantities. To calculate the number of BEs needed for walls of other thicknesses, use multiples of the values given in Table III. These same values can be applied to curved walls by using the outside wall dimensions.

Classifications for insulating firebrick — **Table II**

Group	Reheat* change, °F	Bulk density,† lb/ft^3
16	1,550	34
20	1,950	40
23	2,250	48
26	2,550	54
28	2,750	60
30	2,950	68
32	3,150	95
33	3,250	95

Source: ASTM C 155-70
*Not more than 2% when tested at indicated temperature.
†Not greater than indicated density.

Mortar and mortaring methods

An incorrect selection of mortar not only slows down bricklaying, but can compromise the strength and life of the resultant structure. The mortar must have good water-retention properties, and be a strong air-setting one. Good air-setting properties give the refractory wall proper strength and stability. Brick not strongly bonded will loosen and bulge, making it necessary to rebuild the lining. Furthermore, IFB is most frequently used in furnaces where the temperatures are not high enough to develop a good bond in a heat-setting mortar. Even in furnaces operated as high as 2,600°F, the temperature drop through the IFB is so great that sufficient temperature for a ceramic bond may exist only in the first ½ to 1 in. of the hot face of the brick. The remainder of the wall will not be bonded.

The most common and best mortaring practice is to use "dipped joints." The mortar is thinned to a creamy consistency, just pourable, and the brick dipped into this slurry before being placed in the wall. Mortar supplied wet to trowel consistency will commonly require additional water to reach this creamy consistency. One precaution in dipping is to minimize the amount of mortar deposited on the surface of the brick that will become the hot face of the furnace lining. This coated face is prone to spalling.

Troweled joints are often used but should be discouraged because they require a degree of skill mastered by only a few brickmasons. However, troweling is often used on crowns.

Insulating aggregates

Insulating aggregates are a comparatively unfinished form of refractory. There are two common types: expanded and burn-out:

Expanded aggregate is made by expanding a clay in such a manner that aggregate particles are formed having internal porosity and light weight. Such aggregates have very fine insulating properties. However, expanded insulating aggregates are not readily available. Most producers use their production in-house. An exception is haydite, a clay that bloats upon being heated. It is a low-temperature refractory of limited use. A common application involves mixing it with calcium aluminate cement and vermiculite, and gunning the resulting castable into low-temperature petrochemical heaters.

Burn-out aggregate is made by the burn-out IFB processes. Both reject-brick and dobies [molded blocks of ground clay or refractory materials, crudely formed, and fired] are crushed into aggregate. Various temperature grades are available. The crushed aggregate will often have excessive fines that can detract from its insulating value.

Insulating aggregates are used as fills in places such as tunnel-kiln roofs, flat suspended arches, and car-bottom furnace floors. The major use of the insulating aggregates is in the form of castables in which the aggregate is combined with calcium aluminate cements. Refractory manufacturers formulate these castables into numerous combinations of aggregate sizing, aggregate temperature grades, and types of calcium aluminate cement.

Refractory fibers

Many insulating fibers will block radiation, stop convective currents, and still have a minimum amount of thermal conduction. For high-temperature applications (1,000 to 3,000°F), fibers classified as ceramic or refractory are used. In general, these refractory fibers are considered in four broad categories:

- High-silica-leached and fired-glass fibers.
- Flame-attenuated pure silica fibers.
- Alumina-silica fibers.
- Pure metal-oxide fibers.

The silica fibers have their major application in the aerospace industry. They are very expensive and currently have limited use in the industrial field. Chemically produced metal-oxide fibers are also quite expensive, but their high-temperature capabilities are meeting some very critical needs. For example, pure alumina fibers can be used for insulations up to 3,000°F.

The bulk of the refractory fibers in use are the alumina-silica ones because of their relatively low cost and good thermal properties. Our discussion will primarily deal with alumina-silica fibers, but the principles generally apply to the other fiber types. The easiest way to understand the differences between refractory fibers is to examine their temperature limits (see Table IV).

Characteristics of refractory fibers

The features most important for standard applications are fiber diameter and thermal stability. Fiber diameter is influenced by complex manufacturing considerations. The alumina-silica fibers have average diameters ranging from 2.2 to 3.5 microns. The finer fiber produces a lower thermal conductivity at light density and high mean temperature, but has a marginal effect at higher densities.

The thermal stability of refractory fibers is, in most cases, the more important characteristic. The maximum temperature limit of any fiber is set by the manufacturer to indicate the point at which the fiber becomes thermally unstable. A phenomenon known as devitrification begins to set in at approximately 1,800°F. This crystal-forming process is responsible for shrinkage, loss of strength, and general physical degradation that occurs in such fibers at high temperature. It is a time/temperature phenomenon that rapidly accelerates at 2,300 to 2,400°F.

Let us consider the amount of shrinkage at 2,300°F. Products from different manufacturers, all rated for 2,300°F, have linear shrinkages ranging from 3.5 to 8.8%. Obviously, the thermal stability of the product depends on the amount of shrinkage. Hence, a user of refractory fibers should be aware of the difference in shrinkage of the various refractory fibers and should determine the appropriate design criteria for acceptability. The most efficient insulating product is of no value if it does not have structural integrity.

No matter which fiber is chosen, there are several characteristics inherent to all. Thermal conductivity of refractory fiber is very low compared with brick and castable refractories at the same temperature range. Low density allows a lighter steel-support structure in furnace design. Low heat storage means faster heating and cooling time in periodic kilns. Resiliency of the fiber allows it to be compressed and be packed in various areas, and also makes it resistant to mechanical shock and vibration.

Refractory fibers are completely resistant to thermal shock. They are used in severe applications such as the furnace doors of heat-processing equipment. The fibers are chemically stable except for certain acids and strong alkalies that attack them. After wetting with oil or water, the properties will return when the product is cleaned and dried.

Most products made from refractory fibers use a high percentage of fiber. Hence, the characteristics for the fibers generally apply to finished products.

Health and safety

The airborne particulates generated by handling a refractory fiber are classified as a nuisance dust that does not produce a significant toxic effect. The fiber can cause throat irritation if inhaled, and mechanical skin irritation on contact. After working with the fibers for a period of time, most people no longer react to the skin-irritating effects. It is recommended that workers exposed to the dust wear long sleeves, gloves and safety goggles. Respirators are also recommended where airborne concentrations exceed the limit for a nuisance dust.

Refractory-fiber products

Refractory fibers are processed into products such as felts, blankets, vacuum-formed shapes, sprays and paper. Almost all of these products are available for use from 1,600 to 2,600°F.

The basic material is the refractory fiber in bulk form. It is inorganic and can operate continuously at

Brick equivalents for different wall thicknesses — Table III

Wall thicknesses, in.	Brick equivalents, BE/ft^2
2½	3.6
4½	6.4
9	12.8

1 ft^3 = 17.07 BE
BE = 101.25 in.^{3}

Temperature limits for refractory fibers — Table IV

Refractory fiber	Temperature range, °F
Fluxed alumina-silica	1,600 to 1,800
Alumina-silica	2,300 to 2,400
Modified alumina-silica	2,600 to 2,700
Pure metal oxides	3,000 to 3,100

the maximum recommended service temperature. In single-use applications, the fiber can be taken well above this temperature limit.

There are three varieties of bulk fiber: cleaned or uncleaned, lubricated or unlubricated, and long or short length, or a combination of these.

Uncleaned fiber still has the "shot" or unfiberized material in it, and normally has the longest fiber length. It is not reprocessed before being used as packing or loose fill in expansion joints, furnace-crown cavities, or wall cavities. The long fiber length tends to tie the bulk material together. Lubrication added to bulk fiber makes handling easier. This uncleaned fiber is the most inexpensive type.

Cleaned fiber has been processed in order to separate the shot from it. Certain applications such as aerospace and paper making cannot tolerate a large amount of unfiberized material. The process of shot separation also tends to break the fiber into shorter lengths.

Lubricated fibers contain an organic lubricant that is added in the fiberizing process to give the fibers a thin coating. This allows them to be more easily worked when further handling is required, such as in packing joints. The lubrication allows the fibers to slip on one another, and also significantly reduces dusting. However, a lubricant must not be used on fiber intended for wet mixing since such fiber will not properly disperse in water.

All bulk forms of refractory fiber are highly resilient and are often used to reduce brick cutting in the field, to make emergency repairs, and to pack void spaces in refractory construction. Bulk-fiber density will range from a loose fill of 6 lb/ft^3 to a maximum for packed material of about 12 lb/ft^3. Packing to higher densities will cause the fibers to break up and lose resiliency.

Refractory-fiber felts and blankets

Refractory fibers in the form of felts and blankets are used as linings for furnaces, kilns, and other high-temperature equipment.

Felts are organically bonded mats in roll or sheet form. The product has good cold strength and is used when a semirigid insulating sheet is required. Upon firing, however, the phenolic binder burns out. This causes two problems. First, the binder emits an objectionable odor and irritating fumes. Second, the fired product without the binder is quite weak, which can be serious if strength is required for the material to stay in place.

High-density felts (10 lb/ft^3 or higher) are formed in the same manner as regular-density felts, but the binder is cured in a pressure press to achieve felt densities of up to 24 lb/ft^3. Such high-density felts are one answer where a lower-thermal-conductivity felt of exceptional strength is needed or where space (i.e., thickness) for a felt is limited.

Blankets are completely inorganic (i.e., no binder is used with the refractory fiber). Originally, the blankets were not "needled." In one process, the refractory-fiber mat was sent directly from the collection chamber through a set of heated compression rolls in order to heat-treat the mat into a blanket. In another process, the fiber was water-washed to remove the shot, and then refelted into a blanket. With either process, the resulting product was weak. The product from the first process was subject to laminar failure when flexed, while the weakness in the product from the second process was due to the short fibers.

The un-needled blanket is being replaced by an inorganic needled product. Here, the mat is run from the collection chamber through a needling machine. The needler has thousands of barbed needles that move up and down through the blanket. As the needles penetrate, the barbs hook some of the fibers and pull them through the blanket, binding the blanket together. Needling laces the blanket together with its own fibers. As these fibers become interlocked, the product becomes an integral unit having excellent flexibility. After needling, the blanket is heat-treated to relax the fibers and bring them into intimate contact with each other.

After firing, the strength of the blanket is as great as before because the interlocking mechanism is not affected by heat. This is very important when lining furnaces and kilns where the anchor spacing for the blanket may be as wide as 18 in.

Needled blankets have many applications such as (a) removable insulating blankets for turbines where the blankets must be flexible and resistant to physical abuse, (b) insulation wraps for investment-casting molds and for stress-relieving field welds, and (c) high-temperature sound-absorption systems. Since blanket properties such as strength vary with the manufacturer, detailed product performance data should be studied.

Another material made predominantly of refractory fiber is a board for expansion joints in brick constructions. This board is a refractory felt having a high phenolic-binder content. As brick structures are laid up, a space must be provided for the brick to expand, but this space should also be packed so heat will not escape.

The standard practice is to carefully gauge the joint spacing, and build a free-standing wall away from the first wall. This is tedious and time-consuming, and requires the stuffing of the space at a later time if insulation is desired. An expansion-joint board provides the necessary spacing in this type of construction, as well as the fiber to fill the joint. It is rigid enough to build against, yet resilient enough to spring back and refill the joint when the structure cools. Much time and installation cost can be saved by using such board.

A thin and flexible paper is made from refractory fibers in thicknesses ranging from $\frac{1}{32}$ to $\frac{1}{8}$ in. The fiber is held in a latex-binder matrix. This paper is expensive but its qualities are desired in high-temperature gaskets, linings for combustion chambers, and expansion-joint materials for molten-metal troughs. It can also be used for thermal insulation wherever small clearances are encountered.

Applications for refractory fibers

Perhaps the largest single application of refractory fibers is for the lining of furnaces and kilns. Table V lists the several categories of furnaces that use these fibers as the primary insulation in the roof and walls. Since service conditions for each application vary, different types of fibers will be required. For lining heating equipment with refractory-fiber materials, there are two basic

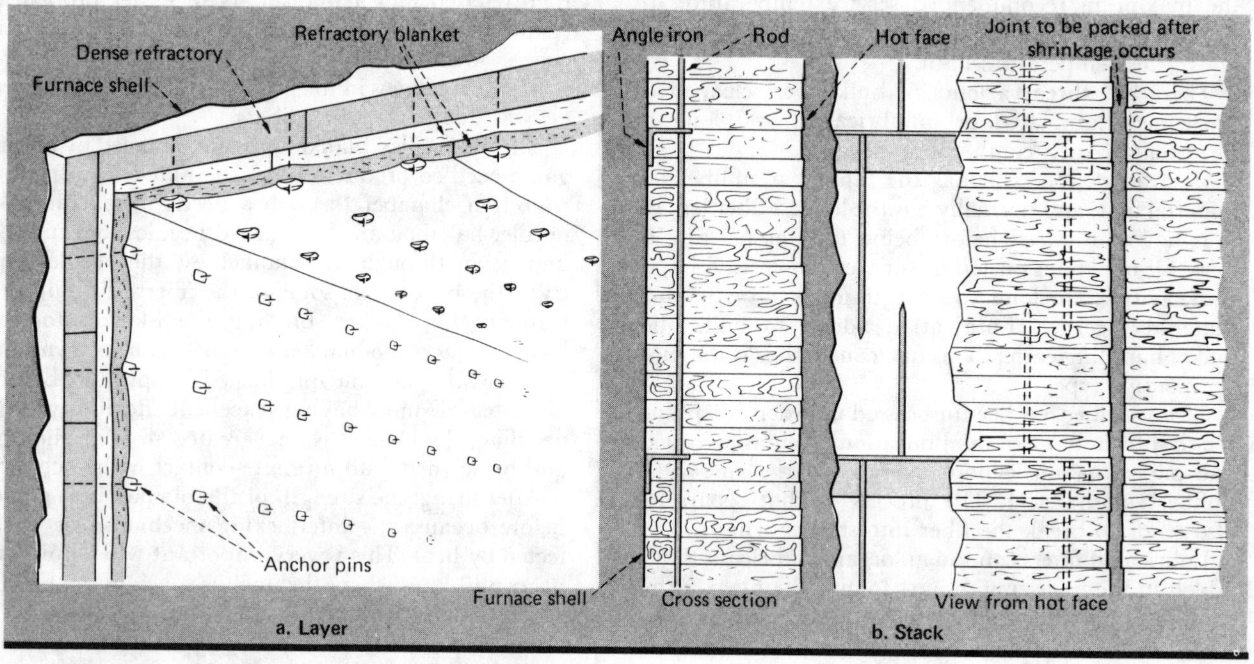

Installation methods for applying refractory-fiber materials Fig. 2

methods—layer or stack construction. Typical construction features for each method are shown in Fig. 2.

Layer construction involves impaling one thickness of insulation on top of another on anchors attached to the furnace shell, until the required thickness is achieved. However, major differences exist in layer construction for furnaces operating up to 2,250°F and those up to 2,600°F in the anchor system and in the type of hot-face fibers. The layer method also enables the use of various types of refractory felts or blankets in combination. For example, a needled blanket (useful to 1,600°F) becomes the backup insulation for the hot-face refractory-fiber material. The needled blanket is less expensive than the hot-face alumina-silica lining, so cost savings are possible.

Stack construction uses refractory fibers or blankets cut into strips that extend from the hot surface to the cold surface of the furnace. These pieces are stacked upon each other in the field or formed into a module that is subsequently mounted as a single unit.

Anchoring systems

In layer construction, a variety of anchors are used to meet varying temperature and operating conditions.

Applications for refractory fibers Table V

Type of furnace	Temperature, °F
Brick firing	1,800 to 2,200
Ceramic firing	1,800 to 3,250
Forging	2,250 to 2,450
Heat treating (metal)	1,500 to 2,100
Petrochemical heaters	1,600 to 2,100

All forms of alumina-silica fiber require anchors that will match, or exceed, these conditions.

Metallic anchors are available in a variety of lengths and alloys for temperatures to 2,250°F. These anchors are welded to the furnace shell, and refractory-fiber insulations and blankets are impaled on them. An anchor washer is then installed to hold the insulation in place.

Ceramic anchors are used where temperature limits exceed 2,250°F. One system uses a Type 304 or 310 alloy stud, welded to the furnace shell. A ceramic pin is threaded over the metal stud. A ceramic washer locks onto the head of the ceramic pin. A major advantage for this system is that a less expensive alloy can be used for the stud as long as the ceramic pin extends into the refractory lining far enough.

Another method uses a ceramic locking cup. The cup extends into the insulation 2 in., and locks onto the standard metal anchor. The cup is then filled with either bulk fiber or insulating cement. The ceramic cup does not penetrate the lining very far, and requires the use of a high-alloy stud. Even then, the temperature on this stud could easily exceed 2,250°F in high-temperature furnaces.

Ceramic anchors are required in carbon-rich reducing atmospheres. The nickel in alloy-metal anchors catalyzes the reduction of carbon monoxide, and carbon accumulates around stud locations where the temperature is 1,000 to 1,200°F, to cause fiber deterioration.

Modular construction

The modular system is a method of prefabricating the refractory blanket into 12-in.-square modules that can be rapidly installed. The modules are attached to the furnace steel and eliminate the layer-by-layer buildup.

In one patented modular system, the strips of blanket

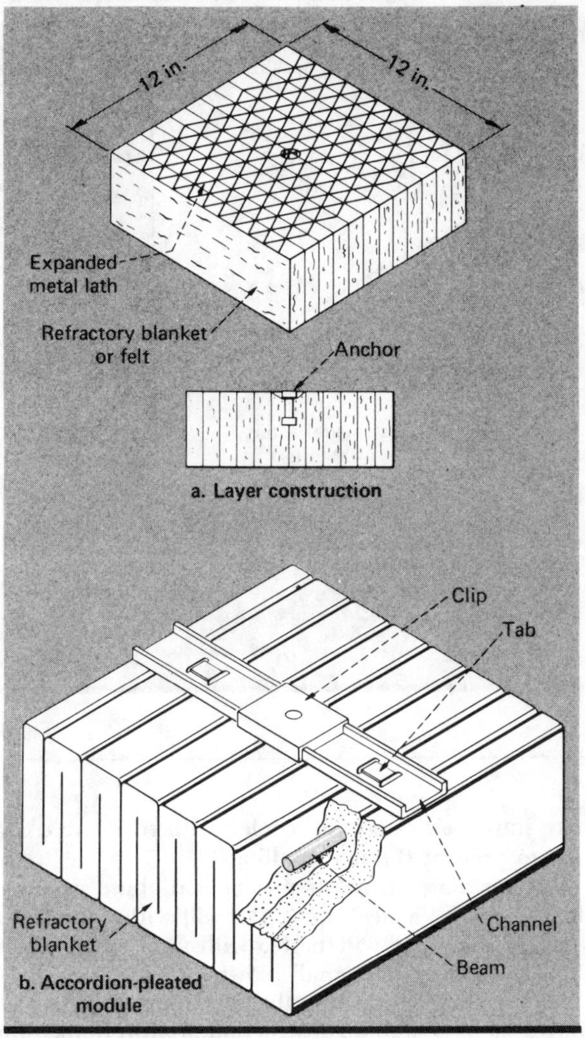

Modular construction saves time and labor during installation Fig. 3

or felt are stack-constructed. The stacks are attached to a 12-in.-square expanded-metal lath by using a refractory cement. The module is then fastened to the furnace shell with a one-step blind-welding and threading operation. In another version, the lath is welded directly to the furnace shell. The strips are compressed when mounted to the metal lath. The module tends to expand to fill the shrinkage joints that occur at high temperatures. However, the joints in this system do open, and must be stuffed with bulk fiber or blanket strips during the first cooldown cycle.

Another patented module is made with a single piece of inorganic, needled, refractory-fiber blanket 12 in. wide. This blanket is accordion-pleated so each section is as deep as the final thickness of the lining (between 5 and 12 in.). Stainless-steel support beams are buried in the fiber pleats next to the cold face of the block. A stainless-steel suspension channel attached to the beams fits into a stainless-steel clip that is fastened to the furnace wall.

Banding and kraft-covers keep the pleated blanket in compression prior to and during application. After the module is attached to the furnace shell, the compression restraints are removed and the refractory fiber expands. Since shrinkage occurs at high temperature, the accordion pleats expand and offset any shrinkage.

Economic considerations

There are several tradeoffs to consider when comparing layer methods to stack or module ones. In layer construction, less-expensive backup insulation is easily used. Stack and module construction employs one material that extends from the hot surface to the furnace shell. Material costs, of course, will be higher for the stack method.

The opening of shrinkage joints at high temperatures is another factor. Layer methods offset the joints from layer to layer, and provide a small amount of compression between the blanket edges. In the modular method, the joints fill during fiber expansion. The success of this depends on how well the job is installed. In some cases, the joints may need to be packed after the initial firing.

The undisputed advantages of modular construction are the large decreases in installation time and labor costs. The buildup of layers and the intricate pin-layout required in layer construction are eliminated. Installation time is saved because the modules are light and more easily handled than large rolls of blanket or felt. Essentially, two-thirds of the onsite labor time is replaced by shop prefabrication. Thus, installation costs and furnace downtime are reduced.

Installation advantages also come from the hidden anchoring system of the modules. Since each module has the hardware located close to the insulation's cold face, lower-cost alloys may be safely used at furnace temperatures that would normally require ceramic anchors. Atmospheric deterioration of the anchors is eliminated because the hardware is located in a relatively cool zone. When repair or replacement is necessary, only the damaged modules are removed. With blankets or felt rolls, such spot repairs are difficult, and large areas must often be replaced.

Layer construction has a definite advantage over modular in relation to thermal efficiency. Fiber orientation in layer construction is perpendicular to the direction of heat flow; whereas in the stack module (Fig. 3a), the fibers are parallel to the flow. This orientation can increase thermal conductivity 20 to 40% as indicated by manufacturers' tests. Hence, more insulation is needed in stack-module construction than in layer construction to achieve the same thermal resistance. The accordion-pleated module (Fig. 3b) will have a thermal efficiency that is greater than the stack module but not as good as the layered construction, due to the orientation of the blanket. Here, the fiber orientation is both parallel and perpendicular to the heat flow.

Retrofit applications

Rising fuel costs are causing a closer look at kiln and furnace operations in relation to heat losses. If the unit is inefficient but the refractory lining is still serviceable, its efficiency can be increased without replacing the entire refractory lining. This can be done by installing a refractory-fiber blanket directly to the hot face of the existing surface. Since refractory fibers do not soak up a lot of heat, they do not require large amounts of heat to

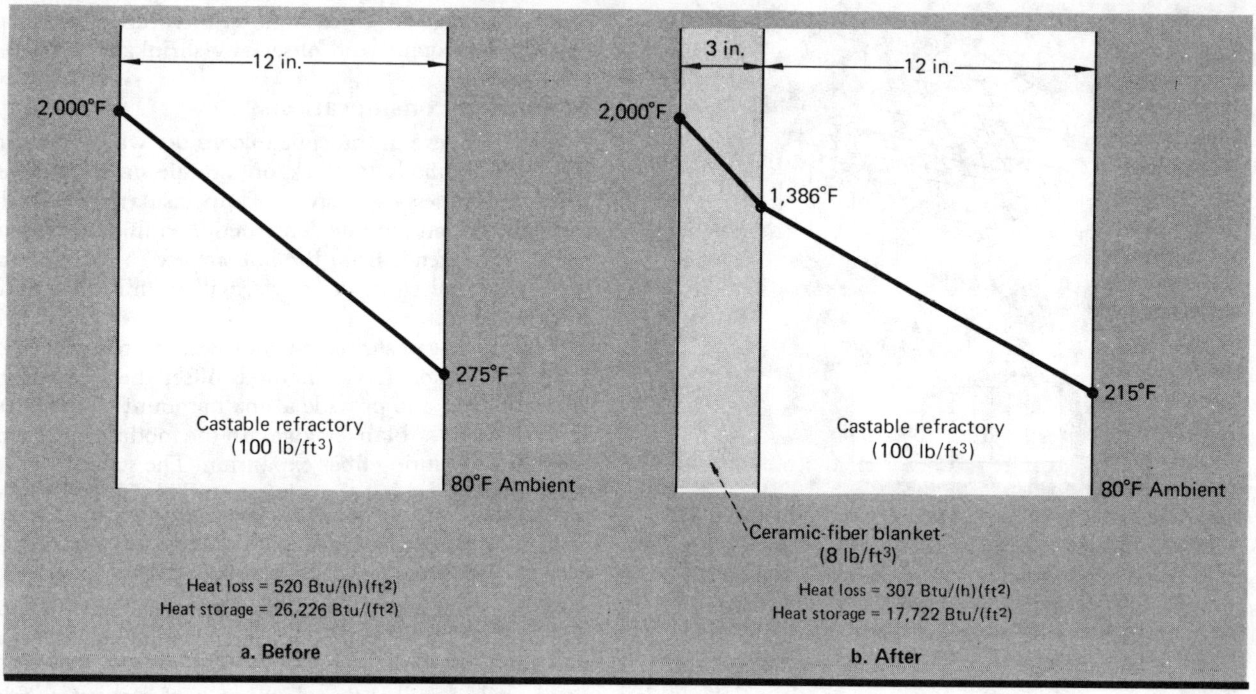

Effect of ceramic fiber on heat loss and heat storage **Fig. 4**

reach equilibrium conditions. In this application, the overall heat loss will be reduced because of additional insulation and the elimination of cracks on the hot face that allow heat to escape. Generally, fiber insulation can be attached to the existing refractory surface by using the wallpaper or modular-veneering technique.

Wallpaper construction—Major cracks or spalled areas should be repaired. A desired stud pattern is laid out. Layers of blanket are then impaled on the studs to the desired thickness. The reduction in overall heat loss for a typical installation is shown in Fig. 4.

Veneer construction—A module is made from strips of refractory fiber 12 in. long, turned edge-grain up and held together with an open-mesh organic cloth. Using an air-setting refractory mortar, the modules are installed over refractories such as dense brick, insulating firebrick, bubble alumina, semisilica brick, castables of any density, and plastic and ramming mixes. The module's thickness can vary from 1½ to 5 in. The organic cloth burns off when the furnace is first fired.

Performance

Insulating firebrick (IFB) and refractory-fiber materials are used in heating equipment for many process operations. Such heating equipment does not normally undergo aggravated chemical attack such as that found in a basic-oxygen steel furnace, or the aggravated physical abuse found in a rotary cement kiln.

Insulating firebrick and refractory fibers have very long service lives. It is not uncommon to find IFB linings in steel-annealing furnaces last more than 20 years. Refractory fibers are completely resistant to thermal shocking—so cyclic heating equipment can be elevated and reduced in temperature as quickly as the burner system will allow. On the other hand, an IFB-lined furnace must be heated and cooled with reasonable caution to prevent thermal spalling.

Neither material resists mechanical abuse such as a direct blow by a steel ingot. IFB will withstand higher air velocities and will better withstand minor abuse such as gouging and small blows.

Compensating for the thermal movement of IFB makes the design and construction of an IFB-lined furnace more difficult. Refractory fibers experience none of this thermal movement.

Refractory fibers are not considered a good backup insulation to dense refractories, because they are resilient and compressible. On the other hand, IFB is thoroughly compatible with dense-refractory construction.

Refractory fibers are the optimum material for periodic heating equipment up to 2,250°F. IFB will be favored above 2,250°F. For cyclic heating equipment operating at any temperature, it is doubtful that dense refractories would be specified unless abuse of the lining or chemical resistance were a major factor.

In a very strong thermal-shocking environment (to 2,600°F), refractory fiber would be selected. Its low heat storage favors periodically operating heating equipment, and the lining will not deteriorate from thermal spalling. In such an environment, the choice is not between IFB and refractory fibers, but between refractory fibers and dense refractories. IFB is too prone to thermal spalling in viscous situations.

Until 1973, dense refractories were used almost exclusively for continuous-operating heating equipment such as tunnel kilns. Now, IFB in these applications saves energy, and reduces foundation requirements and plant space. Refractory fibers are seldom used because low heat storage is not sufficient in such equipment. Only at low mean temperatures (below 1,800°F) is a refractory fiber more cost-effective than IFB. Above 1,800°F, IFB is the optimum material for most applications.

There is an exception as regards continuous-heating equipment such as the fired heater common to refineries and petrochemical plants. In recent years, some furnace builders have made portions of these units of refractory fibers. The reasons are twofold:

■ Downtime for these units is expensive. With refractory-fiber construction, the unit can be shut down and cooled rapidly if repair is needed; the repair is done quickly, and the unit returned to service with minimum downtime.

■ The amount of and cost for the supporting steel is lower. Typically, these units are tall, and the savings in structural steel make refractory fibers cost-effective for portions of the equipment.

Almost any heating equipment that operates above 2,500 to 2,600°F is not practical for refractory fibers. The higher the temperature, the less cost-effective they are. This applies to fibers that are useful to 2,800 to 3,000°F.

Refractories in furnace design

A basic knowledge of refractories is not enough to enable one to specify a refractory for a given installation. Furnace design is a highly specialized field, and few aspects of it are more specialized than refractories. Hence, refractory manufacturers can be invaluable for selecting the optimum material. They do not design heat-processing equipment but do furnish technical data and suggestions on applications.

Specific information is required regarding a kiln or furnace and its operation. Before an intelligent recommendation can be made for a refractory, one should have a complete understanding of the unit, including operating data such as function, cycle time and temperature range; and type of fuel, including its characteristics and impurities. After analyzing the need for a refractory, choosing the optimum product, and engineering and building the refractory system, the next step is performance evaluation.

Refractory application skills are largely developed through on-the-job experience. Moreover, many decisions are based on judgment—carefully made after evaluating a great deal of information. Although innumerable factors influence refractory performance, most come under one of the following subjects: heating-equipment design, refractory-system design and type of refractory, operating practice, refractory quality, and workmanship.

Heating-equipment design

Furnaces are designed to accomplish consistently and economically certain results, such as production of high-pressure steam, reduction of an ore, refining of a metal, or heat treating. The efficiency of heat-processing equipment is normally reported as the ratio of production to fuel consumption.

The effect of furnace design on refractory life too often receives only secondary consideration until after the furnace is in operation, and high refractory costs force a redesign. There are cases where an otherwise efficient and successful furnace design was abandoned because no satisfactory lining material could be found or developed. Furnace design is necessarily the first requirement. Ideally, the refractory-system design is then done, and the furnace design altered as needed and if possible. Thus, refractory and furnace people must work as a team to optimize performance.

The refractory members of the team are not expected to be furnace designers, but should be able to recognize design features that are inherently dangerous from the standpoint of refractory service. Such features are most frequently found in home-made furnaces and in well-designed furnaces that have been altered to reduce fuel consumption or increase production.

Insufficient combustion space (frequently referred to as "a bottled-up condition") is a common cause of refractory failure. More heat is released within the combustion space than is absorbed by the charge, dissipated through the furnace walls by heat flow, or carried out with the flue gases. As a result, refractory-wall temperatures can approach flame temperatures (above 3,000°F). Coupled with impurities, such temperatures will quickly destroy even high-quality refractories.

It would be convenient to say that a heat release of X Btu/(h)(ft^3) of combustion space were the upper limit for the satisfactory use of a given refractory. Unfortunately, it is not this simple. Many other factors—including refractory-wall thickness and height, effect of the amount of insulation, fuel type, amount of excess air, and especially the amount of heat absorbed by the charge or the water tubes—influence the amount of heat that can be released without disastrous effects. Judgment based on experience must be the guide, rather than a formula.

Theoretically, a single combustion space operates at a uniform temperature. This rarely occurs in practice. Only one thermocouple can control temperature in a firing zone, and it is usually placed to control the temperature of the ware or load.

Furnace drawings should be carefully studied to determine the location of control and shutdown thermocouples, and of furnace areas that could be expected to exceed the thermocouple settings. One must study the intended thermocouple settings to determine the maximum controlled-upset condition in the furnace. From this, factors of safety for refractory selection can be established, including possible zones where different refractories can be used.

When portions of the furnace lining (division walls, bridge walls, piers, door jambs, or the nose of an arch) are exposed to high temperatures on more than one side, they frequently fail before the rest of the lining. Failure may be due to load, shrinkage, excessive vitrification and spalling, slagging, or a combination of these. Sometimes, a change in the design of the refractory lining or in the refractory material will eliminate, or at least improve, conditions in the vulnerable spots.

Many burner designs are in use, and some can affect refractory selection. For example, a long-flame front burner can cause flames to lick the opposite wall. A bag wall or muffle in front of a burner can reradiate heat onto the refractories around the burner. A top-mounted flat-flame burner can direct excessive heat to a roof or onto a skew.

Refractory-system design

The study of a refractory system should involve three specific areas, whether one is designing a system or evaluating a refractory. These are: (1) thermal—the refractory system must resist the attack of the heat that is expected; (2) structural—the refractory structure must be mechanically sound; and (3) chemical—the system must resist the attack of whatever chemical materials are present.

Thermal considerations

Temperature—Obviously, the refractory system must withstand the expected temperature of the heat-processing equipment in which it is to be used. However, this is much too simple. The hot-load strength of the refractory, while it might appear to be a mechanical consideration, must be studied as a thermal consideration because it is the thermal design of the system that will determine the refractories to be used.

Load-bearing strength—The refractory's load-bearing strength, measured as hot-load deformation, is not usually of major importance where the refractory is exposed to heat on only one face. In most instances, there is a fairly steep temperature gradient through the lining, and the load is largely carried by that portion of the lining cool enough to be below the temperature at which any softening may occur.

Load deformation—In other refractory constructions, hot-load deformation is the primary factor for refractory selection. Potential trouble spots would arise when the refractory was exposed to high temperature on more than one face, or where the lining would be heavily insulated, thus decreasing the temperature drop through the wall. The ceramic process of why deformation takes place is covered by Norton,* and others.

There is no ironclad rule or formula to predict the amount of deformation that occurs in actual service. A fireclay refractory may experience hot-load deformation when the refractory is exposed to soaking heat, or when the temperature gradient through the refractory is very slight. Although it is risky to make a general rule, it is safe to say that hot-load deformation may occur in a fireclay brick, including IFB, when the refractory will experience anywhere within it a temperature of 2,250°F, or above, with a load of more than 10 psi. At higher temperatures, say above 2,400°F, even lower temperatures within the refractory, and lower loads, may cause trouble via hot-load deformation over an extended period of time.

Backup insulation—Since a load-bearing refractory system might use dense refractory, it is necessary to consider the effect of heavy backup insulation. As an example, let us consider a wall built of 13½ in. of firebrick, and backed with 1 in. of insulating block. With a hot-face temperature of 2,600°F, about 4 in. of the inner brick wall will be exposed to a temperature above 2,250°F. With a concentrated heavy load on the inner portion, this part of the wall could deform. However, with a uniformly distributed load over the entire 13½ in. of brick, there would be little chance of trouble because the relatively cool 9½ in. of brick in the back would carry the entire load, and the wall would remain structurally stable.

This example illustrates the occasional necessity for trial-and-error approaches in refractory-system design. Hot-load deformation properties are used to compare one manufacturer's refractory brick to another's. Experience must be added to hot-load properties to determine which refractories tend to work in given situations, and which do not.

Expansion—Expansion can cause a wall to bow, cause pinch spalling, and create other problems. Many failures have been identified as due to not providing for expansion. The correct procedures, allowing for expansion spaces, are available in refractory manufacturers' installation and construction manuals.

Thermal spalling—IFB has been used in some thermal-spalling situations but is not considered as resistant to spalling as dense castables, fireclay brick, or fireclay-ramming mixes.

Structural considerations

Roof construction—Heat-processing equipment generally requires a roof. The options include a sprung arch or dome, a flat suspended arch, and variations of these. A sprung arch is the least expensive approach, but is limited to small spans. Otherwise, one must use a flat suspended arch, or division walls so that several arches can cover the required roof area. Arches and domes are more successful with low-rise (down to 2 in./ft or even 1½ in./ft of span) roofs, particularly at higher temperatures. This helps to eliminate pinch spalling.

Most insulating refractories require some sort of anchoring. This is one of the most critical aspects of castable and IFB constructions. Therefore, each application requires careful study in order to get the optimum anchoring system.

Abrasion/erosion resistance—Usually, this is a straightforward problem because the materials' limitations are known. What is not appreciated, however, is that erosion occurs more rapidly in turbulent-flow zones such as a corner than in nonturbulent zones such as a straight flue. Thus, the design must recognize that if erosion can occur, it will certainly occur in the turbulent areas. These areas might benefit from a more abrasion/erosion-resistant construction.

Permeability—An important factor with IFB is its very high permeability. Insulating-refractory constructions commonly use lower-temperature grades of refractory as the cold face is approached. If a flow of hot gases can take place through the lining, then damage to the backup insulation can be expected. Hence, we should be very cautious in furnaces where a high positive pressure exists within the furnace, and flow to the outside is possible.

Mechanical spalling—This occurs from high differential stresses placed on the refractories. Refractory fibers, being resilient, do not encounter this problem.

Chemical considerations

Refractories are affected by the action of the atmosphere in the heating equipment, and by chemical attack on permeable materials.

In heat-treating equipment, nonoxidizing atmo-

*Norton, F. H., "Refractories," 4th ed., McGraw-Hill, New York, 1968.

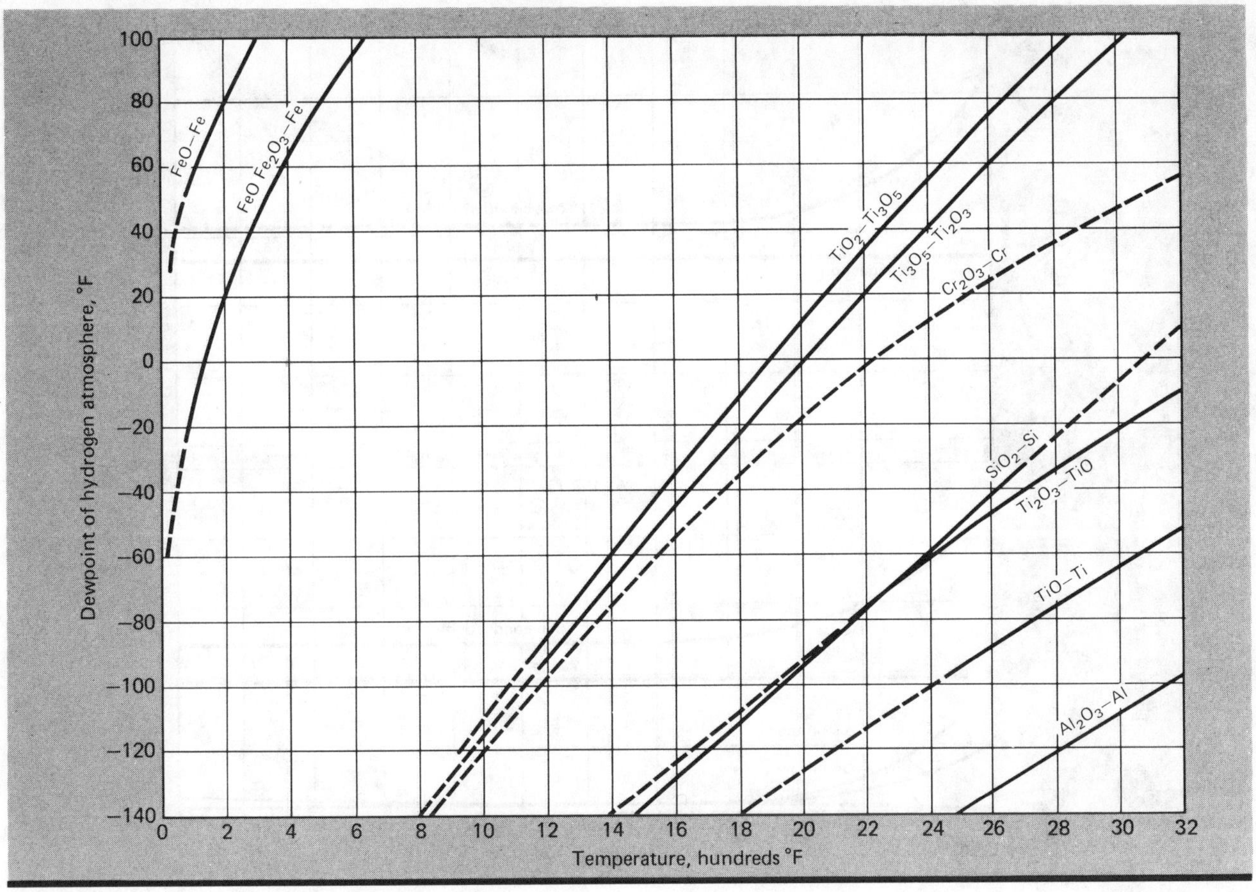

Metal/metal-oxide equilibria in a hydrogen atmosphere **Fig. 5**

spheres create some problems for insulating refractories. Low oxygen pressure (i.e., concentration) reduces Fe_2O_3 to FeO. Since FeO is less refractory than Fe_2O_3, this condition promotes hot-load deformation of the refractory at temperatures above 1,800°F. This effect will also reduce SiO_2 to SiO (a gas), which also causes refractory failure. Upon cooling, the SiO tends to form deposits on surfaces in heat exchangers, boilers and reformers.

Equilibria conditions for a number of metal/metal-oxide combinations in a hydrogen atmosphere are shown in Fig. 5 for the relationship between dewpoint and temperature. As long as operating conditions are maintained to the left and above a particular curve, the metal's oxide will be stable. If the conditions are to the right and below a curve, the oxide will be reduced. This chart is true only for IFB and high-density brick. Refractory fibers should not be used above 900°F, or at a dewpoint lower than −20°F.

As the dewpoint is lowered, the service temperature must also be lowered to prevent a particular oxide from being reduced. For example, if the dewpoint for SiO_2 is −60°F, the temperature must be maintained below 2,400°F to prevent reduction of SiO_2.

A disintegration triggered by the catalytic decomposition of carbon monoxide or hydrocarbons such as methane also occurs in prepared atmospheres. Let us review the mechanism for failure due to this reaction. Ferric oxide (Fe_2O_3) is present in the refractory in localized concentrations. This is converted to iron carbide (Fe_3C) that catalyzes the decomposition of carbon monoxide at 750 to 1,300°F to carbon dioxide and carbon. The carbon is deposited on the Fe_3C. Carbon builds up on the catalytic surfaces that are under stress, and ultimately causes disintegration of the refractory. In many cases, such stresses are severe enough to burst the steel shell of the furnace.

A similar condition exists in hydrocarbon atmospheres, which persists up to 1,700°F.

The risk of carbon disintegration can be minimized by using refractories above the carbon-deposition temperature, and by using products having low iron content. This effect is more noticeable in dense-brick or castable refractories.

Thermal conductivity

Furnace atmospheres also have an effect on the thermal conductivity of insulating refractories. Let us review how the thermal conductivity of the furnace gases affects these refractories. Fig. 6 shows the thermal conductivities for air and hydrogen. The component of the thermal conductivity affected by changing the gas constituents in the furnace atmosphere is the gas conduction. It is quite apparent that the gas (normally air) in the pores of an insulating refractory can be readily replaced by other gas constituents found in the furnace atmosphere.

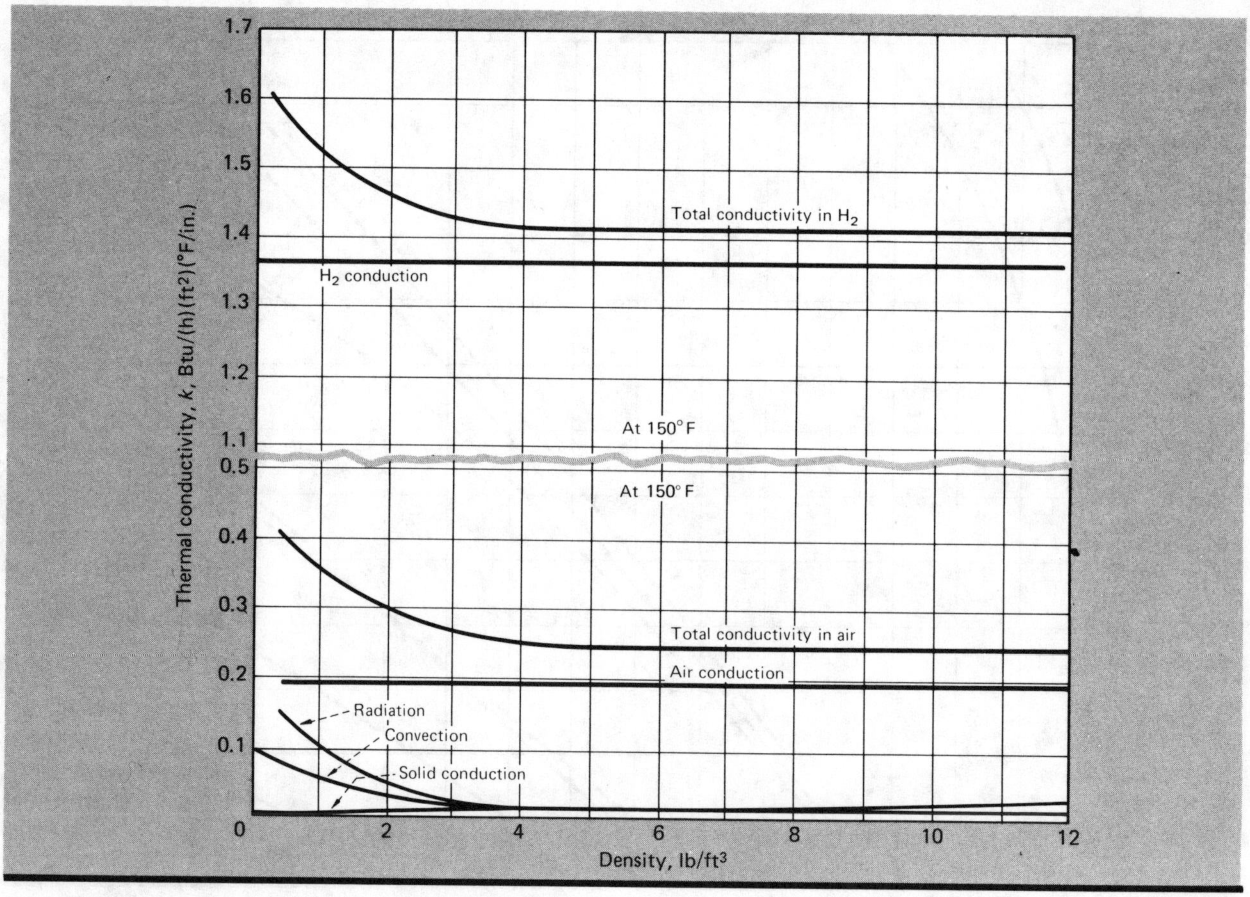

Furnace atmosphere affects thermal conductivity of refractory **Fig. 6**

Replacement or dilution of this air constituent by other gases will change the insulation's k value. The amount of change is dependent on the k value of the replacement gas, and the porosity of the insulation. Most gases involved in a heating atmosphere have essentially the same k value as air (Fig. 7). However, hydrogen has a very high k value, causing a significant change in insulation effectiveness.

Other compounds can be found in small quantities in many furnace atmospheres. They can originate in the fuel, the refractories, or even the charge in the kiln or furnace. One of these is vanadium pentoxide (V_2O_5) from low-grade fuel oils such as Bunker C. Another vanadium compound, sodium vanadate, may be found in oil flames as droplets. It appears to decompose and cause alkali attack at about 1,240°F.

Sulfur occurs in fuels and in some clays. Depending on its content and chemical form, it can become part of a furnace atmosphere as sulfur oxides.

Alumina-silica (45 to 54% Al_2O_3 range) maintains the highest hot strength of the several refractory materials cycled at 1,400 to 1,800°F in the presence of an SO_3 atmosphere.

Test data show that disintegration of the refractory may occur if Na_2SO_4, $MgSO_4$, $Al_2(SO_4)_3$, or $CaSO_4$ are formed in a sulfur dioxide atmosphere.

Insulating refractories are seldom used in installations where chemical attack is expected. Unfortunately, it often arises unexpectedly. Primarily, such attack is due to the permeability of insulating refractories. Many times, dense refractories resist chemical attack, not because of chemical resistance but because their very high density and low permeability prevent damaging materials from entering.

Slagging is defined by the American Soc. of Thermal Manufacturers as "the destructive chemical reaction between refractories and external agencies at high temperatures, resulting in the formation of a liquid with the refractory."

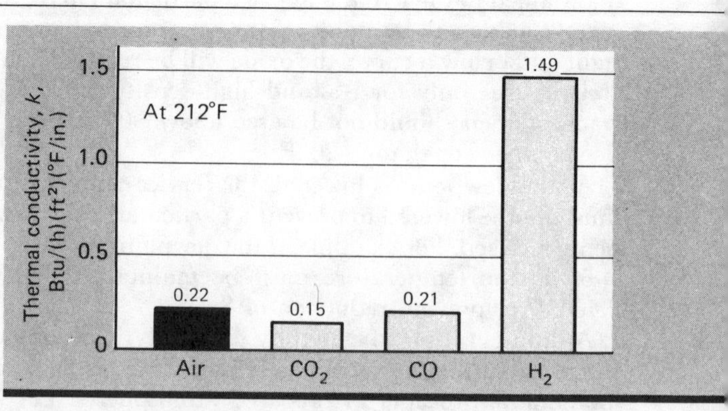

Hydrogen conductivity affects insulation effectiveness **Fig. 7**

Attack from mill scale occasionally causes slagging. As ferrous-metal objects are heated, their surfaces oxidize, and the resulting iron oxides flake. These oxides fall onto the hearth, and will readily attack an alumina-silica refractory because the iron oxides act as a fluxing agent. Should the oxides become airborne, they can contaminate refractories on the furnace wall.

Other materials heated in a furnace can throw off other oxides, considered to be fluxes for an alumina-silica refractory. Incinerators are the worst of all; they burn or oxidize everything put into them.

Refractory fibers are troublesome in fluxing situations. Experience dictates extreme caution when applying such fibers above 1,800°F. However, these fibers have worked well in incinerator afterburners above 1,800°F, perhaps because the fluxes by then have become so oxidized that they are no longer able to attack the refractory.

Chemical attack of refractories has become much easier to diagnose. Chemical analysis of refractory specimens can be made quickly and inexpensively by most refractory manufacturers. The methods used include spectrographic analysis, x-ray diffraction, and wet chemistry. The best approach is to analyze specimens of the damaged refractory, and of the original refractory before firing. On the basis of these data, a decision is made. One comparison may be between the damaged refractory and its original composition. If a fluxing element is found that should not be present, its source is searched for. If the source cannot be eliminated, an alternative refractory must be used.

Direct reactions such as a flux from a ceramic glaze being fired and an alumina-silica IFB are relatively easy to evaluate. However, there are some secondary reactions in which a material in the furnace atmosphere will act as a catalyst, reacting with a second element in the atmosphere and a component in the refractory. These are considerably more difficult to evaluate.

Operating practices

Premature refractory failures may arise because of unintentional changes in furnace conditions. These changes, while unobserved by the operator, may be of long enough duration to destroy a refractory lining but not long enough to be considered a change in practice. The changes may arise from: burners being out of adjustment, furnace atmospheres that are highly reducing for a period of time, furnace temperatures becoming abnormally high, or variations in composition of the raw materials or fuels.

Such changes in practice are usually difficult to discover. They leave no record other than their effects on the refractory lining. Failure of the refractory may not become apparent until weeks or months after the actual damage has been done. This makes investigation difficult, and often raises credibility problems.

Sometimes, the premeditated abuse of heat-processing equipment arises in order to get increased furnace throughput and/or greater return on assets. In almost every case, operating a furnace and refractory system beyond its capacity will reduce refractory life. This abuse is common, and there are really only two alternatives. One, stop doing it; or two, expect a shorter life from the refractories.

Workmanship

It is easy to blame brickmasons for every failure of refractory brick. By the time failure occurs, the evidence is practically destroyed. While good brickwork is extremely important, there are other factors involved. These probably occur more frequently than either poor brickmasonry, or refractory products that are not up to standard quality.

These other factors include the proper size and location of expansion joints, the rise of a furnace arch, and the thickness of a wall. These are not factors of masonry workmanship, and should be taken care of in the design of the refractory system. In practice, many design details are left to the brickmason doing the work. In some cases, this is the best way to handle such details.

Sometimes, however, brickmasons called on to do refractory work have only had experience in the building trades. Such masons make mortared joints up to ½ in. thick. This practice is not compatible with IFB where commonly the mortar may not even be visible. The problem seldom happens in large companies employing their own masons. Smaller companies should keep this in mind, and make sure that the person in charge of the masons understands high-temperature-furnace work.

The authors

James E. Neal is manager of refractory engineering and technical services for Johns-Manville Corp., Ken-Caryl Ranch, Denver, CO 80217. He has had more than 35 years of experience with refractories. A graduate of Rutgers University, he has a B.S. in ceramic engineering. Mr. Neal is a member of the American Ceramic Soc., Tau Beta Pi and Keramos. He is chairman of the refractory-fiber committee for the American Soc. for Testing and Materials.

Roger S. Clark is merchandising manager of refractories for Johns-Manville Corp., Ken-Caryl Ranch, Denver, CO 80217. He has been associated with the refractory industry for many years. Prior to joining Johns-Manville in 1972, he was sales manager of Babcock & Wilcox's ceramic fiber division. He has a B.S. in mechanical engineering from Texas A&M University, and is a member of ASME and the American Ceramic Soc.

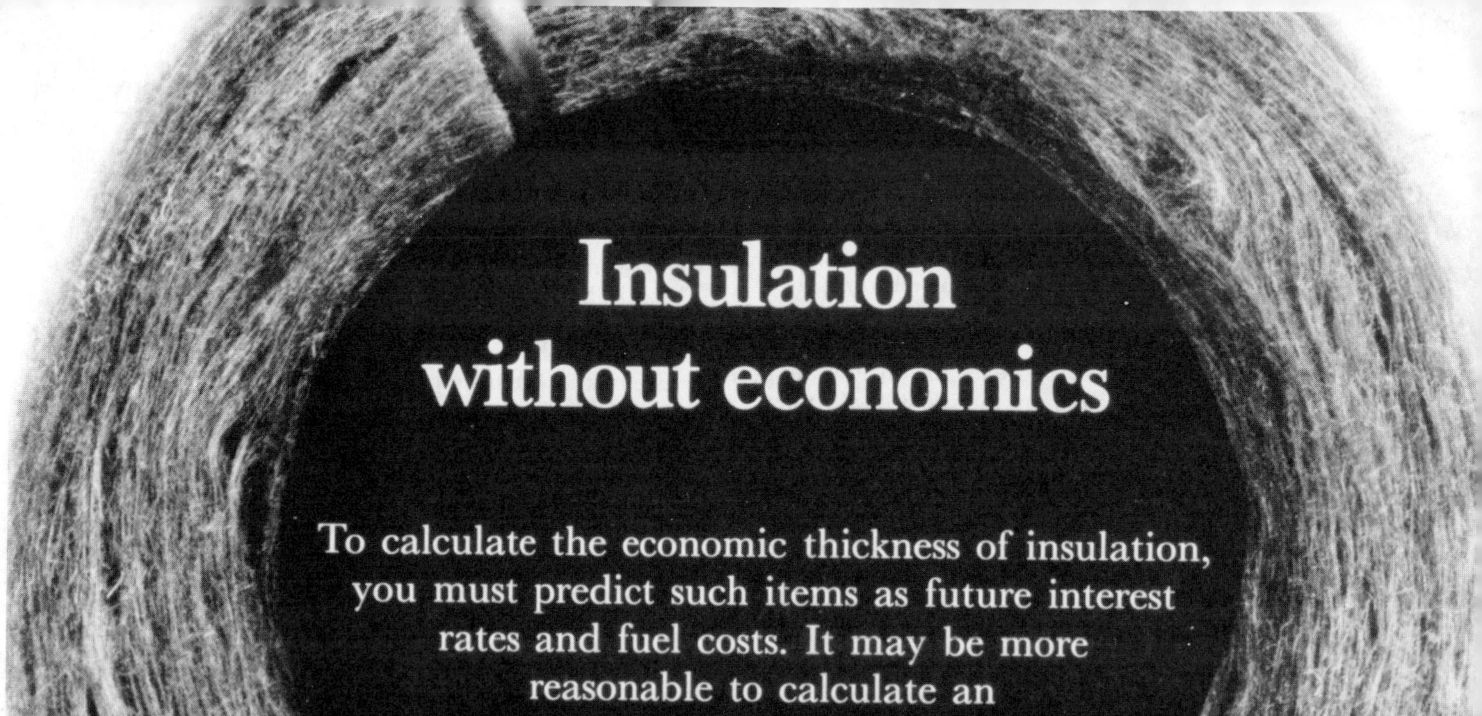

Insulation without economics

To calculate the economic thickness of insulation, you must predict such items as future interest rates and fuel costs. It may be more reasonable to calculate an "acceptable heat-loss" thickness.

M. McChesney and P. McChesney, Fuel Save Associates

☐ Thermal insulation should be the simplest, most generally accepted, cost-effective method of saving energy immediately available to a plant owner. However this clearly is not the case to many—perhaps too many—owners, because to them it is just another capital investment; as such its purchase must ensure a return on the investment.

Their argument is simple and direct: if the plant owner can make a 17% profit on his salable products whereas insulation returns only, say, 12%, then surely it makes no economic sense to buy the insulation—since the more profitable thing to do is to expand plant capacity to produce more salable products.

Having reached this conclusion, the plant owner then can salve his conscience by reducing energy wastage solely by "good housekeeping" involving minimal or even no capital expenditure. After all, there is no shortage of advice and guidance available on how to save 10% of plant energy bills without spending *any* money!

Economic thickness of insulation (ETI)

Because of this "economic" attitude toward insulation, it became necessary to evaluate how much was actually "economic," and this gave rise to the concept of the "economic thickness of insulation." This concept has been extensively discussed on both sides of the Atlantic, being analyzed in the technical literature over the years. There are now lengthy books and manuals available that show the (sometimes bemused) plant owner how to calculate this economic thickness by using tables, graphs or computers. All the plant engineer needs to do is to input data into the tables, etc., and out comes the "economic thickness."

In the U.S.A., the history of the tables highlights their problems. In 1949 a committee was established by Union Carbide Corp. and West Virginia University to reduce the economic-thickness calculation to a simple procedure in which, following the McMillan analysis [*4*], the heat factors were separated from the cost factors. A manual was produced by a forerunner of TIMA (Thermal Insulation Manufacturers Assn.), with nomographs and charts; TIMA itself apparently produced its own version around 1960.

Since that time the data and their presentation underwent modification and, around 1973, programs called ECON-I (tables, charts, worked examples),

Originally published May 3, 1982

ECON-II (marketing manual) and R-ECON (retrofitting) became available. However it was felt by some that these programs had averaged too many variables to nominal values and, in addition, the minimum insulation depreciation (amortization) period was too long. This last objection is important since it showed that plant owners were *not* prepared to regard insulation as a long-term investment—an attitude that is still common today.

Insulation, if properly maintained, can have a very long lifetime (30 years or more) but the pipework around which it is placed may be part of a plant that rapidly becomes obsolescent. Since in many cases insulation is not recovered, but scrapped, it makes no sense to amortize it over a longer period of time.

Accordingly, around 1976 a refinement of these tables was produced by York Research Corp. for the Federal Energy Administration and called ETI (Economic Thickness for Industrial Insulation). It consisted of 10 sections covering fuel costs (which had become a major consideration since the 1974 OPEC price rise), insulation costs, condensation control and retrofitting. All the information was given in tabular, graphical or nomographical form and also in a mathematical appendix.

However, this apparently was not suitable for all needs, and from 1976 to 1980 numerous suggestions were made to make the ETI program more useful. A recent revision has been made incorporating these suggestions, as a joint venture between TIMA, NICA (National Insulation Contractors Assn.) and Louisiana Technical University, the resulting manuals being called ETIH (Economical Thickness of Insulation for Hot Surfaces) and ETIC (Cold Surfaces).

Difficulties with calculating ETI

Over the years, the senior writer has used these tables and watched the brave attempts to present the calculational procedure in a digestible form without sacrificing too much accuracy. Undeniably the use of discounted-cash-flow analysis does complicate the calculation, and the incorporation of future fuel-price increases adds further difficulty. Possibly because of the complexity of these economic factors (but this is only a guess), in the U.K. the Department of Energy produced a slim book of graphs showing the heat loss from unit lengths of pipes of various diameters carrying various thicknesses of different types of insulation and giving a calculational procedure for obtaining the economic thickness of insulation.

However, the book makes no allowance whatever for either discounted cash flows or future fuel-price increases. These graphs have the merit of simplicity, but the writers have found them inaccurate because they are plotted on log-log scales.

In this article we use several of the commonly adopted methods for calculating the economic thickness, and show that a real dilemma arises in trying to decide what is the actual thickness that is *economic.* In fact, we wonder whether there is such a quantity at all!

Put another way, we have a sneaking suspicion that the "economic" thickness of insulation is more or less what any plant engineer wants it to be. The reason is twofold: there are inevitably uncertainties attached to the heat-input data, but these can be controlled to a considerably greater extent than the uncertainties (or, put more euphemistically, the greater range of choice) in the economic-data inputs. Thereafter we propose a simple criterion for determining the *sensible* thickness of insulation (not economic thickness, since economics is taken right out of the problem altogether, at least in explicit form), that puts the decision-making back where it really belongs—in the hands of the plant engineer and not of the company accountant!

A full analysis of insulation economics requires that at least 20 input-data variables be assigned; these can be grouped under four headings for the case of pipe insulation.

Insulation factors

1. Cost of installed insulation, of thickness t, per linear foot $C_{I(t)}$.
2. Thermal conductivity of insulation, k_I.
3. Thermal resistance of insulation surface, R_S.
4. Pipe diameter (nominal), d_1.
5. Ambient temperature, θ_3.
6. Ambient wind speed.
7. Pipe temperature, θ_1.
8. Amortization period of insulation, n.
9. Pipe-complexity factor.
10. Maintenance and insurance costs.

For simplicity we shall assume a pipe-complexity factor of unity, ignore maintenance and insurance costs, and assume zero wind speed.

Fuel factors

11. Type of fuel and cost, C_F.
12. Expected annual price rise of fuel expressed as a decimal, f.

Heat-producing-plant factors

13. Efficiency of conversion of fuel to heat, E.
14. Number of hours of operation per year, N.
15. Capital investment in heat-producing plant.
16. Amortization period of heat-producing plant.

For simplicity we shall completely ignore the economic aspects of the heat-producing plant. Normally the existence of insulation can reduce the size of the needed heat-producing plant, and this represents an incremental positive cash flow. We make the simplifying assumption that the decision to insulate does not affect the plant capacity—as would be the case for insulation retrofit.

Economic factors

17. Cost of money, i.e., return on investment in insulation required, i.
18. Tax rate.
19. Cost of money to finance heat-producing plant.

Nomenclature

A_I External surface area of insulation jacketing, ft^2

A_P External surface area of bare pipe, ft^2

C_F Cost of fuel, 2.83×10^{-6}, \$ per Btu

$C_{H(t)}$ Total cost of heat loss per linear foot of pipe covered with insulation of thickness t, for an entire operational year, \$ ft^{-1} yr^{-1} for money valued at this moment of time

$$= \frac{NC_F}{E}\frac{Q_{(t)}}{L}$$

$C_{I(t)}$ Cost of installed insulation per linear foot of pipe, of thickness t_I inches, \$ ft^{-1} for money valued at this moment of time

d_1 Nominal diameter of pipe, ft

d_2 Nominal outer diameter of insulation jacketing, ft

E Efficiency of conversion of fuel to heat expressed as a decimal, 0.83

f Annual increase in the cost of fuel expressed as a decimal, 0.1

F Heat-loss rate per ft^2 of insulated pipe divided by heat-loss rate per ft^2 of same pipe bare of insulation

h_C Convective heat-transfer coefficient of insulation surface jacketing for still air, Btu h^{-1} ft^{-2} $°F^{-1}$

h_I Total surface heat-transfer coefficient of insulation surface jacketing for still air, Btu h^{-1} ft^{-2} $°F^{-1}$

h_P Total surface heat-transfer coefficient of bare pipe for still air, Btu h^{-1} ft^{-2} $°F^{-1}$

h_R Radiative heat-transfer coefficient of insulation surface jacketing for still air, Btu h^{-1} ft^{-2} $°F^{-1}$

i Cost of money; return on investment required expressed as a decimal

k_I Thermal conductivity of the pipe insulation, Btu ft h^{-1} ft^{-2} $°F^{-1}$

L Length of pipe, ft

n Amortization period of the insulation (number of years over which the insulation economics is to be evaluated), years

N Number of hours of operation of the fuel-to-heat conversion equipment per year, 8,760 h

$Q_{(t)}$ Rate of heat loss from pipe covered with insulation of thickness t_I inches, Btu h^{-1}

r_1 Nominal radius of the pipe, ft

r_2 Nominal radius of insulation plus jacketing, ft

R_I Thermal resistance of the insulation and its jacketing, h ft^2 °F $(Btu)^{-1}$

R_P Thermal resistance of the bare pipe surface, h ft^2 °F $(Btu)^{-1}$

R_S Thermal resistance of the insulation jacket surface, h ft^2 °F $(Btu)^{-1}$

t Thickness of insulation and jacketing, in.

[USPWF] Uniform-series-present-worth factor, $\frac{(1+i)^n - 1}{i(1+i)^n}$

ε Surface emissivity of the insulation surface jacketing or of the bare pipe, 0.9

θ_1 Temperature of steam in pipe: also temperature of pipe wall, °F

θ_2 Temperature of insulation surface jacketing, °F

θ_3 Temperature of still air (70°F)

σ Stefan's radiation constant, 0.171×10^{-8} Btu h^{-1} ft^{-2} $°R^{-4}$

20. Economic model used for determining the economic thickness of insulation.

It must be stressed that the simplifications that we make in no way detract from the overall conclusions drawn. What they do is avoid obscuring the important issues in a morass of arithmetic—they alter number values but not decisions.

The heat-loss equations

Referring to Fig. 1, consider a horizontal pipe of nominal diameter d_1, covered with insulation of thickness t, and of thermal conductivity k_I, carrying dry saturated steam at temperature θ_1. Even with insulation present, the pipe will lose heat, and in the steady state, when the surface temperature of the insulation does not change, the rate at which heat is lost per linear foot of the pipe is:

$$\frac{Q_{(t)}}{L} = \frac{\theta_1 - \theta_3}{\frac{1}{2\pi k_I}\ln(d_2/d_1) + \frac{1}{\pi d_2 h_I}} \quad (1)$$

Strictly, this equation is a simplification because:

- It assumes that the convective heat-transfer coefficient from steam to the inner wall of the pipe is infinitely large compared with h_I, the sum of the convective and radiative heat-transfer coefficients from the insulation surface to the ambient air. This is an excellent approximation because any heat loss from the steam will cause it to condense, and the condensation heat-transfer coefficient (for film-wise condensation) lies in the range 1,000–2,000 Btu hr^{-1} ft^{-2} $°F^{-1}$, which is about one thousand times greater than h_I.
- It ignores the thermal resistance of the pipe wall, which is usually an excellent approximation in most heat-transfer cases other than thick-walled vats.

In Eq. (1) the value of h_I is given by:

$$h_I = h_C + h_R$$

where:

$$h_C = 0.270\,(\theta_2 - \theta_3)^{0.25}\, d_2^{-0.25} \quad (2)$$

$$h_R = \varepsilon\sigma[(\theta_2 + 460) + (\theta_3 + 460)] \times [(\theta_2 + 460)^2 + (\theta_3 + 460)^2] \quad (3)$$

Clearly, to evaluate Eq. (2) and (3) we need to know the value of the insulation surface temperature θ_2. The calculation and importance of θ_2 have been described

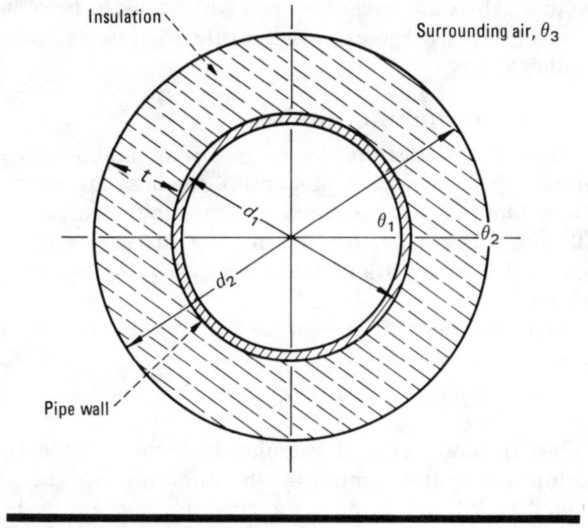

Insulated pipe showing diameters and temperatures **Fig. 1**

500° F, 8-in.-dia. horizontal pipe with calcium silicate insulation **Table I**

Thickness of calcium silicate insulation, t, in.	Surface temperature of insulation θ_2, °F	Heat loss rate per linear foot $Q_{(t)}/L$ Btu $h^{-1}ft^{-1}$	Dollar value of heat lost per linear foot per year of operation $C_{H(t)}$ (\$) $ft^{-1}year^{-1}$	Installed cost of insulation per linear foot $C_{I(t)}$ (\$)ft^{-1}
0	500	3,708	111	–
½	210	756	22.7	2.0
1	160	479	14.5	4.1
1½	136	358	10.7	6.2
2	122	292	8.74	8.4
2½	113	249	7.47	10.5
3	107	220	6.58	12.6
3½	102	198	5.93	14.7
4	99	181	5.42	16.8
4½	95	167	5.02	18.9
5	92	156	4.69	21.0
5½	90	147	4.42	23.1
6	89	140	4.18	25.2

by McChesney and McChesney [*1*]. Using the methods described in that article, the writers have solved Eq. (1), (2), (3) for the specific case of a straight horizontal pipe of nominal 8-in. dia., carrying dry saturated steam at 500°F, located in still air, lagged with various thicknesses of calcium silicate insulation. In addition, the boiler is assumed to be 83% efficient, operating for a full 8,760 hours per year, using fuel costing \$2.84 per million Btu.

Results are shown in Table I, in which the installed cost of insulation is an acceptable simplification. Installed cost is much discussed in the technical literature, and different methods suggested for determining it. The writers have examined manufacturers' quoted costs for an 8-in. pipe with rigid calcium silicate and noted that the cost per linear foot of the actual insulating material is essentially linear with increasing thickness. However, when the cost of layering, jacketing and installation is added, this linearity disappears and a "kinked" straight line is obtained. The values used in Table I are simply those from the "best" straight line through the actual installed costs. Once again, this simplification only affects the numbers calculated from the economic models but not the conclusions drawn from them.

In the article already referred to, [*1*], we have shown that it is desirable purely from the safety-to-personnel point of view to ensure that the insulation surface should be either a canvas jacket (if indoors) or a metal jacket with a very thick layer of high-temperature-resistant matte ("flat") oil-bound paint; either of these will give the high radiation surface emissivity used in this article (0.90). In addition, the insulation surface-temperature should not exceed 140°F for canvas or 113°F for metal, to prevent skin-burn if the surface is touched. This means that the safety thickness of insulation—and this is the minimum that *must* be fitted to our 8-in. pipe—is 1½ in. (canvas cover) or 2½ in. (metal jacket).

In some of the economic models we use discounted cash flows—specifically the uniform-series present-worth factor:

$$[\text{USPWF}] = \frac{(1+i)^n - 1}{i(1+i)^n} \quad (4)$$

This factor is used to discount—to the present time—a uniform series of expenses or receipts occurring annually for n years into the future. It is often referred to as the Net Present Value of future cash flows, and is sufficiently well known not to require further explanation.

What is less well known is how to incorporate fuel price increases into discounted cash flows, i.e., to allow for both the discount factor for money and the increase in the price of fuel over the amortization period. We have already defined and evaluated in Table I the quantity $C_{H(t)}$, which is the dollar value of the heat lost per linear foot per year of operation. This annual cost increases as fuel costs increase, by a multiplication factor $(1+f)$ for the first year, but it must be discounted by an interest factor $(1+i)$. For succeeding years we write Present Value of the total dollar value of the heat lost per linear foot for the n years as:

Present value

$$= C_{H(t)}\left[\left(\frac{1+f}{1+i}\right) + \left(\frac{1+f}{1+i}\right)^2 + \cdots \left(\frac{1+f}{1+i}\right)^n\right]$$

$$= C_{H(t)}\left[\frac{\left(\frac{1+f}{1+i}\right)^{n+1} - \left(\frac{1+f}{1+i}\right)}{\left(\frac{1+f}{1+i}\right) - 1}\right] \quad (5)$$

since the series is a simple geometrical progression of n terms. We have assumed that the value of f is constant, so that the quantity in the square brackets in Eq. (5) is a modified [USPWF] uniform series present-worth factor that allows for a uniform increase in the price of fuel over the lifetime of the insulation.

In addition there are two quite distinct methods of

Typical calculations for Economic Model 1 — Table II

Thickness of calcium silicate insulation, t, in.	Value of $n\,C_H(t) + C_I(t)$, n = 2 years	Value of $n\,C_H(t) + C_I(t)$, n = 5 years
½	47.29	115.23
1	32.79	75.83
1½	27.67	59.87
2	25.78	52.00
2½	25.33*	47.74
3	25.66	45.40
3½	26.45	44.23
4	27.54	43.80*
4½	28.83	43.59
5	30.28	44.39
5½	31.83	45.07
6	33.47	46.01

*The minimum value in each column is the ETI.

ETI as a function of amortization period — Table III

Model 1: Insulation amortization period, n years	Model 1: Economic insulation thickness, t, inches	Model 2: Insulation thickness fitted, t, inches	Model 2: Amortization period, n years
2	2½	2	1.30
3	3	2½	1.80
4	3½	3	2.31
5	4	3½	2.84
6	4½	4	3.38
7	5	4½	3.93
		5	4.62
		5½	5.03
		6	5.59

Typical calculations for Economic Models 3 and 4 — Table IV

Insulation thickness inches, t,	Model 3, Value of [USPWF] $C_{H(t)} - C_{I(t)}$: n = 2 years, i = 8%	Model 3: n = 5 years, i = 12%	Model 4, Value of [USPWF] × $[C_{H(t_a)} - C_{H(t_b)}] - [C_{I(t_b)} - C_{I(t_a)}]$: n = 2 years, i = 8%	Model 4: n = 5 years, i = 12%
½	38.4	79.6	12.7	27.8
1	21.5	47.6	4.34	10.9
1½	12.9	32.5	1.45	5.09
2	7.26	23.2	0.169*	2.49
2½	2.92*	14.4	−0.519	1.10
3	−0.765	11.2	All negative	0.262*
3½	All negative	2.84*	All negative	−0.278
4	All negative	−0.714	All negative	All negative
4½	All negative	All negative	All negative	All negative
5	All negative	All negative	All negative	All negative
5½	All negative	All negative	All negative	All negative
6	All negative	All negative	All negative	All negative

*The ETI is given by the smallest positive number in each column.

looking at increasing insulation thickness, these being the nonincremental method and the incremental method, the differences between which should be obvious from the worked examples for the various economic models used.

Various economic models

Model 1 for economic thickness—In this method a table is drawn up showing the total cost of the heat lost for the entire lifetime of the insulation, added to the cost of the insulation (both per linear foot), for increasing thicknesses of insulation and the minimum value found by inspection.

Algebraically this is expressed as finding the minimum in the quantity:

$$n\,C_{H(t)} + C_{I(t)}$$

Table II shows typical calculations, while Table III, Columns 1 and 2 summarize the minimum values obtained as a function of n, the chosen amortization period of the insulation.

Model 2—If it is felt that an incremental method would be more accurate, i.e., that each additional thickness (increment) of insulation must pay for itself in terms of fuel-cost savings, then the criterion becomes simply:

$$n[C_{H(t_a)} - C_{H(t_b)}] = C_{I(t_b)} - C_{I(t_a)}$$

where t_a and t_b are any two thicknesses of insulation commercially available (not necessarily consecutive). Table III, in Columns 3 and 4, shows values of the amortization period obtained from this equation for given insulation thicknesses.

Model 3—Neither of the above methods allows for the time value of money (discounting the cash flows) nor the inevitable increase in the price of fuel. If the time value of money only is considered, then in a nonincremental analysis, although each thickness of insulation will result in a fixed amount of heat lost per linear foot, the dollar value of this heat lost will vary over the amortization period of the insulation because of the varying cost of money. The criterion in this nonincremental case now becomes:

$$[\text{USPWF}]\,C_{H(t)} - C_{I(t)} > 0$$

In evaluating this inequality we must decide not only on the length of the amortization period but also on the value of money. We consider the two cases for which i has the values 8% and 12%.

Table IV, in Columns 2 and 3, shows typical calculations while Table V, Columns 2 and 3, summarizes the economic thicknesses resulting from this method of analysis.

Model 4—The equivalent incremental method of analysis is obtained from the inequality:

$$[C_{H(t_a)} - C_{H(t_b)}][\text{USPWF}] - [C_{I(t_b)} - C_{I(t_a)}] > 0$$

Table IV, Columns 4 and 5, shows typical calculations while Table V, Columns 4 and 5, summarizes the economic thicknesses resulting from this method of economic analysis.

The final step is to include the factor given in Eq. (5) for the increase in the cost of fuel.

Model 5—In the nonincremental case, the economic thickness of insulation is given by the inequality:

$$\frac{\left(\frac{1+f}{1+i}\right)^{n+1} - \left(\frac{1+f}{1+i}\right)}{\left(\frac{1+f}{1+i}\right) - 1} \quad C_{H(t)} - C_{I(t)} > 0$$

Table VI, in Columns 2 and 3, shows typical calculations while Table V, Columns 6 and 7, shows economic thicknesses resulting from this method of analysis. It is clear that in addition to having to specify the amortization period of the insulation and the value of money, we also need to specify the annual increase in the cost of fuel—which, if anything, is even more imponderable than the value of money. Values of f have changed in a fluctuating manner since 1974, and there is much in the economic literature about these fluctuations. We shall assume a value of 10%—with little justification for doing so!

Model 6—The final method is to make an equivalent incremental analysis allowing for the fuel price increase, and this is given by the inequality:

$$\left[\frac{\left(\frac{1+f}{1+i}\right)^{n+1} - \left(\frac{1+f}{1+i}\right)}{\left(\frac{1+f}{1+i}\right) - 1}\right] [C_{H(t_a)} - C_{H(t_b)}] - [C_{I(t_b)} - C_{I(t_a)}] > 0$$

Table VI, Columns 4 and 5, shows typical calculations, and Table V, Columns 8 and 9, shows economic thicknesses resulting from this method of analysis.

A comparison of ETI as a function of amortization period given by Economic Models 3, 4, 5, and 6 Table V

Insulation amortization period n years	Economic thickness of insulation, t, in.							
	Model 3 Nonincremental		Model 4 Incremental		Model 5 Nonincremental		Model 6 Incremental	
	i = 8%, f = 0	i = 12%, f = 0	i = 8%, f = 0	i = 12%, f = 0	i = 8%, f = 10%	i = 12%, f = 10%	i = 8%, f = 10%	i = 12%, f = 10%
2	2½	2½	2½	2½	3	3	2½	2½
3	3½	3	3	3	4	3½	3	3
4	4	3½	3	3½	4½	4½	4	3½
5	4½	4	3½	3½	5½	5	4½	4
6	4½	4	4	4	5½	5	4½	4
7	5	4½	4½	4	>6	5½	5	4½

Typical calculations for Economic Models 5 and 6 Table VI

Insulation thickness, t, in.	Model 5 Value of $B[C_{H(t)}] - C_{I(t)}$†		Model 6 Value of $B[C_{H(t_a)} - C_{H(t_b)}] - [C_{I(t_b)} - C_{I(t_a)}]$†	
	n = 2 years, i = 8%	n = 5 years, i = 12%	n = 2 years, i = 8%	n = 5 years, i = 12%
½	44.6	105	15.0	37.2
1	25.4	63.9	5.33	15.0
1½	15.8	44.6	2.00	7.35
2	9.57	33.0	0.516*	3.93
2½	4.85	24.9	−0.277	2.10
3	0.929*	18.6	All negative	1.01
3½	−2.52	13.4		0.295*
4	All negative	8.88		−0.193
4½		4.87		All negative
5		1.22*		
5½		−2.18		
6		−5.38		

*The ETI is given by the smallest positive number in each column.

†Where: $B = \left[\frac{\left(\frac{1+f}{1+i}\right)^{n+1} - \left(\frac{1+f}{1+i}\right)}{\left(\frac{1+f}{1+i}\right) - 1}\right]$

ETI calculations and models

If we examine Tables III and V, we see how dependent is the economic thickness on the choice of the economic factors n, f and i. For example, as n varies from 2 to 7 years, then several of the models show that the economic thickness varies from 2½ to 4½ in. The difference in cost between those two thicknesses is $8.40 per foot. Also considering an amortization period of 5 years, then the economic thicknesses are:

3½ in. from Model 4 (i = 8%, f = 0)
4 in. from Model 1
4½ in. from Model 3 (i = 8%, f = 0)
4½ in. from Model 6 (i = 8%, f = 10%)
5½ in. from Model 5 (i = 8%, f = 10%)

The difference in cost between these two extremes is again $8.40 per foot and these cost differences will represent very substantial sums of money for several hundred feet of insulation.

In addition it must be remembered that the values of i and f used here are no more than "reasonable" values chosen from a range of "equally reasonable" values, all of which would give different thicknesses of insulation. The amortization period has been deliberately restricted to a maximum of seven years and it might be thought that in view of the durability of insulation and its cheap maintenance (usually taken as an annual cost of about 1% of the original capital investment) a longer period should be chosen. As far as this article is concerned, the reason for short amortization periods is merely to ensure that the insulation thicknesses would lie within the usual single-layer and likely double-layer costing.

However, there is the more fundamental reason already referred to at the beginning of this article—many plant owners treat insulation as just another investment to be judged by the same criteria as all other plant investments. In addition they argue that quite often energy costs are less than 10% of total manufacturing costs and if there is any money to spare it should go to reducing the remaining 90% of costs. But, there are fallacies in this point of view.

Dollar savings resulting from insulation are totally predictable and without risk; they are guaranteed, whereas investment in more plant or new products is certainly not so predictable.

Energy savings through insulation can directly release cash flows to be spent where they can do most good, which is certainly not in heating the local atmosphere and drains! Every unit of heat saved is a direct addition to income and thus available for investment.

Although energy costs may be only 10% or less of total manufacturing costs, they represent a much higher proportion of controllable costs (actual production costs), unlike costs of buildings, capital and labor.

If this reasoning is accepted, then plant owners should cease to treat insulation as just another investment judged on a simple Payback Period or Return on Investment (ROI) basis, since neither of these gives credit for dollar savings that will continue to accrue long after the insulation has paid for itself. The only way they can do this is to use the discount methods shown here, provided—and this is an important proviso—they know their economic factors and prefer a specific economic model. After all, a reasonable question to ask is "which economic model is right?"

In fact, there is no single answer to this question, since all the models are "right"—being no more than different methods of mathematically modeling the economic influence of money and time. Economists do argue about the accuracy of mathematical models and there seems little consensus among them; this is one reason why we turned back to purely engineering considerations in proposing alternative criteria for assessment of insulation.

The purpose of this article is not to undermine confidence in the computer programs that exist for evaluating the economic thickness, because these programs are a great help if the economic factors are known (a point we return to at the end of this article). However it must be admitted that this is not always the case.

It has been the personal experience of the senior writer that in some cases the plant owners or their chief engineers have been undecided as to the exact value of n and i to use and therefore have tried a range of "reasonable" values. They inevitably ended up with a "reasonable" but confusing range of "economic" thicknesses and were undecided how to determine which one to use—as exemplified in Table V.

In addition, there are other uncertainties, such as the rising cost of insulant material and increasing labor costs, both of which will result in higher installed costs. Also in many new-plant (as opposed to retrofit) situations there will be a variety of pipe sizes, lengths, bends, flanges and valves, all of which must be allowed for. If the fluid temperatures are very high, then costly multilayered insulation is required.

Finally, if an outside insulation contractor is employed, he will likely give a lump-sum cost estimate for the entire insulation work, leaving the plant engineer with little hope of checking its reasonableness by using computer methods.

In fact, the senior writer has had the benefit of frank talks with insulation contractors who, when "put in the hot seat," admit that their costing procedures do not go by the book, because they do not have the staff or time to use the computer programs for every pipe length or flange. One remarked that the time and cost of using the programs "would be commercial suicide"!

Insulation without economics

How, then, can a decision be made on the appropriate insulation thickness to use if the economic factors are not known accurately? One very direct way of resolving this dilemma is to remove all these economic factors from the problem. This is quite possible because it has long been known that the heat-loss factors and the economic factors appear separately in the computer programs; so that if the economic factors are thrown out, then the insulation thickness is decided by heat-loss considerations alone. The result then becomes not an "economic" thickness of insulation, but what could very well be termed an "acceptable heat-loss" thickness of insulation.

The acceptable heat-loss thickness

It should be recalled that although pipe insulation is sold per linear foot, the heat loss from an insulated (or bare) pipe is a *surface-area effect,* measured in Btu h^{-1} ft^{-2}.

It is easy to show that Eq. (1) when written on this surface-area basis becomes:

$$\frac{Q_{(t)}}{A_I} = \frac{\theta_1 - \theta_3}{\dfrac{d_2}{2k_I}\ln\left(\dfrac{d_2}{d_1}\right) + \dfrac{1}{h_I}} \tag{6}$$

$$= \frac{\theta_1 - \theta_3}{R_I + R_S} \tag{7}$$

while the heat loss per unit area from a bare pipe (zero insulation thickness) is:

$$\frac{Q_{(0)}}{A_P} = \frac{\theta_1 - \theta_3}{1/h_P} \tag{8}$$

$$= \frac{\theta_1 - \theta_3}{R_P} \tag{9}$$

Let us propose a criterion:

$$\frac{Q_{(t)}}{A_I} = F\left(\frac{Q_{(0)}}{A_P}\right) \tag{10}$$

Eq. (10) is the algebraic way of stating that the actual heat loss per square foot from the insulated pipe is F percent (where F is expressed as a decimal) of the heat loss from the same pipe when bare of insulation. On this basis Eq. (10) becomes:

$$\frac{1}{R_I + R_S} = \frac{F}{R_P}$$

or:

$$R_I = \frac{R_P}{F} - R_S$$

or:

$$\frac{d_2}{2k_I}\ln\left(\frac{d_2}{d_1}\right) = \frac{R_P}{F} - R_S$$

Whence, replacing diameters by radii:

$$r_2 \ln\left(\frac{r_2}{r_1}\right) = \frac{k_I R_P}{F} - k_I R_S \tag{11}$$

The writers have made a numerical study of the right-hand side of Eq. (11) and shown that (in all but the most unusual circumstances):

$$\frac{R_P}{F} \gg R_S$$

so that we have the accurate approximation equation

500° F, 8-in.-dia. horizontal pipe with calcium silicate insulation — **Table VII**

Thickness of calcium silicate insulation, t, in.	Surface temperature of insulation, θ_2, °F	Heat-loss rate per unit area, $Q_{(t)}/A$, Btu $h^{-1}ft^{-2}$	Heat-loss rate per unit area, compared with that from a bare pipe, %
0	500	1,770	100
½	210	321	18.1
1	160	183	10.3
1½	136	124	7.0
2	122	92.8	5.2
2½	113	73.2	4.1
3	107	59.9	3.4
3½	102	50.3	2.8
4	99	43.2	2.4
4½	95	37.6	2.1
5	92	33.2	1.9
5½	90	29.6	1.7
6	89	26.7	1.5

$$r_2 \ln \frac{r_2}{r_1} = \frac{k_I R_P}{F} \tag{12}$$

Since the insulation thickness is given by

$$t = r_2 - r_1 \tag{13}$$

then, provided that we can solve Eq. (12) for given values of k_I, F and R_P, we can determine this acceptable-heat-loss thickness of insulation.

It is often written in the technical literature that the solution of an equation of the form of Eq. (12) is "difficult," and accordingly extensive tables of presolved equations are given or graphical solutions drawn. Both of these packaged solutions are unnecessary, since the equation can be quickly and easily solved on the simplest of hand-held programmable calculators without the necessity of either interpolation or extrapolation.

Choosing values for F and k

Before discussing these simple programs, it may be of value to discuss appropriate values of F and k_I. If we return to our numerical study of the 8-in. steam main at 500°F, then we can readily determine the percentage heat loss per unit area from it, when it is covered with different thicknesses of insulation. Table VII shows these values, clearly illustrating the law of diminishing returns below about 3% heat-loss rate. It is suggested that F should never be greater than 5%, while 3% appears satisfactory in many situations—however, there is a free choice of F for whoever uses the program.

Choosing insulation materials

The value of k_I is determined by the choice of insulant and its average operating temperature. It is surprising how many plant engineers are still specifying the same pipe-insulation materials that they did 10 years ago. One consequence of this is that they may be paying more for their insulation than is necessary. In fact the writers, in order to draw attention to this point, have quite deliberately used the "wrong" (possibly more expensive) insulation in their calculations on the 8-in.-dia. pipe at 500°F, since calcium silicate is quite likely to be more expensive than rigid preformed glass fiber.

There is, in fact, a choice of insulants at 500°F, such as the mineral wools (slag, rock and glass), foamed glass, 85% magnesia, and diatomite. A discussion of suitable insulants for various temperature ranges has been given by Harrison and Pelanne [2] and also by Probert and Giani [3], and need not be repeated here. In many cases the choice seems to be between calcium silicate and glass fiber, the latter having several advantages over the former. Calcium silicate is widely available up to 3 in. and sometimes 4 in. and additionally up to 5 in. by special order, but apparently is not available up to 6 in.

Thick calcium silicate is heavy and in a retrofit situation on existing pipework that is close to a wall or ceiling or in a duct, the sheer bulk of the additional insulation may be too great. In addition, the existing pipe supports or hangers may not be strong enough to hold the extra weight.

Glass-fiber preformed rigid pipe section is available up to 6-in. thickness as a single layer; it has a low moisture absorption and its low chloride content makes it compatible with stainless steel. Additionally, it has better chemical resistance to both acids and alkalis than does calcium silicate. Another important advantage is that it has a lower thermal conductivity than calcium silicate, which means smaller insulation thicknesses and a more light-weight installation because of its lower density.

The writers were interested to read of a one-mile-long 14-in. steam main at 445°F recently installed in a U.S.A.F. base. It was insulated with 6½ in. of preformed single-layer glass fiber having a factory-applied all-service jacket underneath and field-applied smooth aluminum jacket. It was claimed that if calcium silicate had been used, it would have had to be double-layered, which would have increased contract costs by at least 20%.

The actual value of k_I for a specified insulant to be used in the TI-57 program given in this article is best obtained from manufacturers' data sheets, because not only does the value of k_I depend upon the average temperature $(\theta_1 + \theta_3)/2$, but also upon the method of production, and only the manufacturer can supply this information. However for a *rough* "ball-park approximation" the following are suitable average values of the thermal conductivity, k_I.

Calcium silicate, k_{CaSi}:

$$k_{CaSi} = 3.33 \times 10^{-2} + 8.75 \times 10^{-6} \frac{\theta_1 + \theta_3}{2} + 2.38 \times 10^{-8} \frac{\theta_1 + \theta_3}{2}^2 \tag{14}$$

Glass fiber, $k_{gl\,fib}$:

$$k_{gl\,fib} = 1.25 \times 10^{-2} + 3.95 \times 10^{-11} \times \left(\frac{\theta_1 + \theta_3}{2} + 460\right)^3 \tag{15}$$

For example at $\theta_1 = 500$°F and $\theta_3 = 70$°F the above formulas give:

Program 1: Bare pipe resistance — **Table VIII**

Location	Code	Instruction	Location	Code	Instruction
0	331	RCL 1	25	43	(
1	75	+	26	331	RCL 1
2	04	4	27	65	–
3	06	6	28	07	7
4	00	0	29	00	0
5	85	=	30	44	)
6	23	x^2	31	45	÷
7	75	+	32	332	RCL 2
8	05	5	33	44	)
9	03	3	34	35	y^x
10	00	0	35	04	4
11	23	x^2	36	25	1/x
12	85	=	37	55	X
13	55	X	38	83	∘
14	43	(	39	02	2
15	331	RCL 1	40	07	7
16	75	+	41	85	=
17	09	9	42	25	1/x
18	09	9	43	55	X
19	00	0	44	333	RCL 3
20	44	)	45	45	÷
21	55	X	46	334	RCL 4
22	335	RCL 5	47	85	=
23	75	+	48	81	R/S
24	43	(	49	71	RST

Program 2: Insulation thickness — **Table IX**

Location	Code	Instruction	Location	Code	Instruction
0	333	RCL 3	25	84	+/–
1	321	STO 1	26	75	+
2	331	RCL 1	27	331	RCL 1
3	55	X	28	85	=
4	43	(	29	323	STO 3
5	331	RCL 1	30	65	–
6	45	÷	31	331	RCL 1
7	330	RCL 0	32	85	=
8	44	)	33	40	|x|
9	13	ln x	34	76	x⩾t
10	65	–	35	71	RST
11	332	RCL 2	36	333	RCL 3
12	85	=	37	321	STO 1
13	45	÷	38	65	–
14	43	(	39	330	RCL 0
15	43	(	40	85	=
16	331	RCL 1	41	55	X
17	45	÷	42	01	1
18	330	RCL 0	43	02	2
19	44	)	44	85	=
20	13	ln x	45	81	R/S
21	75	+	46	322	STO 2
22	01	1	47	482	Fix 2
23	44	)	48	71	RST
24	85	=			

$$k_{CaSi} = 3.33 \times 10^{-2} + 8.75 \times 10^{-6} \times [(500 + 70)/2] + 2.38 \times 10^{-8}[(500 + 70)/2]^2 = 0.0377 \text{ Btu ft h}^{-1}\text{ ft}^{-2}\text{ °F}^{-1} \quad (16)$$

and: $$k_{gl\,fib} = 1.25 \times 10^{-2} + 3.95 \times 10^{-11} \times \left[\left(\frac{500 + 70}{2}\right) + 460\right]^3 = 0.0288 \text{ Btu ft h}^{-1}\text{ ft}^{-2}\text{ °F}^{-1} \quad (17)$$

It may be of value to note that in this article all values of k_I have units of Btu ft h^{-1} ft^{-2} °F^{-1}. However, many technical journals and manufacturers' data sheets give values of k_I in Btu in. h^{-1} ft^{-2} °F^{-1}. The relationship between those two units is very simple and involves a factor of 12; thus:

- Multiply a k_I value in Btu in. h^{-1} ft^{-2} °F^{-1} by 1/12 to convert it to Btu ft h^{-1} ft^{-2} °F^{-1}
- Multiply a k_I value in Btu ft h^{-1} ft^{-2} °F^{-1} by 12 to convert it to Btu in. h^{-1} ft^{-2} °F^{-1}

For example, the above value for glass fiber is

$$k_{gl\,fib} = 0.0288 \text{ Btu ft h}^{-1}\text{ ft}^{-2}\text{ °F}^{-1} = 0.0288 \times 12 = 0.346 \text{ Btu in. h}^{-1}\text{ ft}^{-2}\text{ °F}^{-1}$$

Computer programs description

Eq. (12) and (13) have been programmed for the Texas Instruments TI-57 calculator. Because of the limited number of steps (50) in this calculator, it has been necessary to break down the calculation of the acceptable-heat-loss thickness of insulation into two programs. Owners of the more powerful TI-59 calculator or its equivalent will be able to combine the two programs given here into one.

Program 1: Bare-pipe resistance

This program, shown in Table VIII, evaluates the right-hand side of Eq. (12), k_IR_p/F, by first calculating:

$$R_P^{-1} = 0.27\left(\frac{\theta_1 - 70}{d_1}\right)^{0.25} + 0.154 \times 10^{-8} \times (\theta_1 + 990)[(\theta_1 + 460)^2 + (530)^2]$$

and thereafter the value of k_IR_P/F, where k_I is to be obtained from manufacturers' data sheets or, failing that, from Eq. (14) or (15), and F is decided by the user.

Before the program is run, the following calculator store locations must be initialized:

STO 1 θ_1
STO 2 d_1
STO 3 k_I
STO 4 F
STO 5 0.154×10^{-8}

The program is run from Location 0 and, after the data have been entered, it can be run by

- Resetting the calculator by pressing the **RST** key.
- Pressing the **R/S** key.
- Waiting for a steady display showing the value of k_IR_P/F that is to be used in Program 2.

*Program 1 will run directly on the TI59, but Program 2 will not. To run Program 2 on the TI59, add LBL A at the start of the program, substitute A for RST in Step 35, and insert the "accuracy required" figure in the t-register (rather than in STO 7). After the data are entered, press **A** to run—Editor.

Program 2: Insulation thickness

This program shown in Table IX solves Eq. (12) and (13) together and yields the insulation thickness t_I *in inches.* Before the program is run, the following store locations must be initialized:

STO 0	r_1
STO 1	1
STO 2	$k_I R_P/F$
STO 3	1
STO 7	Accuracy required (usually 0.001)

The program is run from Location 0 after the data have been entered (see Program 1); the steady display gives the insulation thickness in inches.

When the calculation is complete, in order to repeat the procedure for another value of $k_I R_P/F$ it is necessary to:

1. Enter the new value of $k_I R_P/F$ on the keyboard.
2. Press the **R/S** key.

As an example of the use of the two programs, let us return to the problem of the 8-in.-dia. horizontal pipe, in 70°F still air, carrying dry saturated steam at 500°F, i.e. $\theta_1 = 500$°F; $\theta_3 = 70$°F; $d_1 = 2/3$ ft; and $r_1 = 1/3$ ft.

For calcium silicate, from Eq. (16)

$$k_{CaSi} = 0.0377 \text{ Btu ft h}^{-1} \text{ft}^{-2} \text{ °F}^{-1}$$

Example 1: $F = 5\% = 0.05$

Program 1 gives $k_I R_P/F = 0.183$

Program 2 gives $t = 1.83$ in.

Clearly 2 in. of calcium silicate insulation would be chosen.

If greater heat conservation is required, then we have:

Example 2: $F = 3\% = 0.03$

Program 1 gives $k_I R_P/F = 0.305$

Program 2 gives $t = 2.83$ in.

Clearly 3 in. of calcium silicate insulation would be chosen.

It is interesting to calculate glass-fiber insulation thickness for the same pipe: in this case $k_{gl\,fib} = 0.0288$ Btu ft h^{-1} ft^{-2} °F^{-1} from Eq. (17).

For $F = 5\%$

Program 1 gives $k_I R_P/F = 0.140$

Program 2 gives $t = 1.45$ in.

This means that 1½ in. of glass fiber would give the same percentage heat loss as 2 in. of calcium silicate.

For $F = 3\%$

Program 1 gives $k_I R_P/F = 0.233$

Program 2 gives $t = 2.25$

This result means that 2½ in. of glass fiber would give the same percentage heat loss as would 3 in. of calcium silicate.

These results not only show the simplicity of the proposed criterion for evaluating the insulation thickness but also confirm the thinner and lighter insulation resulting from the use of glass fiber as opposed to calcium silicate. However, it must not be forgotten that glass fiber needs particularly strong jacketing to avoid mechanical abuse since it is considerably less rigid than calcium silicate.

Another benefit of the proposed criterion is that the dollar cost of the heat loss from any thickness of insulation may readily be found from Program 1, since it is easy to show that:

Cost of heat loss per unit area of insulated pipe, dollars

$$\text{per square foot per hour} = \frac{k_I C_F(\theta_1 - \theta_3)}{(k_I R_P/F)E}$$

Conclusion

In conclusion, the writers would like to reaffirm their confidence in the usefulness of the existing manuals and books that use discounted-cash-flow analysis of insulation economics for those cases where the parameters *n*, *i* and *f* are known. For example, company accountants may *dictate* what values of *n* and *i* are to be used, based upon their economics-oriented viewpoint of all engineering investments. Very often the government lays down the value of *f* that must be used if tax credits are to be claimed. In these cases there is no uncertainty, and the discounted-cash-flow tables can, and should, be used.

However, if the plant engineer is allowed some say in the decision-making, he may care to put aside the economic arguments of the accountants and use the very simple method described in this article—and feel that his intuition and experience about what is an acceptable percentage heat loss from a pipe, and its cost, are just as good a way of looking at the problem of specifying insulation thickness as any other.

References

1. McChesney, M., and McChesney, P., *Chem. Eng.,* Vol. 88, July 27, 1981, p. 58.
2. Harrison, M. R., and Pelanne, C. M., *Chem. Eng.,* Vol. 84, Dec. 19, 1977, p. 61.
3. Probert, S. D., and Giani, S., *Applied Energy,* Vol. 2, p. 83 (1976).
4. McMillan, J. B., Heat Transfer Through Insulation, paper presented to American Soc. of Mechanical Engineers, New York, N.Y., Dec. 6, 1926.

The authors

Malcolm McChesney is founder and Senior Associate of Fuel Save Associates, a group of energy consultants. He is also Senior Lecturer in Energy Studies and Thermodynamics in the Dept. of Mechanical Engineering, University of Liverpool, P.O. Box 147, Liverpool, L69 3BX, U.K. He has written over 20 research papers, three books, and articles for the Encyclopaedia Britannica and *Scientific American.*

Peter McChesney is a college student and Associate (computing) in Fuel Save Associates. He is the son of Malcolm McChesney. Since the age of 12 he has published articles on analog and digital circuitry, ranging from analog to digital converters through musical synthesizers to microprocessor-based hardware. He was a U.K. North-West prize winner in the 1979 Young Engineer for Britain Competition, for designing and constructing a multiplexed digital system for monitoring failure in a chain of steam traps.

Effects of insulation on refractory structures

Adding insulation to reduce heat loss may lower a refractory's mechanical strength and result in failure. Here is how to prevent this from happening.

Gary J. Nagl, Air Resources, Inc.

☐ With increased emphasis on energy conservation in industrial processes, more and more refractory installations are being designed with backup insulation to reduce heat loss. Unfortunately, many such structures are engineered without considering the effect of insulation on the structural integrity of the refractory. Thus, many refractory installations fail, which ups maintenance costs and can offset any energy saving due to the use of insulation.

Softening

Most refractories are a combination of silica and alumina, and do not have a definite melting point. There are a few exceptions—such as silica, corundum and magnesite brick—that consist of essentially pure oxides. Typical refractories soften over a wide temperature range, in which both solid and liquid are present. Fortunately, for fireclay and high-alumina refractories the liquid remains extremely viscous at temperatures exceeding 2,000°F.

Under these conditions, the refractory will deform when subjected to compressive loading, the degree of deformation depending on the temperature and the amount of loading. When the brick temperature is increased to a point at which the viscosity of the liquid is very low, the brick will deform under its own weight.

Generally, refractory walls and arches rarely fail because of compressive stress. This is due to the temperature gradient that exists across the refractory. Under certain temperatures and with certain types of refractory, the hot face of the refractory may be pyroplastic (or soft) and unable to support a compressive load. However, due to the temperature gradient, the cooler portion of the brick may have sufficient strength to support the entire load.

An example of this is illustrated in Fig. 1. Here, a refractory wall consisting of 9 in. of superduty fireclay brick is subjected to a hot-face temperature of 2,600°F. Assuming that the refractory becomes pyroplastic above the temperature of 2,200°F, approximately $1^3/_4$ in. of the brick will be unable to support the load, while the remaining $7^1/_4$ in. should be sufficiently strong to carry the entire load.

Heat loss from the wall would be approximately 2,200

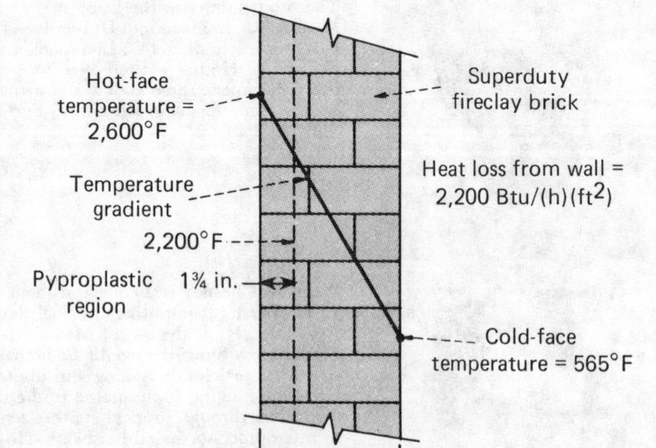

With the proper temperature gradient, the cooler part of the brick handles compression Fig. 1

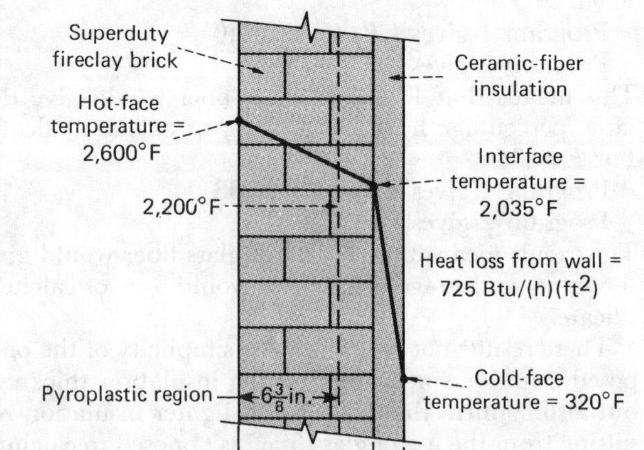

Ceramic-fiber insulation applied to the cold face cuts heat losses but keeps brick hotter Fig. 2

Originally published October 18, 1982

Exceeding these temperatures can result in structural damage of the refractory **Table I**

Refractory	Maximum mean operating temperature, °F
Superduty fireclay	1,950–2,200
60% Al_2O_3	2,225–2,300
70% Al_2O_3	2,400–2,500
80% Al_2O_3	2,425–2,525
90% Al_2O_3	2,600–2,800

In replacing superduty fireclay with higher-grade materials, the cost must be considered **Table II**

Refractory	Relative cost
Superduty fireclay	1.00
60% Al_2O_3	1.82
70% Al_2O_3	2.45
80% Al_2O_3	3.09
90% Al_2O_3	7.59

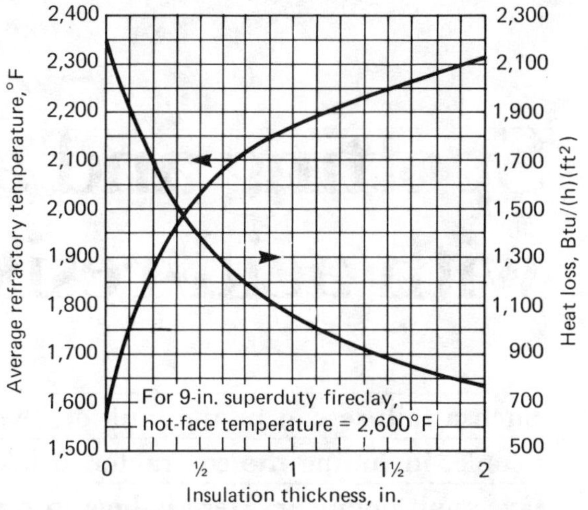

Insulating the refractory lowers heat losses, while increasing the hot-face temperature **Fig. 3**

Btu/(h)(ft²). A considerable portion of this heat can be conserved by applying insulation on the cold face, as shown in Fig. 2. There, 2 in. of ceramic-fiber insulation was applied to the cold face of the refractory, which reduced the heat loss to approximately 770 Btu/(hr)(ft²). However, the interface temperature of the refractory has increased to approximately 2,010°F, resulting in over three-quarters of the brick being pyroplastic.

This installation will not have enough structural strength to give maximum service life. The result is that the money saved by conserving heat may be offset by the increased maintenance cost of replacing the refractory.

Mean temperature

Thus, attention must be given not only to the maximum service temperature of the refractory but also to the mean temperature, which is indicative of the structural strength of the refractory. Refractory manufacturers publish maximum mean operating temperatures that are generally based on a comprehensive loading of 25 psi. Typical maximum mean operating temperatures for commonly used refractories appear in Table I.

In considering these temperatures, it is necessary to make allowances for spalling, slagging and abrasion. Such phenomena reduce the thickness of the refractory and raise the mean operating temperature. Generally, safe practice is to limit this temperature to approximately 100°F below the recommended maximum value.

In Fig. 2, the designer would be required to either reduce the amount of insulation and accept a higher heat loss or to upgrade the refractory. Fig. 3 illustrates the relationship between insulation thickness, average refractory temperature and heat loss, for this example.

If it is assumed that the maximum mean temperature for the superduty brick is 2,200°F, then the maximum amount of insulation that can be applied is approximately 1 in., with a resulting heat loss of 1,050 Btu/(h)(ft²). If the designer desires to decrease the heat loss further, the refractory must be upgraded to 60 or 70% alumina brick, which will cost over twice as much as superduty brick. The cost of the energy saved must offset the increase in material cost. The cost of various grades of refractories relative to superduty brick are shown in Table II.

Tie-backs and suspended walls

The effect of insulation on the operating temperatures of the refractory becomes more critical when tie-back or suspended-wall constructions are used. In both cases, the metallic shelves or clips that are used for structural support of the wall are embedded in the refractory. Thus, these devices are subjected to the operating temperatures of the refractory. As more insulation is applied to a suspended or tie-back refractory wall and the temperature of the brick increases, the maximum service temperature of the embedded metallic may be exceeded, resulting in structural failure of the wall.

For example, consider an overinsulated suspended wall. A 9-in. suspended superduty brick with 1 in. of insulation is exposed to a hot-face temperature of 2,600°F. The metallic support shelf extends 3 in. into the brick, and hence if it is not air-cooled, it will be subjected to the refractory temperatures. Approximately 2½ in. of the shelf will be overheated, resulting in certain failure of the wall. In the design and installation of suspended and tie-back walls, one must ensure that the metallic components are not overheated.

With realization that backup insulation on walls or arches will reduce the structural integrity of the refractory, the proper refractory and insulation thickness can be selected. This saves energy and gives long refractory life.

The author

Gary J. Nagl is vice-president, Energy Systems Div., of Air Resources, Inc., 600 N. First Bank Dr., Palatine, IL 60067. Tel: (312) 359-7810. With the firm since 1973, he is responsible for engineering design and construction, and has developed energy-recovery and particulate-incineration methods for petroleum-coke calcination. Previously, he worked for Universal Oil Products Co. as development engineer, process coordinator and engineering research and development coordinator. He received his B.Ch.E. degree from the University of Illinois and is a member of AIChE and the American Assn. of Cost Engineers, among other societies.

Spotting and avoiding problems with acid-resistant brick

Such a failure can be the fault of several people, including the contractor, bricklayer and engineer. Here is how to track down—and prevent—breakdowns.

Walter Lee Sheppard, Jr., C.C.R.M. Inc.

☐ Most chemical engineers have seen acid brick installations—such as floors, gutters, trenches, linings for waste collection and ducting, or chimney liners. But not many engineers are called upon to design or to install acid brick structures or linings. Fewer still have sufficient experience with these materials to determine the causes of trouble with or failure of such an installation.

When failure occurs, it can almost always be traced either to: (1) a lack of understanding of the limitations and the uses of the materials by the designer; (2) insufficient experience with or understanding of the handling of the materials by the installer; or (3) underbidding of the job by a contractor trying to substitute cheaper materials or cut corners.

Limitations

Acid brick structures have the following limitations:

■ They are strong in compression, but weak in tension and shear. They are also somewhat brittle and nonelastic, so they can be damaged by sudden shock or excessive vibration. Therefore, they are used as either load-bearing structures or as surfacing materials for concrete or steel structures. The structures provide support for them and prevent their bending or flexing.

■ They absorb fluids. Liquids and gases can diffuse into them or slowly pass through them, so they are not in themselves gas-tight, nor will they sustain a hydrostatic head. Therefore, when constantly wet or used to protect the interior of a vessel, they should be used in conjunction with a liquid-tight membrane that acts as a seal to prevent corrosives and/or gases from reaching the supporting steel or concrete substrate.

■ They show thermal elongation and irreversible growth. All acid brick made in North America shows such behavior. Elongation is usually up to 0.16% of any dimension. This growth is slowest in cold, dry, static conditions, and fastest in hot, wet cycling exposures. Growth is not uniform, even for the same batch of brick. The reason for this is unknown. All that we can be sure of is that some growth is almost certain to take place, and design (and of course placement of the expansion joints) must provide for it.

Though acid brick, due to its high compressive strength, is used as liners for chimneys and as supporting structures in acid-contaminated soil, its primary function, in combination with liquid-tight membranes, is to thermally and mechanically protect steel or concrete substrates in corrosive environments. Acid brick linings may also be used to extend the life of a membrane or substrate exposed to an environment that will slowly attack it. Inasmuch as the corrosives traverse the brickwork through winding capillary paths, their attack on the membrane or substrate must be limited to the end of these minute paths. If the rate of attack is small, and the corrosion produces swelling or an insoluble corrosion product, the reaction can seal or partly seal the corrosive's path, and so greatly lengthen the life of the substrate or membrane.

Properly designed and installed acid brickwork can fulfill a great many useful functions. If, however, someone has erred in material selection, design, specification or installation, a failure will result. Our aim here is to provide a key for finding the cause of such a failure.

Cause of failures

These can be grouped under the following:

1. Faulty material selection.
2. Faulty design.
3. Faulty specification.
4. Improper material substitution in erection.
5. Improper protection of materials before, during and after the installation until curing is complete.
6. Improper installation techniques.

The first three items are the responsibility of the architect-engineer who is assigned the job of drawing up the plans and specifications, and is often also charged with listing acceptable installers. Items 4 through 6 are the responsibility of the contractor.

Often, the failure will result not from a single cause, but from a combination of several. Acid brick construction, in most instances, can survive a few minor errors in design and erection, but an accumulation of several practically guarantees early failure.

Unfortunately, few architects and engineers are thoroughly familiar with the design and use of these materials, and only a limited number of contractors have experience in their installation. In most major contracts,

Originally published May 3, 1982

Analyzing failures: Work through the lists here to narrow down the causes of a failure

The **boldface** numbers mean the following:

1 is noted where the cause of the failure can clearly be assigned to the designer.

2 indicates the fault is that of the applicator.

3 means there are two or more individuals who may be at fault.

4 indicates the materials manufacturer is responsible.

5 shows the fault probably lies with the operator.

I—Leaks through acid brick lining or floor. (Detected by "holing" of steel tank, or wet spots, discoloration, or collapse of concrete.)

1. Wrong mortar/membrane for chemicals and/or temperature. **(1)**
2. Wrong mortar/membrane substituted for that specified. **(2)**
3. Sheet membrane material had pinholes in it. **(4)**
4. Mortar/membrane materials were off-spec. **(4)**
5. Membrane applied improperly. **(2)**
6. Hot-asphalt membrane froze, then cracked, after application to concrete. **(3)**
7. Multicomponent membrane/mortar improperly proportioned or mixed. **(2)**
8. Mortar/membrane mixed with foreign material, such as wind-blown dust, or sand. **(2)**
9. Material applied after it passed beyond its worklife or outside of specified temperatures. **(2)**
10. Mortar/membrane not thick enough or in continuous layers; failure to install full-bed joints under or behind brick. **(2)**
11. Damage to membrane during bricklaying. **(2)**
12. Carbon brick and/or carbon-filled mortar in direct contact with lead or stainless steel caused holing of the lead or pitting of the steel. **(1)**
13. Concrete degraded. Anchors in the brickwork, or other penetrations through membrane into concrete, allowed flow of chemicals. **(3)**
14. If a vessel, it was not liquid-tight before being lined. **(2)**

II—Brick wall lining. (rectangular tank, or gutter or trench) falls in. (Wall may or may not carry membrane with it.)

1. Brick lining too thin for height and width. Brick lining must be independent of substrate. **(1)**
2. Concrete wall may have inward bulge at some point. **(2)**
3. Failure to properly install expansion joints at correct locations, or failure to make them large enough. **(3)**
4. Use of improper joint filler, or improper installation of filler. **(3)**
5. Dirt or mortar in expansion joints, preventing functioning. **(2)**
6. See also I, nos. 3, 5, 9, 10 and 11.
7. Dirt or moisture on brick and/or membrane that prevented mortar from bonding properly. Or mortar improperly mixed and applied. **(2)**

III—Damage or loss of mortar joints.

Sulfur mortar:

1. Loss or crumbling can be due to overheating. Check recording thermometer, especially for cleaning cycles. Absolute top service temperature for sulfur is 203°F. Manufacturers say stay under 190°F. **(5)**
2. Empty joints due to poor installation procedures. **(2)**
3. Loss or damage can be caused by chemicals. Look for unnoted trace chemicals or cleaners. Perhaps operating conditions have been changed since design was done. **(1) (5)** (Heat damage usually leaves joint full, but crumbly. Chemical damage involving solvents usually removes some or all material from the joints.)

Resin mortar:

4. Receding or etched joints usually indicate chemical damage. **(1) (5)**
5. Joint damage can also result from putting brickwork into service before cure of the mortar is complete. **(2)** (Overheating may char or crack joint. It rarely shows up as joint loss.)

Silicate and silica mortar:

6. Soft, receding joints result from exposure to steam jets or to neutral or alkaline water. Receding joints can also be caused by HF or acid fluorides. **(1) (5)**
7. Neutral waters or washdowns, before cure was complete, caused loss of mortar. **(2)**

General:

8. Bricks dirty, so mortar did not stick to them and joints fell out. **(2)**
9. Mason never filled joint, laid brick dry. **(2)**
10. Mortar used past work-life, so had no adhesion. **(2)**
11. Unauthorized material mixed into mortar. **(2)**

IV—Damage to shale-fireclay brick in the lining.

1. Fairly uniform surface damage, etching or spalling is almost certainly chemical damage. **(1) (5)**
2. Acid fluorides dissolve off the brick face, leaving an etched surface. **(1) (5)**
   In either of these above exposures, carbon brick is the material of choice.
3. Spalling at edges of brick (at joints), but fairly sound at the middle of the brick. If joints recede **(1) (5)**, or were not full to start with **(2)**, expanding brick are not supported at edges, and corners spall off the brick edges.
4. Spalling of surface in selected areas. This can be due to:
   a. Local overheating from steam impingement. **(5)**
   b. Local exothermic reaction. **(5)**
   c. Use of soft (or underfired) brick. **(4)**

V—Sags or runs in mortar joints, usually accompanied by voids.

Soupy mortar is almost always the fault of the installer. **(2)**

Such joints can be identified by rounded, smooth,

continued

bulging horizontal joints in the brickwork, often glassy, accompanied by pinholes or voids near their tops.

VI—Heaving upward of brick floors and tank bottoms.

Floors:

1. Expansion joints improperly designed, in wrong places, or wrong size or wrong expansion-joint filler. **(1)** (See also II, 3, 4, 5.)
2. Brick underfired or of a clay with excessive, irreversible growth. **(4)**
3. Failure to lay brick void-free. **(2)** (Air voids under brick will pick up liquids, in which crystals can form, causing growth and upward pressure on brick.)
4. Voids or holes in the membrane permit corrosives to reach substrate. **(2)**
5. For epoxy adhesive bed, heaving is due to dirt, dust or faulty application. **(2)**

Tank bottoms (flat bottoms):

Items 1 through 4 apply.

6. Bottom flat, not ventilated, tank hot. **(1)**

VII—Cracking of brick linings of steel tanks.

1. Lining too thin to provide adequate insulation at the operating temperature. **(1)**
2. For lead sheet, or sheet elastomer or plastic, welds (or laps) not padded, brick not notched. **(2)**
3. Seizing of brickwork on seams causes cracking. (See I, 11.) **(2)**
4. "Jacking" of brickwork can cause cracks. **(2)**
5. Failure to provide for expansion at inlets, outlets and other shell penetrations. **(1)**
6. Cure shrinkage of mortar: Strongly bonding mortars, as they shrink on curing, can pull brickwork apart. **(1) (4)**

VIII—Voids in mortar joints.

1. Pinholes and other tiny holes in mortar joints result from air beaten in by a high-speed mixer in the mortar box. **(2)**
2. Pinholes and tiny voids result from insufficient mortar on the trowel. **(2)**

such items are only small parts of the total package, and many engineering and architectural firms do not realize the importance to that total package of having an engineer experienced in this work review material selection, design and specification prior to issue.

Types of failure

The most common type of failure is the heaving upward of a brick floor off the substrate or the bulging upward of a brick bed from the flat bottom of a large tank. Frequently, such heaving or bulging can be traced to the failure of the designer to allow for brick growth and to provide proper expansion joints. But failure can also be due to the improper installation of the joints.

Heaving in a flat-bottom tank may be due to the same lack of provision for brick growth. Or it may be due to the tank's being designed to rest flat upon a concrete pad so that no air can get under it. Such a design effectively insulates the bottom of the tank. In a tank with an internal temperature of 200°F, the steel in the bottom will heat up until it is as hot as the liquid in the tank, while the walls of the tank will be radiating heat outward. The walls will be 30°F to 40°F hotter than the air in the room.

The steel walls, then, will expand only 20 to 30% as much as the steel bottom. The bottom, being restrained by the outer walls and unable to expand downward, will bulge upward, carrying the brick floor with it and causing it to crack.

The corrosives inside the tank can now reach the membrane since it no longer is receiving mechanical and thermal protection from the brick. The membrane is overheated and will fail, and the corrosives can reach the steel bottom and perforate it. Then, the liquid percolates into the concrete base and destroys it. This happens because the designer did not set the tank on I-beams or cribbage. By the time it is discovered, the damage has usually gone too far to be repaired.

Concrete structures

Similarly, carelessness in forming the walls of concrete tanks and trenches can cause the brick lining to fall in. The best design for rectangular vessels is to bow the walls slightly outward at midpoints, giving all the walls a slight outward continuous curve so that the vessel is wider at the center than at the ends. If this is done, the brick laid inside the walls will be in an arched configuration so that as it grows or expands it presses ever more tightly against the supporting structure.

Straight lines are acceptable, but then the lining requires expansion joints, which may not be needed if the walls are contoured. Forms must be set and braced properly so that the weight of the concrete does not bulge the forms inward. The collapse of trench walls or rectangular tanks is often due to improper contouring or inadequate expansion joints. Of course, the breakup or falling-in of a wall may also be due to inadequate support from underdesigned concrete or steel. The concrete may have inadequate foundations and settle and crack, or be inadequately reinforced to support the load. The steel may be thin, or be poorly strengthened, so it moves and permits the brickwork to flex and break.

The designer may have selected the wrong mortar, brick or membrane for the chemical environment or temperature. Thus, the mortar may have been attacked, or the membrane penetrated by corrosives that damaged the supporting structure. The installer may have substituted a wrong material for one specified, stored or used the materials improperly, or allowed dirt or water to enter the structure while it was being built.

When failure occurs through the wall of a tank—by showing up as a leak through the steel or as acid damage on a concrete tank—this is due to membrane failure. If the membrane is a sheet applied with welded or lapped edges (i.e., lead, sheet rubber), and compressible cushions have not been supplied on each side of the lap or weld, the brick lining can seize on the weld or lap and tear it when the brick expands or contracts. Alternatively, such movement can be restrained by the lap, causing the brickwork to crack. This mechanism is explained below.

Lead linings have elevated welded joints between the lead sheets. Sheet rubber (and other sheet) linings are installed with edges lapped, or butted together with a lap strip applied over the butt joints. These joints are elevated to the thickness of the sheet above the level of the surrounding sheet, usually $\frac{1}{4}$ to $\frac{3}{8}$ in. The brick lining inside a sheet membrane, be it lead, rubber or plastic, must be free to move independently of the membrane. Thus, if the bricks are mortared directly against the sheet, the brick and mortar will seize on these elevated sections and be immobilized. Since steel and brick expand differently, either the brickwork will crack or the membrane will be torn apart at the seam. To prevent this, place a strip of compressible foam or other soft packing about $\frac{1}{4}$ to $\frac{1}{2}$ in. on each side of the weld or lap, built up to the same elevation above the surrounding sheet as the weld or lap. Notch the brick and mortar it to go over the lap and the foam on both sides. When the brick expands or contracts, the compressible pad permits the movement without seizing on the lap or weld. For basically the same reason, brick placed around the periphery of a tank (the bottom course of the walls) should be chamfered.

If the membrane linings have pinholes, corrosives can penetrate through them to the supporting steel or concrete, eventually appearing as leaks. Such pinholes may exist in the sheet prior to installation. In installation, bad welds and inadequately sealed laps, or damage to the lining by careless masons installing the brick, can result in the same type of failure.

Also, the membrane may have been applied in a tank that was not liquid-tight. It is a misconception to believe that a tank that is not, in itself, liquid-tight can be made so by lining it with a membrane and brick. Only by removal of the brick and examination of the sheet can the exact cause of failure be determined.

Improper application is often the cause of leaks for fluid-type membranes, such as hot asphalt or neoprene. Failure of the membrane to adhere to the substrate can be due to: (1) failure to apply the primer; (2) failure to clean all dirt from the surface prior to applying the primer; (3) application of primer or of membrane itself to a wet or dirty surface; or (4) improper methods of application. For example, hot asphalt applied to a damp or wet surface will cause surface water to flash into steam, which will pass out through the asphalt surface, leaving pinholes.

To repair such pinholes, it is necessary to cut out an area of at least ten square inches and carefully re-apply the membrane.

Hot asphalt applied with mops, as a roofer does, can leave strings in the asphalt that act as wicks and transmit liquid from one side to the other. Applications should be made only with a smooth-edged squeegee.

Careless bricklaying can also make holes in the membrane and be responsible later for a membrane failure. Bricks are heavy and have sharp corners. Therefore, if the mason hits a corner against the membrane as he seats the brick, he can perforate the membrane, if it is hot asphalt, or an elastomeric or asphalt putty. He is less likely to damage a thick sheet lining.

Pinholing of a lead lining can also result from placing a carbon brick and/or a carbon-filled mortar in direct contact with a lead lining. This results in the creation of a galvanic cell. To prevent this, use a voltage-breaker layer.

Troubleshooting failures

In order to determine the probable cause of a failure, the foregoing box is being provided. By working down the lists, it is usually possible to eliminate many of the possible causes and narrow it down to three or four.

Before using the box, be certain you have the facts regarding the process and the operation. Note the complete process chronology. Include the cleaning-cycle and any idle periods, and the range of temperatures and pressures for all these periods. If the vessel or area is outdoors, include all weather conditions. If it is inside a building that may not be heated when the process is down, then external temperatures must also be included. If there is agitating or mixing equipment, steam or air injection, this must be considered. Look at the plan drawings for the locations of all entries, exits, intercepts, baffles, expansion joints and fixed points, and the relationship of all these points to the failure.

Then check out the specifications to see what materials were specified, and if alternates were used. Examine the manufacturer's literature to be certain that it takes no exception to any of the matters appearing in the process study. Also, compare the literature of all items specified with that of any substitutions installed. Examine the construction to determine if it matches the drawings and specifications.

Check with personnel who were present when the membrane and brickwork were installed, to determine where the materials were stored when they were brought to the site. If they were stored in the open, get the weather reports for the storage and erection periods. When you have accumulated everything you can, turn to the checklist and go through it. With the facts that you have accumulated, you should be able quickly to narrow down the source or sources of the problem.

The author

Walter Lee Sheppard, Jr. is president of C.C.R.M. Inc., 923 Old Manoa Road, Havertown, PA 19083. Tel: (215) 449-2167. C.C.R.M. is a consulting firm, specializing in selection, specification and installation advice on chemically resistant masonry. He is the author of "The Handbook of Chemically Resistant Masonry" and of over 30 articles on the topic, and conducts seminars on the subject. He is a registered P.E. in Delaware and California, and has a B.Chem. degree in chemistry from Cornell and an M.S. in chemistry from the University of Pennsylvania. He belongs to several societies, including the Natl. Assn. of Corrosion Engineers and ASTM.

Ceramic and refractory linings for acid condensation—Part I

Different types of materials and constructions will meet various types of acid service. Here is a guide to the available options.

Robert R. Pierce, *Consultant, and*
Charles E. Semler, *The Ohio State University*

☐ When chemical attack causes the failure of ceramic or refractory linings in a chimney or piece of process equipment, the resulting costs can be high. Although these materials are not necessarily costly themselves, downtime, cleanup and repairs *are.* Attack of these structures—here, by acid—is of increasing concern in the chemical process industries for several reasons:

1. Due to increased energy costs, thermal- and chemical-processing vessel linings are being redesigned to reduce heat loss. As a result, lining temperatures are lower, and the likelihood for acid attack is greater.

2. Chemical wastes are being incinerated more frequently. Such wastes contain a variety of chemicals that can attack linings.

3. Scrubbers are becoming more widely used. Adding a scrubber to a system typically reduces the gas temperature to the range where acid condensation occurs. Use of scrubbers often creates a positive pressure in the system, which complicates the situation by driving the acid into the lining.

4. In the U.S., the supply of low-sulfur coals is being depleted, and the high-sulfur, lower-purity types are beginning to be burned more. Thus, linings will be exposed to a variety of detrimental agents, such as metals, acid and ash.

5. Refuse is being burned more, and refractories and ceramics are used here. A burning mixture of refuse and fossil fuels creates severely corrosive conditions; e.g., sulfuric and hydrochloric acids are produced.

Growing use of ceramics/refractories

Besides the greater potential for increased chemical attack in processing systems, ceramics and refractories will replace high-cost stainless steels and other alloys in these applications for these other reasons:

1. Ceramics/refractories cost less than such metals.

2. Makers of corrosion-resistant alloys obtain some of their raw materials from politically sensitive countries; this could limit the supply. Using ceramics and refractories would avoid such shortages. Also, by using nonmetals now, engineers would develop expertise with them. This would avoid suddenly having to switch from metals to nonmetals (in case of a materials crisis) without having any knowledge of ceramics or refractories.

3. Ceramics and refractories resist the combination of acids and abrasion better than do stainlesses and other alloys.

4. The use of ceramics and refractories (vs. metals) reduces heat losses, and effects energy savings.

Factors that cause lining breakdown

To help in understanding how to design and select linings, the causes of chemical degradation of these structures are discussed here.

Startup and shutdown can create problems, especially when done frequently. Also, as already mentioned, a positive pressure in a system can be a problem.

Acids condensing on a lining at a low temperature will usually revaporize at a higher temperature. For example, sulfuric acid condenses at 185–450°F, depending upon the gas composition. However, revaporization does not occur until the acid concentrates to almost 100%, at above 600°F, when H_2SO_4 dissociates.

Running equipment at reduced gas volumes can result in chemical damage, too. At low speeds, the gas temperature at the lining can be several hundred degrees lower than at the center of the gas stream, and thus fall within

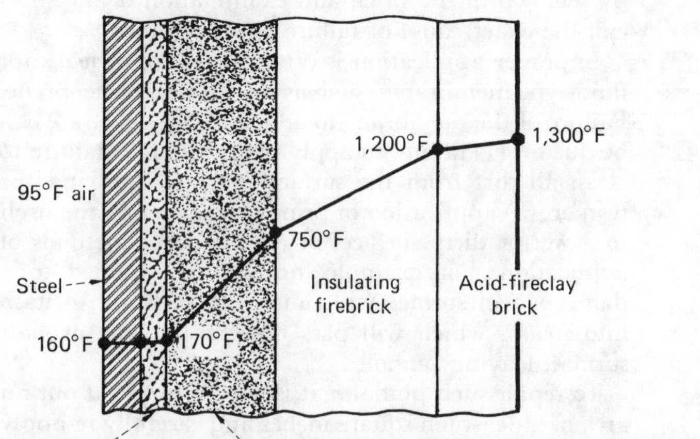

Multilayer ceramic lining has a seamless rubber membrane to protect the steel shell **Fig. 1**

Originally published December 12, 1983

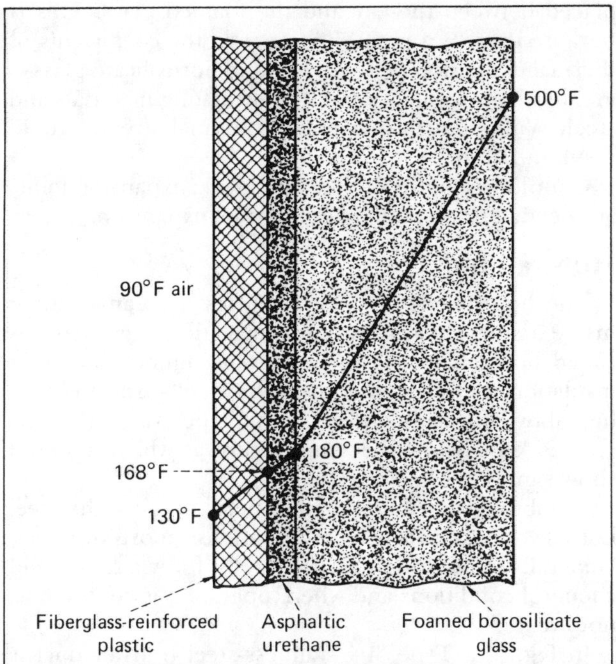

Foamed borosilicate glass protects FRP from high temperatures and corrosive chemicals Fig. 2

the acid-condensation range. Also, at lower flowrates, a normally negative pressure might become a positive one, worsening the situation.

Attack may even happen in the dry state. A refractory lining or a metal shell can degrade without the presence of water; corrosive chemicals alone are enough.

Now, the types of basic constructions that employ ceramics or refractories (along with other materials) will be outlined.

Layers on metal behind a lining

Impervious layers applied to metal structures restrict or prevent acid from reaching the metal (Fig. 1). Selecting the proper membrane or cladding depends upon how process conditions affect the chemistry of the materials to be used. One must determine all possible chemicals that can reach the membrane or cladding surface, and the maximum concentrations of such chemicals. Assessment may be difficult. The designer must be thoroughly familiar with the process, including its chemistry and operating characteristics. All process conditions and material performance should be verified by tests before a design is made final and construction is begun.

For example, in the incineration of chlorinated hydrocarbons, a small concentration of HCl will be formed. However, if the temperature of the membrane or cladding is below the acid dewpoint, the HCl concentration can reach, say, 22%.

The lining must resist this 22% value, rather than the lower concentration that is determined from analyzing the gas stream. If a sulfur-containing fossil fuel is used here, SO_2 and SO_3 will form, become sulfurous or sulfuric acid, and possibly condense.

The maximum temperature at the hot surface of the lining needs to be found, since chemical resistance of such materials is temperature-dependent. Vendors of linings have this information. When obtaining it, advise the firm that the membrane will be used behind a ceramic lining; under the ceramic, a stagnant zone is established that minimizes chemical flow over the membrane's surface.

If the lining design, membrane selection and installation are correct, service life should exceed 15 years. Sheppard *[1]* details commonly used membranes.

Fig. 1 is a cross-section of a lining used in a vessel where the HCl concentration could reach 22%. The operating temperature is 1,300°F, and the maximum summer air temperature is 95°F. A seamless rubber membrane is used, as it tolerates a higher temperature than does a lapped-sheet one.

Metal structure behind a ceramic/refractory

Stainless steels and other alloys, when used alone as structural components, resist both corrosion (of course, within limits) and moderate temperatures (e.g., 1,000-1,600°F) with only a small loss of strength. However, the corrosion rate of metals increases as the temperature increases, often quite dramatically. There are many cases in which unprotected corrosion-resistant metals have failed when the temperature was only 50-100°F higher than the design value *[2]*.

Sometimes, in a hot system, the combined effects of oxidation, corrosion, erosion and abrasion produce degradation or failure of metals. Such problems can usually be avoided by placing a ceramic or refractory over the metal surface.

In the U.S., linings typically are thick (i.e., 4½-15 in.). In many applications, using such thick linings has considerable disadvantages, including higher weight and reduced gas flow. The authors *[3,4]* favor the computerized engineering design of thin ceramic layers, sometimes no more than 1 to 2 in. thick. There are situations, such as in scrubbers and ducts, where thin linings are cheaper and provide operating benefits. Linings certainly can be thick—but the choice should depend upon the application.

The lining must be chosen so that it is compatible with the metal. The physical properties of both components must be considered so that they will not degrade each other under the service conditions of the process. Other factors in metal/ceramic design include summer and winter temperatures, and pressure. If a proper design evaluation is not made, costly repairs, shutdowns and catastrophic failure may result.

For example, the coefficient of expansion of stainless steels and other alloys is usually much higher than that of ceramics. However, in a lining, the average temperature of the ceramic is higher than that of the metal, and so the differing rates of expansion are usually not a problem when design is adequate.

Fiberglass-reinforced plastic

FRP can be used for structural support, as well as for resisting corrosion, both internally and externally. In most cases, using FRP eliminates the need for painting the exterior of equipment and for using a membrane or coating inside the vessel. However, FRP resists only fairly low temperatures; to extend its temperature range, one

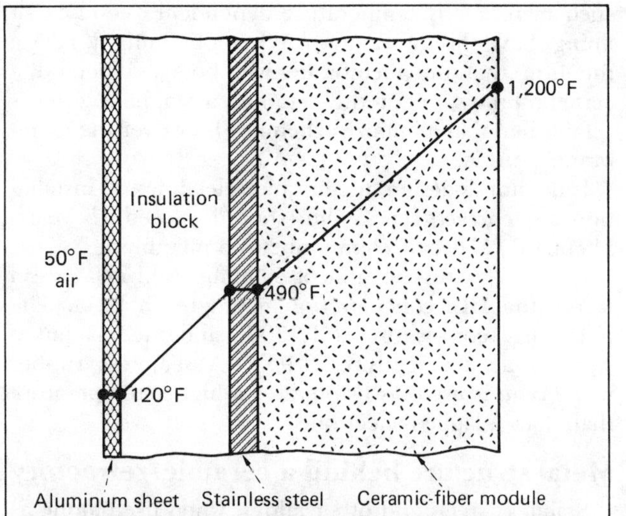

Hot-wall design prevents acid condensation at the surface of the stainless steel **Fig. 3**

should use a protective ceramic lining. Also, FRP is flammable at higher temperatures.

Foamed, closed-cell borosilicate glass is a ceramic that serves as a lining material in FRP systems for temperatures to 950°F. The glass, used with gases or liquids, is available in a variety of shapes. The closed-cell structure provides an insulating effect and helps prevent chemical penetration into the lining, as well as protecting the FRP from high temperatures. If a porous insulation were to be used, acids could possibly reach the FRP surface and degrade it.

Fig. 2 shows a borosilicate-glass-lined FRP structure. The figure is of a scrubber-exhaust gas duct in which temperatures typically range from 120-180°F, but can reach 500°F when the scrubber is bypassed. The bypass temperature is used in designing this system. A 2-in. thickness of the glass reduces the hot-side FRP temperature to a safe 168°F.

Design considerations

There are some important design considerations here:

- The FRP structure must be designed with out-of-roundness tolerances that will be maintained over the full range of operating conditions, including heatup and cooldown. External stiffeners or supports may be needed.
- Sometimes, a layer of ceramic paper may have to be placed between the FRP and the foamed glass. This is because there is a great difference in the coefficients of thermal expansion between FRP and borosilicate glass—greater than that between other insulating materials and steel. Without such a layer, the thermal stresses could result in failure.
- For long lengths of the glass block, expansion joints are needed to accommodate thermal expansion.

Hot-wall lining

The hot-wall design (Fig. 3) uses a ceramic lining inside a steel shell, plus outer insulation. This setup is called hot-wall, because the ceramic lining and outer insulation are selected so that the steel surface is kept hot, above the acid-condensation temperature. Also, the steel is kept below the temperature at which it would show significant loss of strength.

Membranes are seldom used here, because the steel hot-face temperature is too high for most of them. Generally, this design is used only for weak or mild chemical conditions and where operation is continuous, not cyclic.

In Fig. 3, a Type 304 stainless-steel transfer duct is exposed to a 1,200°F gas stream at a velocity of 30-40 ft/s. Corrosion conditions are moderate; there are trace amounts of acid chlorides. Since condensed hydrochloric acid vaporizes at about 300°F, the system was designed to keep the inside metal surface above 350°F during winter. Further, the metal must be kept below 700°F in summer, to avoid loss of metal strength.

There are many hot-wall configurations. Some that can be used inside of a steel shell include: brick over insulating firebrick; monolithic Gunite (a mixture of sand and cement, applied pneumatically with a pressure gun) over insulating firebrick; gunned coating on steel mesh over a ceramic blanket; acid-resistant Gunite over a cast or gunned insulating castable; and refractory-filled hex-grating metal over an insulating castable.

References

1. Sheppard, W. L., Jr., Membranes behind brick, *Chem. Eng.*, Part I, May 15, 1972, Vol. 79, No. 11, p. 122; Part II, June 12, 1972, Vol. 79, No. 13, p. 110.
2. Fontana, M. G., Corrosion Case Histories, paper presented at a seminar, Ohio State Univ., Columbus, Ohio, May 31, 1983.
3. Miller, R. L., A Computer Revolution in Large-Vessel Design?, *Chem. Eng.*, Oct. 2, 1972, Vol. 79, No. 22, p. 26.
4. Pierce, R. R., Engineering Brick-Lined Process Vessels, *Materials Protection & Performance*, Vol. 11, No. 12, 1972, p. 32.

The authors

Robert R. Pierce is a technical consultant, and resides at 412 Robin Rd., Waverly, OH 45690. Tel: (614) 947-7412. He consults with 15 companies on a range of engineering design, failure-analysis and corrosion problems. His experience includes nearly four decades with Pennwalt Corp., where he was senior technical consultant, Corrosion Engineering Div. He is the author of over 50 articles and has two patents. He holds a B.Sc. in chemical engineering from Oregon State University, and executive positions in Amer. Soc. for Testing and Materials, Natl. Assn. of Corrosion Engineers and AIChE.

Charles E. Semler is director of the Refractories Research Center, and director of the Building Research Laboratory, at Ohio State University, 2041 College Rd., Columbus, OH 43210. Tel: (614) 422-7128. Now a consultant to 40 companies, he has worked for Harbison-Walker Refractories Co. as a senior research mineralogist, taught at Washington University, and been a senior research chemist for Monsanto Research Corp. He holds B.Sc. and M.Sc. degrees in geology from Miami University and a Ph.D. in mineralogy from Ohio State. He is listed in "American Men and Women of Science."

Ceramic and refractory linings for acid condensation—Part II

Here is the conclusion of the summary of such types of construction that was begun in Part I, with hints on selection and installation.

Robert R. Pierce, *Consultant, and* ***Charles E. Semler,*** *The Ohio State University*

☐ Part I broadly discussed ceramics and refractories for acid condensation, including the growing need for these materials and the factors that cause their breakdown. Then, several types of acid-resistant materials and constructions were examined. Mentioned here are additional types of constructions, plus hints on selection, properties and construction.

Borosilicate glass over steel

Foamed borosilicate glass block can be used on steel shells, as well as with fiberglass-reinforced polyester (FRP). The glass, whether applied over steel or FRP, often has an asphaltic-urethane jacket that serves as a protective membrane. For steel, a potassium silicate or asphaltic urethane mortar is used for the side-joints.

The glass block begins to deform under load at about 850°F, so when it is used behind an inner (hotside) refractory lining, the temperature must be kept below this value. When the block is used alone, a maximum temperature of 950°F can be tolerated, since there is no induced stress from a covering layer.

The crushing strength of the glass must be considered; it may be necessary to include a cushioning layer to prevent crushing and cracking of the glass block.

Use of this glass makes it possible to design linings to resist strong chemicals at low to moderately high temperatures (i.e., 200–2,200°F). Fig. 1 shows a lining designed to contain a gas stream with high concentrations of SO_2 and SO_3 at 1,500°F. The steel shell temperature is 130°F, but there should be no problem with condensation of sulfuric or sulfurous acid, since chemicals cannot penetrate the foamed-glass layer unless it cracks, separates at the joints, or otherwise breaks down.

Monolithic concrete

Different types and chemistries of monolithic products are available, and process conditions must be carefully considered during selection. The types range from low-density insulating products to high-density materials that

Because of acid gases, this chimney lining failed

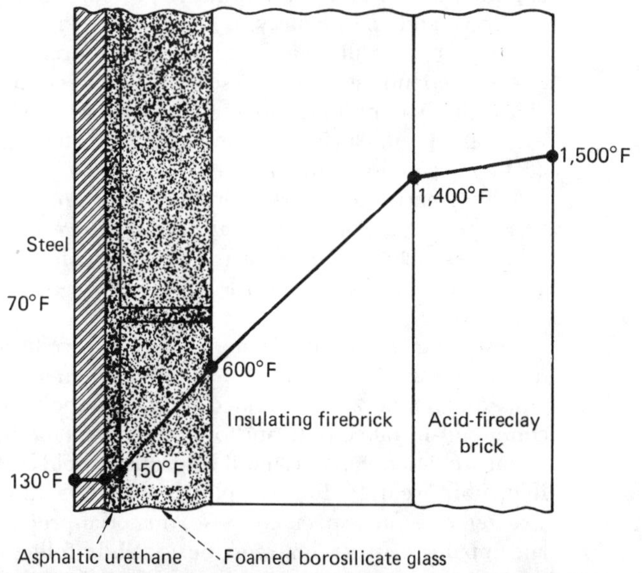

In this multilayer cold-wall design, acid is prevented from reaching the steel shell **Fig. 1**

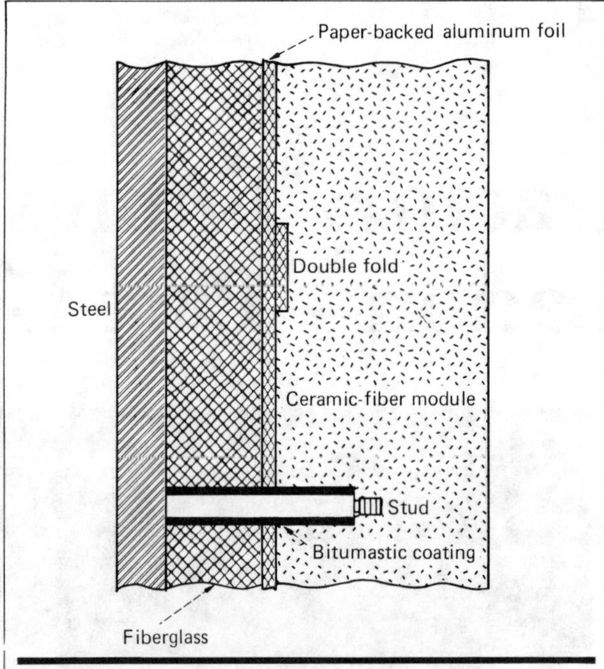

A paper-backed aluminum foil prevents acid from reaching and corroding the steel shell **Fig. 2**

are erosion-resistant. Several installation procedures are used, including gunning, casting, vibrating (there are thixotropic materials that become flowable only when they are vibrated), troweling and ramming.

There are four types of monolithic concretes, characterized by their bonds:

1. Portland-cement types can be used where the pH ranges down to 6½ – 7.
2. Calcium aluminates are useful down to a pH of 4.
3. Phosphate-bonded types do not have well-documented acid resistance but can be used under weakly acid conditions.

These three types generally can be used in weakly acid conditions where there is a negative pressure. If the acid is strong and, particularly, if the pressure is positive, these materials will deteriorate. Breakdown can be slow to fast, depending upon the severity of the conditions. These three monolithics are used in sub-stacks and ducts where the lining-surface temperature is greater than the acid-condensation temperature.

4. Polymeric silicas are more acid-resistant than the first three types, and can be used down to a pH of 0. Examples include fluegas ducts (for burning fossil fuels) and chimneys. These materials have fair to good insulating properties.

Designing monolithic linings requires careful attention to physical attachment, as well as chemical resistance and temperature. Gunned and cast monolithics are commonly installed onto anchors, studs, steel mesh, etc., to ensure that the material will be properly held in place. If anchors are used, for example, their type, coating and spacing must be specified. Also, material-preparation and mixing instructions must be carefully followed so that a good, uniform lining results. Once installed, the lining must be cured and, sometimes, heat-treated.

For all of these concerns, the manufacturers' instructions must be followed carefully. There are countless examples in which departures from recommended procedures resulted in immediate or premature lining failure.

If the lining surface or its interior is subjected to constant or intermittent wetting, a membrane is required on the steel surface. This is because monolithics are fairly porous. A membrane will be especially needed if the gases are under positive pressure.

Foil barrier behind a ceramic

A fairly new method of protection is to install a reinforced-paper-backed metal foil behind a ceramic or refractory lining (Fig. 2). The lining is designed so that the foil surface will be at 300 – 350°F, which is generally above the condensation temperatures for SO_3 and H_2SO_4, which forms from SO_3 and water reacting.

One such system uses aluminum foil in oil-fired equipment, where the oil has a high sulfur content. This method is not widely used for strong chemicals, so little information is available on performance of the various foil types. Until there is more service experience, this foil should be used when only trace or small concentrations of corrosives are present and when pressure is negative, or positive but not high. Gaps and openings in the foil require careful sealing or closure to achieve an effective barrier. Because of this, installation is especially important to successful performance.

Fig. 2 shows a foil lining behind a ceramic-fiber module. Here, cold-shell design is employed, and the amount of acid reaching the shell is restricted. Studs and holding devices attached to the steel shell hold the modules in place. The studs are coated with bitumastic (a carbonaceous organic coating material), and where the studs penetrate the foil, the paper is sealed. At the foil edges, the two pieces are double-folded.

Chemically resistant masonry

For over a century, this has has been the most widely used ceramic or refractory system for acid service (Fig. 3). This lining can be used to protect steel or concrete equipment. Acid-resistant brick is placed over a barrier membrane on steel or concrete. Linings commonly chosen are for low-to-moderate temperatures (up to 500°F). In addition to temperature resistance, masonry linings can be designed for high-pressure service, to 200 psi.

Fig. 3 shows a masonry lining used in the digestion of bauxite by sulfuric acid to manufacture alum. Acid-fireclay brick is installed with resin and silicate mortars, and the brick is backed by a ceramic paper and a lead membrane.

Chemically resistant masonry linings can be used with excellent results if the proper-quality ceramics are selected and the system is designed and installed correctly. Knowledgeable, experienced contractors and maintenance people must be chosen. Such linings are not easy to design, since there are many types of masonry, mortars and membranes. The brick types include acid fireclay, red shale, carbon, fused silica, high-alumina, and silicon carbide. Membranes include synthetic elastomers, rubbers, special metals or asphaltics, and resins.

A membrane must be selected to protect the substrate from the maximum acid concentration that can develop.

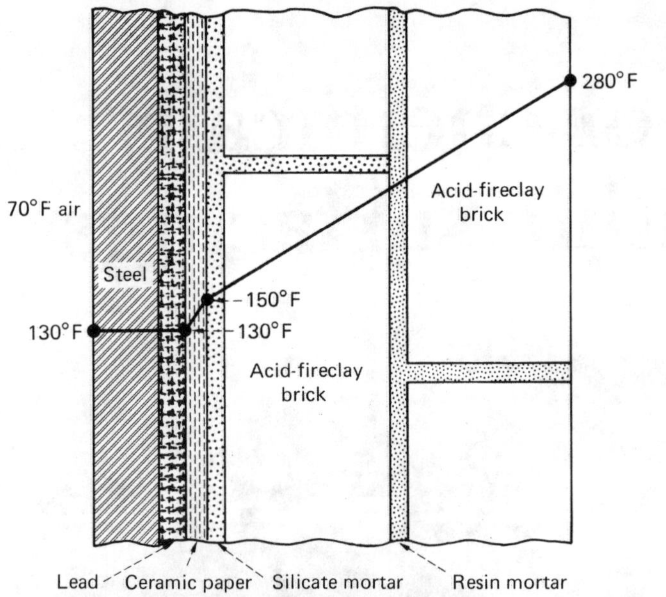

Chemically resistant masonry is used in the sulfuric-acid digestion of bauxite **Fig. 3**

The brick lining, made up of one or more layers of one or more types of materials, is chosen to provide:

1. Needed chemical resistance to the hot-surface conditions.
2. Desired temperature drop through the lining to protect the membrane.
3. Stability so that the lining will not crack.
4. Thermal expansion characteristics that avoid mismatch damage resulting from uncontrolled differential movement of the lining and structure.

Closed-cell borosilicate glass and other foamed ceramics have expanded the temperature range in which masonry linings can be used. These linings are increasingly employed for chemical resistance with higher temperatures, in the range of 500 to 2,300°F.

For example, insulating firebrick is now being used in masonry linings in acid service at higher temperatures, because its temperature can be kept above the acid-condensation range. In the past, the use of insulating firebrick in high-temperature, corrosive systems was limited because of the vulnerability of this material to chemical degradation. This was especially true under positive pressure and cycling conditions.

Selection, properties and installation

In designing a lining for chemical-containing hot environments, material selection is as important as lining design. In some cases, the materials selected dictate the lining method(s) to be used. So, if there are materials limitations, then there will be restrictions on the types of lining designs that can be made.

If a lining is designed on the basis of thermal protection only, and low levels of unsuspected chemical attack occur, degradation can be rapid. Even catastrophic failure can take place. Of course, chemicals need not condense to cause problems; dry-state reactions can degrade lining materials.

Once the lining materials have been selected, ordered and delivered, they should be checked to make sure that they provide the required properties. Many ceramics have properties that vary within established limits, depending upon the production steps for each lot.

There are published specifications for all commercial lining materials. Also, a few pieces or pounds of each lot of material should be labeled and stored indoors for reference later on in case unexpected problems develop.

A problem could develop, for example, if the properties of an insulating firebrick are determined from a brochure published ten years ago. The composition or firing process may have been changed since then, and the new supply may have a different expansion rate or heat flow. Another example: A manufacturer reduces the firing time of a brick, which reduces its strength and chemical resistance.

Even if the utmost care and planning go into specifying the lining design and materials, and the shipments are checked carefully, all can be lost if installation procedures are poor. Bad installation can result in failure. Some common installation mistakes are:

- Incomplete or improper filling of mortar joints in a brick lining.
- Use of a mortar that has been frozen or stored past its shelf life.
- Use of too much mixing water in preparing a silicate solution, to ease the placement of a monolithic refractory concrete.
- Onsite addition of an unspecified filler material, such as sand or coarse aggregate, to a mortar or refractory concrete. This is done to make placement easier.
- Use of cracked or otherwise defective brick.
- Improper mixing or curing of a membrane material.

The authors

Robert R. Pierce is a technical consultant, and resides at 412 Robin Rd., Waverly, OH 45690. Tel: (614) 947-7412. He consults with 15 companies on a range of engineering design, failure-analysis and corrosion problems. His experience includes nearly four decades with Pennwalt Corp., where he was senior technical consultant, Corrosion Engineering Div. He is the author of over 50 articles and has two patents. He holds a B.Sc. in chemical engineering from Oregon State University, and executive positions in Amer. Soc. for Testing and Materials, Natl. Assn. of Corrosion Engineers and AIChE.

Charles E. Semler is director of the Refractories Research Center, and director of the Building Research Laboratory, at Ohio State University, 2041 College Rd., Columbus, OH 43210. Tel: (614) 422-7128. Now a consultant to 40 companies, he has worked for Harbison-Walker Refractories Co. as a senior research mineralogist, taught at Washington University, and been a senior research chemist for Monsanto Research Corp. He holds B.Sc. and M.Sc. degrees in geology from Miami University and a Ph.D. in mineralogy from Ohio State. He is listed in "American Men and Women of Science."

Failure analysis of chemically resistant monolithic surfacings

Factors such as proper mixing of materials, correct curing temperatures and thorough removal of residues on substrates ensure successful application of these surfacings.

Walter L. Sheppard, Jr., *C.C.R.M., Inc.*

☐ Earlier, the author presented a checklist to determine the probable causes of failure of acid-resistant brick and mortar structures (*Chem. Eng.*, May 3, 1982, p. 107). Here is a similar tabulation that will help one to spot and avoid problems with monolithic surfacings. These are surfacings of a single composition that are applied to give a uniform surface.

The checklist primarily deals with resinous toppings of ⅛ (or less) to ⅜ in. thickness. A monolithic less than 1/16 in. thick is considered to be a coating. Monolithics are put over substrates—usually, concrete flooring. Other substrates are steel decking and stairs, steel aprons and, occasionally, wood structures. The most frequent use of these toppings is on concrete floors subject to light traffic. Acid brick is used in heavy-traffic areas. Monolithics are often employed to line trenches and pits and, in some cases, vessels. Whatever the application, the purpose is the same: to protect the substrate from attack by corrosives.

The checklist can be used for failure analysis of thicker linings, such as latex-modified hydraulic surfacings ½ in. or more thick, or polymer concretes 1 in. or more thick. The items in the checklist apply to resinous, siliceous and sulfur formulations, whether gunned or placed by casting and troweling. However, the list should not be considered as a total or absolute diagnostic tool to investigate such failures.

The photographs

Top row: (left) Severe cracking due to weak substrate and poor substrate preparation; (center) Concrete bottom of a thickener surfaced with topping of glass-cloth-reinforced polyester resin. Subsurface moisture 10 ft below surface resulted in disbonding and surface breakup; (right) Delamination and cracking up of vinyl ester topping. Glass fabric too thick, wrong surface treatment used and resin did not bond to fabric. **Bottom row:** (left) Expansion-joint material at top of picture. Repair of crack at middle that "telegraphed" through substrate; (center) Flaking off of resin from glass cloth, probably applied to wrong glass, which did not wet properly and was mixed too thinly; excessive shrinkage resulted; (right) Arrow shows expansion joint in original topping. Concrete wet during application. Possible incomplete, nonuniform mixing of ingredients.

Originally published July 9, 1984

Troubleshooting chemically resistant monolithics

I Failure to cure—material either remains soft or hardens only partially; alternatively, there may be both hard and soft spots

1. Moisture (e.g., rain, spillage) or another liquid has settled on the surface. This may be due to inadequate weather protection, flooding of the surface from one or more sides, cooling of the air below its dewpoint on the substrate, or cold-air drafts over the surface (from external doors, etc.).
2. Surface was wet or chemically contaminated when the surfacing was applied. Cause may be inadequate cleaning, drying or neutralization.
3. Concrete mix contained an admixture (such as an air-entrainment agent, water reducer or curing agent) that either reacted with and depleted the hardener in the surface material or inhibited the cure of the surfacing.
4. Concrete may have contained too much water (i.e., it did not dry enough), so that when the topping was applied, water collected under it to inhibit curing. (Thin sections may fail to cure completely.)
5. Resin and hardener were not mixed in correct proportions, or not mixed uniformly. (If cure is spotty, incomplete mixing is often the cause.)
6. Substrate or materials were too cold to cure properly, or the temperature of the substrate was allowed to fall below the curing temperature after its placement.
7. Overheating beyond their thermal limits will soften some monolithic toppings. If so, these toppings may appear to be partly cured, and show marks and indentations. They will usually re-harden when they are cooled. (See also Item II-11 and Item VI-3.)
8. The material was exposed to intense sunlight.

II Disbondment—material separates from the substrate or does not adhere

1. Surface was inadequately cleaned and dried. The bond is to dirt on the surface. (If dirt is on the surface, it can often be seen adhering to the underside of the delaminated surfacing material.)
2. For new concrete, the contractor may have used a curing agent or sealer that acted as a bond breaker.
3. For new concrete, cement finishers may have dipped trowels in, or wiped them with, a silicone cleaner.
4. See also Item I-4.
5. Where the day-to-night thermal gradient is ± 50°F, water under—not in—the slab may have been drawn up under the topping from as far as 15 ft down. If this happens, apply the monolithic on the 4 p.m. to midnight shift. The cure will be advanced enough to prevent such a problem by the time the temperature rises in the morning.
6. Oils, other release agents, or foams were used, and these left residues on concrete surfaces.
7. Concrete has inadequate surface strength. This can be the result of: (1) inadequate design specifications; (2) failure of the bulk plant to follow design instructions; (3) too much water in the mix; (4) excessive troweling in finishing the concrete; (5) failure to clean the surface or remove "laitance" (fine particles of lime or portland cement that come to the surface upon finishing). The surface of the concrete may be cleaned by brush blasting or by etching with hydrochloric acid. Proper design strength calls for a compressive strength of 3,000 psi at the time of application and a bond strength of 300 psi, using the specified surfacing material as a bonding agent.
8. Either the resin and the hardener were mixed to incorrect proportions or mixing was incomplete.
9. Substrate was too cold for adequate cure, but air temperature was high enough to cure surface.
10. Materials were used that were partly set.
11. Top temperature was exceeded. Topping has a higher coefficient of expansion than does the concrete. When the bond strength is exceeded, the topping will disbond, bulge, then crack. Cleaning by boiling water or steam can produce this. (See also, Item I-7 and Item VI-3.)

III Cracking

1. Topping was applied over an expansion joint, construction joint, or other point of movement in the substrate. (An expansion joint in the topping is required in such areas.)
2. The distance between expansion joints or stress-relief joints was too great. (Cure shrinkage results in accumulation of stresses in the topping. When these exceed the bond strength, the topping disbonds and cracks.)
3. For larger sections of concrete substrate, stress-relief joints were not placed at 20-ft or smaller intervals. (See comment under No. 2.)
4. Materials were improperly proportioned or incompletely mixed.
5. Materials were applied after working life had expired.
6. Note that if disbonding occurs, cracking due to stresses from cure shrinkage will almost certainly follow.
7. If overheating happens, disbondment will take place. If the topping is hard and does not soften appreciably at higher temperatures, it will crack.
8. Area had been frozen at particularly low temperatures. The material becomes more brittle and develops excessive shrinkage stresses at very low temperatures.
9. During cure, materials were exposed to intense sunlight—this is especially seen in spotty areas.

IV Penetration

All resinous, siliceous or sulfur cements, toppings and monolithics are porous to some degree. Diffusion through these surfacings is faster for small molecules. Expect eventual penetration in continuously wet conditions (e.g., puddling, retaining sumps and trenches in continual use). All areas surfaced with such cements, toppings and monolithics should be sloped to prevent puddling.

Penetration will result successively in: (1) disbondment if the penetrant is a chemical that attacks the substrate at all rapidly; (2) cracking of the topping; and (3) curling upward of the edges of the crack. After disbondment and cracking occur, the substrate will show signs of chemical attack. To verify this, peel off an adjacent topping that appears sound, and check the substrate's surface with pH paper.

V Chemical attack

Evidence of chemical attack can be: (1) surface softening; (2) surface discoloration that cannot be removed by cleaning; and (3) surface etching, by either destruction or removal of the aggregate (e.g., silica aggregate removed by HF exposure). For etching, the surface can remain hard and porous or become powdery and crumbly. Discoloration is followed by deterioration. Other signs are softening and swelling.

VI Blisters or bubbles

Part A. Appearing after application and while the material is still soft and before it has hardened. Bubbles or blisters are small; some may break, leaving small indentations

1. Materials have been mixed with a high-speed mixer,

entrapping air in the resin. After the surface has been troweled in place and finished, air bleeds out as blisters or tiny bubbles. A paddle mixer moving at more than 350 rpm can cause this.
2. A breeze over 10 mph can cause this.
3. A highly alkaline surface (pH > 10) can result in "gassing" and subsequent blisters.
4. Application on porous concrete: Air in the pores expands as the daytime temperature rises, and pushes through the topping, resulting in blisters. This can be avoided by applying the monolithic after 4 p.m.
5. Application over concrete that has moisture in it or below it. The effect is the same as in No. 4, but here the underside of the topping also fails to cure properly. (See also Item II-5.)
6. Exposure to varying intense sunlight, heat or cold during installation and cure.

Part B. Appearing weeks after cure has taken place. Bubbles or blisters are thick-skinned and usually rather large

1. Possible diffusion of corrosives through topping attacked the substrate at the bond surface and caused gassing or concrete "growth," with resulting delamination.
2. Possible chemical attack from underneath. Corrosives may have entered the substrate elsewhere and traveled along the rebars.
3. Overheating may have caused disbonding and bulging, similiar to blistering, if the topping softened when overheated. (See also Item II-2 and Item I-7.)

VII Expansion joint failure

1. Loss of bond to sides of joint. This can be due to: (1) poor design—edges of joint were not at right angles to the surface; (2) sealant bonded to bottom of joint—joint opened and material was dragged off its sides; (3) sides of joint were dirty or specified primer was not used when joint was filled; (4) low-level chemical attack occurred—usually caused by a solvent; (5) elastic limit of the sealant was exceeded; (6) sealant was heated beyond its thermal limit; (7) subfreezing of the surface took place; and (8) material suffered thermal shock, resulting in too-rapid movement of adjacent surfaces.
2. Swelling of joint. The causes can be: (1) chemical attack, which results in swelling or gassing—this is detectable by presence of crumbling, a porous condition, or spongy cavitation and loss of strength; and (2) failure to use a compatible material under the sealant—as joint closed, noncompressible material below it extruded upward, pushing the sealant out of the joint.
3. Breakup of joint filler, usually due to either (1) excessive movement beyond the elastic limits of the sealant, (2) chemical attack, or (3) excess hardener in mix.
4. Brittleness, hardness, crumbling and loss of elasticity, due to (1) chemical attack or (2) overheating beyond the maximum service temperature.
5. Uneven cure. This is usually due to (1) exposure to moisture during cure, but can also be due to (2) wind blowing over the surface, (3) excessive sunlight, or (4) nonuniform mixing of the ingredients.

Types of materials

The most important and common monolithics are made from synthetic resins and are called resinous toppings or surfacings. The most frequently used of these synthetics are epoxies, vinyl esters and polyesters, in that order. Resinous monolithics are used from 0 to 210°F, and for a variety of chemical exposures, depending on the resin, filler and formulation. And there are other monolithics—silicates, sulfur formulations (applied hot), asphaltics and hydraulic mixtures.

A distinction should be made between polymer-modified hydraulic cements, which are not very chemically resistant, and true polymer cements, which contain no hydraulic components. The latter cements consist of a binder that is only polymerized by chemical action, an inert filler and a hardening or curing agent. The formulation must be free of any hydraulic cement materials, such as portland cement or Lumnite (an aluminous cement). The polymer-modified hydraulic cements contain ordinary portland cement or any other such cement that is activated with water.

How to use the checklist

The list tries to classify separate visual effects and to identify them by their causes. One cause may be responsible for different effects. However, many of the effects are sequential, and thus one should try to determine which effect appeared first, and which followed it.

For example, if a monolithic seemingly first cracked, then curled along the crack's edges, it might seem that corrosion of the substrate was the cause. The substrate would grow, pushing upward. However, the actual order of events may have been (1) disbondment, (2) cracking of the topping due to cure-shrinkage stresses, and (3) curling due to stresses from the surface shrinking slightly faster than the underside, which was still bonded to the substrate.

Though most of the items in the list refer to concrete substrates, similar effects are sometimes seen for other substrates. For example, for Item II-6, if steel were the substrate, a coating of oil applied at the mill and not removed could cause the same problem.

The author

Walter L. Sheppard, Jr., is president of C.C.R.M., Inc., 923 Old Manoa Rd., Havertown, PA 19083. Tel: (215) 449-2167. C.C.R.M. is a consulting firm, specializing in selection, specification and installation advice on chemically resistant masonry. He is the author of "Chemically Resistant Masonry" and of over 30 articles on the topic. He is a registered P.E. in Delaware and California, and has a B. Chem. degree from Cornell University, and an M.S. in chemistry from the University of Pennsylvania. He belongs to several societies, including the Natl. Assn. of Corrosion Engineers and the Natl. Soc. of Professional Engineers.

INDEX